PLANT TISSUE AND
CELL CULTURE

BOTANICAL MONOGRAPHS

PLANT TISSUE AND
CELL CULTURE

EDITED BY

H. E. STREET
D.Sc. (Lond.)

Botanical Laboratories,
University of Leicester,
England

BLACKWELL SCIENTIFIC PUBLICATIONS

OXFORD LONDON EDINBURGH MELBOURNE

© 1973 Blackwell Scientific Publications
Osney Mead, Oxford,
3 Nottingham Street, London W1,
9 Forrest Road, Edinburgh,
P.O. Box 9, North Balwyn, Victoria, Australia.

ISBN 0 632 09010 3

First published 1973

Distributed in the U.S.A. by
University of California Press
Berkeley, California 94720

Printed in Great Britain by
Alden & Mowbray Ltd at the Alden Press, Oxford
and bound by
Weatherby Woolnough Ltd,
Thrapston, Northamptonshire

CONTENTS

Contributors   vii

Preface   viii

1   Introduction   1
    H.E.STREET

2   Laboratory Organization   11
    H.E.STREET

3   Tissue (Callus) Cultures—Techniques   31
    M.M.YEOMAN

4   Cell (Suspension) Cultures—Techniques   59
    H.E.STREET

5   The Isolation of Protoplasts   100
    E.C.COCKING and P.K.EVANS

6   General Cytology of Cultured Cells   121
    M.M.YEOMAN and H.E.STREET

7   Nuclear Cytology   161
    N.SUNDERLAND

8   Single-Cell Clones   191
    H.E.STREET

9   Pollen and Anther Culture   205
    N.SUNDERLAND

10   Growth Patterns in Tissue (Callus) Cultures   240
    M.M.YEOMAN and P.A.AITCHISON

11   Growth Patterns in Cell Cultures   269
    P.J.KING and H.E.STREET

12 Aspects of Organization—Organogenesis and Embryogenesis   338
    J.REINERT

13 The Origins, Characteristics and Culture of Plant Tumour Cells   356
    D.N.BUTCHER

14 Growth of Plant Parasites in Tissue Culture   392
    D.S.INGRAM

15 Old Problems and New Perspectives   422
    H.E.STREET

    Literature Cited (covers all chapters)   433

    Subject Index   485

CONTRIBUTORS

P.A.AITCHISON, Department of Botany, University of Edinburgh, the King's Buildings, Mayfield Road, Edinburgh 9

D.N.BUTCHER, Unit of Developmental Botany, Agricultural Research Council, University of Cambridge, 181A Huntingdon Road, Cambridge CB3 0DY

E.C.COCKING, Department of Botany, University of Nottingham, University Park, Nottingham NG7 2RD

P.K.EVANS, Department of Botany, University of Nottingham, University Park, Nottingham NG7 2RD

D.S.INGRAM, Unit of Developmental Botany, Agricultural Research Council, University of Cambridge, 181A Huntingdon Road, Cambridge CB3 0DY

P.J.KING, Botanical Laboratories, University of Leicester, Leicester LE1 7RH

J.REINERT, Institute of Plant Physiology, The Free University of Berlin, Königin-Luise-Str. 12-16a, 1 Berlin 33 (Dahlem), West Germany

H.E.STREET, Botanical Laboratories, University of Leicester, Leicester LE1 7RH

N.SUNDERLAND, John Innes Institute, Colney Lane, Norwich NOR 70F

M.M.YEOMAN Department of Botany, University of Edinburgh, The King's Building's, Mayfield Road, Edinburgh 9

PREFACE

This book is an attempt to survey the rapid developments in plant tissue and cell culture techniques which have occurred in the last decade and to assess critically the contributions these techniques are now making to our knowledge of the growth, metabolism and differentiation of plant cells and of the factors controlling morphogenesis in vascular plants. Such a task has demanded the participation of a number of specialists and I feel most fortunate in the enthusiastic collaboration I have received from a group of authors each distinguished in the field of study he has covered. No attempt has been made to achieve any uniformity of approach or style in the separate chapters nor to completely eliminate overlap between them (this aspect is covered by numerous cross-references between chapters). Nevertheless, all the chapters seem to me to cover basic techniques and current researches in a way which not only will introduce established scientists, post-graduates and undergraduates to new fields of enquiry but will also provide details of current work (quite often previously unpublished work) of interest to those already using plant tissue and cell culture techniques in their work.

As Editor I should like to acknowledge my indebtedness not only to my co-authors but to Professor J.H.Burnett (General Editor of the Botanical Monographs Series) and to Robert Campbell (Blackwell Scientific Publications Ltd.). As will be seen in the legends of many text-figures and tables, we have also to acknowledge permission to publish new data and previously published data from many authors and journals. In the general preparation of the manuscript and particularly of the Literature Cited and Index I have been greatly helped by my secretary, Miss Daphne Roberts. Although many people have been involved in preparing the text-figures and plates, particular acknowledgement must be made for photography to Mr. G.G.Asquith (Illustration Service) and for line drawings to Mr. E.M.Singer, Miss S.Pearcey and Miss S.Duffey (Botanical Laboratories), all of the School of Biological Sciences, University of Leicester.

September 1972

H.E.Street

CHAPTER 1

INTRODUCTION

H. E. STREET

Some terms defined 2
Scope of the present work 5
Some landmarks on the way 5

'*The value of continually advancing technique is inestimable so long as it is not allowed to become an end in itself, and thus foster delusive industry of a pointless kind*'
Agnes Arber—*The Mind and the Eye*, p. 13. University Press, Cambridge, 1954

The last decade has seen a very rapid rise in the number of plant scientists using the techniques of organ, tissue and cell culture in plant physiological researches. This is in part due to important development and refinements of these techniques which now make possible an increasing range of reproducible and quantitative experiments involving plant cultures. It is also in part due to the demonstration, by a long sequence of pioneer investigations, that many problems in plant physiology otherwise inaccessible to study can be tackled by the imaginative use of plant culture techniques. Even more widespread use of the techniques is now probably inhibited primarily by a feeling that they are intrinsically difficult to carry out successfully and also that they are demanding of very special facilities and an exceptionally high level of skilled technical assistance. The main purpose of the present volume is therefore to describe clearly current techniques and to assess their potential value and current limitations. This should enable research workers to assess how far they could be of value in their own field of study and to make an informed judgement of the technical problems they are likely to encounter in their use.

There is, at present, in all plant culture work still a strong empirical element; a necessity to arrive at technical solutions by trial and error. In consequence the various protocols, presented in the chapters which follow, can only form a sound starting-point. For successful work with a particular plant material it may be necessary to vary manipulative procedures, methods of sterilization, composition of culture medium, conditions of incubation and so on. Therefore, following upon the various basic protocols are summaries of some of the variations in the basic procedure which have proved valuable with particular plants, and the references from which details of the variations can be obtained. It should therefore be possible for an investigator new to the

1

field to plan a systematic approach to the problem of establishing cultures from his chosen material. Nevertheless, even the best planned approach may fail; there are problems, seemingly formidable problems, still to be solved before it can be claimed that cultures can be successfully established from just any higher plant.

Each aspect of plant culture technique has evolved in an attempt to solve not only a technical problem but to open up or extend a line of biological investigation. This volume is therefore not simply a technical manual but an account of the present state of knowledge regarding a number of aspects of plant physiology to which the culture techniques have made a substantial and often a unique contribution. Here inevitably it has been necessary to select from a now very extensive literature, some of which could not be consulted in the original by the authors. The literature citations are therefore incomplete but nevertheless, it is hoped, properly representative of the main lines of enquiry.

SOME TERMS DEFINED

There is at present considerable confusion and lack of uniformity in the terminology of aseptic plant culture. The term 'plant tissue culture' although commonly used as a blanket phrase to cover all types of aseptic plant culture should now be used in a more restricted sense. It is possible to distinguish the following types of aseptic cultures of plant origin: The culture of seedlings or larger plants (*plant cultures*), of isolated mature or immature embryos (*embryo cultures*), of isolated plant organs (*organ cultures* including cultures derived from root tips, stem tips, leaf primordia, primordia or immature parts of flowers, immature fruits), of the tissues arising by proliferation from segments (explants) of plant organs (*tissue* or *callus cultures*), of isolated cells or very small cell aggregates remaining dispersed as they grow in liquid media (*suspension cultures*).

Tissue or callus cultures do not correspond with any normal tissue of the whole plant. The use of the term *callus cultures* derives from the fact that involved in their origin is proliferation induced in the explant by injury of cells caused by excision, suggesting that the tissue formed can be equated with wound callus. Such injury is, however, only one and in some cases probably not the principal cause of the proliferation; the removal of the explant from controls imposed upon its tissues by the whole plant and the provision to the explant of appropriate nutrients and growth-regulating substances may be among the determinative factors leading to proliferation (see Chapter 3). The established culture may differ very much from young wound callus. The alternative and possibly to be preferred description of such cultures as *tissue cultures* arises from the fact that they can be derived from different tissues in the parent explant and can differ very markedly from one another (either as a consequence of their different specific or varietal origins or the conditions under which they are cultured) and from the primary wound callus in morphology, cellular structure, growth and metabolism. Such cultures always contain both dividing and non-dividing cells, and the non-dividing cells may be of one or several distinct cell types within the tissue mass. The term tissue culture can, therefore, be appropriately applied to any culture growing on solid medium (or attached to a substratum and fed with liquid medium) and which consists of many cells in protoplasmic continuity. It does not imply any structural or functional homogeneity of the constituent cells or equivalence with any normal plant tissue.

The term *suspension culture* is self-explanatory in so far as it implies cells and cell aggregates growing dispersed in a moving liquid medium. At present no such cultures consist entirely of separate cells (such a culture would be described as a free-cell suspension culture). Suspension cultures have also been termed *cell cultures* on the grounds that they represent a distinctly lower level of organization than tissue cultures; at least their free cells can properly be described as 'cells in culture'. The term cell culture has descriptive value even if it can be objected to on the grounds that cell aggregates are never absent and may transiently be very prominent even in the most highly dispersed suspensions currently in culture. The term cell culture has also been applied to work on the growth of single-cultured cells at least initially out of physical contact with other cells. Such work is preferably described as *single-cell cloning* (see Chapter 8).

Where root cultures are all derived from a single initial root tip or where a tissue culture (or suspension culture) derived from a single explant is maintained in culture by repeated subculture of a many-celled piece of the parent culture the term *clone* (root clone, tissue clone, suspension clone) seems appropriate (although the term *strain* is often used in this sense). Occasionally the tissue cultures of such a clone may not be of uniform morphology; they may show limited regions (sectors) differing in texture, colour and/or growth rate. By selective subculture of such sectors, the clone may be split to give two or more lines which appear to remain permanently distinct by propagation of the sector characters. These lines may then be termed *clonal variants*. The fact that such variants differ in recognizable characters does not necessarily imply that the variants are genetically distinct (though it may in fact be so). There is some evidence (almost entirely unpublished) that some such variants represent rather stable 'states of differentiation' which are potentially reversible by appropriate cultural treatments. It should also be borne in mind that the initial proliferation does not occur from a single cell of the primary explant nor even necessarily from a single tissue of the explant. The primary tissue culture will, therefore, frequently contain cells of different previous histories or even of different levels of ploidy. The variant may therefore contain a different balance between the cell types present in the initial clone. Even when variants are not separated the cellular composition of the clone may alter as subculture proceeds by selection of cells at a growth advantage under the chosen cultural conditions.

To distinguish tissue clones as described above from clones derived from single cells the latter are described as *single-cell clones* and, where such clones (derived from a tissue clone) differ from one another, they can be described as *single-cell variants*. The term *mutant* rather than variant should be reserved for cases where there is an established genetic difference or where at least the new culture has arisen as a consequence of the application of an effective mutagenic treatment (preferably to single cells from which the mutant tissue culture has been developed).

It is important to distinguish between (1) culture of primary explants derived from plant organs (*cultured explants*), (2) culture for a single incubation period (passage) following excision of the newly proliferated tissue from the original organ explant (*1st passage cultures*), (3) culture through many successive passages (usually indicative of a potential for indefinite growth in culture provided an appropriate subculture procedure is adopted). Cultures of this last type have been described as continuous cultures. Since, however, they are propagated by successive batch cultures it is preferable to refer to them as *established or permanent cultures* and reserve the term *continuous cultures* for suspension cultures propagated over long periods of time without subculture by constant addition of new medium and a balancing harvesting of the suspension (see Chapters 4 and 11).

Tissue cultures are subcultured by transferring to new medium a fragment of the parent culture. Such a fragment has been termed an explant but it is preferable to term it an *inoculum* (the same term can then be applied to the aliquot of a suspension culture transferred to effect its subculture). The term *explant* should be reserved for the fragment of a plant or tissue (e.g. endosperm) used to initiate a culture clone.

The use of the term *aseptic* is preferable to sterile as a prefix—aseptic tissue culture—since such cultures are grown from surface-sterilized explants under aseptic conditions (conditions designed to exclude contaminants) and should be periodically tested for sterility (by microscopic examination and transfer to appropriately enriched media) but cannot be claimed at all times to be absolutely free from any contaminating organisms (see Chapter 2).

Organogenesis is used to describe the origin of shoot buds or roots from tissue cultures or suspension cultures. Plantlets can arise from these by the formation of adventitious roots from the shoot bud or of a shoot bud from tissue formed by proliferation at the base of the rootlet. *Embryogenesis* is used to describe the origin of plantlets by a developmental pathway closely resembling the normal embryology from the fertilized ovum. That this development in culture involves somatic (body) cells and not the zygote can be indicated by using the term *embryoid* or *adventive embryo* rather than embryo. This does not preclude employing the recognized description of stages in embryology, e.g. proembryoid, globular embryoid, torpedo-shaped embryoid (see Chapter 12).

SCOPE OF THE PRESENT WORK

The present work is confined to a consideration of tissue (callus) and suspension (cell) cultures as defined above, and only those of either flowering plant or gymnosperm origin. The omission of organ cultures and embryo cultures does not imply that their potential has been exhausted, or that they are not being used in interesting current researches. Simply some restrictions of the field covered has been essential lest this book be too long. It is, however, pertinent to the omission both that the techniques of organ culture are not in a period of rapid change and development, and that those techniques and their more important contributions to botanical research have been covered in a number of recent review chapters (Butcher & Street 1964, Street & Henshaw 1966, Street 1969). The anther cultures discussed in Chapter 9 are not organ cultures; the immature anther is being used as an explant which, under appropriate conditions, can yield from its microspores haploid embryoids or a haploid tissue culture.

SOME LANDMARKS ON THE WAY

Haberlandt, in a frequently quoted paper published in 1902 (and now available in translation with an interesting commentary—Krikorian & Berquam 1969) stated clearly the desirability of culturing the isolated vegetative cells of higher plants: 'To my knowledge, no systematically organized attempts to culture isolated vegetative cells from higher plants in simple nutrient solutions have been made. Yet the results of such culture experiments should give some interesting insight into the properties and potentialities which the cell as an elementary organism possesses. Moreover, it would provide information about the inter-relationships and complementary influences to which cells within the multicellular whole organism are exposed.' Experiments along these lines had been started by Haberlandt in 1898, using single cells isolated from the palisade tissue of leaves, pith parenchyma, the epidermis and epidermal hairs of various plants. 'In my cultures . . . cell division was never observed. It will be the problem of future culture experiments to discover the conditions under which isolated cells undergo division.' Haberlandt's failure did not deter others in his laboratory from attempts, over a number of years, at similar experiments but, although the cells in some cases remained alive for prolonged periods, and even expanded in the culture media, they did not divide to give rise to cell aggregates (Winkler 1902, Thielmann 1924, 1925, Küster 1928).

The great interest of Haberlandt's paper rests on its clear concept of the problem and its foresight as to what would be achieved. Although Haberlandt was working 30 years before the discovery of auxins, he was struck by Winkler's observations on the stimulation of ovule development and the swelling of ovaries which occurred when the pollen grains germinated and he suggested 'it would be worth while to culture together in hanging drops vegetative cells and pollen tubes, perhaps the latter would induce the former to divide'. He also proceeded as follows: 'One could also add to the nutrient solutions used an extract from vegetative apices or else culture the cells from such apices. One might also consider utilization of embryo sac fluids.' 'Without permitting myself to pose further questions, I believe in conclusion, that I am not making too bold a prediction if I point to the possibility that in this way, one could successfully cultivate artificial embryos from vegetative cells.'

Very little progress was made during the 30 years following Haberlandt's paper. Various workers in other laboratories reported further unsuccessful attempts to culture cells (Schmucker 1929, Scheitterer 1931, Pfeiffer 1931, 1933, La Rue 1933), and Kotté (1922a, b) and Robbins (1922a, b) reported some progress towards the culture of excised root tips. However, in 1934 the pioneer work on root culture reached fruition when White (1934) was able to report the establishment of an actively growing clone of tomato roots. In the same year Gautheret reported that pieces of cambium removed under aseptic conditions from *Salix capraea*, *Populus nigra* and other trees continued to proliferate for some months, giving rise to alga-like outgrowths if placed on the surface of a solidified medium containing Knop's solution, glucose and cysteine hydrochloride. White's (1937) discovery of the importance of the B vitamins for the growth of cultured roots and the increasing recognition of the importance of auxin (indol-3yl-acetic acid, IAA) in the control of plant growth (Went & Thimann 1937) led Gautheret (1937, 1938) to include these growth factors in his culture medium with the result that he obtained greatly enhanced but still limited growth of the *Salix* cambium. At this time Nobécourt (1937, 1938a, b) was obtaining some cell proliferation in culture using explants from carrot roots. Then, in 1939, Gautheret also reported studies on small explants (which included cambium and functional phloem) from carrot roots using a culture medium containing the modified inorganic salt mixture employed by Nobécourt: glucose, vitamin B_1 (thiamin), cysteine hydrochloride and IAA. White, reviewing the field in 1941 wrote of Gautheret's work as follows: 'On 9 January 1939, Gautheret presented before the French Academy of Sciences the results of studies constituting a combination of his own previous work with that of Nobécourt. Using a Knop solution supplemented with Bertholot's mixture of accessory salts, glucose, gelatine, thiamin, cysteine hydrochloride and indole acetic acid . . . he had cultivated fragments of carrot . . . they showed little or no differentiation beyond the formation of occasional lignified cells . . . they grew slowly and without any indication of diminution of growth rate. The record, though brief, if taken with Gautheret's earlier work, is sufficient to justify the conclusion that he has obtained cultures satisfying both major criteria of a plant tissue culture—potentially unlimited growth and undifferentiated growth—so that there need be no further doubt as to the real success of his efforts.' The emphasis by White on the significance and success of Gautheret's work is particularly commendable for within months of Gautheret's paper, White (1939a) himself reported the formation of a similar tissue culture from the procambial tissue of segments of young stems of the hybrid *Nicotiana glauca* × *N. langsdorffi* and showed that it also could be repeatedly subcultured using a medium containing 0·5% agar but otherwise identical with that he had developed for the growth of his root clone of tomato. The basic technique of tissue culture described in these pioneer papers by Gautheret and White has subsequently resulted in the establishment of such cultures from many species.

Almost immediately, tissue cultures from more species were described (including some from bacteria-free crown-gall tissues—Braun & White 1943) and the cultures were submitted to anatomical study. Although young, actively growing cultures were found to consist of dividing cells (often localized in layers or nodules) and of parenchymatous non-dividing cells, the cultures as they aged showed an increasing degree of organization as exemplified by the development of primitive vascular tissue often composed entirely of tracheidal cells. Many studies on this histogenesis were undertaken but progress towards identification of the factors controlling such cellular differentiation had to wait till much later (Gautheret 1957, 1966). Nobécourt's pioneer studies on carrot had shown that tissue cultures could differentiate roots, and White (1939b) had described the development of leafy buds when his tissue culture of *N. glauca* × *N. langsdorffi* was transferred to liquid medium. Here again, however, no immediate progress was made towards identifying the factors controlling such organogenesis.

During the period from 1939–50, experimental work with root cultures drew attention to the role of vitamins in plant growth and advanced our knowledge of the shoot–root relationship (Street 1957, 1959, 1966a). In the history of plant tissue culture, however, it is a period not inaptly described by Honor Fell (1959) commenting on the first phase in the development of animal tissue culture: 'When tissue culture first appeared on the biological horizon, great things were expected of it. The technique was so elegant, the growing cells were so aesthetically pleasing, that most people felt that tissue culture would immediately lead to major discoveries. Unfortunately, however, too many workers were attracted to the fashionable new field, who were not equipped with the necessary basic knowledge of cytology, physiology and biochemistry that would enable them to make practical use of the method once they had learned it, with the result that tissue culture became a closed world of its own and the mass of descriptive paper that appeared had little relation to current scientific issues.'

However, towards the end of this period and during the early 1950s a number of lines of enquiry were initiated which were to lead to a period of new interest and activity. The studies by Camus (1949) on the induction of vascular differentiation resulting from grafting buds into tissue culture masses led on to important studies on factors controlling vascular tissue differentiation by Wetmore & Sorokin (1955), Wetmore & Rier (1963) and Jeffs & Northcote (1967) (this is further discussed in Chapter 10). The work of Miller & Skoog (1953) on bud formation from cultured pith explants of tobacco led on to the discovery of kinetin (Miller, Skoog, Okumura, van Saltza & Strong 1956). In 1952, Steward initiated work on cultured carrot explants (Steward, Caplin & Millar 1952) which for the first time involved analysis of culture growth in quantitative terms and was to lead to the wide use of coconut milk as a nutrient, and to the discovery of embryogenesis (Steward 1958, Steward, Mapes & Mears 1958, Reinert 1958, 1959, Pilet 1961) (this is further discussed in Chapter 12). In 1953, Muir reported that if fragments of callus of *Tagetes erecta* and *Nicotiana tabacum* are transferred to liquid culture medium and the medium is agitated on a reciprocal shaker, then the callus fragments break up to give a suspension of single cells and cell aggregates and that this suspension can be propagated by subculture (Muir, Hildebrandt & Riker 1954). Similarly, in 1956, Steward and Shantz reported that the supernatant medium bathing their carrot root explants became turbid, due to the presence in it of free-floating cells and small groups of cells. Here again the suspension grew and could be serially subcultured. In 1956 Nickell reported that he had maintained for 4 years by serial subculture a suspension rich in free cells and derived from the hypocotyl of *Phaseolus vulgaris* (these techniques are further discussed in Chapters 4 and 11).

Muir, in his Ph.D. thesis of 1953, reported a further important observation which was to open up the possibility of realizing Haberlandt's objective of culturing single vegetative cells. He carefully isolated, from his suspension cultures and from friable calluses, both of transformed bacteria-free crown-gall tumour cells, uninjured single cells, and succeeded in obtaining from a small proportion of these cells growing tissue cultures (single-cell clones).

This he did by placing the single cells on the upper surface of squares of filter paper whose lower surface made intimate contact with an actively growing 'nurse' crown-gall tissue culture (paper raft nurse technique) (see Fig. 8.1A, p. 192). This arrangement provided the single cells not only with the known nutrients of the culture medium (transmitted via the tissue culture) but also with growth factors synthesized in the massive tissue culture and essential for the induction of division in the isolated cell. Although Muir's success here was due to his choice of the right species and the use of crown-gall tumour cells (see Chapter 13) instead of normal cells, it is from these experiments that later techniques of obtaining single cell clones have developed. One of these techniques is that of agar plating first tested with cells from suspension cultures by Bergmann (1960). By filtering suspension cultures of *Nicotiana tabacum* var. 'Samsum' and *Phaseolus vulgaris* var. 'Early Golden Cluster' he obtained suspensions, 90% of the cells of which were free cells. He then incorporated such suspensions into 1 mm layers of solidified medium (0·6% agar) in petri dishes. The dishes were then sealed and incubated in diffused light. A proportion of the free cells divided and gave rise to visible colonies which could be built up into tissue cultures. This Bergmann technique, modified in a number of small but together important ways, is currently being used in a number of laboratories concerned with single-cell cloning (as discussed in Chapter 8).

In 1959 Melchers and Bergmann cultivated tissue derived from a haploid shoot of *Antirrhinum majus*. The tissue retained its haploid state during several subcultures but then increased in ploidy. Haploid tissue and suspension cultures are clearly of particular interest for those interested in studying mutations, and this subject has been activated again by the demonstration that haploid embryoids and haploid tissue can be obtained by using as explants tobacco anthers excised at the right stage of flower development (Bourgin & Nitsch 1967, Nakata & Tanaka 1968). The species whose anthers are known to behave in this way are very limited at present but if their number can be increased a wide range of haploid tissue cultures will become available (this subject is further discussed in Chapter 9).

The release of protoplasts from root tip cells using a fungal cellulase in 0·6 M sucrose was reported by Cocking in 1960. Protoplasts released by cell-wall degrading enzymes have now been prepared from many plant tissues, including suspension culture cells (Eriksson & Jonasson 1969). The importance of protoplasts to those interested in tissue and cell cultures became apparent when it was shown that such protoplasts could be cultivated, would reform their cell walls and divide. Furthermore, it has already been demonstrated that macromolecules can be readily introduced into protoplasts and that protoplast fusions can be promoted and controlled (this new development is discussed in Chapter 5).

Listing the more important individual steps which have occurred in the development of plant tissue and cell culture techniques fails to bring out the qualitative expansion of potential which results from combining these advances in a given experimental programme nor does it indicate their applications to such fields as plant breeding (see particularly Chapters 5, 8 and 9) and plant pathology (see Chapters 13 and 14). This is particularly so because such a brief outline omits to emphasize the originally gradual, and now rapid increase in sophistication which has been introduced into each aspect of the culture technique. These developments and the ways in which techniques are being combined in current experimental work will emerge clearly in the chapters which follow.

CHAPTER 2

LABORATORY ORGANIZATION

H. E. STREET

Introduction 11
General layout 12
Washing-up facilities and preparation of distilled water 13
Preparation of culture media 16
Sterilization 19
Aseptic manipulations 21
Incubation of cultures 25
Harvesting, growth measurement and biochemical analysis . . . 29
Conclusions 30

INTRODUCTION

An effective laboratory organization for plant tissue culture must include certain essential elements irrespective of whether it is used by a single scientist or a team of workers. In a large unit it is advantageous to separate the individual elements or components of the production line in separate rooms arranged appropriately in relation to one another. However, all the individual components are available in compact form (incubators, incubator shakers, inoculation hoods and so on) which enable an effective small culture unit to be housed within a single laboratory area. In designing any unit, large or small, the overriding consideration in planning must be order and cleanliness, to achieve a high standard of asepsis. The high standard of asepsis is needed because cultures have to be maintained over long periods of time and because many culture media now in use do not promote vigorous growth of contaminants. The culture which collects a contaminant capable of rapid growth can be detected rapidly and rejected; the culture, however, which becomes contaminated with a slow-growing organism may be used to propagate further cultures so that the contamination may not be detected before it is widely distributed in the stock cultures. To avoid this catastrophe it is necessary to adopt very rigorous asepsis and to regularly search for contaminants (test for sterility). The emphasis which will recur in this chapter on cleanliness and purity of reagents and distilled water is not only important as part of the approach to asepsis but also to exclude as far as possible chemical contaminants and to ensure that culture media are always accurately prepared from solutions which have not suffered microbial modification.

The achievement of the high standard of cleanliness requires not only a high standard of work on the part of technicians and scientists, but also the correct design of the laboratories (their siting and ventilation, avoidance of dust traps, choice of hard washable surfaces, good illumination and so on). The atmosphere of a good plant tissue and cell culture laboratory should almost resemble the aseptic streamlined atmosphere conjured up in television science fiction—although we can dispense with the unnatural behaviour and stilted speech of their scientific elite!

GENERAL LAYOUT

Whenever the opportunity presents itself to plan an area for plant tissue and cell culture work, the individual component facilities should be arranged as a production line—there should be a flowline from one room or facility to another in the natural sequence of operations. The kitchen area used for the cleaning of glassware should lead on to the facilities for culture medium preparation, then to the facilities for sterilization. Material should move out of the sterilization area to the facilities for aseptic manipulation, and these should be conveniently placed in relation to the incubation areas. The incubation areas should be placed conveniently in relation to the harvest area, microscope area, and chemical analysis facilities. Cultures or samples from cultures will frequently be moved from the incubation facilities to these areas. The spent cultures and dirty glassware should then easily flow back to the kitchen area. Not only does all this imply careful thought on how the separate facilities are arranged, but emphasizes the need for adequate circulation space between the facilities, and for appropriate storage at each location (clean glassware ready for charging with media, and sterile culture vessels ready for aseptic handling).

A plant tissue and cell culture laboratory should be well serviced with electric power; if there are any facilities for maintained electricity supply (stand-by generator) in case of mains failure, it is essential that this is available to all areas where cultures are being incubated, shaken or stirred. Coal gas is to be avoided but natural gas is less objectionable and can be very useful in small supply in the areas where aseptic manipulations are involved. Compressed air and vacuum are always valuable, and are essential where more sophisticated continuous culture equipment is to be operated. Piped compressed air should be rigorously filtered and maintained at constant pressure by high-grade reducing valves. Piped steam is not essential and, unless absolutely reliable, should not be used for automatic autoclaves; such autoclaves should have their own steam boiler operated by maintained electric supply.

Supply failures or failures of control gear can be disastrous. All temperature-controlled spaces should have safety cut-out devices, in addition to normal controls, to prevent over-heating which can destroy irreplaceable cultures overnight. All essential circuits should have warning lights to attract attention in case of failure; warning bells should be installed to draw attention to critical failures. The closest liaison between night watchmen and scientists can often avert tragedies even at the expense of a night's sleep.

WASHING-UP FACILITIES AND PREPARATION OF DISTILLED WATER

Even a single worker can use a large amount of glassware, particularly of culture vessels. Where more workers are gathered together, the turn-round of glassware must be expected to utilize a large share of available space, and the solution to this problem must be carefully planned. A very high standard of glassware cleaning is essential.

The traditional method of cleaning glassware involves treatment with chromic acid-sulphuric acid mixture, followed by very thorough washing with stiff jets of tap water (at least 5 minutes), and subsequent rinsing with distilled water and finally with double-distilled water. Clean glassware should be protected from dust and, if not used immediately after cleaning, should be rinsed with double-distilled water before use. Dry glassware can be effectively sterilized by heating in an oven to 150°C for 1 hour. All glassware should be of Pyrex or similar boro-silicate glass and should be 'broken in' by filling with distilled water and autoclaving at least twice, being cleaned between auto-clavings with chromic acid-sulphuric acid mixture. To carry out this chromic-sulphuric cleaning the laboratory should have high-pressure water points, lead-topped benches and an acid-resistant floor (Fig. 2.1). The washing out of the chromic acid mixture is most effectively carried out by washing machines constructed of a hard grade of polyvinyl chloride (PVC) which not only provide strong uniform jets of water but effectively dilute the chromic acid as it enters the plumbing system. For immersion of pipettes, test-tubes, petri dishes and small vessels in the chromic acid mixture, polythene or PVC buckets (with handles of the same acid-resistant material) are suitable. Where workers are handling chromic acid mixture they should wear acid-resistant plastic clothing, goggles and hats and rubber boots.

The highly corrosive nature of the chromic acid-sulphuric acid mixture has led to its replacement in many laboratories by modern detergents. Such detergents are used in laboratory glass washing machines and the best of these do follow the detergent wash with an effective hot water wash and finally a wash with cold distilled water. Where such detergents are used, the chromic acid-sulphuric acid mixture is only used to deal with particularly resistant dirt which may survive the detergent wash. The water washes should be just as thorough for removal of detergent as for removal of chromic acid mixture. Chromic-acid mixture should still be used for the breaking in of new glassware. Although washing machines can speed up glassware washing a great deal, it is very important to test for the effectiveness of the water washes in such machines; certain vessels may not be effectively washed internally and will require to be subsequently washed with water jets on jet

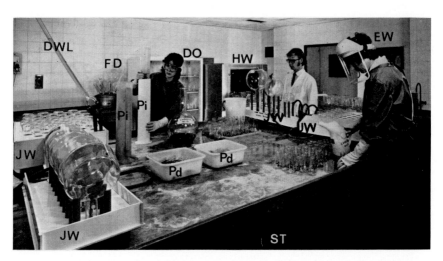

FIG. 2.1. *Special laboratory for the cleaning and drying of glassware used in tissue and cell culture.* Note lead-topped bench with centre gulley and draining to acid-resistant plumbing. The vessels for cleaning pipettes (Pi), petri dishes (Pd) and other glassware with chromic acid are of high-grade polyvinyl chloride (PVC). The various washing machines (JW) are also of hard PVC and supply powerful jets of mains water and distilled water (DWL) to the interior of culture vessels previously treated with chromic acid. Note the protective clothing of the workers handling chromic acid (they are also wearing rubber boots—not shown—and the floor is acid resistant and provided with floor drains). Other items shown include a washing machine (HW) which uses detergent and incorporates a fine jet wash with distilled water; a drying-oven (DO), a hot-air flash dryer (FD) and the emergency first aid (EW). The concentrated sulphuric acid is stored on a large, lead-lined shelf-tray (ST) below the bench. Botanical Laboratories, University of Leicester. (Photograph by G.G.Asquith.)

washers designed for work with the chromic-acid technique. Whenever unusually high variability of growth between replicate cultures or between culture growth on different occasions is encountered, the effectiveness of the glass washing routine should be suspected.

Large quantities of high quality distilled water are needed. A double Pyrex glass distillation unit can be constructed which gives simultaneously a supply of single and double-distilled water and which can run continuously automatically (Fig. 2.2 for a general view of such an assembly; Fig. 2.3 for a flow

FIG. 2.2. *General view of automatic twin still system for the production of single and double distilled water* (for key see Fig. 2.3).
Botanical Laboratories, University of Leicester. (Photograph by G.G.Asquith.)

diagram and details of the floats and electric control gear). Each of the two stills should be effectively baffled to prevent droplet entrainment in the vapour as it moves to the condensers. The first still should be cleaned regularly (as soon as the contents of the boiler show colour or scale). If 'piped' distilled or ion-exchange purified water is available to the laboratory, this should be used to feed the first still; it will greatly extend the period of operation before there is need to dismantle the assembly for cleaning. Where mains water is used, a filter, such as is used to clarify water used in photography, can be fixed at the mains outlet. Large reservoirs collect separate supplies of single and double distilled water and are interconnected by U-tubes. The reservoirs should provide effective protection from dust contamination of the water. Distilled water should not be stored for long periods and the reservoirs should regularly be cleaned with chromic-acid mixture because during storage there can occur a detectable build up of bacterial protein even in double-distilled water in Pyrex bottles. Ammonia vapour can seriously contaminate distilled water.

PREPARATION OF CULTURE MEDIA

Culture media should be prepared from the purest chemicals available; analytical grade chemicals should be used if possible. Significant contaminants can occur in such analytical grade chemicals; hence, to ensure uniformity of medium composition over a long period it is desirable to buy a large amount of each batch of each standard constituent, reserve it exclusively for

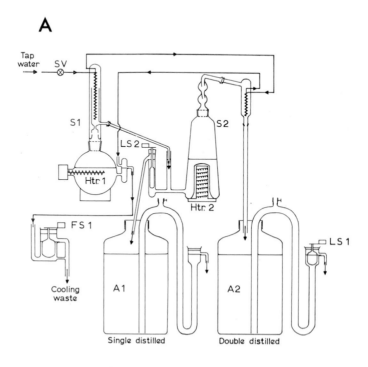

B

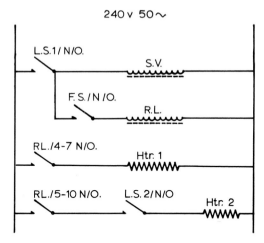

240 v 50 ∿

L.S.1/ N/O.
S.V.
F. S./N /O.
R.L.
RL./4-7 N/O.
Htr. 1
RL./5-10 N/O.   L.S. 2/N/O
Htr. 2

L.S. = Level switch
F. S. = Flow switch
R L. = Relay
S.V. = Solenoid valve
Htr. = Heater

Fig. 2.3. A. *Line drawing of the twin still system* shown in Fig. 2.2. B. *Electrical circuitry of the same still.*

Operation:

1. Level switch (LS 1) is closed by the fall in level of double-distilled water when this is used to prepare culture medium. This energizes the solenoid valve (SV) on the tap water supply and thus admits mains water. The flowswitch (FS) is energized simultaneously.

2. When an adequate supply of mains water is flowing the float rises and closes the flowswitch (FS), and energizes the relay (RL) leading to supply of current to the 3 kw heater (Htr 1) of the first still (S 1, Loughborough Glass Co still) and to activation of the level switch (LS 2) on the second still.

3. As distillation proceeds in S 1, single distilled water collects in the second still (a modified Scorah still) eventually reaching the level which closes LS 2. Excess single distilled now overflows into the collecting aspirator (A 1). When A 1 and other aspirators connected to it by glass siphons are full then excess single-distilled runs to waste (this rarely happens in practice).

4. The closing of the level switch (LS 2) energizes the heater (Htr 2) of S 2 and production of double-distilled water commences and is collected in Aspirator A 2 until the level causes level switch LS 1 to open. This de-energizes the relay (RL) and the solenoid valve (SV) and shuts down the whole still system. Shut-down also occurs automatically if the mains water supply fails to be adequate and restarts when the flow from the mains is adequate. If level switch LS 1 fails, an overflow fitted to aspirator A 2 allows double-distilled water to flow to waste.

media making and check carefully for any change in culture growth when a new batch of any particular chemical has to be used. Details of this sort should always be carefully recorded and dated. The greatest care must be exercised to prevent any cross-contamination of chemicals by dirty spatulas. Chemicals should be weighed into appropriate glass containers; the only exceptions are the weighing of sugars and agar where a hard smooth-surface paper may be used. Wherever possible all weighings should be checked by two persons. Media constituents as they are added should be ticked off on a previously prepared list. More false trails have been laid by mistakes in media preparation than by any other fault of technique. An exact, step by step, routine of media preparation should be worked out.

Often one or only a few basic culture media are in general use in any given laboratory. Such media are required to be made up frequently and often at short notice. It is bad practice and excessively wasteful of glassware to stock-pile sterilized culture medium. It is desirable in most cases to use media within 14 days of their sterilization. The preparation of basic culture media can be facilitated by the use of a concentrated stock solution of the inorganic salt mixture and a concentrated stock solution of vitamins and other chemically defined growth factors stable in aqueous solution. These stock solutions should if necessary be clarified from solid contaminants in the chemicals (cellulose fibres, dust, etc.) by filtration through a No. 3 sintered Pyrex filter. The vitamin and growth factor solution can be dispensed conveniently into Pyrex tubes in small measured volumes appropriate to the normal volumes of media prepared and then frozen and stored at $-20°C$. The inorganic stock solution should be stored at 2–4°C, regularly checked for any precipitate or visible biological contaminant and used within a month of its preparation. Consequently, all such stock solutions should be dated and referenced to the medium preparation book. Some supplements to media, such as yeast extract and casein hydrolysate, should be weighed out on each occasion. Some substances may be used from stock solutions provided these are stored in the dark at low temperature and are prepared fresh every few days (e.g. indol-3yl-acetic acid, IAA). Since the active constituents of coconut milk are reported to be thermostable, the milk drained from the nuts should be immediately deproteinized by boiling, filtered, autoclaved and then stored at $-20°C$.

Agar used to solidify culture media is available in various grades, the most highly purified grades often being very expensive. For preference the agar used should be purified by an ion-exchange process and can be recommended for use in defined media. In our experience there is no substitute for agar as a solidifying agent. Generally it is used within the concentration range 0·6–1·0%.

Radioactive constituents and radioactive media (and the glassware involved in their manipulation) should be kept quite distinct and always labelled 'radioactive'. It should not be overlooked that a significant amount of the radioactive carbon in a culture medium is likely to be released as radioactive carbon dioxide during culture respiration.

STERILIZATION

The standard technique for the sterilization of culture media and culture apparatus is autoclaving at 15 lb/in^2 (121°C) for 15 minutes. When large volumes of medium are autoclaved in bulk two problems arise. It is essential to extend the period at 121°C to ensure that all the solution reaches the required temperature (this should be checked by using 'indicator' ampoules suspended in the centre of the liquid), and a sufficiently slow rate of pressure fall is even more critical if boiling is to be avoided during the cooling phase of the cycle. The autoclaving of large volumes of medium (e.g. 2, 5, or 10 litres) also means that the solution is at high temperature for much longer than when small culture vessels are used (e.g. containing volumes of medium in the range 20–100 ml). A medium relatively stable to autoclaving in small individual volumes may be further modified by autoclaving in bulk, e.g. sugars may show visible caramelization not visible when the normal shorter autoclaving cycle is employed.

The whole question of changes in culture media brought about by autoclaving is in need of further study (Landbouwhogeschool Wageningen 1971). Substances which are stable when autoclaved in pure aqueous solution may be chemically modified when autoclaved in the presence of other medium constituents. Thus hydrolysis of sucrose and the production of sugar acids from monosaccharides proceeds to a greater extent in culture media than in pure aqueous solution (Ferguson, Street & David 1958). Autoclaving a constituent separately and then adding it aseptically to the remainder of the medium (also autoclaved separately) may avoid the necessity of sterilization by filtration. This technique is particularly valuable when testing carbohydrates as carbon sources.

Where a constituent is thermolabile in solution (or is suspected of being so) it may be incorporated in one of two ways. First, by treatment of the dry substance with Analar ethyl ether, removal of the ether at a temperature below 30°C and solution of the solid in sterile water and aseptic addition of the solution to the remainder of the medium, previously autoclaved in the culture vessels (Ferguson, Street & David 1958). Secondly, it may be incorporated by filtration of a solution of the substance through a bacteria-proof filter. Solutions to be filter-sterilized should first be clarified by passing through a No. 3 porosity sintered glass filter. Suitable sterilizing filters are Porosity 5 sintered glass filters, Millipore MF filters pore size = 0·22 μm

(Millipore Filter Corporation, Bedford, Massachusetts) or Gelman Triacetate Metrical (Gelman Instrument Co., P.O. Box 1448, Ann Arbor, Michigan, U.S.A.), or Sartorius Regular filters (via V. A. Howe & Co., Peterborough Road, London S.W.6, U.K.) of similar pore size. When very dilute solutions of growth-active substances are to be sterilized by filtration it is desirable to reject the first portion of the filtrate which may have been reduced in concentration by adsorption. Some thermolabile medium constituents are commercially available as filter-sterilized solutions, e.g. urea (Oxoid Ltd., London S.E.1). When such solutions are to be aseptically added to previously autoclaved medium, the latter should be prepared at more than normal strength so that they are correctly diluted by the added 'ethyl ether' or filter-sterilized solution.

Culture vessels should be fitted with contamination proof closures before sterilization. Traditionally this was by plugs of non-absorbent cotton wool wrapped in a layer of open-weave bandage (previously freed from dressing by boiling in water). Such plugs should be submitted *in situ* to a preliminary autoclaving before use not only to set them but also because they yield waxy material to the sides of the culture vessels during the first autoclaving. The plugs can be used many times provided they are stored in dust-tight containers when not in use. Any plugs which become saturated with water or contaminated with medium should be rejected (and this also applies to plugs from cultures which reveal contamination during incubation). Cotton wool plugs should be covered by inverted loose-fitting aluminium beakers. These keep the exposed rim of the vessel mouth free from dust, protect the plugs from wetting by condensation in the autoclave, and may also be colour marked for identification.

An alternative closure to the cotton plug and one now widely used, particularly for shake suspension cultures, is aluminium foil (0·25 mm thick, soft temper). Appropriately sized squares of foil are placed over the culture vessel opening—test-tube or Erlenmeyer flask—and the overlap pressed neatly to the sides of the tube or neck of the flask by a downward movement of the circle formed by thumb-tip to first finger-tip. These foil caps can be removed and effectively replaced several times to allow for introduction for the inoculum or for sampling of the culture, but are used for one culture only and then rejected. The sterile squares of foil are kept available in the aseptic area to replace cotton plugs used as closures during autoclaving or any covers which may become torn, distorted or inadvertently contaminated. The squares are initially handled into position with sterile forceps. A little experience is needed to make these closures effective and yet not seal them down so hard that the culture becomes partially anaerobic. Presumably air movement between the culture atmosphere and the external atmosphere takes place through fine channels in the folded foil round the neck of the vessel. These closures are effective not only with static cultures but also with suspension cultures on shaking platforms.

Culture apparatus, apparatus for aseptic filtration and individual items of equipment should be wrapped in cellophane sheet (British Cellophane Ltd., Twickenham, Middlesex, U.K.) Grade MAFT 300 or similar; or aluminium foil to protect cotton wool or other filters from saturation and to keep external surfaces sterile in transport from the autoclave to the aseptic area. Bungs to carry glass tubing should be of silicone rubber, as should all flexible tubing in culture apparatus. This silicone rubber is inert and can be repeatedly autoclaved.

Antibiotics have not proved useful in maintaining sterility in plant tissues and cell culture media.

To initiate tissue cultures from seedlings or parts of plant organs it is necessary to effect complete surface sterilization without injury to the cells whose proliferation will generate the initial new tissue. 1% (w/v) bromine water is a very effective sterilizing agent but can only be used on dry seeds to sterilize them prior to germination under aseptic conditions. Bromine water is injurious to many seeds, particularly when the embryo is not heavily protected, e.g. cereal grains. 1% chlorine (as a hypochlorite solution) is widely used due to its lower toxicity. Aqueous (0·1%) or alcoholic mercuric chloride has also been used successfully. Washing with ethanol or with detergent (or incorporation of detergent into the sterilizing solution) may be very valuable where surfaces are not easily wetted (as with surfaces covered with cutin). The sterilizing procedure—choice of sterilizing agent, duration of application —has to be worked out by trial and error. Sometimes an organ from a particular plant may be impossible to sterilize effectively whereas the same organ from another plant grown under different conditions readily yields a sterile culture. The whole process is dependent upon the internal tissue which will proliferate being itself sterile; the sterilizing procedure is designed only to achieve surface sterilization. The organ should if necessary be carefully cleaned and its surface thoroughly wetted before applying the sterilizing agent and subsequently the sterilizing agent must be completely removed by repeated washings with sterile water (see also Chapter 3, section 'Sterilization of material', p. 32 *et seq.*).

ASEPTIC MANIPULATIONS

It is possible and often necessary to carry out certain aseptic manipulations in the incubator rooms or open laboratory; operations such as aseptic sampling of large liquid cultures or the bringing together of ground glass cones and stoppers (using a hand blowpipe) to connect a medium reservoir to a culture vessel. Wherever possible, however, aseptic procedures should be carried out in an area reserved for this purpose where a high standard of air filtration is achieved.

An aseptic area in a multipurpose laboratory can be best achieved by using a Laminair Flow Transfer Cabinet sited at a corner position or along a bench run which is as free from sudden draughts as possible. Cabinets of this kind are manufactured in many countries. Suppliers in the United Kingdom are John Bass Ltd. (Crawley, Sussex), Microflow Ltd. (Fleet, Hants) and Clean Room Construction Ltd. (Sevenoaks, Kent). In such a cabinet (Fig. 2.4) the whole of the back is occupied by a sub-micron particle filter of glass-fibre paper. Air having passed a pre-filter, enters the sub-micron filter and as

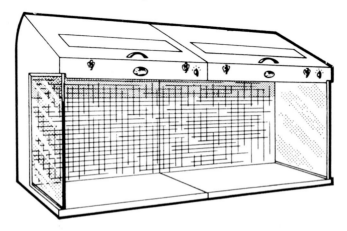

FIG. 2.4. *Laminair Flow Transfer Cabinet for bench mounting.* The entire body of air within the cabinet moves forward along parallel flow lines with uniform velocity (*c.* 90 ft/min) from the whole of the back of the cabinet which is the filter surface. Filtration is achieved in two stages, the second stage being sub-micron. Lighting by fluorescent tubes in the roof gives uniform illumination at the working surface (75 lumens/sq ft). The cabinet shown is made up in two standard units produced by John Bass Ltd., Crawley, Sussex, U.K.

it is released into the working area causes the entire body of air in the cabinet to move forward with uniform velocity (80–100 ft/mm to ensure particles cannot settle in the working area) and along parallel flow lines. Such cabinets can be built to any size specification but the standard units are suitable for bench mounting (33 in. or 84 cm deep, 38–40 in. or 97–102 cm high) and individual units can be mounted side by side (two units are involved in the cabinet in Fig. 2.4).

A completely separate room is very desirable for aseptic work. This room should have a positive pressure of filtered air from a sub-micron filter or should be fitted internally with a recirculating sub-micron filter. All its surfaces should be smooth and washable; an internal windowless room with white walls and a good intensity of artificial light is suitable. Overheating in such a room must be avoided (concentrated work over several hours is often necessary) and it is desirable to have thermostat-controlled refrigeration available for use when ambient temperatures are high. Sterile culture vessels can be placed in this room on formica shelves ready for inoculation. Along

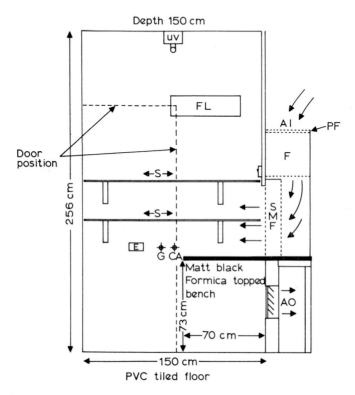

FIG. 2.5. *Section through an aseptic transfer room* (showing services and shelving on wall parallel to section). The fluorescent light (FL) is matched by a similar tube on the opposite wall and they are set at a height to give good shadow-free illumination at the working bench. The filter is a laminiar air-flow (sub-micron) module incorporating a microflow GAA/GA particulate filter. The floor is unbroken PVC and the walls are plastered and covered with light washable paint.

Key: CA=compressed air (20 lb/in²); G=natural gas; E=electric power; UV= Hanovia Bactericidal ultra-violet tube (peak emission 2537 Å) or Phillips TUV Germicidal tube; S=removable plate glass shelves 8 in. deep; AI=air intake to filter; PF=prefilter; F=filter drive fan; SMF=sub-micron filter; AO=air outlet relief vent (adjustable).

Dotted outline shows position of door in near wall (cut away).

(See also Fig. 2.6.)

one side of this room can be sited a series of Laminair Flow Transfer Cabinets or Walk-In Transfer Rooms. Walk-In Transfer Rooms (Figs. 2.5 and 2.6) not only facilitate individual workers starting and completing their work without disturbing others but are particularly valuable where filter-sterilized solutions

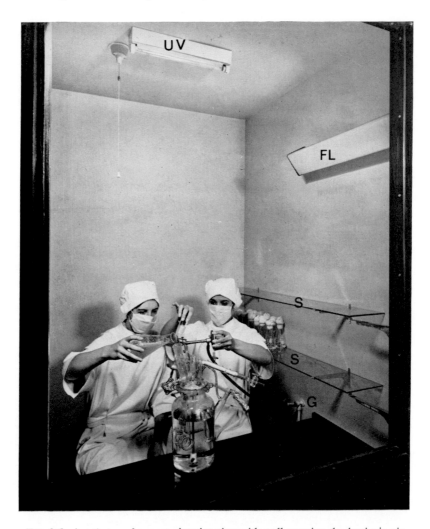

FIG. 2.6. *Aseptic transfer room*—interior view with wall carrying the laminair air-flow filter removed. A 4-litre batch culture vessel (see Chapter 4, p. 71 is being in-oculated with a suspension culture via the inoculation port which is being flamed during the transfer. Note special gowns and hats and face masks being worn. Key as Fig. 2.5.

Botanical Laboratories, University of Leicester. (Photograph by G.G.Asquith.)

are to be added to many culture vessels, where a large number of cultures are to be inoculated, or where large culture equipment has to be handled. A positive forward flow of filtered air over the working bench is maintained in these walk-in rooms throughout their use. They have a well-illuminated matt black-formica work surface, a supply of electricity, natural gas and vacuum and a Hanovia Bactericidal Ultra-violet tube (peak emission 2537 Å) should be fitted at the ceiling. The materials to be handled should be conveniently arranged in the room, the bactericidal tube and the inflow of filtered air should then be switched on and the room closed for at least 30 minutes. The door should form a tight seal. An exit grille (adjustable louvres) below the working surface allows air to escape. The bactericidal tube is switched off not later than 4 minutes after aseptic transfer commences. Unless the operator is sensitive to the radiation he should deliberately expose his hands to it before commencing work (but he should not look directly at the live tube). The wearing of thin sterilized rubber gloves may in some cases be warranted (e.g. to remove and replace a sterilized silicone rubber stopper). Between work sessions the working bench should be washed with alcohol. At least once a week all internal surfaces should be washed over with antiseptic or spirit. A room of this kind can be easily blacked out if it is desired to effect transfer of dark-grown cultures in the light of a green photographic safelight. Dark grown cultures can be incubated alongside light-grown cultures by wrapping the outside of the culture vessels of the former in aluminium foil.

Various techniques can be used to maintain instruments (scalpels, forceps, scissors) in a sterile condition. The working part can be dipped in alcohol, flamed in a spirit lamp or natural-gas microbunsen and cooled in a rack or returned to boiling water after each manipulation. Sterile instruments should be quenched in the medium remote from the culture when handling tissue cultures. Pipettes and syringes used for aseptic handling of solutions or suspension cultures are sterilized by autoclaving either wrapped or in metal containers which can be sealed as they are removed from the autoclave. A new syringe should be taken each time a new stock suspension culture is used for subculture. When performing aseptic manipulations either in a multipurpose laboratory or in a special aseptic room, the operator should carefully scrub his hands and lower parts of his arms, put on a clean surgeon's gown and hat (reserved for such work in a special locker) and wear a disposable face mask. The hat should completely enclose the operator's hair.

INCUBATION OF CULTURES

Tissue cultures growing on the surface of solidified medium in test-tubes or Erlenmeyer flasks should be transported in racks or trays and incubated either

B

in a bacteriological incubator or in a temperature-controlled room. It is usual for such temperature-controlled rooms to be continuously illuminated with artificial light at a good working intensity. If this is the case, cultures which it is desired to grow in the dark are kept in vessels wrapped in aluminium foil or other suitable black-out material. Bacteriological incubators with an illuminated ceiling are available and thus also permit of culture in light and darkness. Many plant tissue cultures grow well (often equally well) in either light or darkness but it is advantageous to be able to test both regimes. Too little is, however, known about the effect of low light intensities on culture growth to suggest any optimum intensity level or spectral composition for general culture maintenance. To study more critically the effects of light on culture growth and morphogenesis, it is necessary to have recourse to special growth chambers where light intensity, light quality, and alternation of light and dark periods can be controlled. If it is desired to obtain maximum greening of cultures, higher light intensities are required than would be desirable in a general purpose temperature-controlled room or incubator; of particular interest for such work with shaken suspension cultures are the Controlled Environment Incubator Shakers of the New Brunswick Scientific Co. and the 'Clim-o-shake' of Adolf Kühner A.G., Basel, Switzerland.

The air temperature of the incubator or room should not vary beyond $\pm 1°C$ of the desired temperature; for general use a temperature of 25°C is usually adopted (although increase in culture growth up to 30°C can be demonstrated with some tissue cultures). If an internal room is used, elaborate lagging of the walls is not required to get good temperature control and evenness of temperature. Much more important is good internal air circulation (a low speed, large blade fan may be valuable here), low temperature heaters and sensitive air thermostats. When such a room is being commissioned, care should be taken in the positioning of the thermostat and accurate maximum–minimum thermometers around the room. Although satisfactory temperature-controlled rooms can often be operated with only a heating cycle it will be impossible to maintain 25°C unless the ambient temperature is several degrees below this; there is always a small continuous heat output from the illumination. The heat output in the room is greatly increased if electric motors are running (when it is essential to have both refrigeration and heating). The simplest arrangement is to have constant (and only just excessive) refrigeration and intermittent heat (thermostat controlled). The refrigeration plant should be in a well-ventilated space as close to the room as possible and readily accessible for maintenance. With incubators, either refrigeration or at least a cold water circulating coil are needed for effective control during hot weather.

One serious risk with all temperature-controlled space is overheating due to failure of refrigeration or failure of thermostats and associated control gear. Overheating is much more dangerous than cooling below the desired temperature. Consequently, a second thermostat should therefore always be installed, set to operate when the temperature rises 2·5–3·0°C above the desired temperature and to effect a complete shut down of the heating cycle until it is reset. This safety thermostat can be linked to a special warning light or alarm bell outside the room. An externally mounted temperature recorder enables the operation of the room to be constantly monitored and undue temperature fluctuations detected (such fluctuations are often a warning of impending thermostat failure).

Cell suspension cultures can be satisfactorily incubated in a multipurpose laboratory by using platform-shaker incubators (Fig. 2.7). These are now available from a number of suppliers: Orbital Controlled-temperature Incubator (L.H. Engineering Co. Ltd., Stoke Poges, Bucks., U.K.), Orbital Incubator (Gallenkamp & Co. Ltd., P.O. Box 290, London E.C.2, U.K.); Incubator Shakers (New Brunswick Scientific Co. Inc., P.O. Box 606, New Brunswick, New Jersey, U.S.A.). Again, it is preferable for them to have a built-in cold water cooling coil so that the temperature can be held at 25°C in hot weather. The lid of the incubator has a glass panel for observation of the cultures and to allow of their illumination. Where a temperature-controlled room is available, platform shakers without an enclosing incubator can be used. Such shakers are available in various capacities and the capacity of single platform models is easily doubled by building a second platform of perspex (Fig. 2.8). Clips are available to fit 100, 200 and 500 ml Erlenmeyer flasks (the larger clip will also hold litre pyrex bottles). With low speeds of agitation, clips can be dispensed with and the surface of the platform covered with studded rubber sheet such as is used on table-tennis bats. These platform-shakers should have a variable speed control; the optimum shaking speed must be determined for the culture in question and for each culture vessel and culture volume in the vessel (Rajasekhar, Edwards, Wilson & Street 1971). Maintained electric supply to such shakers is important for growth of the suspension ceases very quickly after shaker failure and in the matter of a few hours the cultures become partially anaerobic. Platform shakers suitable for plant suspension culture work are manufactured by L.H. Engineering Co. Ltd., Stoke Poges, Bucks., U.K.; New Brunswick Scientific Co. Inc., P.O. Box 606, New Brunswick, New Jersey, U.S.A., the Northern Media Supply Ltd., Hull, U.K., and Adolf Kühner A.G., Basel, Switzerland.

Culture systems such as that designed by Steward (Steward, Caplin & Millar 1952) whereby explants or callus pieces are alternatively bathed in culture fluid and exposed to air must operate in temperature-controlled rooms. This also applies to large spinning cultures such as those described by Lamport (1964) and Short, Brown & Street (1969a).

Batch and continuous culture systems which are stirred by a magnetic

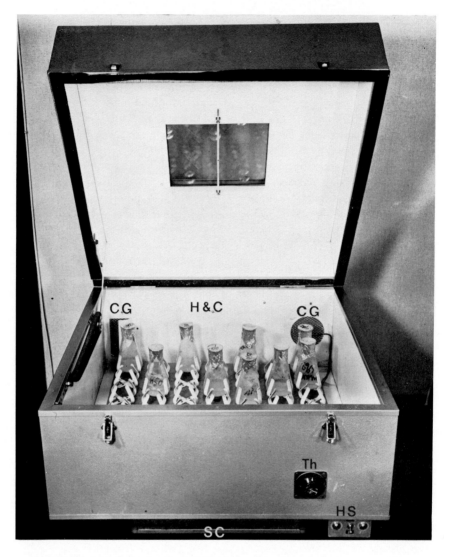

FIG. 2.7. *Platform-shaker incubator* (L.H. Engineering Co. Ltd., Stoke Poges, Bucks., U.K.). Note glass panel and thermometer mounting in lid.
Key: CG=air circulator grills; H+C=panel behind which is mounted heater, cooling coil and fan; Th=Temperature sensor adjustment; HS=Heater switch; SC=Position of access to speed control on shaker platform.

stirrer and a piped air supply (Melchers & Bergmann 1959, Miller, Shyluk Gamborg & Kilpatrick 1968, Wilson, King & Street 1971) can conveniently be operated in a temperature-controlled room provided compressed air is available, but can also be operated in the open laboratory by being fitted with a temperature-controlling internal water circulating coil (Wilson, King & Street 1971).

FIG. 2.8. *Thermostatically-controlled incubation room* containing three large platform shakers (S 1, S 2, S 3) and facilities for larger scale batch cultures. Uniform light over the shakers is obtained by three, round, fluorescent lights (FL) adjustable in height. Note that each shaker carries an upper platform of perspex. The shakers are carrying 100 ml and 250 ml wide-mouthed Erlenmeyer culture flasks closed by aluminium foil and containing various cell suspension cultures.

Key: T + HR = temperature and humidity recorder; BC = batch cultures (41) as described in Chapter 4, Figs. 4.7 and 4.8; CA = compressed air-reducing valve and filter; FM = flow meter controls for air supply to the batch cultures; PB = perspex boxes with inlets and outlets which enable the cultures to be incubated in an experimental atmosphere (which can be intermittently or continuously renewed); TS = temperature sensor.

HARVESTING, GROWTH MEASUREMENT AND BIOCHEMICAL ANALYSIS

The general laboratory facilities associated with a Plant Tissue and Cell

Culture laboratory may be simple or elaborate according to the scope of the work. A microscope area is essential for cell counting and cytological study. The most generally used cell counting procedure involves cell separation by a controlled treatment with 5% aqueous chromic acid followed by violent agitation of the suspension. A protective renewable bench covering is needed where the chromic acid is handled and the flask shakers must not transmit their vibrations to the microscope bench. Rapid accurate balances are needed for fresh and dry weight determinations. A controlled-temperature oven (up to 100°C) is used for drying tissue cultures or pellets of harvested cells. A low speed centrifuge is used to determine packed cell volumes (volume of cells per unit volume of culture). A pH meter is needed for routine checking of the pH of media.

Beyond such basic facilities are also those involved in the analysis of culture media, analysis of cells, enzyme assays and in handling and assaying radio-isotopes. Wherever metabolic studies are undertaken there should be available a rapid freeze-dryer so that lyophilized tissue and cells can be accumulated, particularly for bioassays and chemical detection of growth-regulating substances. A system for effective cell rupture is required for making cell extracts or preparations of cell particles, in our experience cell rupture can be effected by the Willems Polytron homogenizer (available in the U.K. from the Northern Media Supply Ltd., Hull) or the French Press, (French & Milner 1955), or by adpression of frozen cells against a rotating (1500 rpm) abrasive (carborundum) disc (Leis & Ralph 1960).

CONCLUSIONS

Much of what has been said in this chapter may seem obvious to laboratory biologists, even more so to those with experience in plant tissue and cell culture. Nevertheless, principles of good laboratory planning and practice cannot be too often set down. Time and time again setbacks have occurred in the work of my own laboratory through failure to take precautions or follow carefully the procedures outlined. The successful running of a plant tissue and cell culture laboratory is not difficult, but the nature of the work does not allow for short cuts to valid results. Constant vigilance against taking such short cuts is required; the consequences of slipshod work are often not immediate but may be disastrous in the long run.

CHAPTER 3

TISSUE (CALLUS) CULTURES—TECHNIQUES

M. M. Yeoman

Sources of material	31
Sterilization of material	32
Isolation of plant tissues	33
Preparation of explants	35
Standard nutrient media	37
Basic mineral nutrients	37
Organic growth factors	38
Culture techniques	43
Solid media	44
Stationary liquid media	45
Agitated liquid media	46
Subculture and the preservation of cultures	47
Histological and histochemical techniques	48
Choice of parameters for the measurement of growth	50
Biochemical techniques	54
Concluding remarks	58

SOURCES OF MATERIAL

Few plant tissues fail to respond to treatments designed to induce the formation of a callus and it now seems clear that the isolation and successful establishment of a callus largely depends on the culture conditions employed and not on the source of plant material. For a variety of reasons attempts to isolate calluses have been largely directed towards flowering plants with the major emphasis on dicotyledonous materials (Gautheret 1959, Butenko 1964). There are however, reports of successful callus initiations in mono-cotyledons (Straus 1954, Webster 1966, Carter *et al.* 1967, Trione *et al.* 1968), gymnosperms (Reinert & White 1956), ferns (Kato 1963, 1964) and mosses (Ward 1960) although these groups of plants are as yet relatively unexplored. Therefore it may be said that all multicellular plants are potential sources of callus cultures.

The preference for dicotyledonous plants as a source of material is in part due to the free availability of a wide range of specimens and especially because they excel as experimental material in providing a rapid response to callus inducers. Tissues from various parts of the plant can be grown in culture and produce callus cultures: vascular cambia (White 1963) storage parenchyma (Skoog & Tsui 1948, Steward, Caplin & Millar 1952, Nitsch & Nitsch 1956), pericycle of roots (Goldacre, Unt & Kefford 1962), endosperm (La Rue 1947, Straus 1954), cotyledons (Witham 1968), leaf mesophyll (Joshi & Ball 1968a) and provascular tissue (Venketeswaran 1962).

STERILIZATION OF MATERIAL

Plants are inevitably contaminated with a wide range of micro-organisms. The retention of viable contaminants in association with a plant tissue in contact with a nutrient medium containing a source of carbon leads to rapid proliferation of the contaminants and seriously affects the growth of the callus. Therefore the need for complete asepsis is vital. Occasionally anti-biotics have been employed to prevent growth of bacteria (Montant 1957) but this is only effective in very short-term cultures and has the added danger that the antibiotics themselves may seriously impair the growth of the tissue. In addition, only the bacteria within the spectrum of the antibiotic will be affected and other bacteria and fungi will proliferate unmolested. It should be recognized that virus-infected plants are likely to yield a callus in which the virus will persist (Hirth & Durr 1971, Streissle 1971).

In rare circumstances it is possible to obtain completely sterile material from a plant without previous sterilization. Indeed, it is possible to grow completely sterile plants. However, for the majority of isolation procedures, sterilization of the plant or plant part is a vital prerequisite to the successful isolation of a callus. These sterilization procedures will now be described and discussed.

There are a variety of chemical agents in common use for the surface sterilization of plant material. The choice of agent and the time of treatment depends on the sensitivity of the material to be sterilized. Frequently it is discovered that over-zealous sterilization leads not only to the complete removal of all micro-organisms but is also lethal to the plant tissue. It is, therefore, important to determine the optimal conditions for each tissue.

The sterilizing agent should be easily removable, because the retention of such noxious chemicals will seriously affect the establishment of the callus. Repeated washing with distilled water will remove most chemicals. Some sterilizing agents break down and become less toxic and the products can be easily washed away. For example, sodium hypochlorite breaks down to give chlorine, the active agent, and sodium hydroxide, which can be removed, while others like hydrogen peroxide decompose to give harmless components, which evaporate. Other compounds, for example silver nitrate, may be inactivated by the addition of a second chemical (sodium chloride) and therefore, like hydrogen peroxide the sterilizing agent is rendered harmless to the tissue. Dilute mercuric chloride is a satisfactory sterilizing agent but difficult to remove.

The effectiveness of most sterilizing chemicals can be enhanced if a small amount of detergent (0·05%) like Teepol or Lissapol F is incorporated into the sterilizing solution. The addition of the detergent wets the tissue surface and allows the chemical to penetrate and destroy the micro-organisms. Another method frequently employed is to rinse the object briefly in absolute alcohol before placing in the sterilizing agent. In Table 3.1 comparison is made of the more important characteristics of several sterilizing agents in common usage.

TABLE 3.1. A comparison of the effectiveness and properties of several sterilizing agents

Sterilizing agent	Concentration used	Ease of removal	Time of sterilization minutes	Effectiveness
Calcium hypochlorite	9–10%	+++	5–30	Very good
Sodium hypochlorite	2%*	+++	5–30	Very good
Hydrogen peroxide	10–12%	+++++	5–15	Good
Bromine water	1–2%	+++	2–10	Very good
Silver nitrate	1%	+	5–30	Good
Mercuric chloride	0·1–1%	+	2–10	Satisfactory
Antibiotics	4–50 mg/litre	++	30–60	Fairly Good

* 20% v/v of a commercial solution

Details of sterilization procedures for a range of plant structures are included in Table 3.2. It must be emphasized that a procedure should be established for each tissue.

ISOLATION OF PLANT TISSUES

The induction of a callus from a part of a plant occurs when the sterile explant is brought into contact with a nutrient medium known to induce and support cell division. In certain instances the explant may appear to be a uniform piece of tissue largely composed of one cell type, such as an explant from the tuber of the Jerusalem artichoke, the secondary phloem of the carrot, or the pith of the tobacco plant. However, even in a tissue in which the cells appear to be similar, there may be marked differences in the DNA content of the nuclei of the cells and this may be the consequence of endo-polyploidy. This is particularly true of tobacco pith (Patau, Das & Skoog 1957). The callus produced from a morphologically uniform tissue will, at least in the initial stages, give a relatively uniform callus. On the other hand,

TABLE 3.2. Sterilization procedures for different plant organs

	PROCEDURE			
TISSUE	Pre-sterilization	Sterilization	Post-sterilization	Remarks
SEEDS	Submerge in absolute ethanol for 10 sec and rinse in sterile distilled water	Seeds with intact testas submerged for 20–30 min. in 10% w/v calcium hypochlorite or for 5 min. in a 1% (w/v) solution of bromine water	Washed three times in sterile water and germinated in sterile water. Washed five times with sterile distilled water and germinated on damp sterile filter paper	Root and shoot tissue for callus culture. Excellent for tomato seeds (Street & Henshaw 1968)
FRUITS	Rinse briefly with absolute ethanol	Submerge for 10 min. in 2% (w/v) sodium hypochlorite	Wash repeatedly with sterile water, dissect out seeds or interior tissue	Good source of sterile seedlings
PIECES OF STEM	Scrub clean in running tap water and rinse with absolute ethanol	Immerse for 15–30 min. in 2% (w/v) sodium hypochlorite, remove ends	Wash three times in sterile water	Plant vertically in agar medium or dissect tissue out and culture in isolation
STORAGE ORGANS	Scrub clean in running tap water	Submerge for 20–30 min. in 2% (w/v) hypochlorite	Wash three times in sterile water. Dry with sterile tissue paper	Remove tissue from the inside of the structure
LEAVES	Rub surfaces briefly with absolute ethanol	Immerse for about 1 min. in 0·1% (w/v) mercuric chloride	Wash repeatedly with sterile water. Dry with sterile tissue paper	Difficult material to sterilize. Choose very young leaves. Lamina laid on agar or petiole inserted in agar

explants which contain a variety of different cell types will tend to produce a mixed callus. Frequently callus cultures are established from roots or stem sections and owe their origins to a variety of different cells present in the original tissues. Sometimes it is possible to dissect out a particular tissue from a piece of root or stem and culture this fragment in isolation. Occasionally it may be possible to encourage one group of cells to divide at the expense of the rest and establish a callus of one cell type (Torrey 1961). However, it has been shown that in instances where this has been achieved a simple change in the culture medium can alter the whole nature of the callus by promoting rapid division in a minority population (Matthysse & Torrey 1967a, b).

PREPARATION OF EXPLANTS

The size and shape of the initial explant is normally not critical, although proliferation may completely fail to occur with explants below a critical size (see below). In general, fairly large pieces of tissue are favoured because the large number of cells present increases the chance of obtaining a viable culture. In investigations of this type explants may be pieces of stem, leaf, root, flower, fruit or seed or tissues dissected from such structures. Where the aim is the establishment of a callus culture the successful establishment of one explant may be enough.

In contrast, where quantitative studies on the development of the callus are the aim, it is essential to prepare and select explants which are, as far as possible, equivalent in size, shape and composition. Clearly, only certain structures can be used for such investigations for they must be large enough to provide a sufficient amount of sterile uniform tissue suitable for culture. Large organs such as the storage root of carrot and parsnip and the underground stems of potato and the Jerusalem artichoke have proved to be popular sources of material for such study (Steward, Caplin & Millar 1952, Yeoman, Dyer & Robertson 1965). A typical example of such a technique is in constant use in the author's laboratory for the preparation of tissue from the dormant tuber of the Jerusalem artichoke. This technique has also been successfully employed with carrot, white turnip and parsnip.

The sterile tuber is cut into slices 25 mm deep and cylinders of tissue are removed from the storage parenchyma region. These cylinders are lined up parallel to one another on a cutter (Fig. 3.1) and the tissue cut into explants $2\cdot4 \times 2\cdot0$ mm by the razor blades fixed to the upper part of the cutter. The choice of size and shape of the explant was decided after careful consideration of a number of factors. A cylinder is convenient because it is a simple shape to produce with accuracy in very large numbers and also possesses a high surface area to volume thus facilitating exchange of gases and nutrients (Caplin 1963). It is also clear from a series of investigations (Yeoman, Naik & Robertson 1968, Yeoman & Mitchell 1970) that some substance emanates from the damaged surface of the explant and is active in inducing the divisions which lead to the development of a callus. Therefore, a high surface

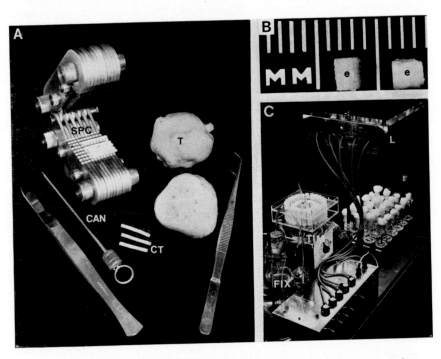

FIG. 3.1. *Technique for the isolation and culture of explants from the Jerusalem artichoke tuber.* (A) Cylinders of tissue (CT) have been removed from the parent tuber (T) with a stainless steel cannula (CAN) and sub-divided into individual explants with a special cutter (SPC). (B) Isolated explants (E) photographed against a millimetre scale (MM). (C) Explants and a magnetic stirring bar encased in PTFE placed in 15 ml of nutrient medium in a 100 ml conical flask. Tissue and medium are stirred by the coated magnet which is revolved by a synchronous electric motor revolving beneath the flask at 250 rpm. Provision is made for the illumination of the cultures (L) and for killing the contents of each flask at predetermined times with a fixative (FIX). Automatic timer (TIM).

area/volume ratio is desirable to provide a maximum amount of this growth substance. The explant should be as small as possible so that a maximum number of explants may be obtained from a single source. The minimum size of the explant is determined by the average cell size of the parent tissue. For example explants removed from the secondary phloem of the carrot root are made up of cells about one-third the size of similar explants removed from the Jerusalem artichoke tuber. A carrot explant of 3·8 mg which contains approximately 25,000 cells is a viable entity, an artichoke explant of similar size in which the constituent cells are much larger tends to be less viable because of the damage produced during excision. The minimum size of the artichoke explant is about 8 mg and contains approximately 20,000 cells. Caplin (1963), however, using a different method of excising tissue reports the successful culture of much smaller explants of artichoke.

STANDARD NUTRIENT MEDIA

So far, in plant tissue culture research the establishment of an autotrophic culture has not been achieved. The basis of all nutrient media is a mixture of mineral salts combining the essential macro- and micro-elements together with a source of carbon which is almost always a sugar and usually sucrose. Very few plant tissue cultures can, however, be established or maintained on such a simple medium and for the majority of tissues various additives are essential. The usual supplements are vitamins, amino acids, sugar alcohols, auxins and related plant growth regulators, gibberellins, a chelate such as EDTA, kinetin or other cytokinin and various natural extracts such as coconut milk. The present trend is, however, towards fully defined media and the use of complex mixtures such as yeast extract, tomato juice and coconut milk is losing favour. It is important to recognize that a culture medium suitable for the isolation of a callus from an explant may not be suitable for the maintenance of such callus culture and also, that a mixture which promotes rapid growth in a liquid medium may not be as suitable in a solid medium. In the following pages an appraisal will be made of several standard nutrient media in use in laboratories all over the world.

BASIC MINERAL NUTRIENTS

Actively growing plant tissues require a continuous supply of inorganic elements and the composition of the mineral salts mixture must take this into account. The elements required in greatest amount apart from C, O and H are: N, usually supplied as nitrate or ammonium; P, added almost universally as phosphate; S, as sulphate and K, present as the major cation. Ca, Na and Mg are required in smaller amounts, together with Cl. Various mixtures of macro-elements are in use and these have been developed from nutrient mixtures originally devised for the culture of whole plants, such as Knop's solution. Mixtures of macro-inorganic and micro-inorganic nutrients are compared in Tables 3.3 and 3.4. These media have been selected because they are used most widely in basic or modified forms.

The overall concentration of all mineral nutrients is high in the media of Nitsch, and Murashige and Skoog and low in Gautheret's medium. Apart from the medium of Murashige and Skoog, where a mixture of nitrate and ammonia is used, the preferred nitrogen source is nitrate. Potassium levels are high in all media except that of Gautheret. Levels of sodium and sulphur are exceptionally high in the medium of Hidebrandt, Riker and Duggar where the phosphorus level is especially low. It is interesting to note that widely divergent media in terms of concentration of individual ions are reported to support rapid rates of growth of similar tissues. For example the medium of Hildebrandt, Riker and Duggar contrasts with that of Murashige and Skoog, although both are used for tobacco tissue.

In addition to basic macro-inorganic nutrients most culture media contain selected micro-inorganic nutrients. It is inevitable that even the purest available salts will contain traces of contaminating substances and these will provide a hidden supply of micro-nutrients. Those not convinced by this, need only peruse the label of a bottle of some Analar chemical when they will learn just how pure a particular salt may or may not be! Despite these traces, it is considered necessary to add specific mixtures of micro-elements. Several combinations are shown in Table 3.4 from which it is clear that there are considerable differences between recipes in the total concentrations of micro-nutrients. The medium of Murashige and Skoog has a particularly high micro-nutrient content. Gautheret's micro-nutrient mixture contains titanium and beryllium which are not present in other mixtures of micro-nutrients. To ensure the availability of iron to cultures throughout extended periods of growth it is advisable to add this element as ferric-sodium ethylenediamine tetra-acetate (Fe-EDTA). The presence of this compound in the medium also ensures that iron is available over a wide pH range. (Ferguson, Street & David 1958, Sheat, Fletcher & Street 1959, Klein & Manos 1960). Little really critical work has been done on the essential micro-nutrients required by callus cultures and all of the mixtures quoted appear adequate to support rapid and prolonged growth over the limited cultural periods employed in propagating callus.

ORGANIC GROWTH FACTORS

The response of a piece of excised tissue in culture depends on the endogenous growth substances present at the time of excision. For example, relatively small cuttings can produce large callus outgrowths at the cut surface in the absence of added growth substances (Fig. 3.2). Excision alone will produce

TABLE 3.3. Inorganic macro-nutrients present in selected plant tissue culture media (all values expressed as milligrammes per litre)

Constituent	Heller (1953)	Nitsch & Nitsch (1956)	White (1963)	Hildebrandt, Riker & Duggar (1946)	Murashige & Skoog (1962)	Gautheret (1942)
KCl	750	1500	65	65	—	—
$NaNO_3$	600	—	—	—	—	—
$MgSO_4.7H_2O$	250	250	720	180	370	125
$NaH_2PO_4.H_2O$	125	250	16·5	33	—	—
$CaCl_2.2H_2O$	75	—	—	—	440	—
KNO_3	—	2000	80	80	1900	125
$CaCl_2$	—	25	—	—	—	—
Na_2SO_4	—	—	200	800	—	—
$Ca(NO_3)_2$	—	—	—	—	—	—
NH_4NO_3	—	—	—	—	1650	—
KH_2PO_4	—	—	—	—	170	125
$MgSO_4$	—	—	—	—	—	—
$Ca(NO_3).4H_2O$	—	—	300	400	—	500

TABLE 3.4. Inorganic micro-nutrients present in selected plant tissue culture media (all values expressed as milligrammes per litre)

Constituent	Heller (1953)	Nitsch & Nitsch (1956)	White (1963)	Hildebrandt, Riker & Duggar (1946)	Murashige & Skoog (1962)	Gautheret (1942)
$NiSO_4$	—	—	—	—	—	0·05
$FeSO_4.7H_2O$	—	—	—	—	27·8	0·05
$MnSO_4.4H_2O$	0·01	3	7	4·5	22·3	3
KI	0·01	—	0·75	3·0	0·83	0·5
$NiCl_2.6H_2O$	0·03	—	—	—	—	—
$CoCl_2.6H_2O$	—	—	—	—	0·025	—
$Ti(SO_4)_3$	—	—	—	—	—	0·2
$ZnSO_4.7H_2O$	1·0	0·5	3	6·0	8·6	0·18
$CuSO_4.5H_2O$	0·03	0·025	—	—	0·025	0·05
$BeSO_4$	—	—	—	—	—	0·1
H_3BO_3	1·0	0·5	1·5	0·38	6·2	0·05
H_2SO_4	—	—	—	—	—	1·0
$FeCl_3.6H_2O$	1·0	—	—	—	—	—
$Na_2MoO_4.2H_2O$	—	0·025	—	—	0·25	—
$AlCl_3$	0·03	—	—	—	—	—
$Fe_2(SO_4)_3$	—	—	2·5	—	—	—
Ferric tartrate	—	—	—	40·0	—	—

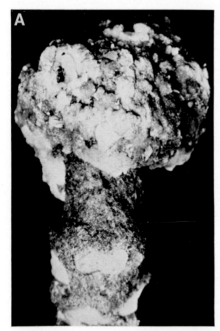

a wound response, accompanied by the induction of cell division and a callus may result. Tissue from the immature lemon fruit (Kordan 1959) and isolated vascular cambia (White 1963) will produce a callus culture in the presence of a sugar and mineral salts alone. However, few excised tissues yield a prominent callus in the absence of added growth factors. These tissues may be placed in one of four categories:

(1) Tissues which require only an auxin or related growth regulator;
(2) Those that require only a cytokinin;
(3) Those that require both an auxin and a cytokinin;
(4) Tissues which will only respond to media containing complex natural extracts.

The first group includes tissue isolated from the dormant tuber of the Jerusalem artichoke and the roots of *Cichorium* and *Scorzonera*. The second group is small and includes tissue from the root of the white turnip while the third group includes tissue from tobacco pith and those from the storage root of the carrot and potato tuber. Group four includes a wide range of tissues for which defined conditions have not so far been achieved.

The maintenance of the callus, once established is, however, another matter and it is here that the requirement for complex additives may become necessary. In anticipation of this requirement many workers use complex media for callus initiation. Some workers (for example White 1963) maintain that the use of media with complex natural extracts is advantageous for the initiation of a callus and the callus, once established, may then be cultured on simple defined media. For all practical purposes a mineral salts mixture with sucrose and 10% (v/v) coconut milk is difficult to surpass. Sometimes the coconut milk may be replaced by $0\cdot5\%$ yeast extract (Straus & La Rue 1954, Torrey & Shigemura 1957) or 5–10% tomato juice (La Rue 1949).

The auxins and related growth regulators normally used in the initiation and maintenance of callus cultures are indol-3yl-acetic acid (IAA), at a concentration of from 10^{-5} to 10^{-10} M, naphthalene acetic acid (NAA) which can be used at a somewhat higher concentration and 2,4-dichlorophenoxy-acetic acid (2,4-D) which is usually active at 10^{-5} to 10^{-7} M. The data presented in Fig. 3.3 show the effect of 2,4-D concentration on the growth of explants isolated from the Jerusalem artichoke tuber. The optimum concentration of 2,4-D both for cell division and fresh weight increase is in the region of 10^{-6} M. Little growth takes place in the absence of this growth substance

---

FIG. 3.2. *Formation of a callus outgrowth at cut surfaces.* (A) Massive callus on a *Rhododendron* cutting. (B) Callus on the stem of *Rubus fruticosus*. (C) A chronological series showing a fresh stem cutting of *Chamaecyparis lawsoniana*, the formation of a callus and the emergence of roots accompanied by the drying out of the callus. All of the cuttings had been exposed to a mist propagation procedure for several weeks.

while high concentrations inhibit growth completely. The optimum concentration of 2,4-D for carrot explants cultured in the presence of coconut milk is $2 \cdot 7 \times 10^{-6}$ M (Steward & Caplin 1952). Murashige & Skoog (1962) showed that an overall concentration of 10^{-6} M NAA or 2,4-D gave satisfactory growth in terms of fresh weight for tobacco pith.

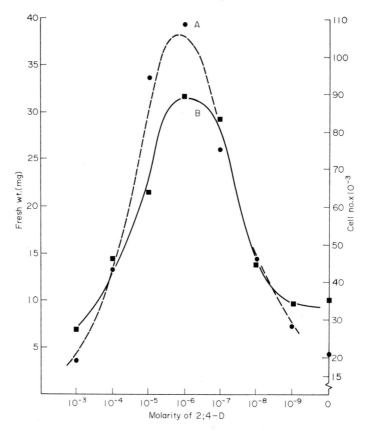

FIG. 3.3. *Effect of 2,4-D concentration on the increase in cell number* (A) *and fresh weight* (B) *of tissue excised from the Jerusalem artichoke tuber.* Cultures grown for 7 days in an agitated liquid medium containing sucrose and mineral salts.

As has already been explained some tissues fail to respond to auxin alone and have a requirement for a cytokinin. Cytokinins are, by definition, naturally occurring substances and are present in extracts from green plants. Few cytokinins have been characterized since Letham (1964) isolated and characterized zeatin from maize fruits. However, a range of synthetic substances have the properties of cytokinins and foremost among these is kinetin, a substance discovered by Miller, Skoog, Saltza & Strong (1955).

Kinetin has been widely used to initiate and maintain growth in callus cultures and is particularly active as an interactant with IAA or 2,4-D in the culture of tobacco pith. The optimum concentration for tobacco pith is 10^{-7} to 10^{-6} M.

The maintenance of a callus in culture may require the presence of selected amino acids such as glycine or a mixture of amino acids usually supplied as casein hydrolysate (Steward & Caplin 1952). Vitamins may also be required and are frequently included in standard media. White's medium (White 1963) contains vitamin B_1 and B_6 at 0·1 mg/litre and nicotinic acid at 0·5 mg/litre as well as 3 mg/litre of glycine. Reinert & White (1956) used

TABLE 3.5. Organic constituents present in selected plant tissue culture media (all values expressed as milligrammes per litre)

Constituent	Heller (1953)	Nitsch & Nitsch (1956)	White (1963)	Hildebrandt, Riker & Duggar (1946)	Murashige & Skoog (1962)	Gautheret (1942)
Sucrose	20,000	34,000	20,000	20,000	30,000	30,000
Glycine	—	—	3	3	2	3
Myo-Inositol	—	—	—	—	100	—
IAA	—	0·18–1·8	—	—	1–30	—
Cysteine	—	- -	1·0	—	—	10
Vit B_1	1·0	—	0·1	0·1	0·1	0·1
Vit B_6	—	—	0·1	—	0·5	0·1
Nicotinic acid	—	—	0·5	—	0·5	0·5
EDTA* (disodium salt)	—	—	—	—	37·3	—
Ca D-pantothenic acid	—	—	1·0	—	—	—
2,4-D	—	—	6	—	—	—
Kinetin	—	—	—	—	0·04–10	—

* Ethylenediamine tetra-acetate

a variety of vitamins for the culture of callus tissue of *Picea glauca*. Gautheret (1959) added cysteine at 10 mg/litre and vitamin B_1 at 1 mg litre to the mixture of mineral salts and sucrose.

The data presented in Table 3.5 show the suitability of particular media for a series of tissues and direct the reader to consult the relevant literature to examine the merits and demerits of any particular culture medium.

CULTURE TECHNIQUES

Techniques for the isolation and cultivation of plant tissues have been adapted from methods used for micro-organisms. Today the most modern procedures for the growth of plant cells in suspension culture using chemostats and turbidostats have been developed from similar devices currently in use in microbiological laboratories (see Chapter 4).

SOLID MEDIA

The first techniques developed for callus culture were simple and employed media solidified with agar, gelatin or silica gel. More recently certain types of acrylamide gels have been used to solidify media. The great merit of this form of culture is its extreme simplicity. Only simple standard laboratory glassware is required, and there is no need for complex mechanical devices or elaborate containers. In addition large numbers of cultures can be accommodated in a small space.

Agar is the most popular solidifying agent and is usually present at an overall concentration of $0·6–1\%$ (w/v). Several commercial agar preparations are available, the purest being Difco Nobel bacteriological agar which is employed for critical work. The other grades are usually acceptable for routine work where the presence of small amounts of interfering substances is not considered to be important. Agar can, of course, be purified further in the laboratory but such procedures are usually not worthwhile unless critical work on micro-nutrients is planned. Gelatin, which is usually employed at an overall concentration of 10% (w/v), also provides an organic base but is not widely used. Recently Biogels have been used and their biological properties are under investigation in France.

A rapid survey of the literature on plant tissue cultures will show that solid media have now been largely relegated to the establishment and maintenance of callus cultures. Much of the recent critical work on nutrition, metabolism and growth has been performed with liquid media. The reason for this change becomes readily apparent when the limitations of culture on solid media are examined. Firstly, only one part of the callus or explant is in contact with the surface of the medium. It is likely, therefore that as culture proceeds, inequalities in the growth response will arise in response to the nutrient gradients set up between callus and medium. Similarly there may be gradients in the exchange of respiratory gases due to occlusion of the base of the explant. Gradients of toxic waste products may be established. Also, in a static system, the callus is subject to polarization by gravity and perhaps variation of incidental light. (Of course retention of the original polarity of the tissue or a permanent orientation with respect to gravity may be desirable in certain morphogenetic investigations.) A further disadvantage observed in the author's laboratory is that cultures grown on solid media cannot be transferred to liquid media without some disturbance to the tissue. This makes it difficult to make measurements which involve immersion of the tissue in liquid because the tissue becomes waterlogged and displays properties quite unlike those shown during growth on the original solid medium. This makes it difficult to make respiratory measurements using the Warburg technique or 'pulse-chase' experiments with radioactive precursors.

Despite these limitations, the culture of calluses on media solidified with agar remains the method *par excellence* for the routine maintenance of cultures (Fig. 3.4) and is still used for experimental investigations.

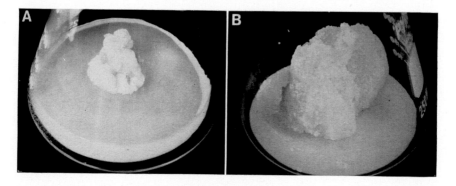

FIG. 3.4. *Established callus cultures* of (A) *Acer pseudoplatanus* and (B) Crown-gall tissue from *Parthenocissus tricuspidata* growing on an agar medium.

STATIONARY LIQUID MEDIA

The culture of tissues in unshaken liquid media has all the advantages of the solid medium methods with the further advantage that interfering substances often present in solidifying agents are absent, so critical nutritional work can be performed. It is therefore not surprising that the outstanding method for culture of tissues in unshaken liquid media was developed by workers with a keen interest in mineral nutrition (Heller & Gautheret 1949). The method described is elegant in its simplicity. The tissue is placed on an ash-less filter paper support held at the interface of the medium with the air in the test-tube. The filter paper acts as a wick providing nutrient while keeping the tissue in the gas phase. The construction of these supports is described in Butenko (1964). White (1953) has also employed a method using a filter paper support in which the test-tubes are inclined at an angle, whereas with Heller's method the tubes are held in a vertical position. These methods are, however, not widely used today and have been replaced by other techniques in which the medium is agitated.

AGITATED LIQUID MEDIA

The culture of explants agitated in a liquid medium eliminates many of the disadvantages ascribed to the culture of tissues in stationary culture. Movement of the tissue in relation to the nutrient medium facilitates gaseous exchange, removes polarization of the tissue due to gravity, and eliminates nutrient gradients within the medium and at the surface of the tissue. As a result the only polarity which remains is that which exists at the time of inoculation. Tissues grown in contact with a liquid medium are also more amenable to manipulation in experiments where radioisotopes are fed as tracers.

Culture techniques in which tissues are agitated in contact with a liquid medium may be placed in one of two categories:

(1) *Continuous immersion:* techniques in which the tissue is always in contact with the culture medium and the mixture is shaken or stirred continuously. In such an arrangement the volume of culture medium must be carefully chosen so as to provide a maximum surface to the gaseous environment and ensure adequate aeration. This includes the use of shaking machines, sometimes reciprocating but usually rotary in action, and comprising many widely used techniques. The tissue is usually placed in a flask which contains a liquid medium occupying about 20% of the total volume. The tissue is suspended in the medium by agitation, at speeds between 50 and 100 rpm. Cultures are usually incubated in darkness or in low intensity white light.

An alternative method of agitation employed in this laboratory is the use of magnetic stirrers. (Fig. 3.1). In this method a small magnet encased in an inert plastic such as PTFE is placed in the culture medium with the tissue and the magnet rotated by a larger external magnet driven by a small synchronous electric motor at a speed of 250 rpm. 15 ml of culture medium in a 150 ml conical flask provides an ideal environment for the growth of tissue up to about 10 days. After this period the rate of growth is reduced but can be restored by subculturing the tissue. This method is especially suitable for short-term culture and has the great merit of simplicity in terms of equipment.

(2) *Periodic immersion:* techniques in which the tissue spends periods immersed in the liquid medium alternating with periods in air. Such an arrangement ensures adequate mixing as well as providing efficient gaseous exchange. This has proved very popular for the culture of calluses and one outstanding technique, although widely known, deserves particular mention. The use of roller tubes and roller bottles in the culture of animal tissue has become standard in animal tissue culture laboratories but the lack of adhesion of plant tissues to glass has reduced their effectiveness. This problem of lack of adhesion has been overcome by the 'tumble tube' method described by Steward, Caplin & Millar (1952) in which the culture tubes rotate 'end over

end' rather than around their main axes, the basic method can be readily appreciated from Fig. 4.1 in Chapter 4 which shows a modified auxophyton. The tubes are fastened with their long axes parallel to the radius of the discs which are about 2 feet in diameter. The discs are rotated at 1 rpm by a shaft held at an angle of about 12°. The medium runs from one end of the tube to the other leaving the tissue 'high and dry' in the other limb of the tube. In practice the tissue becomes distributed between the two limbs of the tube and may occasionally change ends. The effect of this rotation is to gently wash the tissue with nutrients and to provide adequate mixing and gaseous exchange. It is with this apparatus that Steward, Mapes, Kent & Holsten (1964) first showed that a carrot plant may be raised from a single cell thus bringing to reality Haberlandt's dream of totipotency.

The method of culture chosen for a particular investigation obviously depends on a variety of factors including the nature of the tissue, the length of the culture period, and the availability of mechanical devices. Generally much simpler devices may be employed for the culture of callus cultures than for cell suspension cultures (see Chapter 4).

SUBCULTURE AND THE PRESERVATION OF CULTURES

Callus cultures require to be transferred periodically to a fresh nutrient medium. Extensive growth leads to the exhaustion of nutrients, drying out of solid media or concentration by evaporation of liquid media, and the accumulation of tissue metabolites. Cultures maintained on agar at 25°C or above require to be subcultured every 4 to 6 weeks. In the early stages of callus development it may be convenient to transfer the whole piece of tissue but the subculture of established callus demands the frequent sub-division and transfer of separated pieces. In this case it is important to transfer small healthy-looking pieces to the surface of fresh agar medium. Very small pieces of tissue tend not to survive the rigour of transfer. Failure to transfer cultures ultimately leads to the death of the callus while a subculture from necrotic callus tends to grow much less actively than one taken from an actively growing healthy culture. In this situation active growth can often be restored by repeated and sufficiently frequent subculture. Loss in vigour is possibly related to the accumulation of toxic materials which are carried over to the subcultures from the necrotic callus and which are gradually lost during subsequent subculture. Whereas slow-growing callus strains are common, a permanent change in growth activity may be due to the presence of a microbial or virus contaminant present in the callus. Although most contaminants make their presence known immediately and swamp the culture, others are slow growing and less obvious and present a considerable nuisance.

Animal tissue cultures may be preserved at low temperatures ($-70°C$) in glycerol or DSMD solutions, and retain viability for periods up to a year. Plant tissue cultures cannot be preserved in this way (see however references to current work, Chapter 15, p. 428) but can be preserved under mineral oil for many months (Caplin 1959). Special methods of preservation are not widely employed and most workers prefer the technique of continual subculture to maintain viable cultures.

HISTOLOGICAL AND HISTOCHEMICAL TECHNIQUES

It is not intended in this section to cover all the available techniques but merely to refer to methods which are best suited to callus tissue. Because callus tissues are composed of a high proportion of vacuolated cells the techniques must be chosen to cope with the problems associated with the handling of such cells. The choice of fixative is critical in this respect especially if the full detail of the cells is to be retained. If the preparations are made specifically to observe the nucleus under the light microscope, fixatives long established for this purpose are equally suited to vacuolated cells as to non-vacuolated cells. For example a mixture of ethanol and acetic acid in the proportions of 3:1, a fixative suitable for meristematic non-vacuolated cells, is also a good fixative for chromosome preparations in vacuolated cells. Such a fixative does not, however, retain much of the detail of the cytoplasm. Two fixatives stand out as particularly suited to preparations of callus cells in which it is essential to preserve cytoplasmic details and cell walls. The first is a fixative well known to plant anatomists, FAA, a freshly prepared mixture (v/v) of formalin (1), 50% ethanol (18) and acetic acid (1), the second is a fixative commonly used in electron microscopy but well suited to light microscopy, 6% glutaraldehyde. Optimum times of fixation vary from tissue to tissue and must be determined for each specimen. Subsequent handling and sectioning of the tissue is by standard technique (Johansen 1940, Jensen 1962).

Electron microscopy

Preparation of vacuolated tissues for electron microscopy also presents problems of preservation. The fixative *par excellence* for such studies is glutaraldehyde followed by osmium tetroxide.

The procedure reported by Tulett, Bagshaw & Yeoman (1969) is to fix small pieces of artichoke callus tissue in 6% glutaraldehyde in 0·1 M phosphate buffer at pH 6·9. It is important to note that some commercial preparations of glutaraldehyde may require neutralization. The fixation is started at room temperature for at least 2 hours and continued overnight at 5°C. Tissue which does not sink in the fixative is kept under vacuum until it does so. After fixation the tissue is washed over a period of 3 hours in several changes of phosphate buffer at pH 6·9 to remove excess glutaraldehyde. Post-fixation is then carried out for 1 hour in 1–2% buffered osmium solution or 1–2 hours in 2% unbuffered aqueous potassium permanganate. The washing and post-fixation are carried out at room temperature (c. 22°C). Tissue fixed in glutaraldehyde and permanganate is very brittle when embedded and most difficult to cut.

It is clearly desirable to obtain ultrastructural results with a variety of fixatives as the detail of structure may vary with different fixatives. Artichoke callus tissue will not fix satisfactorily with either potassium permanganate or osmium tetroxide alone. Other workers have obtained successful results with these fixatives on callus tissue from carrot (Israel & Steward 1966, 1967), *Andrographis* (Bowes & Butcher 1967) and *Oxalis* (Sunderland & Wells 1968).

Dehydration is carried out in ethanol according to standard procedures and the tissue embedded in araldite. Frequently difficulty is encountered in embedding callus tissue. On many occasions the tissue floats when put into the propylene: araldite mixture even though it does not float in the fixative. When this occurs successful embedding is never obtained. The tissue floats in the embedding medium and remains spongy in the hardened block. The floating is not due to the presence of air bubbles on the surface of the explant. Variations in the size of the tissue, the rate of dehydration, and whether propylene oxide is used or not, fail to overcome this problem.

Sections are cut with glass knives on a Huxley microtome and mounted on uncoated grids according to established procedures. They are stained on the grids with uranyl acetate (saturated aqueous solution) for 60–90 minutes and lead citrate for 5–30 minutes or with lead alone (5–30 minutes) by the methods of Reynolds (1963) and Venable & Coggeshall (1965). Alternatively sections are stained in either potassium permanganate or barium permanganate (20–30 minutes) by the method of Lawn (1960). Sections are examined by conventional electron microscopy (see also Chapter 4, section 'Study of cell structure', p. 94).

Histochemistry

If the choice of fixative is important for histological studies then it is critical for histochemical investigations and such investigations must be preceded by a careful analysis of the effects of the fixative and subsequent processing on the substance being measured. The careful approach of the histochemist and the basic principles involved in measurements of this kind are clearly explained by Pearse (1961). Applications of histochemical techniques to plant tissues generally are described by Jensen (1962). However, for applications of histochemical procedures to callus cells specific papers must be consulted (Adamson 1962, Mitchell 1967, 1968, 1969). In these papers methods have been adapted to measure amounts of various macromolecules in plant cells, DNA, utilizing the Feulgen-staining procedure, total nucleic acid with the gallocyanin–chrome alum technique, and protein using either dinitrofluorobenzene procedure or staining with napthol yellow S. Autoradiography has also been exploited in the study of callus cells using techniques established for other tissues (Rogers 1967). Nowadays liquid emulsion methods are much preferred to stripping film procedures. Techniques for measuring enzyme activities have also been exploited although exact quantification is difficult. The great advantage of the histochemical–cytochemical approach is that differences between individual cells may be measured and specific zones of activity recognized. The value of such investigations will be discussed in Chapter 10.

CHOICE OF PARAMETERS FOR THE MEASUREMENT OF GROWTH

Analysis of growth in callus cultures is usually based on fresh weight measurements, occasionally determinations of dry weight or cell number are also made. Total protein is not widely employed as a parameter of growth. The attraction of fresh weight as the major growth parameter is obvious for it is a quick and simple means of following an increase in mass of the tissue. Measurements may be made without sacrificing the sample and the callus may be weighed under aseptic conditions and returned to the surface of the culture medium. Such measurements have the disadvantage that changes in the growth rate may result from disturbance of the tissue. Fresh weight may also in certain circumstances be used as a means of estimating cell number. Steward has stated that in carrot calluses growing in the auxophyton 1 mg may be taken to represent 10,000 cells. However, great care must be exercised in accepting such relationships for in certain situations such as the early development of the callus in the Jerusalem artichoke no such relationship exists, indeed, rapidly decreasing average cell weight means that the number of cells per unit weight of tissue is increasing rapidly. However, once the callus is established 'Steward's rule' may apply.

Investigation of cell division cannot be conducted satisfactorily without the determination of cell numbers, and a full growth analysis of any situation requires a series of parameters to be measured such as fresh weight, cell number, cell wall and protein. The use of dry weight as a growth parameter must also be considered with caution because contamination of the tissue with sugar from the culture medium and the storage of large amounts of carbohydrate within the cells tend to complicate interpretation of the data.

Techniques for cell counting and the determination of mitotic indices

There are two possible approaches to the problem of estimating the cell number in a callus tissue. First, in a geometrically regular structure it is possible to cut sections, count the cells in the section and after taking measurements and making certain assumptions about average cell length calculate the total number of cells in the tissue fragment. Obviously such a method becomes less and less appropriate as the tissues become more and more irregular and as established callus tissues are by their very nature irregular growth masses such an approach is unsuitable for such tissues. The second approach has proved to be more successful and involves the maceration of the tissue and subsequently counting the constituent cells. This technique is based on that used by Brown & Rickless (1949) for the estimation of cell numbers in roots. Maceration is achieved by treatment of the tissue with a solution of chromium trioxide in water. Immersion of most callus tissues in 5% chromic acid for 16 hours at 20°C suffices. However, maceration may be accelerated by the use of an increased concentration of chromic acid and a higher temperature (see Chapter 4, section 'Cell counting', p. 90). Too vigorous a treatment leads to the breakdown of the cells and a much reduced cell count. In some instances pre-treatment of the tissue with N HCl at 60°C before immersion in chromic acid facilitates maceration. Other macerating agents may be used and include pectinase at 37°C for 1 hour or EDTA at pH 8·0 at 40°C for 2 hours.

Once the tissue has been softened by the macerating agent it can be converted into a cell suspension firstly by breaking it with a glass rod and secondly by drawing the suspension into a Pasteur pipette or hypodermic syringe and expelling it. This is repeated usually two or three times depending on the state of the tissue. The most critical and time consuming part of cell number determination is undoubtedly the counting of the cells in the macerate. Mechanical aids to counting such as the Coulter Counter have not so far proved satisfactory for macerated plant tissue and therefore it is necessary to count aliquots of the cell suspension on a haemocytometer slide. This method can be accurate but is tedious if performed with a conventional microscope and a tally counter. It is possible to semi-automate the process by the use of a projection microscope and an automatic bacterial counter device. The apparatus shown in Fig. 3.5 is designed for this purpose. The haemocytometer slide is loaded in the normal way ensuring even distribution of cells on the grid and the image of the complete grid is projected on to tracing paper on the front of the viewing screen. Remote controls enable the mechanical stage to be moved and the microscope to be focused. A permanent record of the distribution of the cells on the grid and a total cell count are obtained at the same time, for the pen not only marks the paper, but also transmits a pulse to the counter which presents the sum of the pulses. Viewing the cells in this counting apparatus is facilitated by staining

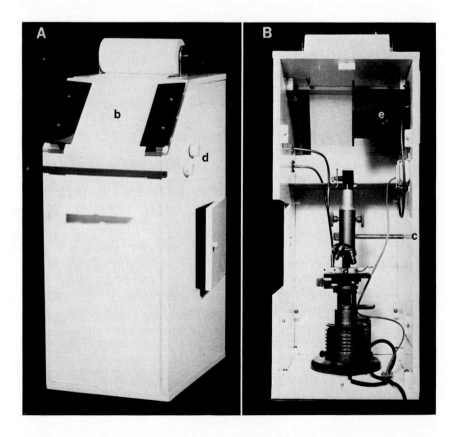

FIG. 3.5. *Apparatus to aid cell counting in tissue macerates.* (A) External view. (B) Internal view. The haemocytometer slide with the macerate is placed on the stage of the microprojector (a) and the image of the specimen projected on to the screen (b) which is covered with a sheet of translucent paper. The position of the cells is marked on the paper with an electronic pen attached to a bacterial colony counter. This gives a record of the distribution and total number of the cells. The image of the cells is focused by remote control (c). The mechanical stage may also be operated by remote control (d). An extractor fan (e) is fitted to reduce the temperature within the box. Counts are made in conditions of low light intensity.

with crystal violet. The procedure for cell counting using this method is as follows.

The tissue is placed in 5% chromium trioxide at 70°C for 25 minutes and washed three times with distilled water. The tissue is stained in 1% aqueous crystal violet made up in 0·03 M EDTA at pH 10 for 10 minutes and then thoroughly washed with distilled water. At this stage the tissue is placed in a small glass tube with 2 ml of distilled water and macerated further with a glass rod followed by sucking up and expelling the macerate with a Pasteur pipette. The macerate is placed on a haemocytometer slide and the number of cells counted. An average of six grids is obtained.

$$\frac{\text{Vol. of macerate}}{\text{Vol. above Haemocytometer grid}} \times \frac{\text{Cell count}}{\text{Number of Explants}} = \frac{\text{Cell number}}{\text{per explant}}$$

Note that careful handling of the tissue is required after the initial treatment with chromium trioxide so as to preserve it intact. The tissue does not stain adequately with crystal violet in the absence of the EDTA treatment.

High mitotic indices in an asynchronously dividing population of cells have been interpreted as representing fast rates of division and low indices, slow rates of division. However, mitotic index alone only reflects the time spent by the average cell in mitosis. Mitotic indices which fluctuate with time may show that the tissue under investigation is synchronous and, indeed, this is the best indicator of synchrony of cell division. Used in conjunction with data on the rates of increase in cell number, the change in mitotic index with time can provide valuable information about the characteristics of dividing cells. Determination of mitotic indices is a simple but time consuming procedure. The simplest approach for callus cultures is to process the tissue fragments according to the Feulgen procedure after ensuring the optimum time for hydrolysis with N HCl at 60°C to obtain intense staining. The softened fragments may then be tapped out on a slide and squashed according to established techniques. Counts of cells in interphase and in various stages of mitosis are made along random transects until a population of 500 cells has been examined. Mitotic index is the percentage of cells in visible mitosis. The number of preparations made for each tissue depends on the mass of the callus (see an alternative technique described in Chapter 4, section 'Mitotic index', p. 91). It is clearly impracticable to measure mitotic indices in large callus masses and if it is considered important to make such measurements then the tissue must be sub-divided. The concept of mitotic index originated from studies on apical meristems where the population of dividing cells can be reasonably closely defined. In a mature callus the situation is much more complex, with only a small proportion of dividing cells, and these may be organized into isolated growth centres or cambia. In this situation the concept of mitotic index is perhaps of limited value and should be regarded with caution. During the early development of callus cultures and in actively dividing suspension cultures mitotic index can provide evidence of division synchrony (Yeoman, Evans & Naik 1966, Yeoman & Evans 1967, Street 1968a).

BIOCHEMICAL TECHNIQUES

Assay of nucleic acids and protein, including isotopic procedures

Limitations of space make it impossible to do more than begin to explore this subject, and provide reference sources for methods. However, some description of the application of established techniques to callus systems, with the problems they entail will, it is hoped, be helpful to those approaching this subject for the first time (see also Chapter 4, p. 91 *et seq.*).

The determination of total nucleic acid (RNA+DNA) is relatively straightforward as long as care is taken to remove substances which interfere with the final assay procedure. Mature plant tissues and calluses contain a variety of substances which interfere with the standard techniques used for the determination of nucleic acids. Some of these substances such as storage polysaccharide and sugars can be removed relatively easily by mild acid hydrolysis and efficient washing but other substances such as pectins and derivatives are less easy to remove and if present in substantial amounts can seriously interfere with the determinations (Evans 1967). Old-established callus tissues contain large amounts of these materials and in consequence are most difficult to handle.

A rapid and reliable method of following the total amount of nucleic acid (RNA and DNA) in developing calluses is described in Yeoman & Mitchell (1970). The method is based on the washing procedure of Kupila, Bryan & Stern (1961) and the extraction procedure of Schneider (1945). After the removal of interfering substances by a careful washing procedure (Kupila, Bryan & Stern 1961), the ether-dried pellet is extracted with successive aliquots of 0·5 N perchloric acid and the absorption of a suitably diluted extract measured at 260 nm. In all cases the absorption at 235 mm was determined to check the amount of contamination in the extract. This method, with slight modification, is applicable to a wide range of callus tissues. This procedure has the great merit that simultaneously with the determination of RNA, DNA content may be measured on the perchloric acid extract using the diphenylamine procedure of Burton (1956). It is at this stage that individual pectins may seriously interfere with the development of the blue colour. Measurement of the absorption of the extract at 600 nm and 650 nm and subsequent calculation of the difference is essential to correct for the green colour due to pectins and related compounds. Failure to recognize this interference and correct for it may lead to inflated values for DNA and subsequent errors in interpretation (Yeoman & Mitchell 1970).

Protein nitrogen determinations using the Conway microdiffusion technique may be carried out on the residue from perchloric acid extraction after digestion in a micro-Kjeldahl apparatus. Protein determinations may alternatively be carried out on macerated tissue using the Lowry procedure (Lowry, Rosebrough, Farr & Randall 1951). Separation of proteins in homogenates may be made with conventional polyacrylamide gel electrophoresis (Davis 1964, Ornstein 1964).

Characterization of the RNA and DNA requires quite a different approach and is a most complex procedure. Standard techniques of extraction using the phenol-detergent method (Loening 1965, Kirby 1965) and separation and identification with the polyacrylamide gel technique (Loening 1967, Richards & Lecanidou 1971, Richards & Temple 1971) have been used successfully with developing callus cultures from the Jerusalem artichoke tuber (Fraser 1968).

Isotopic procedures

Callus cultures growing on media solidified with agar can be supplied with radioactive tracers from the substrate. However, such an arrangement is undesirable for pulse-chase experiments. Transfer of cultures grown on an agar medium to a liquid medium for the application of tracers is complicated by the reaction of the tissue to the sudden change in environment.

Callus cultures growing in a liquid medium provide near ideal material for studies of the incorporation of labelled intermediates. Radioactive tracers may be introduced into the culture medium without disturbing the tissue and the termination of the labelling period is simply achieved by replacing the radioactive growth medium with a non-radioactive culture medium containing high concentrations of the non-radioactive intermediate; a pulse and chase procedure. Although there is no difficulty in applying the tracer, difficulties of penetration are encountered. For example, highly ionized molecules such as acids fail to penetrate plant tissues. Such a difficulty can be overcome by lowering the pH of the incubation mixture and thereby suppressing the ionization of the molecule. This usually leads to increased penetration of the substance. Generally, however, the tracers employed should be selected not only for their metabolic properties, but also for their ease of penetration. Sugars, amino acids, nucleotides and inorganic phosphate penetrate rapidly and have been used widely for the investigation of different facets of metabolism. ^{3}H-Tritiated thymidine is used routinely to detect the onset, duration and extent of DNA synthesis (Harland 1971). Dougall (1965) has employed ^{14}C-amino acids to investigate protein synthesis in growing callus cultures of rose, while Fraser (1968) made extensive use of ^{32}P-phosphate to investigate the metabolism of RNA in developing callus cultures of the Jerusalem artichoke.

It is important to choose the specific activity of the radioactive tracer with care so as not to excessively irradiate the cells and bring about severe changes in metabolism. High specific activities are best used only for short exposure and it is important to be able to check the affects of the tracer. This is particularly true when an isotope is fed continuously to a culture, in which case the results must be considered with caution.

Enzyme techniques, including respirometry

There are a limited number of accounts in the literature of enzymic studies with callus cultures (Jaspars & Veldstra 1965a, b, Newcomb 1951, 1960, Dougall 1970, Harland 1971, Fowler 1971, Goh 1971). In general terms the extraction procedures used and the assay techniques employed are based on methods developed for animal and whole plant tissues. However, especial difficulties are frequently encountered when mature callus tissues are used as a source of enzymes. The physical disruption of the tissue is usually achieved with a pestle and mortar, a high-speed blender or a glass/glass homogenizer. Once the cells are ruptured, care must be taken to ensure that the proteins present in the extract are not inactivated by polyphenol oxidases and hydrolytic enzymes. Interference from such inactivating agents is not always encountered. Jaspars & Veldstra (1965a) report no such difficulty in the preparation of α-amylase from tobacco crown-gall tissue cultures. Active preparations of ascorbic oxidase can be obtained without special precautions from cultured tobacco pith (Newcomb 1960). Some callus tissues are very difficult to handle and enzymes may only be preserved in the homogenate by the addition of protective agents such as β-mercaptoethanol, dithiothreitol (Cleland's reagent), polyvinyl pyrolidone (PVP) or cysteine. Aitchison (1972) has shown that the activity of glucose-6-phosphate dehydrogenase in an extract prepared from a developing callus of the Jerusalem artichoke is affected by the concentration of β-mercaptoethanol present in the tissue homogenate (Fig. 3.6). In the complete absence of this reducing agent no activity can be detected. A similar relationship holds for DNA polymerase, thymidine kinase, dTMP kinase and DNAase with this tissue (Harland 1971). Dithiothreitol is as effective as β-mercaptoethanol but at much lower concentrations. PVP is only effective at low concentrations in the presence of β-mercaptoethanol. High concentrations of PVP alone prevent browning of the extract but do not always give active enzyme preparations.

The gaseous exchange of callus tissue grown in a liquid medium can be determined by manometry or with an 'oxygen' and 'carbon dioxide' electrode system. Values reported in the literature for oxygen uptake and carbon dioxide output of callus tissues have almost all been obtained using the Warburg respirometer (Umbreit, Burris & Stauffer, 1957). Usually during these determinations the tissue is agitated in a liquid medium or buffer (Yeoman, Dyer & Robertson 1965, Barnes & Naylor 1958) but measurements can be made with the tissue resting on top of absorbent paper soaked in the

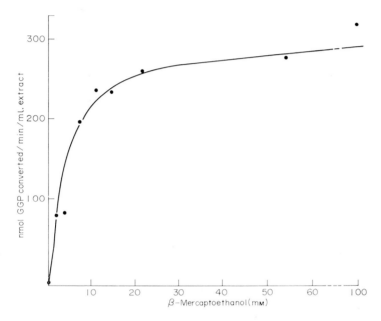

FIG. 3.6. *Relationship between the concentration of β-mercaptoethanol in the enzyme extraction medium and the level of glucose-6-phosphate dehydrogenase activity in developing callus cultures of the Jerusalem artichoke.*

liquid phase. Such a procedure is essential if the tissue has been cultured at the surface of a solidified medium because the transfer of such tissue from a solid to a liquid medium leads to waterlogging of the tissue and a change in the pattern of gaseous exchange. The uptake of oxygen or release of carbon dioxide by submerged callus tissues may be determined using an 'oxygen' or 'carbon dioxide' electrode in conjunction with a physiological gas analyser, such as the Beckman Model 160. This device measures the levels of dissolved oxygen or carbon dioxide in the agitated liquid medium and can be coupled with a pen recorder to give a continuous record of changes in dissolved gases (Davidson 1971). It is essential in such determinations to maintain the level of dissolved oxygen in the respirometer flask by periodic aeration.

C

CONCLUDING REMARKS

The techniques described in this chapter for the initiation and isolation of plant tissue cultures are still in continuous use in many laboratories throughout the world and have not become outdated. In contrast some of the techniques described for the cultivation of tissues have become obsolete and are frequently relegated to the establishment and maintenance of cultures. These older traditional methods are being replaced by more sophisticated techniques utilizing cell suspension cultures. Even here batch-propagated culture procedures are being superseded by chemostats and turbidostats (see Chapter 4, section 'Stirred cultures and continuous culture systems', p. 67 *et seq.*). However, with all of these techniques the desired end-product is a viable culture and this provides the first step in a variety of investigations into the growth and metabolism of cells (see Chapters 10 and 11).

CHAPTER 4

CELL (SUSPENSION) CULTURES—
TECHNIQUES

H. E. STREET

Introduction 59
Culture systems 60
Culture media 79
Maintenance of stock suspension cultures 84
Conditioning of media—concept of minimum effective density . . . 87
Measurements of growth and metabolism 89
Petri dish plating of cell suspensions 95
Growth of cell suspensions in microchambers 97

INTRODUCTION

A suspension culture consists of cells and cell aggregates dispersed and growing in moving liquid medium. During incubation the amount of cell material increases; this increase occurs only for a limited time and the culture reaches a point of maximum yield of cell material. If the culture is at this point diluted back (by subculture) to the same initial cellular content, as established at the beginning of the previous and now completed culture passage, it will in a subsequent and similar incubation period go through a similar pattern of growth and yield a similar amount of cell material. Thus the culture can be continuously propagated by successive batch cultures of appropriate duration and the number of stock cultures increased at regular intervals.

Such suspension cultures are usually initiated by placing pieces of a friable tissue (callus) culture in moving liquid medium. They have also been started from sterile seedlings or imbibed embryos by breaking up the soft tissues in a hand-operated glass homogenizer and then transferring the homogenate, containing intact living cells, dead cells and cell debris, to moving liquid medium. The first passage suspension thus obtained after a suitable period of incubation may contain residual pieces of inoculum as well as more finely dispersed aggregates and free cells. It should therefore be subcultured using a pipette or syringe with an orifice or cannula sufficiently fine to exclude these residual pieces of inoculum. It may even be desirable to let the culture settle for a few seconds and take off suspension only from the upper part of the culture to obtain as dispersed as possible a culture in its second passage.

The different culture systems now to be described achieve movement of the liquid medium in different ways. The movement of the liquid medium serves to maintain the cells and cell aggregates evenly distributed and promotes adequate gaseous exchange between the culture medium and the culture air.

CULTURE SYSTEMS

The Steward apparatus (Auxophyton)

Steward, Caplin & Millar (1952) described a culture apparatus (their Auxophyton) designed for rapid proliferation of their small (3 mg) explants from the storage root of the carrot. The culture vessel (tumble-tube) (Fig. 4.1B) of this apparatus is a tube 12·5 cm long and 3·5 cm diameter with a side neck (1·7 cm diameter) at the midpoint which serves to introduce the inoculum and which when plugged with a cotton plug permits gaseous exchange. This culture tube normally carries 10 ml culture medium and is mounted along a radius and near to the outer edge of an 18–24 in. disc (some twenty-four tubes can be mounted on a disc) rotated at 1–2 revolutions per minute by a shaft at a slight angle (10–12°) to the horizontal (Fig. 4.1A). When small explants of carrot or other plant organ are introduced into these tubes the explant is alternately exposed to air and immersed in liquid (at 1 rpm the explant spends about two-thirds of its time exposed). As the medium flows along the tube from end to end an equilibrium gaseous exchange is maintained; the slight turbulence caused by the flow of liquid probably aids gaseous exchange across the cotton plug of the side neck. The whole apparatus is placed in a temperature-controlled room, and fluorescent light arranged to equally illuminate the culture tubes. Some details of construction and some growth data from carrot root explants growing in the Auxophyton are given in the paper by Steward, Caplin & Millar (1952).

In 1956 Steward and Shantz described the replacement of the culture tubes described above by nipple flasks in order to culture larger numbers of explants for biochemical examination. Nipple flasks of 1000 ml with ten nipples or of 250 ml with eight nipples (Fig. 4.1C) are constructed in such a way that the explants distribute themselves in these projections and hence are alternately exposed to air and bathed in culture medium.

The Auxophyton was designed for the growth of carrot-root explants but Steward & Shantz (1956) reported that when these were cultured in the nipple flasks the culture medium often became turbid owing to the release of free cells and small cell aggregates from the explants into the bathing medium and that this suspension itself was capable of growth and subculture in the apparatus.

Growth of cell suspensions in the Auxophyton involves slow movement of the culture medium but this medium is always in part distributed as a thin film over a considerable area of the inside of the culture vessel and hence a high rate of gaseous exchange is effected between the liquid and gas phase of the cultures.

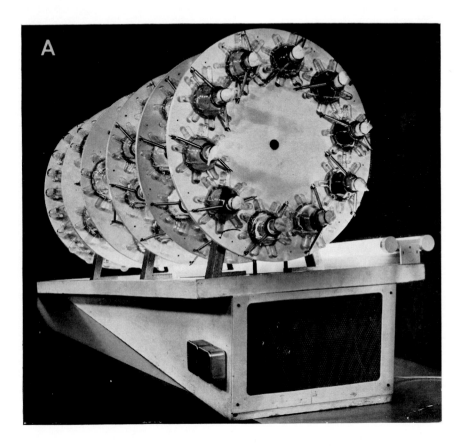

FIG. 4.1. *A form of the auxophyton* based upon designs of Steward, Caplin & Millar (1952) and Steward & Shantz (1956).

A. Total apparatus showing drive shaft (10–12° to the horizontal) carrying six discs which rotate at 1–2 revolutions per minute. In this picture each disc is carrying ten nipple flasks of 250 ml. Note fluorescent lights mounted parallel to the drive shaft.

B.(p. 62) Close-up of disc carrying tumble-tubes (twenty-four per disc) each containing 10 ml culture medium.

C.(p. 63) Close-up of 250-ml nipple flasks each with eight nipples and showing spring-loaded rings which keep each culture vessel in place.

(Photographs by G.G.Asquith.)

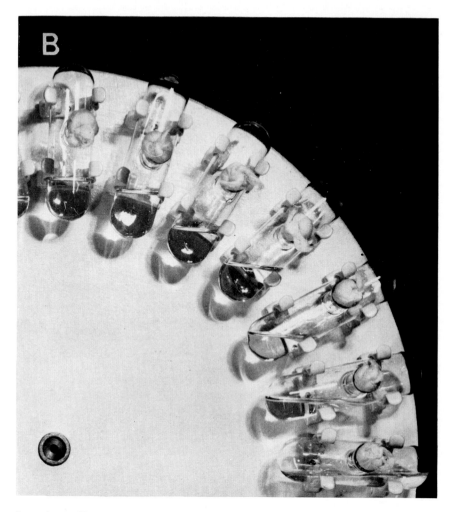

Legend page 61

Platform shakers—orbital shakers

Muir (1953) used a platform shaker to impart a circular motion to liquid culture medium contained in Erlenmeyer conical flasks when he first demonstrated that friable tissue culture pieces could be used to obtain a suspension of dispersed cells capable of continuing growth. Such platform shakers have since been widely used for the initiation and propogation of plant suspension cultures and a list of some manufacturers who offer a range of such shakers is given in Chapter 2. By fitting appropriate clips culture vessels ranging from

Legend page 61

100-ml Erlenmeyer flasks to 1000-ml cylindrical culture bottles can be shaken on such platforms (Figs. 2.7, 2.8). Although the culture vessels can be plugged with cotton plugs these are now usually replaced with aluminium foil caps (see Chapter 2) because the latter reduce evaporation loss, which can be considerable during an incubation period (often of 21–28 days), without apparently impeding essential gaseous exchange.

Orbital shakers should have variable speed control (a range from 30–150 rpm is likely to be adequate; speeds above 150 rpm are unsuitable) and the stroke should be in the range of $\frac{3}{4}$–$1\frac{1}{2}$ in. orbital motion. Where 3-phase electric supply is available this is to be preferred to single phase for long continuous running of the electric motors. The optimum shaking speed (as judged by growth rate, total cell yield, and best possible cell separation) will depend upon the particular culture, culture medium, type of culture vessel, and volume of culture relative to culture vessel size and shape (Rajasekhar, Edwards, Wilson & Street 1971).

For the propagation of stock cultures and for experimental work 100 ml (volume of culture 20–25 ml) and 250 ml (volume of culture 70 ml) wide-mouthed Erlenmeyer flasks have been used as culture vessels on orbital shakers. These shake culture flasks have been modified in various ways to produce closed system cultures in which modified gas phases can be estab-

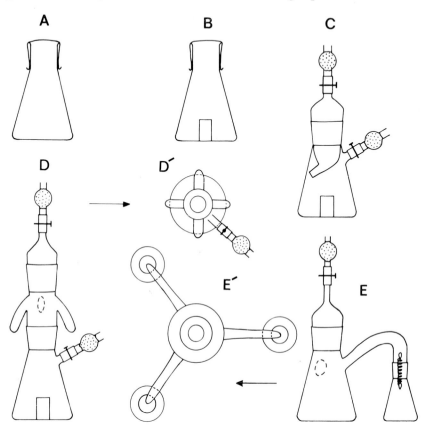

FIG. 4.2. *Culture vessels used for the growth of cell suspension cultures on platform shakers* (see Figs. 2.7 and 2.8, pp. 28 and 29).
A=Standard culture vessel, 250-ml Erlenmeyer, wide-mouth flask closed with aluminium foil. B=same culture vessel but with centre well for introduction of a gas absorbent. C=closed-system vessel in which flask air can be replenished or replaced by another gas at intervals during culture. D=vertical section, and D′=view from above, of closed-system vessel with provision for additional gas absorbents by use of three side arms; provision for gas change as in vessel C. E=vertical section, and E′= view from above, of closed-system vessel providing for inclusion of three gas absorbents placed in each of the three 100-ml Erlenmeyer side flasks, thereby permitting the use of larger volumes of gas absorbents and allowing, via the tapped stopper, the replacement of oxygen consumed by the culture.
All after Rajasekhar, Edwards, Wilson & Street (1971).

lished (Fig. 4.2) and to grow two cultures within the same vessel (Fig. 4.3). By building perspex boxes with a gas inlet and outlet it is possible to maintain cultures on orbital shakers in a controlled atmosphere (Fig. 2.7, p. 28).

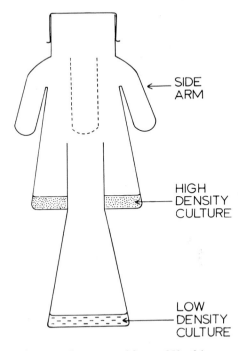

SIDE ARM

HIGH DENSITY CULTURE

LOW DENSITY CULTURE

Fig. 4.3. *Two-tier culture vessel* constructed from a 250-ml (upper compartment) and a 100-ml (lower compartment) Erlenmeyer flask. Volume of low density culture= 20 ml, volume of high density culture= 35 ml. The three side arms can each carry 2 ml of a liquid gas absorbent. Closure shown by aluminium foil. Alternatively by use of a silicone-rubber stopper with entry and exit tubes the culture atmosphere can be renewed in a 'closed-system' culture. After Stuart & Street (1971).

Spinning cultures

Spinning cultures have been described by Lamport (1964) and Short, Brown & Street (1969a) (Fig. 4.4). The apparatus shown in Fig. 4.4 will take 2×10 litre Pyrex bottles each carrying a culture volume of 4·5 litres. These are at 45° to the horizontal and are rotated at 80–120 rpm. The neck of the culture bottle is sealed by a cotton plug carrying a sample tube (sealed except when the culture is to be aseptically sampled). This type of closure allows adequate gaseous exchange. With sycamore cell suspension cultures (Henshaw, Jha, Mehta, Shakeshaft & Street, 1966) the growth pattern and cell yield per unit volume of culture is very similar to that obtained in the 250-ml Erlenmeyer flask cultures on the horizontal shaker. Short, Brown & Street (1969a) recorded, after 21 days' incubation, $1 \cdot 0 \times 10^9$ cells, $1 \cdot 4$ litres packed cell volume and 40 g cell dry wt per $4 \cdot 5$ litres culture.

Fig. 4.4. *Rotator used to spin two 10-litre, Pyrex bottles each carrying a suspension culture of 4·5 litres.* Angle to the horizontal 45°. Normal speed of rotation 80–12 rpm. After Short, Brown & Street (1969a). (Photograph by G.G.Asquith.)

Stirred cultures and continuous culture systems

Large batch cultures (culture volumes 1·5–10·0 litres) have been devised in which the cells are maintained distributed throughout the culture and adequate gaseous exchange effected either solely by forced aeration (Tulecke & Nickell 1960, Tulecke 1966, Graebe & Novelli 1966, Kurz 1971) or by aeration combined with internal magnetic stirring (Melchers & Engelmann 1955; Melchers & Bergmann 1959, Miller, Shyluk, Gamborg & Kirkpatrick 1968, Veliky & Martin 1970, Wilson, King & Street 1971). These stirred cultures because they are stationary can be readily instrumented and connected to reservoirs and gas supplies. By an internal coil or water jacket the temperature of the culture can be accurately controlled in the open laboratory.

The more recently developed of these culture units are designed also to be the culture vessel for continuous culture systems of the chemostat or turbidostat type and have been tested, albeit as yet often in only a preliminary way in such systems. Features which need to be examined much further before the different units can be properly compared and evaluated are; ease of maintaining sterility over long periods, freedom from mechanical failures during long periods of operation, degree of automation and hence level of personal supervision required for successful operation, versatility in regard to the growth conditions which can be established and simply and quickly changed (temperature, stirring, aeration, illumination, nutrient supply), ease with which extra culture chambers for multistage (sequential) treatments can be linked, facilities and space for monitoring equipment (oxygen electrode, pH electrode, density measurement). However, over and above these engineering considerations are the growth performances in these systems of cultures derived from different species and varieties and the effect of the regime of movement and aeration upon cell aggregation.

Very active development work on continuous culture systems for work with plant suspension cultures is now proceeding in a number of laboratories and the present account is therefore confined to describing the basic principles of the few systems which have so far been described in any detail.

The Kurz (1971) system (Fig. 4.5) has been run as a chemostat with a suspension culture of soybean and is reported to support a mean generation time of 30 hours and to maintain the culture predominantly in the form of single cells and cell pairs. Its culture vessel is a Pyrex glass cylinder (55×75 cm), flat bottomed with rounded corners and a working capacity of 1800 ml. There are medium input and exit, sampling and air ports and air enters in pulses (of 0·1 second duration at 2–3 second intervals) by a central pipe at the base via a Skinner Magnetic valve (Skinner Electric Valve Division, New Britain, Connecticut, U.S.A.). The compressed air (at 5–10 psi) as it enters the culture vessel expands into a large gas bubble of the diameter of the vessel

Fig. 4.5. *A chemostat based upon the cylindrical culture vessel (CV) and aeration system of Kurz (1971).*
Key: CV = culture vessel (2000 ml) containing 1500 ml culture fitted with air outlet (AO), condenser, and input (medium) and output (culture) points. RV + CF = reducing valve and carbon.

(see Fig. 4.5 insets A and B) and as it moves slowly upwards causes the medium to form a thin layer in contact with the air. This expansion of the incoming air causes a vibration in the culture and this is considered to prevent cell aggregation. More precise data must be awaited regarding cell aggregation in this system compared with that in the same culture when grown in shake flasks and other stirred vessels. Data is also needed on the behaviour of other cultures than soybean in this system.

The culture vessel described by Veliky & Martin (1970) is constructed for an inverted 3- to 6-litre Erlenmeyer flask. A basically similar culture vessel (Fermenter Type FG) is produced by Biotec, P.O. Box 16152, Stockholm, Sweden. The Veliky and Martin vessel (Fig. 4.6) has two interesting features (1) a teflon-coated double bar magnet stirrer supported by and rotating (200–300 rpm) on a short glass rod located at the bottom of the vessel, (2) an inlet consisting of a 17-gauge stainless steel or teflon hypodermic needle. These two devices avoid the difficulty encountered by the tendency of plant suspension cultures to build up dense masses of cells on tubes or rods just above the medium level. This culture vessel has been used for batch culture and for intermittent renewal of medium and harvesting of culture. This 'drain and refill' technique tends initially to preferentially harvest the heavier aggregates and after several harvests the culture stabilizes at a highly dispersed state and then can be made to yield regular uniform harvests of actively growing culture to inoculate experimental shake flask cultures or continuous culture systems.

The systems of Miller, Shyluk, Gamborg & Kirkpatrick (1968) and of Wilson, King & Street (1971) use a similar culture vessel and both have been developed as continuous systems and fitted with automatic needle sampling valves. Basically similar all-glass culture vessels are manufactured by Jencon

---

filter at input from compressed air system; air pressure to MV $c.$ 7·5 lb/in². MV = magnetic valve (Hoke, Inc., Cresskill, N.J., U.S.A., Cat. Ref. S90A, 180R) which every 2–3 seconds releases a pulse of air (0·1 seconds) under the control of electronic solenoid valve timer. The air as it enters the base of the culture vessel expands violently (see inset A which is taken at the level A′) to give a bubble of the diameter of the vessel (at B′ this appears as inset B) which moves slowly upwards causing the medium to flow past it in a thin film. NM = new medium input; HP = Hughes Micro-Metering Pump controlling rate of input of new medium; CM = reservoir of new medium; IMR = intermediate medium reservoir; OS = outlet solenoid controlling harvesting of culture to the culture receiving vessel (CRV) and controlled electronically (COS) so that harvesting balances new medium input. HgCl₂ = mercuric chloride solution used to sterilize outlet tubing and taps of CRV after removal of harvested culture; SR = sample receiver for collecting of samples to monitor growth of the culture; SWL = sterile water line to wash out sample receiver; F = miniature air-line filters (Microflow Ltd., Fleet, Hants).
(Photograph by G.G.Asquith.)

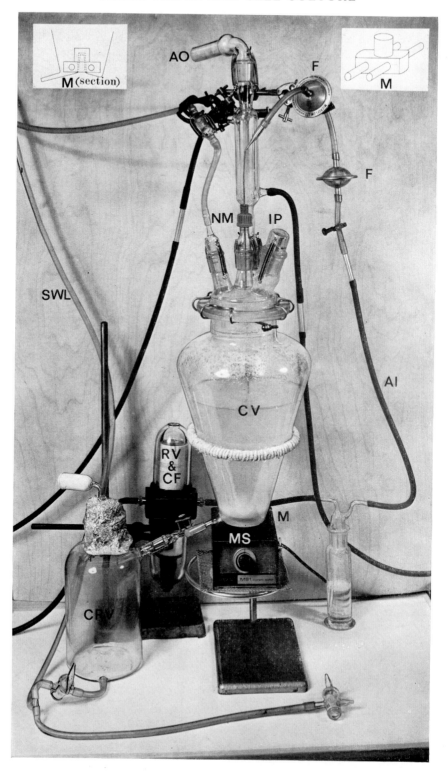

(Scientific) Ltd., Hemel Hempstead, U.K. and by Bellco, Vineland, New Jersey, U.S.A. The basic batch culture unit of the system of Wilson *et al.* is shown in Fig. 4.7, its flow-diagram in Fig. 4.8. The culture vessel is a 5-litre round-bottom flask fitted with a large diameter flat-flange joint and closed by a matching reaction flask lid with five ports (these and the cone and socket joints are commercially available, e.g. as Quickfit items of J.A.Jobling & Co. Ltd., Stone, U.K.). The magnetic stirrer revolves (range 200–600 rpm) on a glass cone worked into the base of the culture flask, the stirrer is a 6-cm PTFE-coated magnetic bar and the shaft is of 6 mm stainless steel rod capped with a PTFE sleeve. The air supply passes a vapour trap, a carbon filter and two microflow miniature sub-micron line filters (Microflow Ltd., Fleet, Hants, U.K.) and enters the vessel via a No. 2 porosity sintered glass aerator. By adjusting stirring speed and rate of air flow, oxygen absorption coefficients up to at least 50 nmoles $O_2/ml^{-1}/min^{-1}$ can be obtained. Air outlets are plugged by non-absorbent cotton plugs and protected from water condensation by electric heating tapes. The system is air-tight so that carbon dioxide or ethylene production can be determined by analysis of the exit air stream. Culture sampling is effected by temporary closure of the air-outlet leading to a positive pressure in the vessel which causes culture to flow into the sample receiver. The air-outlet is then opened, the culture sample run off and the sample receiver and sample line washed with sterile water. Ports in the lid can be used to introduce a thermometer or an autoclavable glass electrode (Activion Glass Ltd., Kinglassie, Scotland). A separate port into the side of the culture vessel can be blown to admit an autoclavable dissolved oxygen electrode (the NBS Electrode of New Brunswick Scientific Co., New Brunswick, New Jersey, U.S.A., is very suitable).

For studies on the cell cycle of a synchronously dividing suspension

---

FIG. 4.6. *Batch culture system based upon a culture vessel as described by Veliky & Martin (1970) and also manufactured by Biotec (Stockholm).*

Key: CV = culture vessel manufactured from an inverted, 3-litre Erlenmeyer flask and fitted with a Quickfit reaction flask lid (MAF 3/52) with five ports, with an outlet tube for harvesting the culture, and with an upright glass pivot rod at the centre of the base to carry the magnetic stirrer (M) (see also insets showing the form of the magnetic stirrer—a teflon-coated double bar magnet and the stirrer in section showing how it sits on the pivot rod). MS = magnetic stirrer; AI = air input line, receiving air from the reducing valve and carbon filter (RV + CF) and passing via two miniature air-line filters into the culture via a 17-gauge stainless steel tube. AO = air outlet via condenser; CRV = culture receiving vessel with wash line of sterile water (SWL). The culture is grown as a batch culture but at any time the greater part of the culture can be drawn off via CRV and then brought back to its original volume by the addition of new medium (NM = new medium input) and incubation continued. IP = inoculation port for introduction of stock suspension to initiate the culture.

(Photograph by G.G.Asquith.)

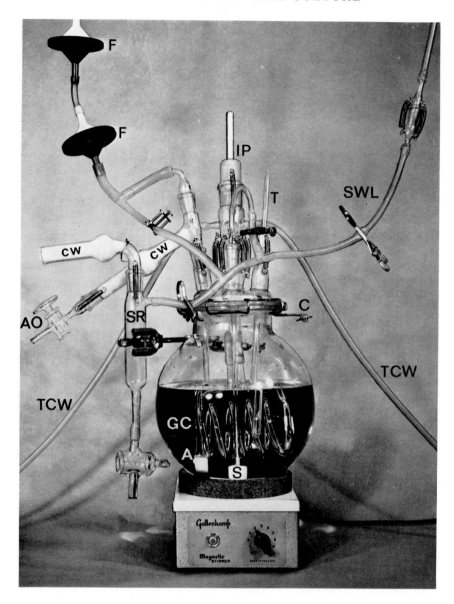

FIG. 4.7. *Basic batch culture unit as described by Wilson, King & Street* (1971). The culture vessel shown here is a 5-litre, Quickfit, wide-neck, reaction vessel (FR5LF) but alternatively can be a 5-litre culture vessel (FV5L) (this alternative vessel is shown in Fig. 2.6, p. 24 and Fig. 4.11, p. 78). The closure of the vessel in either case is by a Quickfit reaction vessel lid (MAF3/52) with five ports, one of which is a B34/35 socket for the introduction of cell suspension (IP). A cone at the base of the culture

culture (see Chapter 11, p. 297) the culture unit can be sampled automatically and at short intervals by a needle valve constructed of stainless steel and similar to that first described by Miller *et al.* (1968) (Fig. 4.9). The valve is fitted through a ground circular hole in the side of the culture vessel and

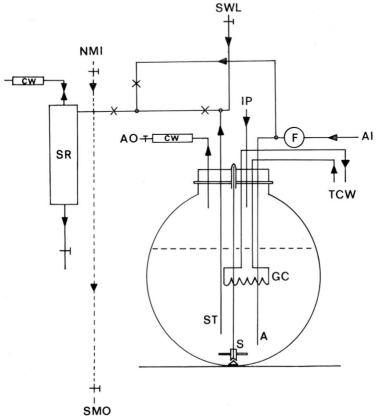

FIG. 4.8. *Flow-diagram for batch culture unit shown in Fig. 4.7.*
Key: as Fig. 4.7 plus ST=sample tube connecting to sample receiver (SR). This batch culture unit can be adapted to permit of removal of part of the culture and replacement by an equal volume of new medium. This is indicated by the vertical broken line: NMI=new medium input; SMO=outlet for spent culture.
After Wilson, King & Street (1971).

vessel carries the PTFE-coated magnetic bar (S) carried on a stainless steel rod. Key: A=aerator (No. 2 sintered aerator); AO=air outlet; C=flat flange lid clip (Quickfit, JC 100F); CW=cotton wool filter; F=miniature airline filter; GC=glass-coil through which circulates water at controlled temperature (TCW); SR=sample receiver; SWL=sterile water line used to wash-out SR; T=thermometer.
(Photograph by G.G.Asquith.)

operated by solenoid valves controlled by an electronic circuit linked to a multiset time clock. A gravity operated fraction collector moved on by solenoids completes the assembly (Fig. 4.10, p. 76). Each time a sample is to be collected the valve opens and a predetermined volume of culture passes into the sample detector. From the detector the sample is passed to the next sample tube (containing appropriate fixative) in the turn-table. The needle valve dead

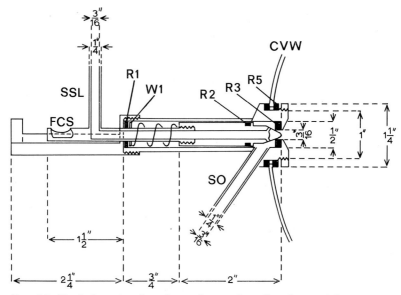

Fig. 4.9. *Vertical section through stainless steel needle valve used for automatic sampling of a suspension culture.* Design based upon that of Miller, Shyluk, Gamborg & Kirkpatrick (1968).
Key: CVW = culture vessel wall; FCS = flexible cable socket; R 1, 2, 3, 5 = silicone rubber 'O' rings; SO = sample outlet; SSL = sterile saline line; W 1 = stainless steel washer (see also Fig. 4.10).
After Wilson, King & Street (1971).

space and the sample detector are then automatically washed with sterile saline (drains into the next wash water tube of the turn-table) and blown free of liquid with a pulse of sterile air and the system set to collect the next sample when the next signal comes from the time clock. The flow-diagram of the system and the electronic control circuit are described in Wilson *et al.* (1971).

   This kind of culture vessel can easily be modified for a *closed continuous culture system.* By adding a port for controlled entry of new medium and a wide siphon for *balancing removal of spent medium* free from cells (Fig. 4.11 and Fig. 4.12, pp. 78, 79). Such a system permits of a significant prolongation of the exponential growth phase so that this can be accurately assessed under different culture regimes (temperature, nutrient supply, aeration). For the controlled supply of new medium a micro-metering pump is required. This may work either on the peristaltic principle (Watson Marlow Ltd. Flow-Inducer Mark 55, Fig. 4.15) or involve the use of a miniature pump unit and pulsation damper (the Hughes Micro-Metering Pump manufactured by Metering Pump Ltd., Ealing Broadway, London W.5, U.K., Fig. 4.5). The Hughes pump is particularly suitable for the prolonged operation often needed in plant cell suspension continuous cultures. All peristaltic pumps involve wear on the silicone rubber tubing with the attendant risk of burst tubing and consequent infection of the culture.

Open continuous systems involving regulated new medium input and *balancing harvest of an equal volume of culture* are of particular interest because they allow of the establishment of steady states of growth and metabolism, transition from one steady state to another and determination of the factors controlling growth rates in such steady states. These open continuous systems may be either *chemostats* (fixed rate of new medium input) or *turbidostats* (intermittent new medium input so as to maintain the culture at a fixed optical density of suspension). The essential modifications needed to the basic culture unit for it to function in such continuous systems are (1) the addition of a constant level device, the electrodes of which operate the solenoid valve controlling harvesting of the culture, and (2) of a loop through which the culture is circulated (by a peristaltic flow inducer) external to the main culture vessel, from which it is harvested in pulses of 40–60 ml and in which its optical properties can be monitored. The circulating loop is introduced by adding two tubulures to the culture vessel below the surface of the culture and connecting these through glass junctions and a flow-through cell by an appropriate length of silicone rubber tubing (6·3 mm internal diameter).

The flow-diagram for the chemostat system is shown in Fig. 4.13 (p. 80). For accurate cell counting and determination of other growth parameters and for biochemical analysis, samples are withdrawn from the manually operated sample receiver. Culture recovered from the culture receiving vessel can be used to check the dilution rate (new medium addition expressed as culture volumes added per day) and yield of the culture. Stable cell densities are achieved in a chemostat as a result of an equilibrium established between the dilution rate (which determines the limiting nutrient in the culture) and the mean generation time of the cells. The dilution rate (D), which is equal to

$$\frac{\text{flow rate (l day}^{-1})}{\text{culture volume (l)}}$$

and the specific growth rate (μ) of the culture, which is equal to $log_e 2/g = 0·69/g$ where g = mean generation time (doubling time for cell number), are at equilibrium numerically equal. By raising the flow rate and hence the dilution

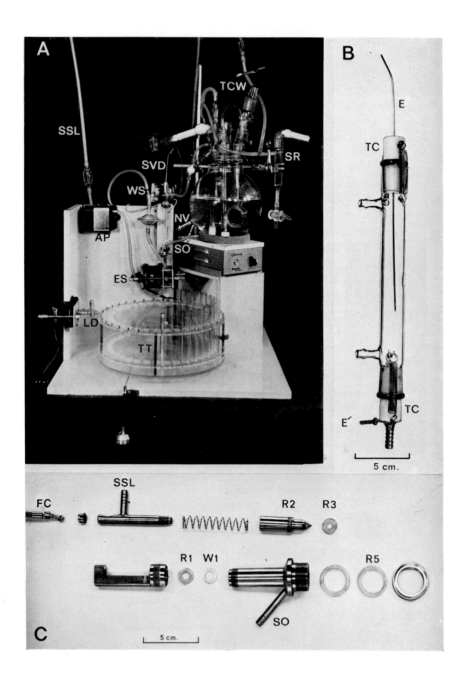

rate, a new, uniform and faster specific growth rate and a new, fixed and lower cell density should result. Eventually, the rising dilution rate reaches the critical dilution rate which is in equilibrium with the maximum specific growth rate of the cells under the particular environmental conditions of the system (temperature, light, nutrient composition of the culture medium). Any further increase in dilution rate will then wash out all the cells from the culture vessel. Clearly it is very difficult to run a chemostat at close to the critical dilution rate, only a small increase in dilution rate at this point will result in wash-out of the culture (for further discussion of chemostat theory and operation see Chapter 11, p. 321 *et. seq.*).

During chemostat studies with the above system on a sycamore cell suspension culture it was shown that at each steady state the light absorption, measured as culture flowed through a cuvette in the circulation loop, varied only within narrow limits and decreased to a new uniform value when a steady state characterized by a lower cell density was established. This enabled a turbidostat system to be operated (flow-diagram Fig. 4.14, photograph of complete assembly Fig. 4.15, pp. 81, 82). Here an electronic control circuit receiving signals from a density detector (light source, cuvette and photocell) maintained a fixed level of light absorption by the growing suspension. Rise in the absorption (as a consequence of culture growth) above a pre-set threshold leads to the admission of a pulse of new medium (via the medium input solenoid valve and of a volume determined by the volume of the loop electrode module, and observation chamber separating the input from the density detector). Such admissions are balanced by automatic harvesting of the culture controlled by the constant level device. This turbidostat system is particularly valuable for work at low cell density and high growth rate because there is no danger of wash-out (for further discussion of turbidostat operation see Chapter 11, p. 317 *et seq.*).

FIG. 4.10. *Assembly for automatic sampling of a 4-litre batch culture.*
A. View of complete assembly.
Key: AP=air pump which clears the sample volume detector (SVD) of liquid by a pulse of air; ES=solenoid valve when open empties sample into collecting tube on turn-table; LD=latching device, operated by a solenoid, which activates movement of turn-table (powered by the brass weight shown hanging down below turn-table); NV=needle valve (see Fig. 4.9 and C this figure); SO=sample outlet line for needle valve; SR=sample receiver permitting manual sampling of the culture; SSL=sterile saline line used to wash out sample volume detector and valve; SVD=sample volume detector electrode; TCW=port carrying the temperature controlling glass coil; TT= turn-table; WS=solenoid controlling the saline wash.
B. Sample volume detector showing details of construction: E and E′=electrodes; TC=teflon cones.
C. Units of the needle valve separated out. Key as Fig. 4.9 which should be consulted to show how the individual components are assembled.
All after Wilson, King & Street (1971). (Photograph by G.G.Asquith.)

The continuous culture systems which have been briefly described above have been developed specifically for the growth of plant cell suspensions. They are designed to operate over the long periods necessitated by the long mean generation times of plant cells as compared with micro-organisms. They

FIG. 4.11. *A closed continuous culture system* based upon the batch culture unit (shown in Fig. 2.6 and detailed in Fig. 4.7).

Key as Fig. 4.7 with the following additional features: ABT = air bleed tap for removal of air which may become trapped in the stilling tube and collects here; NMI = new medium input; SMR = 'spent' medium reservoir; StT = Stilling tube up which spent medium rises and is separated from cells of the culture; STU = siphon tube unit.

After Street, King & Mansfield (1971). (Photograph by G.G.Asquith.)

avoid excessive agitation (such as is intentionally developed by baffles and impellers in systems used for bacterial cultures) and other features which may lead to rupture of the thin-walled vacuolated cells of such cultures. While gentle circulation by peristaltic flow inducers seems to assist cell separation, attempts to achieve very high flow rates by this method may cause a high level of cell destruction. The all borosilicate glass construction of the above systems enables them to be easily modified for a particular kind of experiment or to overcome some particular mechanical problem associated with the culture of a particular cell line. Nevertheless some commercially available chemostat systems (often termed fermentors) developed primarily for the

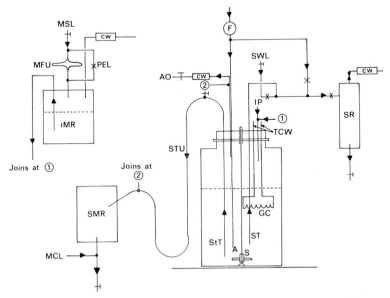

FIG. 4.12. *Flow-diagram for closed continuous culture system* shown in Fig. 4.11. Key as Fig. 4.11 plus units of the new medium line (top left): MSL=new medium supply line; MFU=medium filter unit (sub-micron); PEL=pressure equalizing line; IMR=intermediate reservoir of new medium from which medium is pumped to the new medium input point of the culture vessel; MCL=mercuric chloride wash line to sterilize collecting tube and tap.
After Wilson, King & Street (1971).

culture of micro-organisms may be successfully used for the growth of plant cell suspension cultures. Such systems have been developed by the New Brunswick Scientific Co. Inc. (P.O. Box 606, New Brunswick, New Jersey, U.S.A.) and A. Gallenkamp & Co. Ltd. (P.O. Box 290, London E.C.2, U.K.).

CULTURE MEDIA

General facilities required and some basic rules for preparation of media are outlined in Chapter 2 (p. 16).

This is one of the most confusing aspects of work with plant tissue and

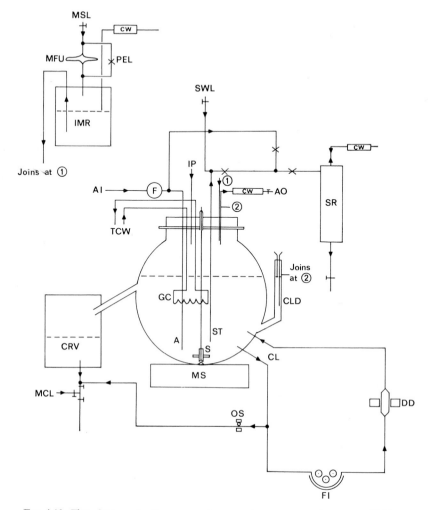

Fig. 4.13. *Flow-diagram for the open continuous culture system—chemostat* of Wilson King, & Street (1971).
Key as Fig. 4.7, 4.8 and 4.11 plus CL=circulation loop; CLD=constant level device; CRV=culture receiving vessel; DD=density detector; FI=flow inducer; MS= magnetic stirrer motor; OS=outlet solenoid valve through which culture harvested in response to signal from the constant level device.

suspension cultures for those newly entering the field. There are a multiplicity of media and often it is very difficult to find out exactly what medium has been used in a particular investigation without working ones way back through previous published papers. Again most of the published media although they have proved capable of supporting the growth of the cultures described can

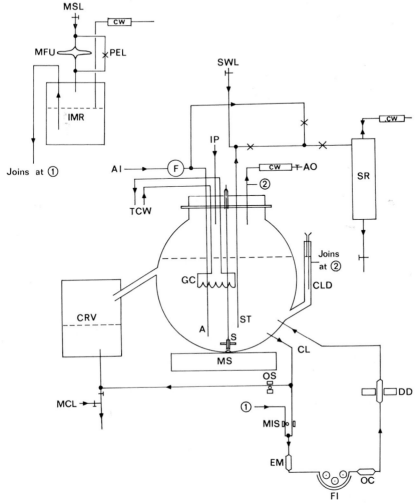

Fig. 4.14. *Flow-diagram for the open continuous culture system—turbidostat* of Wilson, King & Street (1971).
Key as Fig. 4.13 (and earlier flow-diagrams plus: EM = electrode module permitting insertion of pH electrode and/or oxygen electrode in circulation line; MIS = new medium input solenoid valve; OC = observation chamber the volume of which determines the volume of the pulse of new medium.

rarely be regarded either as optimal or as being as simple as possible for maintaining the growth rate recorded. Torrey & Reinert (1961) in work with *Daucus carota* and *Convolvulus arvensis* have reported, using defined media, that the vitamin requirements are different for rapidly growing suspensions than for tissue cultures of the same clones. It may therefore be necessary to

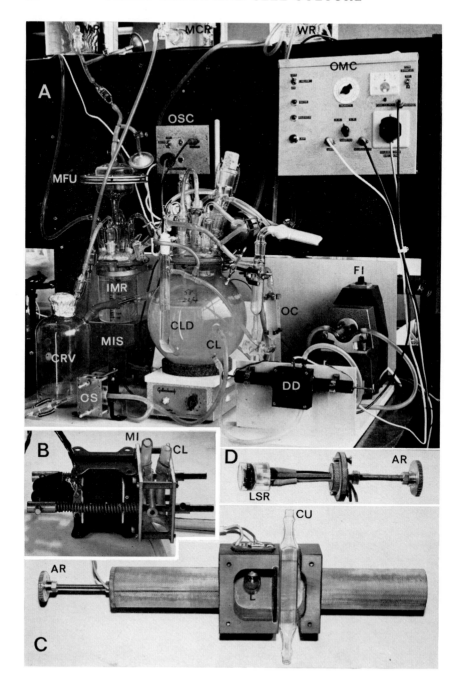

use a different medium for suspension culture growth than has been developed for satisfactory culture of the primary tissue culture. Nevertheless the most widely used basic media developed for callus cultures (Chapter 3, p. 37 *et seq.*) form the starting-point for developing appropriate media for suspension cultures. Whenever it is desired to develop a satisfactory suspension culture from a callus culture the first step is to study the influence of medium composition, and particularly the levels of auxin and cytokinin, on the texture of the callus culture growing on solid medium. A medium which makes the callus culture friable—capable of being easily fragmented—not only gives a tissue mass which will readily break up when transferred to agitated liquid medium, but will probably result in a suspension which remains well dispersed.

When considering media for the growth of suspension cultures it should be recognized that most of the basic media are very lightly buffered and that considerable pH change may occur in response to the introduction of the inoculum and its subsequent growth. Where media initially acid in pH (4·8–5·4) rapidly drift towards a more neutral pH it may be important to incorporate ethylenediaminetetracetate (EDTA) to maintain the availability of iron and other metal ions (Ferguson, Street & David 1958, Sheat, Fletcher & Street 1959, Klein & Manos 1960). The rapid absorption of nitrogen also implies that adjustment of the balance between nitrate and ammonium ion concentrations can be used as a means of stabilizing pH. The use of solid buffers (the sparingly soluble compounds calcium dehydrogen orthophosphate, precipitated calcium phosphate, and calcium carbonate) has also been used as a technique for stabilizing the pH of liquid media used for the growth of suspension cultures (Sheat, Fletcher & Street 1959).

FIG. 4.15. *The open continuous culture system—turbidostat* of Wilson, King & Street (1971).

A. View of complete system.

Key: CL=exit and entry tube-lines of the circulating loop; CLD=constant level device; CRV=culture receiving vessel; DD=density detector; FI =flow inducer (Watson Marlow Mark 55); IMR=intermediate medium reservoir; MCR=mercuric chloride reservoir used to supply solution to maintain sterile the output from CRV; MFU=new medium filter unit; MIS=medium inlet solenoid valve; OC=observation chamber; OMC=optical monitoring control unit; OS=culture outlet solenoid valve; OSC=outlet solenoid control unit; WR=sterile water reservoir used to wash out the manual sample receiver (compare with Fig. 4.14).

B. Detail of a solenoid valve—the medium inlet in the closed position. MI=medium inlet line; CL=circulation loop.

C. Density detector with front plate and right light sensitive resister removed. AR= screw adjustment rod; CU=cuvette (flow through cell in the circulation loop); L= lamp.

D. Right-hand light sensitive resistor (LSR) and screwed adjustment rod (AR) removed from density detector shown in C below.

(Photographs by G.G.Asquith.)

Although as indicated in Chapter 3 (p. 47 *et seq.*) it is desirable to sub-culture callus cultures frequently to maintain them in a state of active growth, nevertheless some callus cultures on solid media will remain viable, even though growing very slowly, when the passage length is extended to many weeks. This is probably due to slow diffusion of nutrients from remote parts of the mass of solidified medium. By contrast cell suspension cultures may reach a very high cell density per unit volume of culture (40–60% of the culture volume represented by cells is not uncommon) and complete exhaustion of essential nutrients may occur by the time the culture enters stationary phase. In consequence (as discussed in the next section of this chapter) it may be critical to effect subculture very soon after the culture has reached its maxi-mum dry weight yield or otherwise extensive cell death often accompanied by cell lysis occurs.

As will be discussed in more detail later (section 'Conditioning of media', p. 87) the question of the required complexity of the culture medium is related to the size of the inoculum used to initiate the culture; this applies to both callus and cell suspension culture. The smaller the inoculum the more exacting are its growth requirements. In developing a simple defined medium for stock culture maintenance it is therefore important to ensure that sufficient cells are transferred at each subculture. There is also considerable experience, albeit not documented by published papers, that the stability of cultures (stability of potential growth rate, cytology and growth requirements) is best maintained by using as simple a medium as possible even if this is not necessarily the medium supporting maximum growth rate. There is clearly need for more critical studies on this aspect of plant cell culture (see Chapter 7).

MAINTENANCE OF STOCK SUSPENSION CULTURES

When cell number in a suspension culture is plotted against time of incubation a curve of the form shown in Fig. 4.16 is obtained (for further discussion see Chapter 11). The general technique of clonal maintenance of cell suspensions is to subculture them early in the stationary phase. With any given suspension it is important to test viability as incubation is continued beyond the beginning of the stationary phase. If viability is well maintained over several days some latitude is available in the timing of the sequential sub-cultures. In some cases it may be necessary to monitor rather carefully and to sub-culture as soon as the maximum cell density is reached or better, at a time when experience shows that the culture will be in the phase of negative acceleration of growth in cell number which preceeds stationary phase. Some cultures have been propagated over many culture passages by much more frequent subculture;

by subculture towards the end of the period of most rapid cell division (the end of the transient period of exponential growth). However, for many suspensions, this technique means subculture at a time when cell aggregation is maximal and in many (though not all) cases leads to progressively increased aggregation. Such a subculture procedure is therefore of limited application for the routine propagation of clones (see Chapter 11).

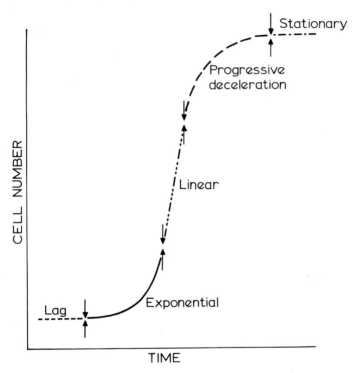

FIG. 4.16. *Model curve relating cell number per unit volume of culture to time in a batch grown cell suspension culture.* Growth phases of the growth cycle labelled. After Wilson, King & Street (1971).

The period from culture initiation to stationary phase is determined primarily by initial cell density, duration of lag phase and growth rate. Generally speaking suspensions are maintained by using an inoculum which establishes a relatively high initial cell density ($0\cdot5$–$2\cdot5 \times 10^5$ cells/ml) and this density rises during incubation often to within the range 1–4×10^6 cells/ml. This means that the increase in cell number is such as would occur if all the cells underwent four to six cell divisions. For many cell clones this increase is achieved in 18–25 days, implying a normal incubation of stock cultures of 21–28 days (the culture being in stationary phase when subcultured). Where subculture is attempted with cells in a very active state of division the passage length is likely to be 6–9 days (particularly as such cultures have no detectable lag phase or a very short lag phase).

A representative sample of a suspension for subculture can be obtained from a shake flask culture by agitating and withdrawing a pre-set volume rapidly and via a long stainless steel cannula of appropriately large diameter using a previously sterilized unit such as the ARH Pipetting Unit (A.R.Howell Ltd., Kilburn High Road, London N.W.6, U.K.) (Fig. 4.17). For routine propagation this sample volume is fixed from knowledge of the average final cell count at stationary phase so as to give a reasonably repeatable initial cell density. When, as in some experimental work, greater accuracy in the initial

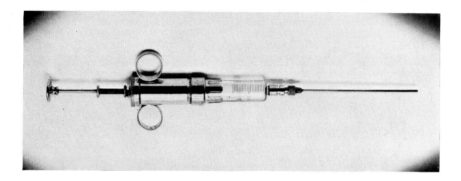

FIG. 4.17. *A.R.Howell Ltd. Pipetting Unit.* This is fitted with a wide bore cannula to admit a cell suspension culture. It can be adjusted to deliver a pre-set volume (in this particular case up to 5 ml).

cell density is required it is necessary to perform a cell count on the stock culture and then adjust the pre-set volume of the pipetting unit so that the desired initial cell density will be obtained. A small sample of at least one of the new cultures should be aseptically withdrawn and submitted to a cell count to check that the required cell density has been obtained. This is particularly necessary where very low initial cell densities are being established.

The use of lower initial cell densities prolongs the period of exponential growth and enables maximum growth rates in different media and under different environmental conditions to be determined by plotting logarithm of cell number, or cell dry weight or cell protein against time. However, for each clone-culture medium combination there is a critical initial density below which the culture will not grow and, at densities approaching this, the lag phase is prolonged and variable as between separate cultures (Stuart & Street 1969). Using a standard synthetic medium the critical initial density for a clone of sycamore cells was found to be $9–15 \times 10^3$ cells/ml. A culture initiated from this density undergoes at least 8 doublings of cell number before reaching a maximum cell density of $c.$ $4 \cdot 0 \times 10^6$ cells/ml.

Stock cultures should be regularly examined microscopically for evidence of contaminating organisms and from time to time plated on enriched media to check for the growth of such contaminants.

CONDITIONING OF MEDIA—CONCEPT OF MINIMUM EFFECTIVE DENSITY

The critical initial cell density or minimum effective density (the smallest inoculum from which a new suspension culture can be reproducibly grown) is a function of the culture clone, the duration and conditions of incubation of the stock culture being used and the composition of the culture medium. Reference was made in Chapter 1 to the discovery that single cells can be induced to divide by being placed in contact with a growing tissue culture or by plating out on solid medium along with a piece of nurse tissue culture. In such plating experiments there is again a minimum effective density and colonies arise more frequently from cell aggregates in the plated suspension than from single cells (Blakely & Steward 1964a, b, Earle & Torrey 1965, Street 1968a). A similar situation has been observed in mammalian cell cultures where it has been shown that a critical number of cells per unit volume must be established at each subculture for self-sustaining growth (Earle, Sanford, Evans & Shannon 1951). Here again successful growth could be obtained from a much lower initial cell density by using, as a 'feeder' system, cells whose own capacity for division had been destroyed by X-irradiation (Puck, Marcus & Ciecura 1956). Further work with cultured animal cells has shown that the action of the 'feeder' cells in promoting growth at low cell density is related to their release into the culture medium of amino acids.

These observations appear to indicate that in order to divide both single cells and a low density of cells require to be supplied with substances not required by a tissue mass of many cells or by a dense suspension and that the substances in question are actually released by the cells into their bathing medium. This implies that cells not only absorb nutrients from a culture medium but release into that medium products of their own biosynthetic activity. This metabolite release from the cells alters (*conditions*) the medium in a way which assists the growth in it of single cells or few cells per unit volume. If this is so, and if the released metabolites are stable in solution it should be possible to induce cell division at very low cell densities by use of 'conditioned' medium and thus separate in time the action of the 'nurse' culture and the induction of division in the single cells or low density suspension.

Stuart & Street (1969) working with a suspension of sycamore cells, prepared such liquid 'conditioned' medium and demonstrated that by its use the critical initial cell density could be lowered by a factor of at least 10 (down to $1 \cdot 0 – 1 \cdot 5 \times 10^3$ cells/ml^{-1}). The techniques adopted in this study are

A   **B**

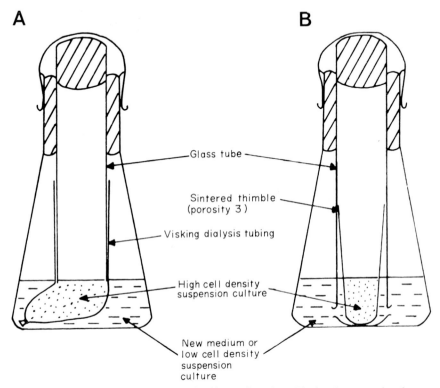

Glass tube

Sintered thimble (porosity 3)

Visking dialysis tubing

High cell density suspension culture

New medium or low cell density suspension culture

FIG. 4.18. *Apparatus used to prepare 'conditioned' medium.* The barrier separating the high density from the low density culture is in A, a Visking dialysis tube and in B, a No. 3 Pyrex sintered thimble. After Stuart & Street (1969).

probably applicable to the preparation of 'conditioned' media from any tissue clone.

The preparation of conditioned medium involves the separation of a growing high density suspension or tissue culture from new synthetic medium (the medium being conditioned) by a barrier permitting the ready diffusion of solutes. This barrier, for instance, may be either Visking dialysis tubing (Fig. 4.18A) or a No. 3 Pyrex sinter (Fig. 4.18B). The extent of the conditioning is determined by the cell count and age of the dense (nurse) suspension or tissue culture, the volume of this suspension or tissue mass relative to the volume of new medium and the duration of incubation on the shaker before the 'nurse' culture is removed aseptically.

The effectiveness of the conditioning can be assessed by determining the minimum effective cell density for the conditioned as compared with new medium. With a given nurse culture and volume ratio of nurse to new medium, there is an incubation time which is most effective; the growth-promoting activity of the medium rises during this incubation period and then declines if the incubation is further prolonged. This decline (over-conditioning) is only in part due to the intervention of nutrient depletion. The contribution to the over-conditioning of nutrient depletion can be tested by mixing conditioned with new medium of either normal or enhanced strength. When medium is over-conditioned neither the adding back of nutrients nor dilution of conditioned medium can fully restore the activity to that of medium conditioned for the optimum time. This appears to be a 'staling' problem but the chemical basis of the effect has not been determined.

Studies with clones of different species have shown that medium conditioned by one species may be effective in promoting the growth from a low initial cell density of a different clone of the same or a different species; in certain cases such medium may be more effective than self-conditioned medium (Benbadis 1965). In all such work it is essential to carefully establish the conditions which yield the most effectively conditioned medium bearing in mind that over-conditioning can occur quickly after the point of optimum conditioning and increase in intensity rapidly as incubation is prolonged.

It has now been shown for a number of clones that appropriately conditioned medium is much more effective in supporting growth from low initial cell densities; particularly so when a simple synthetic medium is used for the successful growth of cultures from high initial densities. Nevertheless for most, if not for all cultures, and as already indicated by the figures quoted from work with sycamore suspensions, such conditioned media do not provide conditions under which individual single cells will normally divide when introduced into a relatively large volume of the medium. This limitation will be further discussed in Chapter 8.

MEASUREMENTS OF GROWTH AND METABOLISM

Growth in plant cell suspension cultures is normally measured by cell counting, determination of the total cell volume (packed cell volume), and by determinations of cell fresh weight, cell dry weight and content of cellular protein.

D

Cell counting

This usually requires prior separation of the cells in cell aggregates and this is usually best achieved by a controlled treatment with chromic acid (5–15% w/v chromium trioxide solution) (Brown & Rickless 1949, Butcher & Street 1960). The chromic acid treatment should be just sufficient to achieve the necessary degree of separation to permit accurate counting and it should be demonstrated that this treatment is not close to that causing significant cell destruction. Our standard treatment for the counting of sycamore cell suspension is to add 1 volume of culture to 2 volumes 8% chromium trioxide, heat to 70°C for 2–15 minutes (according to the phase in the growth cycle at which the cells are harvested), cool, and then shake very vigorously for 10 minutes (on a Baird and Tatlock 'Microid' or similar shaker). This treatment has been found to be effective for a number of cell suspensions. In some cases a controlled treatment with a commercial pectinase is more effective. For instance good cell separation was obtained in a *Parthenocissus tricuspidata* crown-gall suspension by treatment at room temperature for 16 hours with 0·1% w/v pectinase (Nutritional Biochemical Co., Cleveland, U.S.A.) at pH 3·5. The suspensions, prepared as above and suitably diluted (and if desired stained—see Chapter 3, p. 53) are counted on a cell counting slide with three channels. The depth of this slide must be greater than the diameter of the largest cells. A depth of 1200 μ (μm) will accommodate most suspensions. After allowing the cells to settle the cells are counted at a magnification of approx. × 100. Twenty random fields are usually counted per channel and five channels prepared for each suspension. Each field should include not less than 5–10 cells. The mean of the 100 counts and the known volume of each field (e.g. diameter of field 105 μ, depth 1200 μ, volume = 0·99 μl) enables the number of cells per ml of original culture to be calculated. The counting procedure can be semi-automated as described in Chapter 3 (p. 51).

Packed cell volume

This can be determined by transfer of a known volume of suspension to a 15 ml graduated conical centrifuge tube and centrifuging at 2000 × **g** for 5 minutes and expressed as millilitre cell pellet per millilitre culture.

Cell fresh weight

This can be determined by collecting cells on pre-weighed (in wet condition), circular filters of industrial nylon fabric supported in a Hartley funnel, washing the cells with water to remove medium, draining under vacuum and weighing. A large culture sample is needed to get a reasonably accurate fresh weight.

Cell dry weight

This can be determined as described above except that the nylon filters are pre-weighed dry and the nylon plus cells dried for 12 hours at 60°C, cooled in a desiccator containing silica gel and weighed. Cell fresh and dry weights are normally expressed per ml culture and per 10^6 cells.

Total cell protein

This can be determined by collecting the cells on a Whatman GF/C pad, washing them twice with boiling 70% ethanol, drying with acetone and then transferring the pad and cells to a measured volume of N NaOH, hydrolysing at 85°C for 100 minutes and filtering. The measured volume of NaOH should be such as to yield a filtrate (hydrolysate) containing up to 1 mg protein activity ml^{-1}. Protein can be determined by the method of Lowry *et al.* 1951 (via Layne 1957) using bovine serum albumin as a standard.

Mitotic index

Where we are concerned with the relative contributions of cell aggregates and free cells to the growth of cultures or where cultures are being examined for synchrony of cell division the determination of the mitotic index is of value. Determination of mitotic index in randomly dividing cultures enables us to determine the relative lengths of interphase and of the separate stages of mitoses and combined with knowledge of the mean generation time of the culture to calculate the absolute length of these phases of the growth cycle (see also Chapter 11). The technique of determination of the *mitotic index* developed in our laboratory and applied to cell suspension cultures, is as follows: The fresh suspension may be spun down (as described under packed cell volume), pipetted onto a slide and a drop of lactopropionic orcein added and the slide heated gently over a spirit burner flame. The preparation is then covered with a cover slip, the slide tapped gently and then a squash prepared. The cover slip is then carefully eased off with a razor blade and the cells on the cover slip and slide cleared with a drop of ethanol. The cover slip is then mounted on a new slide in Euparol and the cells on the original slide also mounted in Euparol under a new cover slip. Alternatively the suspension can be fixed for 24 hours in 65% formic acid (1 volume of suspension plus 1 volume fixative) and the fixed suspension pipetted onto a slide and a drop of propionic orcein added. The slide is then gently heated, the preparation covered with a cover slip, set aside for 10–15 minutes and then a squash prepared. Two slides are then prepared as described above except that the cells are mounted in 45% acetic acid and the slides sealed with brown cement (such slides can be examined any time within 14 days of preparation).

The slides prepared as above from fresh and fixed suspensions are examined under oil immersion (approx. × 1250) and 1000 cells examined and the number of prophase, metaphase and telophase figures recorded. The sum of these as a percentage of the cells examined constitutes the mitotic index. To obtain consistent results, particularly where several workers are involved, a set of photographs should be prepared to define the beginning and termination of each phase of mitosis in the cells being scored (see also Chapter 3, p. 51).

There is no clear distinction between parameters of growth and those primarily descriptive of metabolic activity. Thus when we construct a carbon balance sheet of a culture we are not only measuring respiration rate and the rate of consumption of the primary source of carbon and energy in the medium but also the rate of synthesis of cellular carbon compounds (a process closely related to the growth rate of the culture). Similarly there is often, and this is particularly so in a culture in a steady state of growth, a fixed relationship between increase in cell number and the accumulation and rate of synthesis of vital cell constituents, for example between growth and increase in protein, DNA and RNA content and between growth and rate of protein synthesis or rate of O_2 uptake per unit volume of culture (see Chapter 11, p. 321). The measurement of such physiological and biochemical parameters therefore serves to describe further aspects of the growth process. Clearly it would be inappropriate to outline here the whole array of physiological and biochemical estimations which can be applied to the study of the growth and metabolism of plant cell suspensions. Attention is therefore directed below only to a very restricted number of easily performed analyses which assess basic aspects of cellular physiology and biochemistry and whose wider use would greatly increase the information yield of experiments involving plant cell suspensions (see also Chapter 3, p. 54 et seq.).

Estimations of DNA, RNA and protein

Freshly harvested cells are collected on GF/C pads, killed with boiling 85% methanol, cooled, washed with ice-cold 85% methanol, extracted three times with ice-cold 10% trichloracetic acid (TCA) and the TCA then neutralized by two washings with ice-cold 80% ethanol saturated with sodium acetate and finally with ice-cold 95% ethanol. Lipid is then extracted from this residue by washing twice and in succession, at room temperature, with chloroform-ethanol (1:3), ether-ethanol (1:1) and ether. The residue, dried in vacuum, can be stored. Protein, RNA and DNA can then be separated from this residue by a modified Schmidt-Thanhauser procedure (Short, Brown & Street 1969b). Protein is removed by extraction with 0·3 N KOH and protein estimated in this extract (Layne 1957). DNA and protein are precipitated from the KOH extract by cold perchloric acid and the RNA in the supernatant estimated by spectrophotometry at 260–290 nm using yeast t-RNA as a standard. The precipitated DNA is then solubilized by hot perchloric acid and estimated by a modified diphenylamine reaction using either deoxyadenosine or calf-thymus DNA as standards.

Feulgen staining and assay of nuclear DNA

Cultured cells may need a particular fixative for subsequent successful Feulgen staining. Often accumulated starch interferes with examination of nuclei and in such cases a technique for clearing the cytoplasm (e.g. controlled heat treatment to disperse the starch) may be necessary. The technique of Hillary (1939, 1940) should be followed for Feulgen staining. The cells are placed in N HCl preheated to 60°C and the hydrolysis time for optimum staining determined and subsequently carefully observed. With cultured sycamore cells 12 minutes at 60°C gives good staining of cells fixed in a non-metallic fixative such as acetic-alcohol (1:3). The Feulgen reagent must be prepared from a suitable basic fuchsin (magenta basic fuchsin-crystals technical, CI 42500, Cat. No. 26120 of British Drug Houses Ltd. has proved suitable and consistent) (see also Chapter 7, p. 165). For DNA determinations by microdensitometry following Feulgen staining, the procedure of McLeish (1963) should be followed. The cells are suspended in an ice-cold solution of M/30 phosphate buffer at pH 7·0 and 2% w/v formaldehyde for 15–45 minutes depending on the material. McLeish then used a mechanical press to extract the nuclei which were collected by centrifugation. A simpler and satisfactory procedure is that described by Bennett (1969). The cell suspension is placed on a glass slide and macerated with a glass tapper. The slide is then flooded with absolute alcohol to stick the nuclei on the slide and then dried in ether vapour in a Copling jar for 30 minutes. The slides are then hydrolysed and stained in the normal way. After alcohol application to fix the nuclei to the slide, all trace of formaldehyde should be removed by washing with water since formaldehyde reacts with Feulgen reagent. Studies involving microdensitometry are further discussed in Chapter 7.

Protein synthesis by study of ^{35}sulphate incorporation

Lescure (1966) working with sycamore cells and Johanneau and Péaud-Lenoël (1967) working with tobacco cells followed incorporation of ^{35}S from sulphate into the trichloracetic acid (TCA) insoluble fraction of the cells as a measure of protein synthesis and indirectly of culture growth. The technique is to add $^{35}SO_4$ aseptically to the culture at about 5 μc/100 ml culture. In a growing culture a steady rate of incorporation is reached after 3–4 days (if it is desired to follow incorporation throughout the growth cycle of a batch culture it is necessary to add $^{35}SO_4$ during the later part of a previous passage in order to have achieved a steady-state level of radioactivity in the cellular pools from which sulphate is incorporated). To measure incorporation, 2 ml of culture is added to 5 ml ice-cold 10% w/v TCA. After centrifuging the supernatant is discarded and the residue washed twice in 5% TCA, collected on a GF/C pad, washed three times with 70% ethanol, dried overnight at 100°C and the radioactivity determined (preferably by scintillation counting).

Similar techniques can be used to study U-^{14}C-leucine incorporation into protein, 2-^{14}C-uridine incorporation in RNA and 6-^{14}C-thymidine incorporation into DNA. These experiments are, however, conducted over short periods (one to several hours) so that the ^{14}C does not become distributed amongst other metabolites and the rates of incorporation are related not only to the rates of synthesis of the particular macromolecules but to rates of entry of the precursors and the size of the unlabelled pools of the precursor. In such experiments the cells at the end of the test period of incorporation are washed with 10% TCA containing 200 μg/ml of cold leucine, uridine or thymidine to remove the radioactive precursors from the free space as rapidly as possible.

Respiration

Oxygen uptake of cultured cells can be measured by removing a sample of the culture and use of a Clark oxygen electrode (Rajasekhar, Edwards, Wilson & Street 1971) or Warburg manometer (also permitting measurement of CO_2 release and calculation of the respiratory quotient). When aeration of the culture is achieved by bubbling air at a known rate through the culture vessel, respiration can be continuously monitored by measuring the CO_2 concentration in the exit air by an Infra-Red Carbon Dioxide Analyser.

Total organic carbon of cells and culture media

Total cell carbon can be conveniently determined in heat-dried or freeze-dried cells by the method of Baker, Feinberg & Hill (1954). To determine total organic carbon in culture medium, a persulphate oxidation is used (Osburn & Werkman 1932, Katz, Abraham & Baker 1954).

Study of cell structure

The techniques of histology and histochemistry described in Chapter 3 (section *Electron microscopy*) are applicable to the study of the cells of suspension cultures. The cells are most conveniently handled in 'transfer tubes' during fixation, staining and embedding (Sutton-Jones & Street 1968). Successful preparations for the electron microscope can be obtained using material very slowly impregnated with araldite/epoxypropane (1·2) mixture but better results are obtained by embedding in methacrylate-styrene (Mohr & Cocking 1968). A successful technique involves pre-fixation in 0·6% glutaraldehyde in cacodylate buffer of pH 7·2, post-fixation in osmium tetroxide, dehydration, pre-staining with uranyl acetate, embedding in methacrylate-styrene, polymerization at 55°C, sectioning and staining with lead citrate (Davey & Street 1971).

PETRI DISH PLATING OF CELL SUSPENSIONS

The technique of plating out cell suspensions on agar plates was first used by Bergmann (1960). This technique is of particular value where attempts are being made to obtain single-cell clones. The objective is to establish, evenly distributed in a thin layer of culture medium, as low a cellular density as is compatible with the growth of colonies from a high proportion of the cellular units (single cells and cell aggregates). Further in order that most, or if possible, all of the colonies are of single-cell origin, the suspension should be freed of large aggregates by aseptic filtration through appropriate mesh bolting cloth or by some other means (see below), so that it contains a high proportion of single cells and that any aggregates remaining contain only a few cells (if possible not more than four to six cells). Such small aggregates are likely themselves to be of single-cell origin. This removal of aggregates is important because they not only undergo division on the plates with greater frequency than do the single cells but division is initiated earlier in aggregates than in the single cells that do divide (Blakely & Steward 1962).

The basic technique of plating is as follows: The suspension culture filtrate is submitted to a cellular unit count (performed as the standard cell count, see p. 90, except that it is not submitted to the maceration stage). This count enables a known number of cellular units to be established per unit volume of plating medium. The counted suspension is adjusted by dilution or concentrated by low speed centrifugation and removal of supernatant so that 2 ml suspension when inoculated into the sterile medium will give the required cellular unit density. The sterile medium contains 0·6% agar (Oxoid Agar No. 3 or similar grade) and is cooled to 35°C before adding the suspension. The suspension and medium are mixed and then immediately dispensed into 9 cm sterile disposable petri dishes (10 ml medium plus cells per dish). The dishes are then normally sealed with parafilm and the cellular units counted directly *in situ* by examining twenty fields of known area per plate using a ×40 magnification stereobinocular microscope. From this the number of cellular units per mm^2 and per plate is calculated. The plates are then incubated at 25°C in the dark for 21 days. After incubation the number of colonies per plate is determined as follows: Shadowgraph prints are made of each plate by resting the plate on a sheet of photographic document paper (such as DR Ilfoprint DR3 5L. No. 3. Projection Document Matt Lightweight) under a photographic enlarger as the light source. In the negative holder of the enlarger is a negative of the ruled area of a Fuchs-Rosenthal counting chamber so that when the shadograph is developed this grid is superimposed on the photograph of the petri dish (Fig. 4.19). The squares of this superimposed grid are numbered and from a table of random numbers co-ordinates are obtained for twenty squares and then the number of colonies in these squares

FIG. 4.19. *Petri dish containing colonies developed by plating out a cell suspension culture.* The grid of a Fuchs-Rosenthal counting chamber is superimposed on the shadowgraph of the petri dish (for counting technique see text). Previously unpublished photograph by K.J.Mansfield.

are counted using a × 10 magnification stereobinocular microscope. From this count is calculated the number of visible colonies per plate. The mean of five such counts on replicate plates is used to determine the *plating efficiency* (PE) thus:

$$PE = \frac{\text{No. of colonies/plate}}{\text{No. of cellular units/plate}} \times 100$$

This photographic method records clearly colonies not easily observed on the standard type of bacterial colony counter.

It is clearly highly desirable to obtain the highest possible plating efficiency with a cellular unit density sufficiently low that colonies can reach a size permitting their subculture before they make contact with other colonies.

To achieve this it is important: (i) to use either a conditioned medium or a synthetic medium specially designed to permit growth from a low initial density, (ii) to avoid the use of cells held too long in stationary phase (cells harvested during the phase of active cell division show the highest plating efficiency), (iii) never to expose the cells to temperatures in excess of 35°C during their dispersion in the agar medium and to plate with minimum delay and also (iv) to incubate plates in darkness or very low intensity light. In our experience whenever plates have been inspected frequently with the binocular low power microscope during the progress of incubation, development of colonies has been inhibited as compared with plates incubated without interruption in the dark.

The necessity of preparing conditioned medium (see p. 88) or even more of developing a medium specially designed to promote growth for a low density ('synthetic' conditioned medium) can be overcome by directly placing a tissue culture mass on the plates (Fig. 4.20). This tissue culture can be confined (and hence prevented from overgrowing colonies) without impeding release of metabolites into the surrounding agar by a special glass ring (Fig. 4.20). As further considered in Chapter 8 (p. 196) colony growth can often be significantly promoted by incubating plates in an atmosphere artificially enhanced in carbon dioxide content. Further consideration is also given in Chapter 8 to techniques of increasing cell separation as an aid to isolating single-cell clones and to the problems of clonal selection.

GROWTH OF CELL SUSPENSIONS IN MICROCHAMBERS

In 1955 De Ropp first tried to culture single cells in hanging drops of medium in microchambers. The single cells did not divide, only cellular aggregates of at least ten cells showed any mitosis. Torrey (1957) used the Maximov double coverslip method (Fig. 4.21) in an attempt to induce division in single cells derived from a pea root callus. His cells were placed around the outer part of a circular film of agar medium in the centre of which was placed a piece of the parent callus to 'nurse' the isolated cells. Some cells quickly died. Others remained alive for up to several weeks and in about 8% of the cells, division occurred. This division, however, did not continue and the largest aggregate derived from a single cell only contained seven cells. The most successful experiment with microchambers were those reported by Jones, Hildebrandt, Riker & Wu (1960). Their microchamber (Fig. 4.22) was sealed with inert mineral oil. Using cells derived from a callus of the hybrid *Nicotiana tabacum* × *N. glutinosa* they observed division when at least thirty cells were present in the chamber, cells which divided eventually became senescent and this was observed to induce division in other cells previously quiescent. By using in

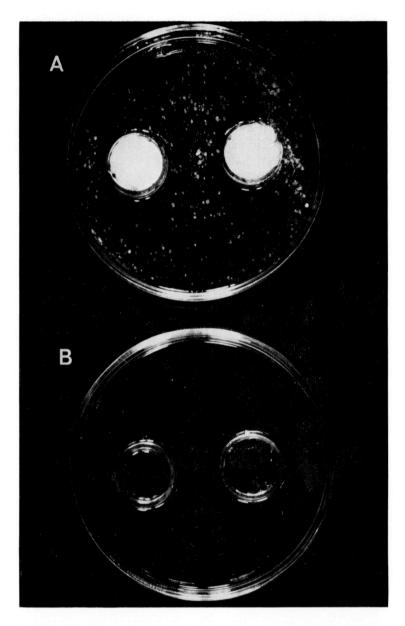

FIG. 4.20. *Petri dishes seeded with a suspension of Acer pseudoplatanus cells and with glass rings, notched around the bottom edge to allow free diffusion.* A. Nurse callus confined within the glass rings. Note colony formation in agar beyond the rings. B. No nurse callus within rings—no colony formation. After R.Stuart (1969).

their microchambers 'conditioned' medium they observed divisions when only individual single cells were present. These microchambers are primarily of interest for phase-contrast observation of mitosis in cultured cells, but colonies have been raised from cells whose division was initiated in the microchamber (see Chapter 8).

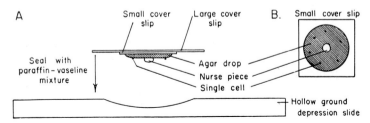

FIG. 4.21 *Maximov's double coverslip method as used by Torrey (1957)* to culture single cells derived from a tissue culture in presence of nurse tissue.

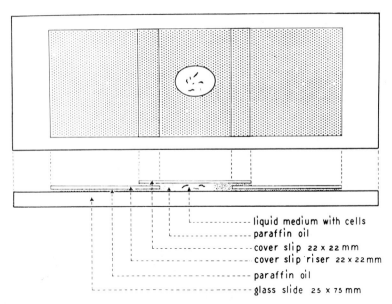

FIG. 4.22. *The microchamber of Jones, Hildebrandt, Riker & Wu (1960)* used to observe the growth of free cells of a hybrid tobacco tissue culture.

CHAPTER 5

THE ISOLATION OF PROTOPLASTS

E. C. COCKING and P. K. EVANS

General introduction 100
Isolation of protoplasts from plant tissue cultures 103
Isolation of protoplasts from leaves 108
Isolation of mature tobacco pollen protoplasts 118
Applications 119

GENERAL INTRODUCTION

Increasingly those interested in isolated protoplasts are concerned with what they can do with such naked cells. With his own particular limited horizon each worker usually perfects, as best he can, the isolation procedure which is required for reproducible isolations. Each tissue or cell system from which it is desired to isolate such naked cells presents its own particular and characteristic problems.

It is important to stress that the procedures used for the isolation of protoplasts can and do have effects on the subsequent behaviour of these naked cells. Consequently, subsequent reproducible studies of the properties of these isolated protoplasts are dependent on a rigidly controlled isolation procedure.

Isolated protoplasts being naked cells present special problems not normally encountered in the usual plant cell culture. They need to be kept in a suitable plasmolyticum whose osmotic potential needs to be carefully related to that of the protoplast system. Irreversible, damaging effects on the viability of isolated protoplasts readily result from the use of plasmolytica of either too high or too low an osmotic potential. When incubations are carried out for several hours the penetration of plasmolytica (for example sucrose) into the protoplasts can be a complicating factor in this respect. Environmental factors, particularly light effects, can also markedly influence the stability of isolated protoplasts. This probably results from light-induced changes in the internal osmotic potential, particularly if the system is potentially photosynthetic.

When protoplasts are isolated enzymatically using crude cell-wall degrading enzymes some of the enzymes present can have deleterious effects on the isolated protoplasts. One reason why certain workers (Pilet 1971, Pilet, Prat

& Roland 1972) have utilized mechanical methods of isolation is that by using such methods possible deleterious effects are largely eliminated. As discussed by Cocking (1972), however, the number of protoplasts which can be isolated mechanically is always small and their isolation is restricted to the cells of tissues which can be readily plasmolysed. Experience over several years extending back to the early work of Tribe (1955), who used mechanical rather than enzymatic methods for protoplast isolation, has indicated that isolated protoplasts are far more resistant to such possible harmful effects when they are in the plasmolysed condition. One possible reason for this has recently become clear from the work of Withers & Cocking (1972). From studies using thorium dioxide as an electron-dense electron-microscopic marker they showed that extensive uptake from the plasmolyticum took place during plasmolysis probably by large-scale membrane invagination. In view of this extensive uptake it would seem advantageous to carry out prior plasmolysis in a suitable plasmolyticum before subjecting the tissue to digestion in a possibly harmful, cell-wall degrading enzyme mixture. As a result, the procedures recommended for the isolation of tobacco leaf protoplasts for induced fusion studies with sodium nitrate involved prior plasmolysis in 25% sucrose (Withers & Cocking 1972). Sometimes prior plasmolysis does not improve the viability of isolated protoplasts. However, when harmful effects of enzymatic isolation are indicated, prior plasmolysis should be carried out since it may be beneficial to survival. Cocking (1972) has recently reviewed critically the isolation and development of plant cell protoplasts. The basic concept is that these isolated protoplasts are naked cells and, being cells, they can be induced to grow and divide under suitable conditions. A special characteristic of these naked cells is that under suitable conditions they will resynthesize a new cell wall (for a detailed discussion of cell-wall regeneration in isolated protoplasts see Cocking 1970 and Cocking 1973a). As a result these naked cells come to behave like the usual cell suspension cultures (Fig. 5.1). They appear to have unique advantages for the cloning of plant cells since they are a true single-cell suspension culture and can sometimes be plated with a high frequency of colony formation (Nagata & Takebe 1971, Frearson, Power & Cocking 1973). They are also well suited to studies on plant cell fusion as a first step towards the somatic hybridization of plants (Power & Cocking 1971). They also promise to be of considerable use in the synchronous infection of single plant cells by viruses (Coutts, Cocking & Kassanis 1972) and for elucidating activity at the plasmalemma in relation to endocytosis (Willison, Grout & Cocking 1971, Grout, Willison & Cocking 1972) and fusion (Withers & Cocking 1972).

As shown diagrammatically in Fig. 5.1, protoplasts can now be isolated either directly, from various plant tissues, or from callus, or suspension culture cells. The range of purposes for which protoplasts are being isolated is ever increasing in scope. For instance, they are being isolated from the aleurone layer as a preliminary step towards the study of the action of gibberellic acid on the secretion of amylase (Taiz & Jones 1971) and from *Avena* coleoptiles to obtain further insight into the basic mode of operation of auxins (Hall & Cocking 1971). We have, therefore, felt it best to describe in detail the actual procedures for the isolation of protoplasts for specific physiological, biochemical and virological studies rather than to generalize on the procedures to be adopted; more general and historical aspects of the

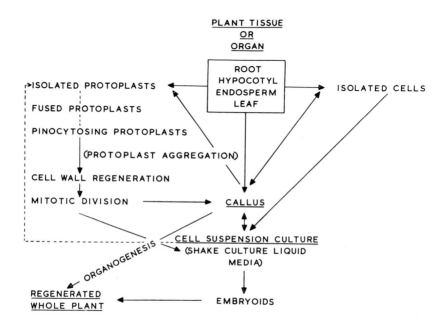

FIG. 5.1. *Schematic viewpoint of the behaviour of higher plant cells in culture* (reproduced with permission of McGraw-Hill).

subject have been fully reviewed recently (Ruesink 1971). As stressed by Professor Street in his introductory chapter, there is in all plant tissue culture work—and work on isolated plant protoplasts is no exception!—still a strong empirical element and, accordingly, a necessity to arrive at technical solutions by trial and error. As of science in general, the mass of necessary information should decrease as knowledge increases. We have not yet quite reached this stage with procedures for the general isolation of protoplasts but the isolation of fruit protoplasts can now be satisfactorily described in a few lines (Willison & Cocking 1972).

ISOLATION OF PROTOPLASTS FROM PLANT TISSUE CULTURES

Introduction

At first sight the advantages of using plant tissue cultures as a source of protoplasts look very attractive. Callus cultures usually grow rapidly and their growth conditions can be rigorously controlled (see Chapter 3). The tissue is normally friable and the cells are loosely packed together allowing ease of penetration of the hydrolytic enzymes. In the case of suspension cultures the need for any cell separating enzyme would be minimal (see Chapter 8). Further, protoplasts arising from callus cultures might be expected to respond more readily to *in vitro* culture than protoplasts arising from plant tissues which have not been conditioned to such culture. By far the most attractive feature of tissue cultures, however, is their sterility, dispensing with the need for surface sterilization of plant material or the growth of whole plants under sterile conditions. It is not surprising, therefore, that in some of the early work on protoplasts, callus or suspension cultures were the source material. Eriksson & Jonassen (1967) chose suspension cultures of *Haplopappus gracilis* from which they were able to isolate protoplasts. Whilst Keller, Harvey, Gamborg, Miller & Eveleigh (1970) have studied the isolation and culture of protoplasts from soybean suspension cultures and have recently (Kao, Gamborg, Miller & Keller 1971) reported that these proto- plasts will regenerate a wall and undergo continued cell division. The range of tissue cultures from which protoplasts have been isolated is now extensive and is steadily increasing. However, in spite of these obvious advantages of callus cultures as a source of protoplasts, they also have some disadvantages. A number of workers (Keller *et al.* 1970, Schenk & Hildebrandt 1969) have found it necessary to desalt the crude commercial enzymes for protoplast isolation from callus material, either to increase the yield of protoplasts, or because protoplasts once isolated using the crude enzymes rapidly deteriorate and die. One procedure (Kao *et al.* 1971) is to suspend the crude enzyme overnight in a small volume of dilute salt solution, such as 15 g of cellulase in 30 ml of 0·5 M NaCl, and then to spin down the undissolved material. This residue, when tested for its ability to liberate protoplasts, is inactive even at high concentrations. The enzyme solution is then loaded into a Biogel P6 Column, $40 \times 2·5$ cm, the enzyme solution washed into the column with 5 ml of water and the column subsequently eluted with water. Fractions are collected and the protein elutes in the void volume whereas the salt elutes in approximately twice the void volume. The fractions which precede the salt are therefore bulked together and freeze dried. This procedure leads to about a tenfold increase in the activity per gramme of enzyme material. Even with the use of these desalted slightly purified enzymes many callus and suspension cultures still fail to produce protoplasts or, in some cases, the yields even after prolonged enzyme incubation are too low for experimental purposes. Lignin present in many suspension culture cells may contribute to this difficulty (Leppard, Colvin, Rose & Martin 1971).

In protoplast preparations from callus material it is often difficult to separate protoplasts from single cells or from cell debris. Chupeau & Morel (1970), however, overcame this problem by using a sucrose/sorbitol density gradient to obtain a clean protoplast preparation from carrot callus material.

In marked contrast to the ease with which leaf protoplasts from a number of sources readily form a new wall and, in the case of tobacco var. Xanthi (Nagata & Takebe 1971), undergo rapid sustained cell division, protoplasts derived from a number of callus and suspension cultures are unexpectedly reluctant to reform a cell wall and enter division. Hellman & Reinert (1971) cultured protoplasts from carrot tissue cultures for up to 5 months without recording wall regeneration or cell division. However, it should be noted that Chupeau & Morel (1970) were able to obtain cell-wall regeneration in their isolated carrot callus protoplasts (see also Grambow, Kao, Miller & Gamborg 1972).

Another disadvantage of tissue cultures as sources of protoplasts, particularly where protoplast fusion is the objective, is the usual high degree of vacuolation of these protoplasts. Ideally for fusion studies protoplasts with little vacuolation and considerable cytoplasm are the best material (Power & Cocking 1971) and few cultured systems yield protoplasts of this type. Giles (1972) has, however, isolated protoplasts from soybean callus grown on a medium with a low auxin and high cytokinin concentration which results in a hard compact growth. These protoplasts are highly cytoplasmic and have been observed to undergo induced fusion with sodium nitrate.

Isolation of protoplasts from crown-gall callus of Parthenocissus tricuspidata

The callus material grown in the dark at 25°C was regularly subcultured at 5-weekly intervals on a medium containing Heller's salt solution (Heller 1953), 0·32 mg/litre thiamine HCl and 2% sucrose, at pH 5·8 solidified with 1% agar. Callus tissue 4 to 5 weeks after subculture is the most suitable material for protoplast isolation; older tissue produces progressively lower yields of protoplasts. For the most rapid release of protoplasts desalted enzymes are required. Thus 4% desalted Meicelase P (Meiji Seika Kaisha Ltd., Chuo-ku, Tokyo, Japan) or Cellulase Onozuka 1500 u/g (activity measured by decomposing activity on filter paper) (All Japan Biochemicals Co. Ltd., Nishinomiya, Japan), 1% desalted Macerozyme (All Japan Biochemicals Co. Ltd., Nishinomiya, Japan) and 10% mannitol dissolved in the liquid culture medium with the omission of sucrose has produced high yields of protoplasts in 4 hours at 25°C. Crude enzyme will also produce protoplasts but only after an extended incubation time (24 hours). Centrifugation at $100 \times g$ for 5 minutes causes the liberated protoplasts to rise and collect at the top of the tube, whereas the cell and cell debris sediment. The isolated protoplasts were washed free of the enzyme mixture by transferring them to culture medium containing 10% mannitol, but without sucrose. On centrifugation at $100 \times g$ they now sediment. These protoplasts have been cultured for up to 5 weeks in the same nutrient medium with the addition of 10% mannitol, and under the same conditions used for the culture of the callus tissue. Although many appear to remain alive and healthy no wall synthesis can be observed optically in this medium nor does cell division occur. A medium specially formulated for cell-wall regeneration rather than for division may be required in the first instance.

Isolation and culture of rose protoplasts

Protoplasts from Paul's Scarlet rose suspension culture cells have been extensively studied by Pearce (1972) and the account here is based on his work. The cells grow rapidly in batch culture on a fully defined medium originally devised by Davies (1971). The cellulase preparations Meicelase P and Onozuka (1500 and 3000 u/g) give satisfactory yields of protoplasts but effectiveness can vary between batches of enzyme. The enzyme mixture was 14% sucrose plus crude cellulase ($0\cdot5\%$ Meicelase P) or $3\cdot0\%$ slightly purified cellulase (see purification details as previously described). The pH was adjusted to between $5\cdot5$ and $6\cdot0$ with NaOH and HCl. Enzymatic damage to the protoplasts also varies between batches. The suspension culture cells form a pipettable suspension composed of single cells and small groups of cells, and thus the necessity for the use of a pectinase is eliminated. Cells from all phases of the growth cycle have been used successfully. One gramme fresh weight of cells filtered from medium was added to every 10 ml of enzyme mixture Twenty millilitres of incubation mixture were placed in a 100-ml Erlenmeyer flask (too great a depth of mixture is deleterious to survival) and placed at a constant temperature of between 25–33°C for up to 24 hours. The incubation mixture was then centrifuged for 1 minute at $300 \times g$ and the protoplasts, which collect at the surface of the liquid, transferred using a Pasteur pipette to 14% sucrose and recentrifuged. This procedure was repeated for the second wash. Live material forms a dense layer at the surface while debris sinks.

It has always been the purpose in isolating rose protoplasts to achieve a floating layer composed entirely of free protoplasts with no wall debris present at all. Complete degradation of the wall was initially achieved when cell suspension cultures recently grown up from callus were incubated for

24 hours at 25°C with a crude cellulase. In subsequent passages it became necessary first to raise the incubation temperature to 33°C and later to use 3% slightly purified cellulase. It was found, however, that if the suspension cells were returned to agar and grown as callus for several passages then the original isolation conditions again became effective. Callus itself, however, did not give good release under these conditions.

Progressive changes in either wall composition or the cells' content of some enzyme inhibitor, with time, in liquid culture, are thus implied. One of

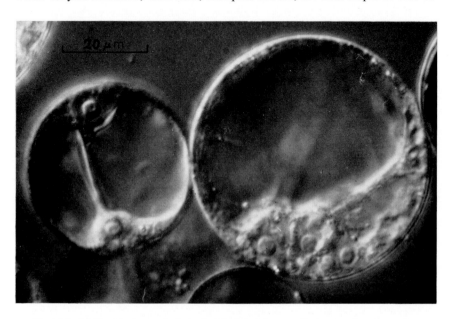

FIG. 5.2. *Multinucleate protoplasts just isolated from actively growing suspension culture cells of Paul's Scarlet rose*. The nuclei can be seen containing prominent nucleoli. (Nomarski interference contrast.)

the more variable components of the suspension cells are phenols (Nash 1968) and hence polyvinylpyrrolidone (360—Sigma, 5% w/v), which adsorbs some phenols (Andersen & Sowers 1968), was added to the various incubation mixtures. This improved release and survival, particularly in the case of cells grown in suspension for many passages. Prior plasmolysis in sucrose for $\frac{1}{2}$ hour or 24 hours reduces release without increasing total survival. The condition of protoplasts in the floating layer is improved but the proportion of the incubation which floats is reduced by about half. Pre-treatment with fresh non-plasmolysing medium gives normal release and survival.

Protoplasts are more readily released from cells which are just entering the growth phase than from mature cells and they also survive the incubation better. Cells from later in the growth phase release protoplasts no more easily than do mature cells, but they tend to survive better. It has been observed by other workers that cells from young, actively growing callus of many tissues most readily release protoplasts (Schenk & Hildebrandt 1969). It might be surmised that cells in this condition contain a minimum of substances deleterious to release or that their walls are then most susceptible

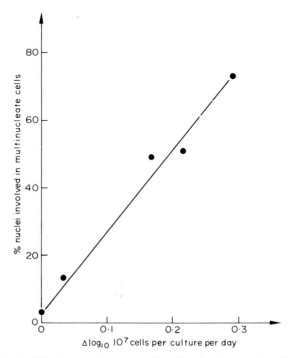

FIG. 5.3. *Relationship between the extent of multinucleations in isolated protoplasts and the division rate of the Paul's Scarlet rose suspension culture cells from which they were isolated.* Cells were taken from early, middle and late growth phase and stationary phase batch cultures. The division rate was obtained from all counts made on daily samples of the cultures.

to the enzyme. Improved survival may be due to a reduced content of such substances or, on the other hand, to greater stability of the more cytoplasmic protoplasts obtained from these cells.

Isolates from mature cells contain very few multinucleate protoplasts. In preparations from the phase of mature cell division, they are frequent and may predominate (Fig. 5.2). Figure 5·3 shows the correlation between growth rate and degree of multinucleation expressed as proportion of nuclei in multinucleate protoplasts. The extent of multinucleation is neither increased by centrifugation nor other physical methods of pushing the protoplasts together. The majority of multinucleate protoplasts are binucleate but numbers as high as nine and ten, while uncommon, are not very rare and even eighteen nuclei in a single protoplast have been observed. A number of the binucleate protoplasts can be attributed to cells in which new cross-wall formation is incomplete. However, the proportion of cells with incomplete cross-walls is far fewer at any one time than the proportion of binucleate protoplasts which can be isolated from those cells. The majority must, therefore, be due to spontaneous fusion (in the sense of Power, Cummins & Cocking 1970, see also Withers & Cocking 1972). The implication here that recently divided cells spontaneously fuse more readily than more mature cells is further supported by the observation that cells kept in rapid division for far longer than is possible in batch culture, by a process of frequent transfer, such that the culture is maintained in the logarithmic phase of growth, also produce a higher proportion of multinucleate protoplasts. Protoplasts isolated from cultures early in the phase of rapid growth enter division very soon after isolation while protoplasts from mature cells do not begin to divide until several days after placing in culture medium.

ISOLATION OF PROTOPLASTS FROM LEAVES

Introduction

The typical leaf with its numerous intercellular air spaces and the loose packing of the tissue is a very suitable material for the isolation of cells. In fact it has been known for some time that cells from the mesophyll tissue of leaves can be isolated mechanically and, in culture, these isolated cells can be induced to divide and form colonies (Kohlenbach 1959, 1966, Joshi & Ball 1968b). Recently, Takebe, Otsuki & Aoki (1968) used an enzymatic procedure for the isolation of viable mesophyll cells from tobacco leaves. Takebe subsequently took this method to the logical next step and treated the isolated mesophyll cells with cellulase to yield isolated protoplasts (Otsuki & Takebe 1969). In 1970 Power and Cocking (1970) found that this two-step procedure was unnecessary and they were able to isolate quantities of tobacco leaf protoplasts by incubating the leaf in a mixture of cellulase and pectinase with an appropriate osmotic stabilizer. In practice the two methods produce protoplast preparations which differ. In the sequential method the spongy mesophyll cells are discarded and the protoplasts arise from the palisade mesophyll whereas the mixed enzyme method produces protoplasts from both palisade and spongy mesophyll. These techniques of protoplast isolation pioneered with tobacco are now being extended, with suitable modifications (see, for instance, Schilde-Rentschler 1972), to other leaf systems.

Isolation of protoplasts from cereal leaves

Mature healthy leaves from 21–28 days old cereal plants were surface sterilized by treatment with 70% ethanol for 2 minutes followed by 30 minutes in 3% sodium hypochlorite (0·3–0·42% w/v available chlorine produced by dilution from a hypochlorite solution, BDH, Poole, England) containing 0·5% Teepol as a wetting agent (BDH, Poole, England). The hypochlorite was removed by three successive washes in sterile water.

Unlike tobacco leaves the epidermis of cereal leaves cannot readily be peeled and so the leaves were cut into narrow longitudinal shreds using a scalpel. The leaf pieces were placed in a petri dish and immersed in 15% sorbitol for 30 minutes to achieve plasmolysis before exposing the cells to the hydrolytic enzymes. This sorbitol solution was replaced by a shallow layer of membrane filter-sterilized (Sartorius Membrane Filter, GmbH, Göttingen, Germany) enzyme solution, approximately 1 g leaf material to 7 ml of enzyme solution. For the isolation of protoplasts from rye leaves this enzyme solution contained 4% Meicelase and Pectinol R10 (Rohm & Haas, Philadelphia, U.S.A.). Pectinol R10 is standardized with diatomaceous earth and a solution of the enzymes was prepared by suspending 20 g in 100 ml of distilled water for 2 hours. The insoluble material was removed by filtration. Since only 6·5% of this Pectinol is water soluble, a 1·3% solution of enzyme results. For the isolation of protoplasts from wheat, oats and barley leaves 1% Macerozyme can be substituted for Pectinol R10. In all cases 1% Potassium dextran sulphate and 14–18% sorbitol or mannitol are added to the enzyme solutions which are adjusted to pH 5·8 with 1 N HCl. After 5 hours' incubation the enzyme solution was removed and the leaf pieces rinsed briefly with a sterile washing solution of the following composition, 13–18% sorbitol, 1 mM KNO_3, 0·2 mM KH_2PO_3, 0·1 mM $MgSO_4$, 1 mM $CaCl_2$, 1 μM KI and 0·01 μM $CuSO_4$ (Jensen, Francki & Zaitlin, 1971) before being returned to a fresh enzyme solution. After a further 12 hours incubation the leaf pieces were teased apart in the washing solution with release of the protoplasts. The protoplasts were sedimented by centrifugation at $100 \times \mathbf{g}$ for 5 minutes and resuspended in sterile 25% sucrose. Further centrifugation at $200 \times \mathbf{g}$ for 5 minutes causes the intact protoplasts to rise to the surface whilst cells, cell debris and free chloroplasts sediment. The floating layer of protoplasts was resuspended in a large excess of washing solution and centrifuged at $100 \times \mathbf{g}$ for 5 minutes. The protoplasts were then suspended in a known volume of washing solution and a small sample taken for counting using a modified Fuchs Rosenthal haemocytometer (see later). Yields of over 1×10^6 protoplasts per gramme of leaf material can be obtained from rye leaves, whereas barley, oat and wheat leaves yield somewhat fewer protoplasts ($1–5 \times 10^5$ per gramme/leaf) (Evans, Keates & Cocking 1972). Actual yields based on the number of cells/leaf may, however, be comparable. The efficiency of the sterilization procedure outlined above is somewhat variable and the addition of the antibiotic Gentamycin at 10 μg/ml to the enzyme incubation mixture and the culture medium has proved valuable in reducing the level of bacterial contaminants during subsequent culture. Some of the wheat and rye protoplasts will reform a cell wall and survive for several weeks when cultured at a low light intensity (70 lux) in White's Medium (White 1963) with the substitution of 15 mg/litre $CuSO_45H_2O$ for the $ZnSO_47H_2O$ (Lamport 1964) and the addition of 1 mg/litre 2,4-D, 1 mg/litre 6-benzyl amino purine and 14% sorbitol, adjusted to pH 5·8 with 1 N NaOH and solidified with 0·6% agar. However, cell division occurs only rarely under these conditions. Other media may be required for division, or it could be that these mature cereal leaf cells are not capable of being so readily cultured as are cells arising from isolated protoplasts of tobacco (Nagata & Takebe 1971) or petunia (Frearson, Power & Cocking 1973) leaf material.

Isolation of protoplasts from leaves of leguminous plants
(protocol worked out by M.R.Davey)

Five to 7 weeks' old pea (*Pisum sativum* var. Little Marvel) and broad bean (*Vicia faba* var. Aquadulce Claudia) at the time of flowering or early fruit set, and 4 months' old garden lupin plants (*Lupinus polyphyllus*) were used for protoplast isolation. Plants were grown in the greenhouse under natural daylight during the summer, and under 10,000 lux of illumination from banks of warm white fluorescent tubes maintained on a 16 hours light/8 hours dark cycle during the winter. Plants were transferred from the greenhouse to the laboratory 24 hours before use and maintained at 22°C under 5000 lux of illumination provided by warm white fluorescent tubes on a 16 hours light/8 hours dark cycle. Fully expanded leaves were removed from the plants, the leaflets excised, and 5–10 g fresh weight of material sterilized (by immersion in 600 ml of 1% sterile Teepol for 2·5 minutes with constant agitation using a flame sterilized long-handled spatula, followed by immersion in 600 ml of 1·5% sodium hypochlorite for 10 minutes). The leaves were then thoroughly washed with 8×400 ml changes of sterile distilled water.

The lower epidermis was then removed from the leaflets using fine forceps (see later), and the peeled leaflets floated on the surface of 25% sucrose with the exposed mesophyll tissue in contact with the plasmolyticum. All incubations were carried out in 9 cm sterile plastic petri dishes. Following prior plasmolysis for 1 hour the leaf material was transferred to the surface of a mixture of 5% Meicelase P with 5% Macerozyme in 25% sucrose (pH 5·8) for 20 hours at 25°C, followed by replacement of the mixture with fresh enzymes at 3 and 6 hours after the beginning of incubation. Protoplasts were released by gentle agitation of the leaf pieces at the end of the incubation period, the protoplast-enzyme mixture strained through fine wire gauze and three layers of muslin to remove leaf debris, transferred to 13 × 100 mm screw-capped tubes, and centrifuged at 225 × g for 5 minutes. After centrifugation, the protoplast layer covering the surface of the enzyme solution in each tube was removed using a Pasteur pipette, and the samples bulked. Protoplasts were then washed free of enzymes by resuspending in 25% sucrose, followed by centrifugation at 225 × g for 5 minutes. This washing procedure was repeated five times. Isolated protoplasts were suspended in a standard volume of 25% sucrose and an aliquot transferred to a modified Fuchs/Rosenthal silver-backed counting chamber with a depth of 0·2 mm (Hawksley Cristalite B.S. 748, Lancing, Sussex, U.K.) for cell number estimations. Protoplast yields for both pea and broad bean were in the range $2·0-3·0 \times 10^6$ per g fresh weight of leaf tissue.

Isolation of tobacco leaf protoplasts for virus infection studies
(protocol worked out by R.H.A.Coutts)

Plant virology studies have been hampered by the lack of a suitable host cell system in which synchronous virus replication can be observed (Seigal & Zaitlin 1964). The introduction of the use of tobacco leaf protoplasts, for studying infection, and replication of tobacco mosaic virus (Takebe & Otsuki 1969), will perhaps provide such a 'one-step' virus replication system. The use of such a system is restricted, unless the procedure can be modified to different host plants and viral strains. The system outlined is such a modification. For details of the virus infection studies see Coutts, Cocking & Kassanis (1972).

Sixty to seventy days' old *Nicotiana tabacum*, var. Samsun, plants were grown in 7 inch pots of 'John Innes' compost at 22–25°C. Fluorescent warm-white tube illumination at a light intensity of 10,000–11,000 lux at the middle leaf surface, and a photo-period of 16 hours was maintained. The relative humidity was kept at 70–75%, and the plants left 24 hours without water before use in an experiment. This allowed a slight loss in turgor of the plants, and facilitated easier epidermal peeling. Leaves not quite fully expanded (20–25 cm in length) gave the more stable and most virus susceptible protoplasts. These leaves are often in position six to eight leaves from the apex of the plant. The leaves were then surface sterilized by immersion in 70% alcohol for 30–60 seconds, followed by further immersion in 2·0% sodium hypochlorite for 30 minutes. A little Cetavlon (ICI Ltd., Macclesfield, U.K.) may be added as a wetting agent. The leaves are then washed three times in sterile distilled water to remove the surface sterilants since these are often inhibitory to protoplast stability (Power 1971) and after this stage all work is carried out under aseptic conditions in a sterile air cabinet.

The lower epidermis was peeled from the leaves, and they were cut into small strips, to give 4–6 g peeled leaf material. The peeled epidermal strips were placed into 20 ml of a maceration medium containing 0·5% Macerozyme, 13% mannitol, and 1·0% potassium dextran sulphate (molecular weight source Dextran 560, sulphur content 17·3%, Meito Sangyo Co. Ltd., Japan). The pH of the medium was adjusted to 5·8 with 2 N HCl, before sterilization, through a millipore filter. The epidermal strips were then shaken in the maceration medium at 25°C on a reciprocal shaker (100–120 cycles/minute, 4·5 cm stroke) and cell fractions isolated after 30 minutes', 1 hour 15 minutes', 2 hours' and 3 hours' shaking. The maceration medium was replaced after each time interval. The final two fractions contained almost pure preparations of palisade parenchyma cells (Takebe, Otsuki & Aoki 1968). These were centrifuged out of the medium (100–200 × g for 2–3 minutes) and washed by two resuspensions in fresh 13% mannitol, and centrifugation as before.

The isolated cells were then placed into 40 ml of 4% Cellulase Onozuka SS 1500 u/g (a crude cellulase) made up in 13% mannitol, adjusted to pH 5·2 with 2 N HCl, before millipore sterilization. The cell suspension was then incubated at 36°C for 3–3½ hours, with occasional gentle swirling to stop clumping of the cells during cell-wall removal, and the protoplasts harvested from the cellulolytic medium by slow speed centrifugation (100 × g for 1 minute) and washed by two resuspensions in 13% mannitol, plus 0·1 mM calcium chloride, followed by centrifugation as before. The protoplasts suspended in 5 ml of 13% mannitol, were counted in a modified Fuchs/Rosenthal silver-backed counting chamber with a depth of 0·2 mm after which they were sedimented by centrifugation and resuspended in fresh 13% mannitol to give a concentration of $1–4 \times 10^6$ protoplasts/ml.

The efficiency of dextran sulphate as a ribonuclease inhibitor has been studied for different viral systems (Philipson & Kaufman 1964) and its inclusion in a protoplast/virus system is advantageous because it reduces the ribonuclease level in the maceration medium to a low level. This is useful because the deleterious effects of ribonuclease on tobacco mosaic virus during initiation of infection are well known (Hamers-Casterman & Jeener 1957). The activity of ribonuclease in a 0·5% medium of Macerozyme was reduced from $5·56 \times 10^{-1}$ Kunitz units/mg of Macerozyme to $8·40 \times 10^{-6}$ Kunitz units on the inclusion of potassium dextran sulphate at 1·0% concentration. Potassium dextran sulphate also aids in stability of isolated mesophyll cells and protoplasts, but whether this is an ionic effect or one of enzyme suppression is not clear.

The effects of poly-L-ornithine (mol. wt 120,000) on protoplasts are not fully understood, but if pinocytosis of virus particles does occur in the protoplast system (Cocking 1970) then the inclusion of a known pinocytic inducer (Ryser 1967), such as a poly-amino acid, would perhaps assist infection. In fact poly-L-ornithine is necessary for infection of protoplasts by tobacco mosaic virus particles (Takebe & Otsuki 1969, Coutts, Cocking

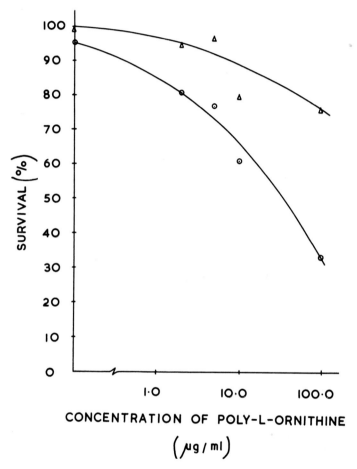

Fig. 5.4. *Survival of Samsun tobacco leaf protoplasts after exposure to various concentrations of poly-L-ornithine, for 1 hour at 25°C, in the infection buffer.* Protoplasts: freshly isolated △–△; 12 hour incubated ○–○.

& Kassanis 1972). However, the effects of poly-L-ornithine depend upon its concentration (Takebe & Otsuki 1969).

Tobacco leaf protoplasts variety Samsun were isolated and exposed to various concentrations of poly-L-ornithine in the same buffer as used for infection of the protoplasts with tobacco mosaic virus (Coutts, Cocking & Kassanis 1972). Intact protoplasts in each aliquot were counted in the Fuchs/Rosenthal counting chamber. The treated aliquots were then incubated under the same conditions as used for the incubation of virus-infected protoplasts (Coutts, Cocking & Kassanis 1972), and intact protoplasts were counted after 12 hours' incubation. The assessment of intact protoplasts is made by observing that chloroplasts are evenly distributed and that the isolated protoplasts are compact spherical entities. On increasing the concentration of poly-L-ornithine above 2 μg/ml, no greater uptake or replication of virus was observed, only an instability of the protoplasts (see Fig. 5.4).

Other poly-electrolytes tested for their effects on uptake of tobacco mosaic virus into protoplasts to cause infection were DEAE-Dextran, and poly-L-lysine (mol. wt 140,000). These had little effect on virus uptake and the infection levels were low.

Isolation of tobacco and petunia protoplasts (with particular reference to spontaneous fusion) (protocol worked out by J.B.Power and E.M.Frearson)

During the enzymatic degradation of the cell walls of tissues, spontaneous fusion of the protoplasts can frequently occur (Power, Cummins & Cocking 1970). As the cell walls are degraded within the tissue, a gradual expansion of the plasmodesmatal connections between adjacent cells results in the spontaneous fusion of protoplasts (Withers & Cocking 1972) with the eventual production of spontaneous fusion bodies consisting of two or more protoplasts. This provides a readily reproducible system for the study and culture of multinucleate protoplasts.

Two methods will be described for the isolation of leaf protoplasts, whereby this spontaneous fusion can be reduced to a minimum, thus providing large quantities of single protoplasts. These methods demonstrate two fundamental approaches to the isolation of protoplasts. First, for tobacco, a short incubation of the leaf material when the cell walls are readily degradable and secondly, for petunia, a longer (overnight) incubation when the cell walls of the material under examination cannot be degraded rapidly. Both techniques release large quantities of viable protoplasts which will in the case of suitable species undergo cell-wall regeneration and cell division when cultured under suitable conditions.

For the isolation of the tobacco leaf protoplasts tobacco seeds (*N. tabacum* var. Xanthi) were germinated in 'John Innes' seed compost and allowed to reach the four-leaf stage. The seedlings were then transferred to 'John Innes' No. 1 potting compost and grown up under fluorescent lights

(7500 lux, top leaf surface) at 20–21°C. The enzyme mixture used consisted of 4% Cellulase Onozuka SS 1500 u/g, and 0·5% Macerozyme in 13% mannitol (pH 5·8, using 2 N HCl); the enzymes dissolved at room temperature and millipore-filtered (0·45 μ pore size). Leaves (14 days' old) of the tobacco plants (50–60 days) were removed from the plants and surface sterilized by successive immersion in the following: 70% ethanol (30 seconds), 2·5% sodium hypochlorite solution (30 minutes), sterile distilled water (five changes).

Working now under aseptic conditions the lower epidermis was removed by horizontally inserting fine jeweller's forceps into a junction of the midrib and a vein and carefully peeling away the epidermis at an angle to the main axis of the leaf. Peeled leaf pieces (approximately 4 cm^2) were placed, exposed surface downwards, on the surface of a 13% mannitol solution. During the subsequent 3 hours, plasmolysis within the tissue occurred, after which the leaf pieces (1 g) were transferred to the enzyme mixture (20 ml), maintained in a closed petri dish and submitted to static incubation at room temperature (22°C) for 4 hours. Incubation was terminated by pouring the enzyme incubation mixture through a coarse wire gauze (or cheese-cloth) and gently washing the residual leaf material with 13% mannitol. The protoplasts were collected by centrifugation (100 × **g**, 5 minutes) and washed free of debris by repeated resuspension in a small volume (10 ml) of the appropriate culture medium containing 13% mannitol. Sucrose (23%) may be substituted throughout for mannitol and in this case protoplasts are collected and freed of debris by flotation.

Under the conditions described above, comparable yields of leaf proto-plasts can be obtained using the following enzyme mixtures as alterations to the one included in the above protocol: (i) 4% Meicelase P + 0·5% Macero-zyme; (ii) 4% Cellulase Onozuka SS 1500 u/g + 20% Pectinol R10; (iii) 3% Cellulase Onozuka 3000 u/g + 0·5% Macerozyme; (iv) 2% Celluluse Onozuka 5000 u/g + 0·5% Macerozyme. The enzymes are used in the presence of 13% mannitol or 23% sucrose and at pH 5·8.

For the isolation of petunia leaf protoplasts the middle and lower leaves of petunia plants (Sutton's, Reading, England, F1 hybrids, mutiflora, var. Blue Dandy and Red Cap, 60 days' old) were removed and surface sterilized as follows: 70% ethanol (30 seconds), 10% 'Domestos' (Lever Bros., London) solution (30 minutes), sterile distilled water (six changes). The lower epidermis was removed (cf. tobacco) and peeled leaf pieces (3 cm^2) placed, exposed surface downwards, on the surface of a 13·6% mannitol solution for 3 hours. The plasmolysing solution was then removed and replaced by the enzyme incubation mixture (1% Cellulase Onozuka SS 1500 u/g, 0·3% Macerozyme, 13·6% mannitol, pH 5·8) and the leaf pieces were incubated statically overnight (18 hours) at 26°C. The protoplasts released were collected and freed of debris as described above for tobacco. Sucrose (23%) may be substituted for the mannitol.

Two methods can be successfully employed for the production of spontaneously fused tobacco leaf protoplasts.

(i) Surface sterilized, peeled leaf pieces were incubated, statically at room temperature and without prior plasmolysis, in an enzyme mixture consisting of 4% Cellulase Onozuka SS 1500 u/g, 0·5% Macerozyme, 13% mannitol, pH 5·8. Degradation was allowed to proceed until some protoplasts were seen to be released from the tissue (usually 4–5 hours), whereupon the leaf pieces were carefully lifted out of the enzyme mixture and transferred to a small volume (5–10 ml) of the culture medium maintained in a petri dish. Using forceps the leaf pieces were slowly agitated or gently pressed against the edge of the petri dish thus facilitating the release of large numbers of spontaneously fused leaf protoplasts, together with some unfused material.

(ii) Large peeled leaf pieces, surface sterilized in the normal manner, were incubated statically overnight (26°C) in reduced levels of the enzyme mixture (1·3% Cellulase Onozuka SS 1500 u/g, 0·3% Macerozyme, 13% mannitol, pH 5·8). Incubation was terminated by gently swirling the incubation mixture, whereupon the whole of the leaf tissue disintegrates, leaving only the upper epidermis on the surface of the enzyme mixture. The upper epidermis was removed, and the protoplasts, together with some of the enzyme mixture, transferred to test-tubes. Protoplast suspensions, enriched with spontaneously fused protoplasts, were collected at the bottom of the tubes by either static sedimentation of the initial preparation (30–45 minutes) or a brief centrifugation ($100 \times g$, 1 minute). These enriched protoplast suspensions can be freed of debris by flotation in 23% sucrose.

Frequently 60% of the protoplasts obtained by methods (i) and (ii) are fused, normally in twos or threes, but sometimes larger masses (termed fusion bodies) often consisting of fourteen or more protoplasts, are produced.

Culture of leaf protoplasts

Nagata & Takebe (1971) described the regeneration of whole plants from isolated mesophyll protoplasts of tobacco obtained using a sequential enzyme treatment. Protoplasts isolated using mixtures of pectinases and cellulases as a 'one-step' procedure readily regenerate into whole plants. Spontaneously fused protoplasts cannot be readily produced using a sequential enzyme treatment since this involves the initial separation of the cells and a subsequent cellulase treatment to degrade their cell walls. Essentially the same technique is employed for the culture of both spontaneously fused and unfused material. Spontaneously fused tobacco leaf protoplasts have been shown to undergo cell-wall regeneration, and cell division with some synchrony of nuclear division (Power, Frearson & Cocking 1971) when maintained in the medium of Nagata & Takebe (1970) containing 23% sucrose instead of mannitol, or in a commercial White's medium (S-3, Biocult Laboratories Ltd., 3 Washington Road, Abbotsinch Industrial Estate, Paisley, Scotland) also containing 23% sucrose.

It has also been shown that multinucleate soybean callus protoplasts (Miller, Gamborg, Keller & Kao 1971) obtained as a result of spontaneous fusion during the enzymatic degradation of the tissue exhibit extensive synchronous nuclear division. In this system nuclear fusion prior to nuclear division was common. Synchrony of nuclear division in such multinucleate protoplasts should afford the opportunity of nuclear fusion provided the nuclei come together (Power, Frearson & Cocking 1971). Such nuclear fusion could in turn result in ploidy changes in the system. Regenerated cell colonies and, subsequently, whole plants, from spontaneously fused protoplasts could when eventually realized result in the production of polyploid plants and hence a greater control over ploidy changes in plants generally.

The media requirements for protoplast culture will vary depending upon the species or even the variety but the basic technique of culture is similar. For instance unfused or fused tobacco protoplasts, prepared as described above, were collected by centrifugation ($100 \times g$, 5 minutes) and washed free of debris and contaminating enzyme solution by resuspension in the culture medium of Nagata & Takebe (1971) containing 13% mannitol so as to give a concentration of protoplasts of 4×10^5/ml. A 2 ml aliquot of this suspension was placed in a tight-lidded petri dish (50×12 mm: Falcon Plastics Ltd) and mixed with an equal volume of the above medium containing 1·2% agar (maintained at 45°C until required) so as to give a final concentration of 2×10^5 protoplasts/ml (Nagata & Takebe 1971). The dishes were then inverted and maintained at 28°C (± 2°C) with a continuous illumination of 2300 lux from white fluorescent tubes. Protoplasts maintained in agar in this way are readily observable during culture and it is possible to follow the development of individual protoplasts or spontaneous fusion bodies during this period. Cultured tobacco protoplasts readily undergo cell-wall regeneration followed by repeated cell division with the production of clearly visible cell colonies. Spontaneous fusion bodies have been observed to divide but it has not as yet been established that they give rise to visible cell colonies. After approximately 6 weeks' culture, colonies (0·5–1 mm diam.) can be pricked out individually and placed on the surface of the B_3 medium of Sacristán & Melchers (1969) (containing 1% agar). Under the same light and temperature regime these colonies grow rapidly and differentiate numerous shoots after approximately 4 weeks. These shoots subsequently produce roots when placed on White's medium (containing 0·1 mg/l thiamine HCl and 2% sucrose).

Individual petunia leaf protoplasts readily form callus in a manner similar to tobacco. The culture medium for petunia consists of the tobacco regeneration medium (Nagata & Takebe 1971) mixed with an equal volume of an earlier cell-wall regeneration medium (Nagata & Takebe 1970) and 13·6% mannitol as the osmoticum (Frearson, Power & Cocking 1973). Petunia leaf protoplasts cultured in continuous illumination at 700 lux and at 26°C undergo cell-wall regeneration and division with the production of cell colonies. A rapidly dividing callus can be produced when individual colonies (0·5–1 mm diam.) are transferred to agar plates containing the B_3 medium of Sacristán & Melchers (1969).

ISOLATION OF MATURE TOBACCO POLLEN PROTOPLASTS
(protocol worked out by J.B.Power)

Using a fine scalpel, pollen grains were removed from the anthers of fully expanded flowers of tobacco (*N. tabacum* var. Bright Yellow), and the pollen of eight flowers placed in a small volume (5–10 ml) of a 15% sucrose solution maintained at 25°C. During the subsequent 5 hours the pollen sank in this medium and, although germination was not synchronous, normally 60–70% of the pollen grains germinated during this period, and the extreme tip of the pollen tube was visible. A further rapid growth of the pollen tubes was prevented by transferring the pollen grains to 10 ml of a non-sucrose containing medium (e.g. 13% mannitol plus the salt composition of Nagata & Takebe 1970), but containing 1·0–1·5% Macerozyme (pH 5·8). The pollen grains were then incubated overnight in this medium maintained in a reciprocal shaking water bath (70 cycles/minute, stroke 4·5 cm) at 30°C, and then centrifuged and the overnight medium replaced by 10 ml of a 15% sucrose solution containing 2% Cellulase Onozuka SS 1500 u/g and 2% Cellulase Onozuka 5000 u/g (pH 5·8) and the preparation statically incubated at 25°C for a further 5 hours.

During the final cellulase treatment, protoplasts are released through the weakened tips of the pollen tubes (Fig. 5.5) or, when pollen tube production is minimal, directly through the exine. The latter normally produces two or three non-vacuolated sub-protoplasts which generally remain attached to the pollen grain, but which can be seen freely suspended in the medium after gentle agitation.

Under these conditions the yield of protoplasts is low, but if germination of the pollen tube is allowed to proceed unretarded, then no protoplasts are produced, since the rate of pollen tube wall production is far in excess of the rate of cell-wall degradation by the cellulases so far tested. Pollen grains placed directly in a cellulase solution either do not germinate, or, if low levels of cellulases ($<1\%$) are employed, the pollen will germinate normally and will yield no protoplasts. Germinating pollen placed directly in cellulase solutions produces maximum tube growth with no protoplast production but the desired intermediate stage of small tube production, lack of vacuolation and minimal tube elongation can be achieved with most of the pollen by the inclusion of a Macerozyme treatment in a non-sucrose medium. Occasionally

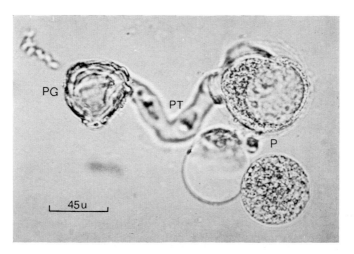

Fig. 5.5. *A mature pollen grain (PG) of tobacco* germinated for 5 hours in 15% sucrose, incubated overnight with 1·5% Macerozyme (in 13% mannitol) and finally treated for 5 hours with a mixture of cellulases (2% Cellulase Onozuka SS 1500 u/g and 2% Cellulase Onozuka 5000 u/g, in 15% sucrose) at 25°C.

A vacuolated protoplast (P) is seen emerging from the weakened tip of the pollen tube (PT). Freely suspended in the medium is a non-vacuolated pollen protoplast together with an isolated vacuole.

during the Macerozyme treatment, a combination of shaking, and pollen expansion as a result of germination, results in a sloughing off of the exine with the production of a pollen cell. Such pollen cells can occasionally be converted, during a subsequent cellulase treatment into relatively non-vacuolated isolated protoplasts.

APPLICATIONS

The areas of experimental investigation in which protoplasts are finding a use are becoming increasingly varied. No longer are investigators satisfied with the isolation of protoplasts from a particular material but are using protoplasts to help answer problems in plant and cell physiology. Hence most of the procedures which have been described have been developed from a particular standpoint. Indeed isolated protoplasts offer in a number of ways a unique approach to many of these problems.

Perhaps the area of investigation which is attracting the most active work currently centres on the possible use of isolated protoplasts to provide experimental material for plant cell modification. Normally, of course, the cell wall is a barrier to both uptake and fusion. Studies on the production of homokaryons and heterokaryons as a result of protoplast fusion were greatly stimulated by the work of Power *et al.* (1970). What is now needed by plant cell biologists are isolated protoplast systems, preferably haploid, having negligible vacuolation suitable for fusion and cytological studies including nuclear characterizations by microdensitometry. One such protoplast system may be that from young pollen tetrads. Bhojwani & Cocking (1972) used an enzyme rich in β1-3 glucanase to degrade the callose wall of the tetrads with subsequent release of naked haploid microspores which were densely cytoplasmic with negligible vacuolation. It is known that highly vacuolated protoplasts do not readily undergo induced fusion to the extent that the nuclei in the homo or heterokaryons are in close association with each other (Cocking 1973b); and indeed as a result the induced fusion of isolated leaf protoplasts and most callus and suspension culture isolated protoplasts may be more suited to the formation of cytoplasmic, rather than nuclear, hybrid cells. Less vacuolated and perhaps in certain instances more meristematic, protoplasts such as those from the pollen tetrads may be more suitable for the production of heterokaryons with nuclei in close association with one another permitting more ready nuclear fusion and hybrid cell formation.

Activity at the plasmalemma which expresses itself in cell-wall synthesis by isolated protoplasts (Willison & Cocking 1972) and in the uptake of particles by endocytosis has already been discussed. The actual isolation of protoplasts is also associated with activity at the plasmalemma. This can result in the uptake of whole micro-organisms such as bacteria into protoplasts as a result of plasmolytic uptake (Davey & Cocking 1972). In this way it may be possible to establish new endosymbiotic associations as well as to investigate the possibility of various forms of transformation in higher plants such as the transfer of genes for nitrogen fixation (Cocking 1973b).

CHAPTER 6

GENERAL CYTOLOGY OF CULTURED CELLS

M. M. YEOMAN and H. E. STREET

Introduction 121
Structure and physiology of quiescent cells 121
Changes in cell structure associated with the induction of division in explants 127
Cytological characteristics of the developing callus 137
Range of cell form and structure in suspension cultures 143
Cell division in suspension cultures 150
Light-microscope studies of cells in mitosis 153
Electron-microscope studies of cells in suspension culture 154

INTRODUCTION

Very nearly all callus cultures are derived from tissues composed of vacuolated cells. Two major types of vacuolated cells are involved: those of the vascular cambium which may already be in a state of active division and a variety of parenchyma cells which are quiescent and have to be induced to divide. Studies on cultured callus cells have often been preceded by a detailed examination of the cells from which they have been derived, in the hope that the differences revealed between mature quiescent cells and those cells engaged in division will help to provide the answer as to how cells are induced to divide. Rapid division of the tissue leads to the callus from which large masses of highly vacuolated cells are formed and these cells may be used to initiate suspension cultures. Considerable changes in cellular structure accompany the activation of quiescent cells and development of calluses which, once formed, exhibit a wide range of cell form and structure characteristic of their mode of growth.

STRUCTURE AND PHYSIOLOGY OF QUIESCENT CELLS

Living differentiated cells are frequently characterized by a large central vacuole and a thin layer of cytoplasm lining the wall. Sometimes cytoplasmic stands traverse the central vacuole. They frequently contain storage products, which may be in the vacuole (e.g. inulin or sucrose) or contained in plastids

E

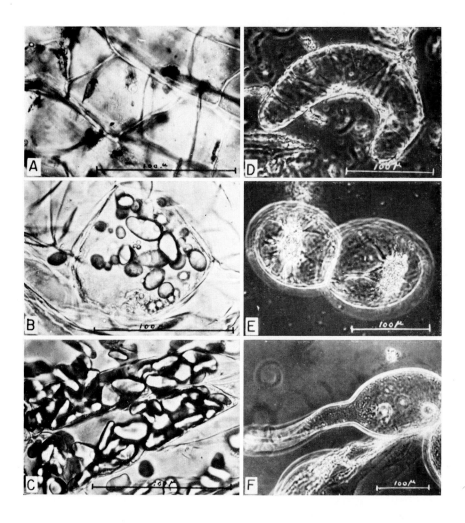

FIG. 6.1. *Contrasts between quiescent cells and their active counterparts growing free.*
A. Resting cell of the secondary phloem of carrot root, showing carotene in chromatophores.
B. Resting cell of potato tuber showing abundant starch grains.
C. Resting cell of banana fruit pulp showing abundant starch grains.
D. Cf. A; free cell of carrot, growing in liquid medium containing coconut milk (10%).
E. Cf. B; free cells of potato tuber, having grown in liquid medium containing coconut milk (10%)+2,4-D (2,4-dichlorophenoxyacetic acid) at 6 ppm.
F. Cf. C; free cells of banana fruit activated to grow by coconut milk and naphthalene acetic acid (all from Steward 1961).

and chloroplasts (starch and occasionally phytoferritin). In addition oil droplets and protein-containing bodies may be present within the cytoplasm. Usually the nucleus is flattened and peripherally located in quiescent cells. In complete contrast nuclei in dividing vacuolated cells (Fig. 6.1) are commonly more spherical and held in a central position by transvacuolar cytoplasmic strands (Bailey 1920, Sinnott & Bloch 1940, 1941, Steward, Mapes & Smith 1958, Jones, Hildebrandt, Riker & Wu 1960, Israel & Steward 1966).

The appearance of the nucleoli also betrays the quiescent nature of the cells (Fig. 6.2A): the nucleolar particles are tightly packed and the granular particles, usually surrounding the central fibrous ones, are less prominent (Rose, Setterfield & Fowke 1972). This is in contrast to the arrangement in dividing cells where during prophase the nucleolar particles are reported to be less compact with the granular and fibrous regions intermingled (Lafontaine & Chouinard 1963, Brinkley 1965, Birnstiel 1967). The arrangement of ribosomes in quiescent cells is consistent with a low rate of protein synthesis as polyribosomes are rarely seen. Such polysomes as are observed in quiescent cells from the Jerusalem artichoke tuber are helical and apparently free in the ground substance (Bagshaw 1969). Wooding (1968) has reported the presence of large helical aggregations of ribosomes in some of the companion cells in the mature secondary phloem of *Acer pseudoplatanus*. He suggested that these aggregations might represent the inactivation and storage of ribosomes in response to some environmental influence. However, most of the ribosomes in quiescent artichoke cells are not found in clusters but scattered singly in the ground substance. Some ribosomes may be seen attached to the endoplasmic reticulum or to the nuclear envelope.

The respiratory activity of quiescent cells is very low (Yeoman, Dyer & Robertson 1965) but the internal structure of the mitochondria appears quite normal. The simplest mitochondrial form observed is a sphere or a short rod. Bagshaw, Brown & Yeoman (1969) have, however, observed cup-shaped mitochondria in quiescent cells from the tuber of the Jerusalem artichoke (Fig. 6.3A).

It has been reported that dictyosomes in quiescent cells are sometimes composed of curved cisternae with few vesicles (Mollenhauer & Morré 1966). Bagshaw (1969) has shown that the dictyosomes in cells from artichoke tissue are frequently curved and associated with some vesicles although some dictyosomes composed of straight cisternae may also be observed (Fig. 6.2B). It is likely that the vesicles were derived from the dictyosomes during an earlier stage of development when metabolism was more active and are not a true reflection of the activity of these quiescent cells. Crystal-containing bodies (CCBs) (Fig. 6.3C) have been reported in a range of quiescent cells; Frederick, Newcomb, Vigil & Wergin (1968) interpreted the presence of these microbodies as being characteristic of metabolically less active cells.

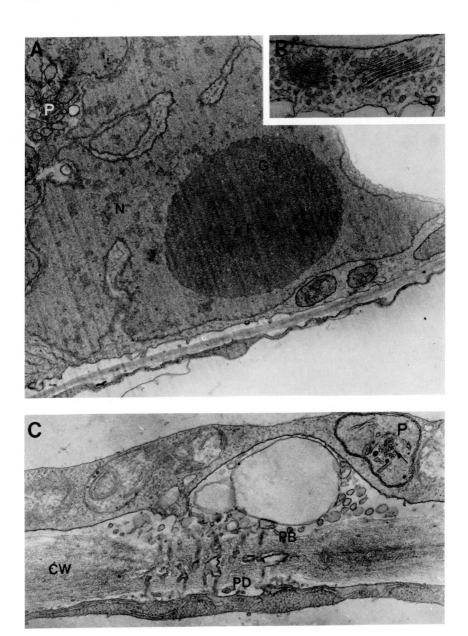

They also pointed out that association between CCBs and the endoplasmic reticulum occurred more frequently in metabolically active cells. This is consistent with the situation in artichoke tuber cells where the CCBs are not associated with the endoplasmic reticulum.

A few microtubules have been observed in quiescent cells adjacent to the cell wall (Bagshaw 1969). This is as expected in cells where cell-wall deposition is proceeding very slowly or has stopped completely. In actively growing cells many microtubules are found in the cytoplasm adjacent to regions of wall growth where they are probably active in the deposition of cellulose micro-fibrils. (Ledbetter & Porter 1963, 1964, Hepler & Newcomb 1964, Wooding & Northcote 1965, Cronshaw & Bouck 1965). Paramural bodies are also present in association with plasmadesmata (Fig. 6.2C).

Plastids are found in a variety of shapes and forms in quiescent tissue (Israel & Steward 1967; Tulett, Bagshaw & Yeoman 1969). They are generally scattered within the cell but may on occasions be found in tight clusters which may become dispersed during subsequent culture (Tulett, Bagshaw & Yeoman 1969). They are not, however, a characteristic of quiescent cells for plastid clusters are also found in the cambial region of developing tubers. Plastids in quiescent cells of the carrot root lack an internal lamellar structure (Israel & Steward 1967). A distinctive peripheral system may be observed, which is formed by invagination from the inner of the limiting membranes. The central stroma may contain starch grains alone or starch and carotene together. Many particulate ribosome-like granules are also present in the stroma. Two distinct membrane-bound systems were recognized in artichoke tuber plastids by Gerola & Dassu (1960). They called the electron-dense sac-like system the *corpo opaco*. Tulett, Bagshaw & Yeoman (1969) showed that this central system or *corpo opaco* was often irregular and connected to branching tubules and cisternae (Fig. 6.3B). The apparently independent second system is peripheral. The differences in appearance of the peripheral and central systems presumably relate to their different functions. The function of the peripheral system is not obvious but considering its position in the plastid it could be suggested to have a function of transport associated with movement of materials through the plastid envelope to which it is connected. The central system is likely to be a store of protein.

FIG. 6.2. *Quiescent parenchyma cell from Jerusalem artichoke tuber tissue* (Bagshaw 1969).

A. A peripherally situated, lobed nucleus (N) containing dense aggregations of chromatin and a nucleolus with tightly packed granular (G) and fibrillar (F) portions. Part of a cluster of plastids (P) lies near the nucleus ($\times 20,000$).

B. Dictyosomes composed of straight cisternae and vesicles ($\times 42,000$).

C. Paramural body (PB) associated with plasmadesmata (PD). A plastid (P) containing phytoferritin (arrowed) is located close to the cell wall (CW) ($\times 36,000$)

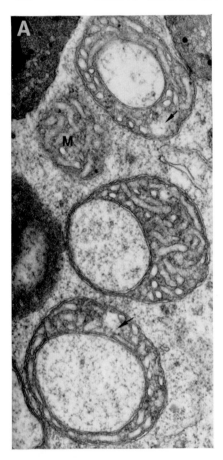

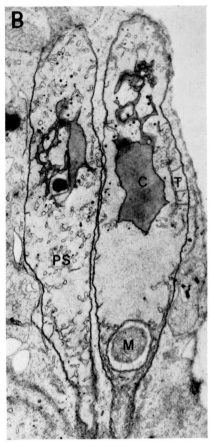

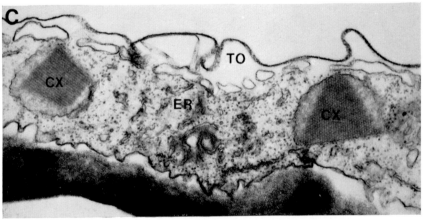

CHANGES IN CELL STRUCTURE ASSOCIATED WITH THE INDUCTION OF DIVISION IN EXPLANTS

The induction of division is accompanied by a transformation of the structure of quiescent cells. These changes largely reflect changes in the metabolism of the cells. The majority of observations have been made on parenchyma cells of the carrot root (Israel & Steward 1966, 1967) and the Jerusalem artichoke tuber (Fowke & Setterfield 1968, Bagshaw 1969, Yeoman, Tulett & Bagshaw, 1970). The earliest detectable change is an increase in the number of polyribosomes (within 1 hour of excision). These assume the form of helices and spirals. The spirals are seen to be associated with the endoplasmic reticulum while the helices are found scattered in the ground cytoplasm. Polysomes increase in frequency with time (Fig. 6.4A and B). There is also an increase in bound ribosomes of which a high proportion are in spiral clusters. Nicholson & Flamm (1965) found an increase in protein synthesis, in cultures of tobacco cells, associated with a rise in the proportion of membrane-bound ribosomes which they isolated from this tissue. Sealey (1973) has also shown that an increase in polyribosomes accompanies the excision and culture of artichoke tuber tissue. Israel & Steward (1966) and Halperin & Jensen (1967), using electron microscopy, have shown a large increase in the amount of free ribosomes in cultured carrot tissue. In contrast, no such increase was detected in cultured artichoke tissue (Bagshaw 1969) where these changes in the machinery for protein synthesis are followed by the appearance of large quantities of unidentified electron-dense bodies in many of the cell vacuoles. Some of these bodies are fibrillar, while others are more solid (Fig. 6.4B). These bodies are also found in the cytoplasm. Some resemble lipid bodies both in their homogeneous appearance and in the fact that they are often shattered. Their nature and significance are not understood.

Gerola & Bassi (1964) detected CCBs which they called proteosomes in artichoke tuber cells cultured in the light. They did not detect them in freshly excised tissue. Bagshaw (1969) found CCBs scattered in the cytoplasm of

---

Fig. 6.3. *Mitochondria, plastids and crystal-containing bodies (CCBs) in quiescent parenchyma cells of Jerusalem artichoke tuber.*

A. Three cup-shaped mitochondria encircle cytoplasm which is less dense than the surrounding cytoplasm. The arrows mark fibres in the mitochondria which might be DNA. One mitochondrion (M) shows a simple profile which may represent a section through a sphere or cylindrical rod ($\times 34,000$) (Bagshaw 1969).

B. Plastids showing central (C), peripheral (PS), and tubular (T) systems. A mitochondrion (M) is shown in the encircled cytoplasm ($\times 23,000$) (Tulett, Bagshaw & Yeoman 1969).

C. Two CCBs containing large crystals (CX). The tonoplast (TO) is partially detached and the underlying endoplasmic reticulum (ER) is apparently swollen ($\times 44,000$) (Bagshaw 1969).

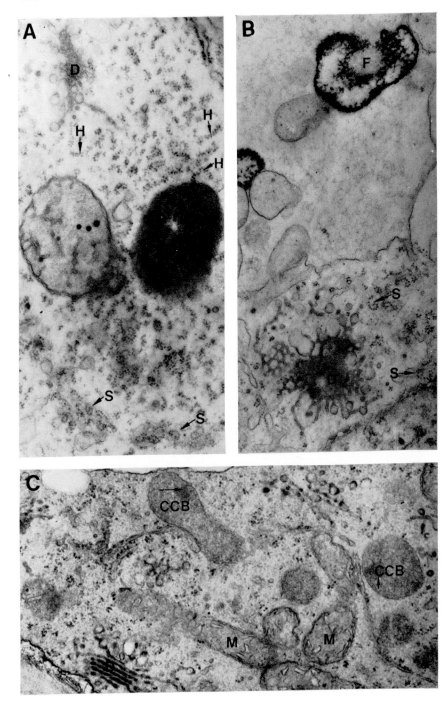

tuber cells, both before (Fig. 6.3C) and after culture (Fig. 6.4C). She also showed that the CCBs decrease in size during culture (Fig. 6.5). It has been suggested (Schnepf 1964, Bouck & Cronshaw 1965, O'Brien & Thimann 1967, Frederick, Newcomb, Vigil & Wergin 1968) that CCBs in plants may be the functional equivalent of lysosomes in animal cells. Frederick *et al.* (1968) suggested that they may, in addition, be the functional equivalent of animal microbodies. It is possible that the CCBs of the artichoke contain at least some hydrolytic enzymes and the reduction in size of these bodies may be a result of the dispersal of these enzymes in preparation for cell division.

Storage products characteristic of quiescent cells tend to disappear during the preparation for division. Starch vanishes from carrot cells brought into culture (Israel & Steward 1966) and phytoferritin is no longer visible in the plastid stroma of cultured artichoke cells. In contrast, small clear circular areas appear in the stroma of artichoke plastids. These clear areas are assumed to contain starch (Tulett, Bagshaw & Yeoman 1969) as they give a positive reaction with periodic acid-Schiff. The number and size of these starch bodies, absent in quiescent tuber cells, increase over the first few days of culture.

A rise in the respiration rate accompanies the excision (Edelman & Hall 1965) and culture of quiescent tissue. (Steward, Caplin & Millar 1952, Steward, Bidwell & Yemm 1958, Evans 1967). Increases in the number of cristae have been associated with enhanced mitochondrial activity and an elevated respiration rate in some plant tissues (Cherry 1963, Simon & Chapman 1961). However, no increase in the number of cristae per mito-chondrial profile was found in artichoke tuber cells during culture (Bagshaw 1969). The mitochondrial matrix was often less electron-dense in cultured cells and there is some evidence to suggest an increase in volume during the early increase in the rate of respiration. Israel & Steward (1966) and Sutton-Jones & Street (1968) reported increases in the number of mitochondria during culture of carrot storage tissue and sycamore tissue respectively. However, details of numbers were not given. Another sign of enhanced metabolic activity is an increase in the number of dictyosomes. The cisternae of the

---

FIG. 6.4. *Tissue from the Jerusalem artichoke tuber cultured in contact with a medium containing 20% coconut milk and 10^{-6} M, 2,4-D*. These cells are approaching cell division (Bagshaw 1969).

A. Spiral (↑S) and helical (↑H) polysomes are seen in profusion. The spiral ones (↑S) are apparently bound to the endoplasmic reticulum but the helical ones (↑H) are free in the ground substance (× 35,000) (cultured for 24 hours).

B. Vesicles of various sizes are connected to the dictyosome (D). A fibrillar dense body (F) is shown in the vacuole (× 38,000) (cultured for 24 hours).

C. Showing sections of flat dictyosomes, (D), simple mitochondria (M) and small CCBs. The CCBs contain small crystals (arrowed) and a larger proportion of matrix (× 43,000) (cultured for 12 hours).

dictyosomes were always flat, in contrast to the predominantly curved ones in quiescent cells, and were associated with many vesicles of assorted sizes (Fig. 6.4B and C). The dictyosomes were often grouped together.

An increasing number of microtubules was detected near the plasmalemma in cultured cells before the first division (Fig. 6.6B) but large numbers such

A

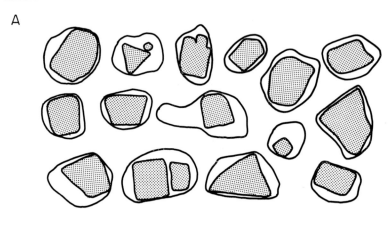

B

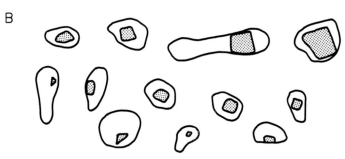

FIG. 6.5. *Tracings of crystal-containing bodies (CCBs) from electromicrographs* (× 40,000) (Bagshaw 1969) of (A) quiescent uncultured cells and (B) cultured cells undergoing cell division. Tissue was dividing synchronously (see Fig. 6.4 for details of tissue and medium). The decrease in the size of the crystals (:::) is quite distinct.

as those found by Pickett-Heaps & Northcote (1966a, b) in their 'pre-prophase band' were not detected. However, as serial sections were not cut through the entire length of these large cells a 'pre-prophase band' could have existed but remained undetected.

In quiescent cells the nucleus is usually flattened and positioned between the vacuole and the cell wall in a very narrow band of peripheral cytoplasm

(Fig. 6.2A). In complete contrast the nuclei of dividing cells are rounded and found away from the wall towards the centre of the cell. Bagshaw (1969) from her study with vacuolated artichoke tuber cells preparing for division has placed a series of profiles of nuclear positions and cytoplasmic arrangements between these two extremes which is thought to represent the sequence of events (Fig. 6.7). However, it is important to remember that the arrangement of some of these profiles may not form a developmental sequence but represents sections cut in different planes through cells with similar arrangements of cytoplasm. It is consistent with the sequence proposed by Sinnott & Bloch (1940, 1941) from light-microscope studies on dividing vacuolate plant cells and also with the cinemicrographic observations of Mota, Hildebrandt & Riker (1964) on dividing tobacco callus cells.

The first change noticed in the nucleus is that it assumes a more rounded profile (Fig. 6.6A) and does not lie as close to the wall as in quiescent uncultured cells. Dense aggregations of chromatin are more frequently observed in the nuclei. The change in shape of the nucleus is accompanied by alterations to the structure of the nucleolus, which becomes less compact and the fibrillar and granular regions become intermingled. Often a very large more electron-transparent central region appears in the nucleolus (Fig. 6.6A) which contains and is surrounded by granular particles similar in size to ribosomes (Fig. 6.6A). Fibrillar material similar to the chromatin outside the nucleolus is also found within this body, and in addition smaller electron-transparent regions are present within the fibrillar zone surrounding the central area. A few karyosome-like bodies can still be seen at this stage.

Similar observations have been reported by Jordan & Chapman (1971) and Rose, Setterfield & Fowke (1972) in nucleoli of cells from ageing tuber discs of Jerusalem artichoke.

Nuclei, whatever their position in the cell, are always surrounded by perinuclear cytoplasm. In many of the early prophase nuclei positioned near the cell wall the perinuclear cytoplasm is connected to cytoplasmic strands on the side away from the wall (Fig. 6.6A). These strands may have been connected to peripheral cytoplasm in other parts of the cell and may have been concerned with the initial movement of the nucleus away from the wall. Profiles of nuclei at varying distances from the cell wall and various cytoplasmic arrangements are shown diagrammatically in Fig. 6.7. Many profiles show connections of cytoplasm between the cell wall and the nucleus in only one area (Fig. 6.8A). It is probable that the nucleus was originally situated in this area. Some of these profiles show a substantial column of cytoplasm between the nucleus and the cell wall (Fig. 6.8A), whilst others show several, often narrower strands separated by sections of the vacuole between the nucleus and the cell wall. In contrast to these profiles many nuclei are found in the centre of a cytoplasmic strand traversing the whole cell (Fig. 6.8B).

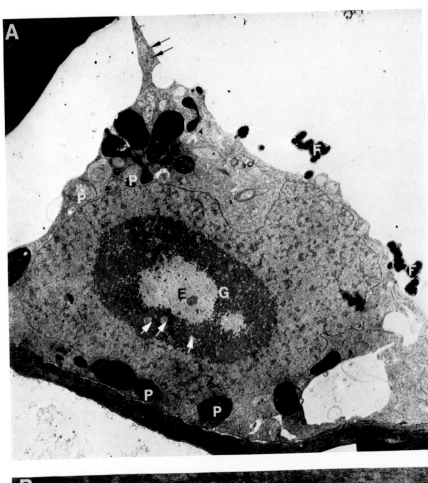

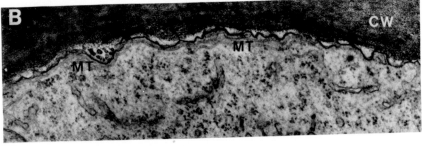

The direction of this strand is nearly always similar to the direction of cell plates and new cell walls in cells which have completed the first division. Similar cytoplasmic strands in other types of vacuolated cells were designated the phragmosome by Sinnott & Bloch (1940). They found that the expanding cell plate followed exactly the course of the phragmosome so that this structure occupied the position where the new cell wall would be laid down. Many vesicles, dictyosomes, endoplasmic reticulum, polysomes, mitochondria, plastids and a few CCBs can be seen in the cytoplasmic strands and in the perinuclear cytoplasm. In the strands all the organelles are orientated so that the largest dimension appears to be parallel with the direction of the strand (Fig. 6.8A).

Another structure exists connected to the nucleus in the perinuclear cytoplasm, and in the cytoplasmic strands up to a distance of 10 μm from the nucleus, which appears to arise separately from the nuclear envelope and may branch (Yeoman, Tulett & Bagshaw 1970). These nuclear extensions are frequently associated with microtubules and are connected to the endoplasmic reticulum. No nuclear extensions can be detected in cells without cytoplasmic strands and they are only present in early prophase. They can be found in tissue cultured for varying lengths of time and are characteristic of the early prophase state in dividing vacuolated artichoke cells. In these cells, the nuclear extensions may provide connections between the nucleus and the cell wall which position the nucleus prior to division. The nuclear extensions and microtubules might also provide support for the developing cytoplasmic strands. The dissolution of the nuclear envelope at the end of prophase ensures the eventual disorganization of the nuclear extensions. Structures similar to the nuclear extensions described here have been reported in algal cells by Pickett-Heaps & Fowke (1969), but it remains to be seen whether nuclear extensions are associated with nuclear movement during early prophase in highly vacuolated cells from other plants. Prophase nuclei are frequently surrounded by microtubules and small vesicles. These microtubules presumably come from the nuclear spindle when the nuclear membrane breaks down. Whether the microtubules which were found in the cytoplasmic strands near nuclear extensions are also involved in spindle formation is not known.

FIG. 6.6. *Cultured artichoke tuber tissue a few hours before cell division.* Medium contained 10^{-5} M, 2,4-D and no coconut milk.
A. Compare the structure of the nucleolus with that of the quiescent cell (Fig. 6.2). The nucleolar material is loosely packed. The nucleolus contains a large central electron transparent region (E) surrounded by granular material (G). Smaller transparent regions (arrowed) are surrounded by fibrillar material. Part of a cytoplasmic strand is visible (double arrowed). Note the variation in the density of the plastids (P) and the presence of electron dense fibrillar material (F) ($\times$ 10,000).
B. Showing microtubules (MT) near the cell wall (CW) ($\times$ 43,000).

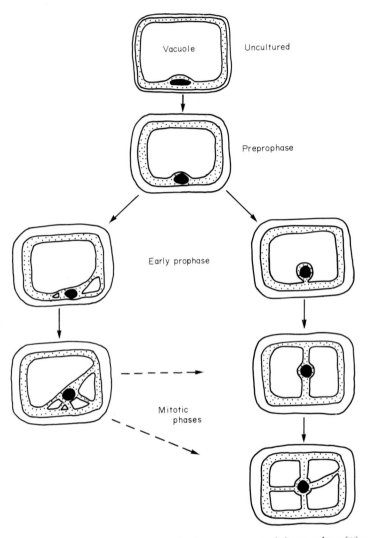

Fig. 6.7. *Schematic diagram of changes in the arrangement of the cytoplasm* (⋮⋮⋮) *and the position of the nucleus* (■) *during the first synchronous cell division in cultured artichoke tuber cells* (Bagshaw 1969).

The sequence of events during mitosis in vacuolated cells appears to follow the usual pattern. Mitochondria, dictyosomes, CCBs and plastids appear to be excluded from the spindle but endoplasmic reticulum and vesicles (which may have been formed from dictyosomes) can be seen within the spindle and between the chromosomes. In a longitudinal section through the spindle at anaphase (Fig. 6.9A) it can be seen that the chromatids have separated and moved to opposite, broad poles. The region of the equatorial plate is magnified in Fig. 6.9B, and can be seen to be composed of microtubules, endoplasmic reticulum, vesicles and ribosomes. The vesicles may be an early stage in cell plate formation. An early telophase nucleus near the cell plate is shown in Fig. 6.10A. The chromosomes have lost their identity and the nuclear envelope has reformed. Early telophase nuclei are irregular in shape and contain spaces surrounded by two membranes (Fig. 6.10A), which may be continuous with the cytoplasm. Myelin-like bodies are also associated with these nuclei (Fig. 6.10B). Nuclear material in early telophase nuclei is very dispersed. There are large aggregates of fibrous material, but the granular material is dispersed between clumps of chromatin throughout the whole nucleus (Fig. 6.10B). Later the granular material presumably becomes compacted around the fibrous material.

Cell plate and cell wall formation proceed from the centre of the spindle along the phragmosome, towards the parent cell wall. The cell plate is composed of variously sized vesicles, multivesicular bodies and microtubules (Fig. 6.11A). Many dictyosomes are found around the cell plate, particularly at the growing end and it is assumed that many, if not all the vesicles, are derived from them.

The participation of dictyosome vesicles in cell plate and wall formation is well established (Whaley & Mollenhauer 1963, Mollenhauer & Morré 1966, Pickett-Heaps & Northcote 1966a, b, c, Pickett-Heaps 1967, Cronshaw & Esau 1968). The autoradiographic studies of Pickett-Heaps (1967) also suggest that the endoplasmic reticulum plays a part in cell wall synthesis. Cronshaw & Esau (1968) reported that endoplasmic reticulum may contribute some vesicles to the plate. In artichoke tuber cells groups of dictyosomes are found outside the spindle region during metaphase and anaphase and near the developing cell plate during telophase. Lengths of endoplasmic reticulum are also abundant near the cell plate.

In the dividing artichoke cell, microtubules, possibly the remains of spindle fibres, are only found at the growing edge of the cell plate. In other tissues microtubules have been observed passing through the cell plate and forming plasmadesmata (Robards 1968). This same function has been attributed to endoplasmic reticulum (e.g. Frey-Wyssling & Muhlethaler 1965). The newly formed cell walls in cultured artichoke tissue were not found to contain many plasmadesmata nor was their subsequent formation observed (Bagshaw 1969). The formation of paramural bodies associated with the new cell wall is of some interest. The membranous structures and multivesicular bodies found near developing cell plates (Fig. 6.11A) are possibly an early stage in the development of paramural bodies. The origin of these is thought to be cytoplasmic and not from the plasmalemma, as at this stage no recognizable plasmalemma has been formed. Multivesicular bodies are reported to

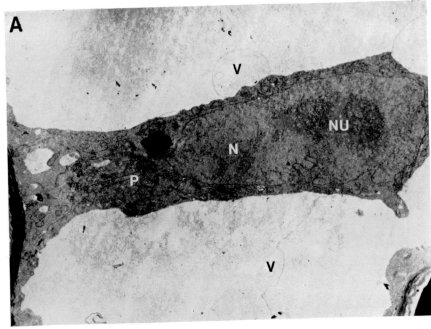

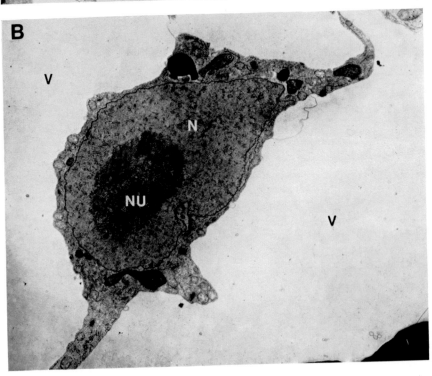

develop from dictyosome vesicles in cultural carrot cells (Halperin & Jensen 1967) and from vesicles of endoplasmic reticulum in cotton nucellus (Jensen 1965). This development was not observed in the tuber cells (Bagshaw 1969). Repeated cell divisions gave rise to a regular wound cambium (Fig. 6.11B) and eventually to the formation of a callus (see Chapter 10).

CYTOLOGICAL CHARACTERISTICS OF THE DEVELOPING CALLUS

The induction of division in a tissue fragment is followed by considerable anatomical changes (Gautheret 1966, Street 1966b). At first the major change is one of dedifferentiation in which a wound cambium is established and the average cell size drops sharply (see Chapter 10). The much smaller cells which result from this active phase of division have small vacuoles and appear quite different from the highly vacuolated cells from which they are derived (Fig. 6.12C). The structure and arrangement of the organelles suggest that the cells are metabolically active and changes may be observed in the mitochondria (Bagshaw, Brown & Yeoman 1969) and plastids (Israel & Steward 1967, Tulett, Bagshaw & Yeoman 1969). This phase of dedifferentiation is followed by a redifferentiation phase in which a new developmental pattern is increasingly superimposed upon the actively dividing wound cambium eventually obliterating it (see Chapter 10). Differentiation may take many forms: a periderm, vascular cambium or meristematic nodules may appear. Often all three growth forms may be found in one lump of tissue. Meristematic nodules are a common feature of developing calluses and they become growth centres which may not differentiate further but produce expanded parenchymatous cells from their periphery. These cells form the tissue universally recognized as a callus. These large cells may also be derived from the wound cambium, periderm or vascular cambium of the developing callus and are similar in size, shape and organization to the cells found in suspension cultures (see below). Not all the division products of the callus develop in this way, some may differentiate into phloem elements or become extensively lignified and form tracheids (Fig. 6.12A). Other cells such as the small meristematic cells of the growth centres remain undifferentiated and similar in appearance to the cells of the apical meristem (Fig. 6.12B). These nodular

---

FIG. 6.8. *Cultured cells in early prophase* (for details of medium and tissue see Fig. 6.6). A. The nucleus (N) is lying in the vacuole (V) within a stout column of cytoplasm. Plastids (P) are orientated so that the largest dimension appears to be parallel with the direction of the cytoplasmic strand ($\times 5000$).
B. The rounded nucleus (N) is suspended in the vacuole (V) by a narrow transvacuolar cytoplasmic strand. The nucleolar material is loosely packed (NU) ($\times 5000$).

structures may of course be induced to emerge as fully differentiated roots or shoots (Skoog & Miller 1957) and exhibit the wide range of cellular types typical of these organs.

Embryoids are usually formed from the superficial cells of a callus mass

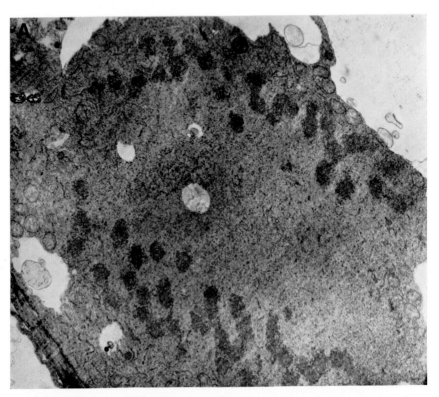

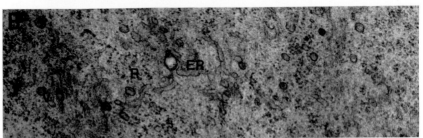

FIG. 6.9. *A cultured cell in anaphase* (see Fig. 6.6 for details of tissue and medium). A. The chromatids have separated and moved to broad poles (×8000). B. Enlargement of the equatorial plate region. Vesicles (V), endoplasmic reticulum (ER) and ribosomes (R) are shown (×43,000).

(Fig. 6.13). However, Thomas, Konar & Street (1972) have shown that embryoids can arise from single callus cells of *Ranunculus sceleratus*.

Chloroplast ontogeny has been studied in a variety of callus cultures (Sunderland & Wells 1968; Israel & Steward 1967). Etioplasts of triploid aspen cells (*Populus tremuloides*) contain an interconnected tubular net, phytoferritin aggregates, electron-transparent vesicles which appear to invaginate from the inner plastid membrane, membrane-bound homogeneous spheroids and starch grains. Illumination with white light or low intensity red causes changes within the proplastid and the emergence of a thylakoidal system from the tubular complex typical of mature chloroplasts. The way in which this transformation is achieved is interesting because it represents an additional pathway for the development of photosynthetic lamellae (Blackwell, Laetsch & Hyde 1969). Similarly in greening plastids of callus cells from

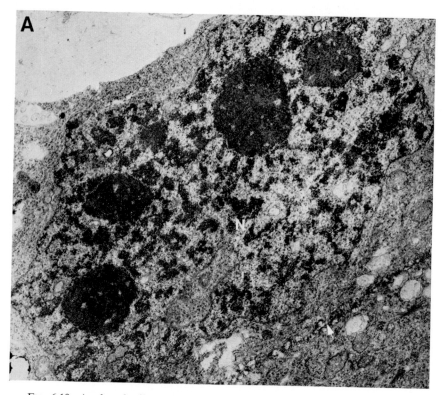

FIG. 6.10. *A cultured cell in early telophase* (see Fig. 6.6. for details of tissue and medium).

A. The newly formed nucleus (N) lies near the developing cell plate (arrowed) (×11,000).

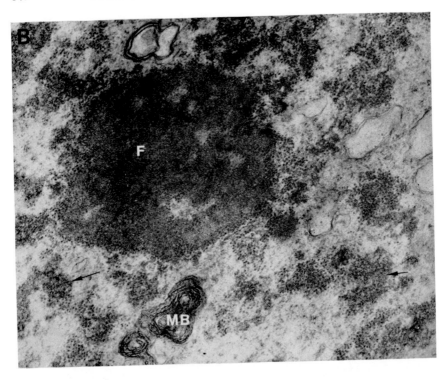

(FIG. 6.10 *cont*.)

B. Enlargement of part of A showing a nucleolus (mostly composed of fibrils (F),
scattered granules around fibrous chromatin (arrowed), membrane-bound spaces
and whorls of membranes (myelin-like bodies (MB)) ($\times$ 36,000).

the artichoke (Tulett, Bagshaw & Yeoman 1969) only the central membrane
system is involved in the formation of thylakoids. The subsequent development of the chloroplast follows the pattern described by Gerola & Dassu
(1960). Israel & Steward (1967) reported that, during greening of callus tissue
from carrots, a thylakoid system develops from the pre-thylakoidal body
analagous to the *corpo opaco* of the artichoke. Invaginations of the inner
membrane of the plastid envelope arise during greening but in appearance
they are distinctly different from the thylakoids. Plastids showing a globular
centre and pre-thylakoid body similar to those of carrot cell plastids have been
described by Thomas, Konar & Street (1972) from *Ranunculus sceleratus*
callus growing in 2,4-D-omitted medium. Sjolund & Weier (1971) have
studied the structure and development of chloroplasts in callus cultures of
Streptanthus tortuosus. They showed that dark-grown cells contain proplastids
without a prolamellar body but with a complex of loosely associated membranes. When these cells are placed in the light chloroplasts develop from the

plastids. If light-grown cultures with mature chloroplasts are transferred to the dark, a dedifferentiation to proplastids is observed. The change in the structure of the plastids is paralleled by a dedifferentiation of the vacuolate cells to a less differentiated meristematic state.

A variety of complex mitochondria is found in cultured artichoke tuber tissue (Bagshaw, Brown & Yeoman 1969) and the range of forms tends to increase during culture. Two major groups are distinguishable. The first group includes long cylindrical rods, branched structures and plates which all contain a matrix and cristae. These are probably derived from simple spheres and rod-shaped mitochondria. The second group consists of plates and bell-shaped forms and have also been reported by Thomas, Konar & Street (1972) in embryogenic callus of *Ranunculus sceleratus*. These have the usual mito-chondrial structure containing cristae around the rim but only a thin strip of matrix containing no cristae in the centre of the plate or dome of the bell. These mitochondria are probably derived from the cup-shaped structures

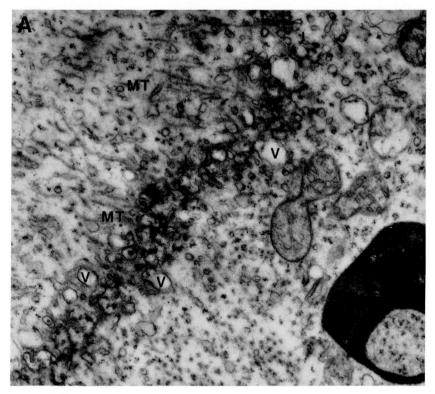

FIG. 6.11.
A. *A cultured cell showing the growing end of a developing cell plate composed of vesi-cles (V) of various sizes.* Microtubules (MT) are present in large numbers (× 28,000).

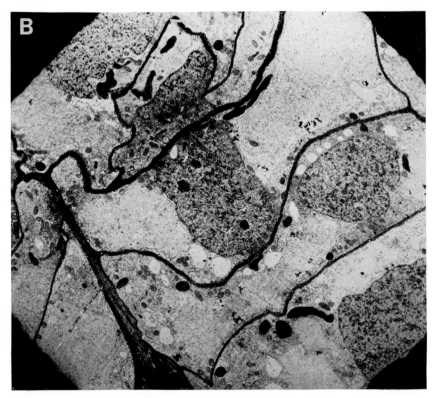

(FIG. 6.11 *cont.*)
B. *Tissue cultured for 6 days* (see Fig. 6.4 for details of tissue and medium) *showing part of a row of newly divided cells resembling a wound cambium.* The divisions are periclinal (× 3500).

characteristic of quiescent tissue (Fig. 6.3A). The development of different mitochondrial forms may be related to the changing metabolic and mechanical conditions in the cells. Also the development of a particular form could possibly influence the subsequent development of the cell (Bagshaw, Brown & Yeoman 1969). Profiles similar to those derived from the rimmed plate and from the bell have been reported by Manton in *Anthoceros* (1961), Albergoni (1964), in *Allium*, and by Bell & Muhlethaler (1964) in *Pteridium* showing that these complex forms are not restricted to callus cells.

Spherosomes are a characteristic feature of cells from normal callus and crown-gall tissue derived from *Helianthus annuus* (Holcomb, Hildebrandt & Evert 1967). These authors have examined the morphology and staining properties of these organelles and report they are consistent with those reported for other plant materials (Sorokin 1955). They are lipid rich and contain acid phosphatase. They may be the plant equivalent of animal lysosomes (these structures are further considered in the next section of this chapter).

RANGE OF CELL FORM AND STRUCTURE IN SUSPENSION CULTURES

The cells of suspension cultures are derived from callus and correspond in basic structure to the vacuolated parenchymatous cells of the parent culture.

Whenever growing suspension cultures have been examined they have been found to contain both cell aggregates and free cells. In many cases a range of cell sizes and shapes has been encountered in the free cell fraction.

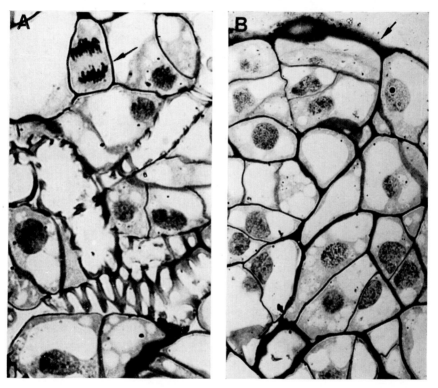

FIG. 6.12. *Tissue from the Jerusalem atrichoke tuber as* Fig. 6.4. A and B, light micrographs of tissue cultured for 17 days. C, electron micrograph of tissue cultured for 6 days.

A. An area where tracheids are differentiating is shown. One cell is in anaphase (arrowed) (× 2300).

B. Meristematic callus cells near the outside (arrowed) of the explant (× 1800).

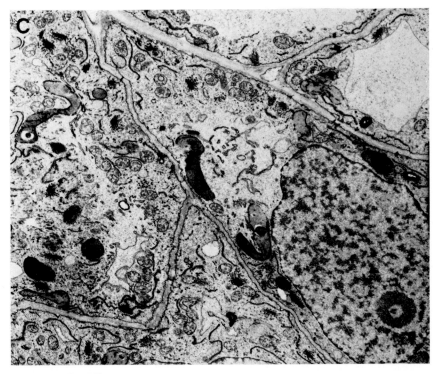

(FIG. 6.12 *cont.*)
C. Small circular and elongate mitochondrial profiles are shown. The dictyosomes and plastids are also abundant and the cell vacuoles are much less prominent (Bagshaw 1969) (× 7500).

Thus Nickell (1956) found in suspension cultures of *Phaseolus vulgaris* mainly spherical cells, 12–40 μm in diameter, but also slipper shaped (30–60 × 10–12 μm) and gourd-shaped cells (95 × 40 μm). Muir, Hildebrandt & Riker (1958) found, in suspension cultures of a number of species, both spherical cells (15–220 μm diam.) and elongated cells. In suspensions of crown-gall tissues of marigold and sunflower these elongated cells ranged from 20–150 μm in diameter and from 40 μm to 1·2 mm in length. In suspensions derived from normal and hybrid tobacco, giant elongated cells (many of which were multinucleate) were observed, up to 300 μm in diameter and up to 2·5 mm long. Steward, Mapes & Smith (1958) have reported the presence in their carrot suspension cultures of spherical cells (50–100 μm diam.), giant cells 100–300 μm diam.) and very elongated tubular cells. Suspension cultures derived from the tuber of *Solanum tuberosum* and from the cotyledon tissue of *Arachis hypogea* also showed a similar range of cell morphology. In some suspension cultures (e.g. Paul's Scarlet rose and sycamore), however, the diversity of cell shapes encountered is much less and the sizes of the majority of their cells, both free and aggregated, fall within a narrow range, the mean of which depends upon the growth phase of the culture. In these cultures, giant cells are only encountered in cultures which have been allowed to remain in stationary phase beyond the normal period of subculture. Jones, Hildebrandt, Riker & Wu (1960) interpreted the giant cells in their tobacco cultures as senescent cells and observed that their mitochondria became joined end-to-end to form filiform aggregates of 2–12 units which accumulated in the parietal sheath of cytoplasm where they exhibited worm-like movements (see also section 'Cytological characteristics of the developing callus', p. 137 *et seq.*).

The very elongated tubular cells mentioned in the earlier papers cited above may develop only in particular culture media; many of the cultures in which they have been prominent have been grown in complex media containing coconut milk and/or casein hydrolysate. Simpkins, Collin & Street (1970) noted that many of the cells of their sycamore suspension became excessively elongated in a medium supplying only organic nitrogen (casein hydrolysate + urea + cysteine) in contrast to the roughly spherical cells developing in a medium in which nitrate was the sole nitrogen source (Fig. 6.16). Similarly, elongated cells can also be observed in sycamore cultures grown in presence of coconut milk. Studies in our laboratory with carrot cell suspensions growing on a synthetic medium (Murashige & Skoog 1962 supplemented with 2,4-D) indicate that cultures rich in elongated and giant free cells are not only of low embryonic potential but that many of these cells are polyploid. The initiation of the original callus in presence of a medium containing kinetin plus 2,4-D gives, in contrast to callus initiated in absence of kinetin, a culture which when grown in suspension is of relatively low embryogenic potential and is rich in diversely shaped free cells. Isolation of clones of carrot cells of different mean ploidy level indicates that mean cell size increases as the ploidy level rises and that cultures which are extensively aneuploid show a high proportion of curiously shaped (mis-shapen) cells and often markedly elongated cells (Fig. 6.14). Clearly the question of the causative agents responsible for diversity of shape and size of cells in suspension cultures is in need of more critical study. The present indications are, however, that important factors here are the composition of the culture medium, the degree of senescence of the culture and the degree of cytological variation.

Although there is this diversity of size and shape, the cells of suspension cultures generally remain parenchymatous (living, thin-walled and vacuolated) (Fig. 6.15). The diversity of cell *types* is much more limited than is observed in callus cultures. The occasional reports (e.g. Muir, Hildebrandt & Riker 1958), for instance, of free-floating, tracheid-like cells relate to highly aggregated 'suspensions' and probably such cells have their origin in floating

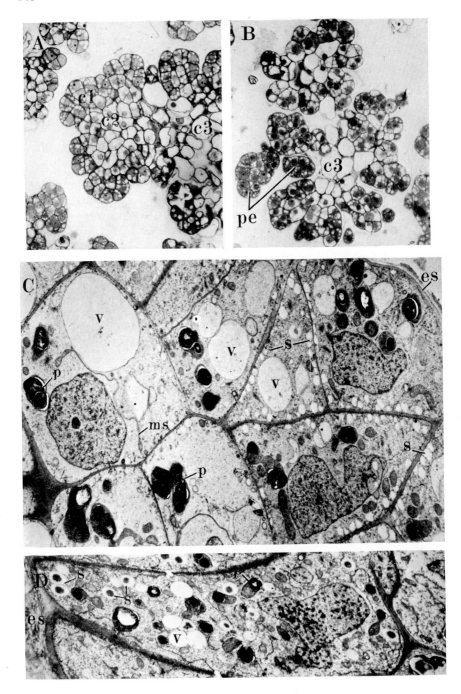

callus fragments. Conditions conducive to lignification (high sucrose and high levels of 2,4-D or kinetin) in more dispersed suspension cultures have been reported to lead to the presence of cells with uniform cell walls and staining for lignin *throughout* their protoplasts (Carceller, Davey, Fowler & Street 1971). Further, much of the lignin-positive material of such cultures is extracellular. Even here the lignin-containing cells appear to arise within small aggregates undergoing regular fragmentation. A large number of callus cultures are now available which turn green when cultured in light and in these cultures many of the cells contain chloroplasts. However, when such calluses are used to initiate light-grown suspension cultures the level of pigmentation is significantly reduced and conditions conducive to maximum cell dispersion still further reduce the pigment level. There is strong evidence that in such suspensions chloroplast differentiation occurs in non-dividing cells within transient aggregates rather than in free cells. A similar loss of pigmentation occurs when callus fragments rich in anthocyanin are grown in suspension culture. The more disperse are cell suspension cultures the more they consist of colourless parenchymatous cells.

The structural features of a limited number of suspension cultures have been examined in some detail by light microscopy choosing that stage in the growth cycle of the cultures when the cells are most actively dividing. Thus, Jones, Hildebrandt, Riker & Wu (1960) and Mota, Hildebrandt & Riker (1964) have used time-lapse phase contrast in the study of a suspension of hybrid tobacco cells. Roberts & Northcote (1970) have examined a sycamore suspension in the differential interference microscope of Nomarski. Holcomb Hildebrandt & Evert (1967) have described in some detail the movement and chemical properties of the spherosomes seen in cultured cells of normal and crown-gall sunflower. Many workers have published single pictures of

---

FIG. 6.13. *Light and electron micrographs of embryogenic callus of* Ranunculus sceleratus. *L. cultured on a medium with 2,4-D* (from Thomas, Konar & Street 1972).
1. Light micrograph of a section through a callus showing a large aggregate composed of superficial cytoplasm—rich cells (c1) and internal more vacuolated cells (c2) and at its junction with the main callus mass senescent or dead cells (c3) ($\times 220$).
B. Light micrograph of a section of callus showing an aggregate undergoing fragmentation at regions of c3 cells. Note the release of presumptive pro-embryoids (pe) ($\times 220$).
C. Electron micrograph of a section of callus showing external embryogenic cells (right) and internal cells (left) with less-dense cytoplasm, larger vacuoles and elongated mitochondrial structures (ms). The embryonic cells show numerous 'empty' spherosomes (s) adjacent to the cell walls. es = external surface of aggregate; p = plastid; v = vacuole ($\times 4600$).
D. Superficial embryogenic cell similar to those in C but containing spherosomes (s) with small, dense, central stroma. es = external surface of aggregate; p = plastid; v = vacuole ($\times 4800$).

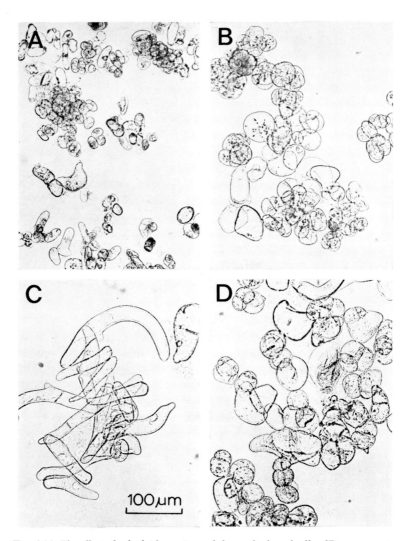

FIG. 6.14. *The effect of polyploidy on size and shape of cultured cells of* Daucus carota. A, diploid; B, tetraploid; C, D, octoploid. C shows the abnormal cell shapes often observed in highly polyploid cell lines (previously unpublished data of M.W.Bayliss).

suspension culture cells as seen under high-power phase contrast (Fig. 6.17).

The nucleus is usually suspended away from the cell wall in cultures undergoing active cell division. The cytoplasm around the nucleus is connected to the peripheral cytoplasm by a variable number of cytoplasmic strands. These strands are actively streaming as revealed by the movement along the strands of mitochondria and spherosomes. Holcomb, Hildebrandt

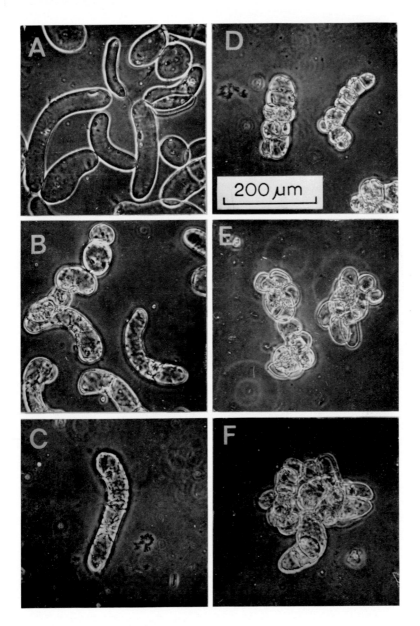

FIG. 6.15. *Cells from batch suspension cultures of* Acer pseudoplatanus *at different stages of the growth cycle.* Phase-contrast.
A. Stationary phase cells used to initiate culture.
B. Towards the end of lag phase showing increased prominence of cytoplasm.
C, D. In early exponential phase; C, a dividing cell; D, beginning of aggregate formation.
E, F. Aggregates from later in the growth phase showing rounding off of cells as they expand.
(From Henshaw, Jha, Mehta, Shakeshaft & Street, 1966.)

& Evert (1967) have concluded that the most conspicuous particles in the cytoplasm are spherosomes (0·5–3·0 μm diam.), identified by their staining with lipid stains, and their golden-yellow fluorescence in ultra-violet following use of Nile blue as a vital stain. The cytoplasm is often flowing in opposite directions in parallel strands; it can flow in opposite directions within the same strand. Some strands appear to arise completely from the peripheral

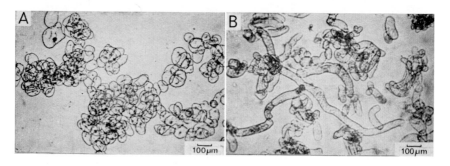

FIG. 6.16. *Cells from 24-day-old batch suspension cultures of* Acer pseudoplatanus.
A. In medium containing nitrate as sole nitrogen source.
B. In medium containing nitrogen as casein hydrolysate, urea and cysteine.
(From Simpkins, Collin & Street 1970.)

cytoplasm and to end blindly within the usually single large central vacuole. The nuclei contain nucleoli with vacuoles. The cells contain amyloplasts, some of which are in the peripheral cytoplasm but others may be associated with the nucleus. The external cell walls are thin and uniform; recently formed dividing walls often follow a wavy course. Extensions of the tonoplast membrane into the vacuole have been observed giving rise to membrane-bound vesicles eventually released into the vacuole.

CELL DIVISION IN SUSPENSION CULTURES

Steward, Mapes & Smith (1958) have described division in the spherical cells of their carrot cultures by an equatorial division wall. They also noted small

FIG. 6.17. *Cells from suspension cultures of different species.* Phase-contrast photographs showing cell walls, cytoplasm, transvacuolar cytoplasmic strands, nuclei and other inclusions.
A. *Linum usitatissimum.*
B. *Nicotiana sylvaticum.*
C. *Haplopappus gracilis.*
D. *Atropa belladonna.*
(Photographs by M.W.Bayliss.)

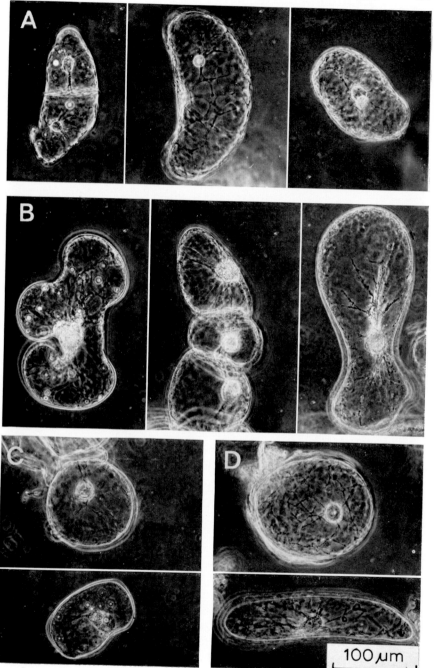

groups of cells arranged in such a way as to suggest that the next division gives rise to a row of three cells, the central cell of which is smaller than the two 'wing' cells and that from such groups divisions both at right angles to and along the axis give rise to 'dense moruloid masses of actively dividing cells'. This led to the suggestion that the 'moruloid' masses have their origin in single spherical cells and that new single spherical cells can arise by release from these masses. The larger elongated cells apparently divide into a row or mass of cells by internal divisions within the original cell wall and without cell expansion taking place. The free cells of stationary phase sycamore suspension cultures undergo internal segmentation in this way and then the parent cell wall is ruptured by cell expansion and the daughter cells give rise to more or less isodiametric aggregates by 'randomly' orientated further division walls (Henshaw, Jha, Mehta, Shakeshaft & Street 1966, Street, Henshaw & Buiatti 1965) (Fig. 6.15). These aggregates continue to increase in mean cell number and their cells remain small until the cell division rate of the culture begins to decline; then cell expansion promotes the break up of the aggregates. In carrot cultures (Steward, Mapes & Smith 1958), superficial cells of the developing aggregates can apparently give rise to pronounced filamentous outgrowths which later show division by transverse walls to give uniseriate filaments. Short uniseriate filaments, whether arising from superficial cells of aggregates or from free cells, are commonly seen in many cell suspensions. Bergmann (1960) and Jones, Hildebrandt, Riker & Wu (1960) in their studies on tobacco cell suspensions not only describe equatorial divisions in roughly spherical cells and repeated transverse divisions in tubular cells but oblique division walls in somewhat elongated cells corresponding closely to the division walls seen in dividing fusiform cambial initials. Where giant cells have been observed, these are often multinucleate and when such cells divide they appear to give rise to uninucleate cells so that the early divisions give rise to very unequal cells (small, uninucleate cells occurring within the old cell wall whilst the 'parent' cell is still multinucleate). An extreme case was described by Jones, Hildebrandt, Riker & Wu (1960) of certain giant cells in stationary phase tobacco suspensions which came to contain first a number of internally free uninucleate cells and ultimately several hundred daughter cells by the development of cell masses from each of the initial uninucleate cells.

Steward, Mapes & Smith (1958) report that infrequently in carrot cultures, but more frequently in their suspension cultures derived from peanut cotyledon tissue, many of the spherical cells have small outgrowths or papillae at their surface and that such papillae have been observed to receive cytoplasm and a nucleus and to become very restricted at the point of attachment to the mother cell. Observations on other cells suggest that such papillae may then enlarge and become 'pinched off' at the base so that new cells arise by a process reminiscent of 'budding' in yeasts. The studies of Blakely & Steward

(1961) indicate that the various forms of division observed in carrot cultures are also paralleled in suspension cultures of *Haplopappus gracilis*.

LIGHT-MICROSCOPE STUDIES OF CELLS IN MITOSIS

Jones, Hildebrandt, Riker & Wu (1960) have described in some detail the mitosis of tobacco cells in micro-culture as observed by phase-contrast. The stages of mitosis were timed as follows: prophase up to 3 hours 48 minutes; metaphase 14–70 minutes; anaphase 3–7 minutes; cell plate formation following anaphase 11–95 minutes; formation of daughter nuclei containing nucleoli following anaphase 6–22 minutes. The longer times for each stage of mitosis were recorded for the larger cells of the culture. Following the disappearance of nucleoli and of the nuclear envelope many of the cytoplasmic strands fused and thickened and from mid-prophase mass flow of organelles within the cytoplasm ceased. This 'fixed tension', considered to be indicative of an increase in cytoplasmic viscosity, started in the strands near the nucleus and extended to the ends of the cell. This stage of 'fixed tension' continued right through to cytokinesis. A transverse plate of cytoplasm became apparent during metaphase; it was described as resembling the phragmosome found in dividing cells by Sinnott & Bloch (1941). This plate of cytoplasm was not involved in a general movement of cytoplasm towards the two daughter nuclei which occurs during cell plate formation. A later study of nuclear division in tobacco cells from the same laboratory (Mota, Hildebrandt & Riker 1964) confirmed the stage of 'fixed-tension' but described the phragmosome as arising at or immediately before prophase and as being the first visible indication that the cell would divide in the near future. This paper figures the chromosomes as arranged as a dark band in the equatorial plane at metaphase and their movement as two compact groups to the poles in anaphase. Bergmann (1960) has also made rather less detailed observations on the division of isolated tobacco cells cultured on agar plates. One interesting difference between the accounts of Bergmann and of Jones, Hildebrandt, Riker & Wu (1960) relates to protoplasmic streaming; Bergmann states that during the cell division process there was very active protoplasmic streaming throughout the entire cell accompanied by movements of the nuclei. Roberts & Northcote (1970), in their observations on dividing sycamore cells also stress, contrary to Jones, Hildebrandt, Riker & Wu (1960), that at no stage during cell division does cytoplasmic streaming completely stop. Roberts & Northcote (1970) observed that the plane of the future cell plate was indicated by a region of cytoplasm (the phragmosome) in which a fibrous phragmoplast developed centrifugally outwards to fuse first with the mother cell wall at the side of the cell where the nucleus was placed during mitosis. The cell

F

plate arose along the line of the phragmoplast. These workers also noted a band of reticulate cytoplasm, arising during prophase, applied to the cell wall and partially or completely encircling the cell. This band persisted until the beginning of telophase (see also this Chapter, p. 131 *et seq.*).

ELECTRON-MICROSCOPE STUDIES ON CELLS IN SUSPENSION CULTURE

The fine structure of sycamore cells in suspension culture has been subjected to detailed study (Sutton-Jones & Street 1968, Roberts & Northcote 1970, Davey & Street 1971, Carceller, Davey, Fowler & Street 1971).

The stationary phase cells have a large central vacuole and a very thin peripheral layer of cytoplasm containing very few mitochondria (mainly of the light matrix type), ribosomes and E.R. membranes (Fig. 6.18A). The nuclei are often extensively lobed. Many of the cells contain prominent amyloplasts packed with starch. An organelle prominent in these stationary phase cells is a crystalloid microbody of (0·5–1·0 μm diam.) bounded by a unit membrane and containing a distinctive crystalloid (Fig. 6.18B). Such bodies have also been described from suspension culture cells of *Eucalyptus camaldulensis* (Cronshaw 1964). Such microbodies have been interpreted as containing deposits of catalase (Frederick & Newcomb 1969) but this has not been shown for the microbodies of suspension culture cells. These microbodies decline sharply in number during lag phase and do not become prominent again until the culture again enters stationary phase.

During lag phase, preparatory to the onset of cell division suspension culture cells undergo changes some of which resemble those observed during the activation of quiescent cells in explants (see this chapter, 'Changes in structure of cells associated with induction of division in explants', p. 127 *et seq.*). There is a massive increase in the volume of cytoplasm per cell (Fig. 6.18C) and associated with this the development of transvacuolar strands and sheets of cytoplasm. The numbers of mitochondria, ribosomes, E.R. profiles

---

FIG. 6.18. *Electron micrographs of suspension culture cells of* Acer pseudoplatanus.
A. Stationary phase cells with large central vacuole and a thin peripheral layer of cytoplasm (× 1755).
B. Microbodies, with crystalloid core, cut longitudinally and transversely, from a stationary phase cell (× 46,000).
C. Late lag phase cell with cytoplasm containing numerous organelles and small vacuoles (× 2400).
D. Profiles of rough endoplasmic reticulum running parallel with the cell wall, in a late lag phase cell (× 32,000).
(A from Sutton-Jones & Street 1968; B from Davey & Street 1971; C, D from Davey 1970.) (Key as Fig. 6.20.)

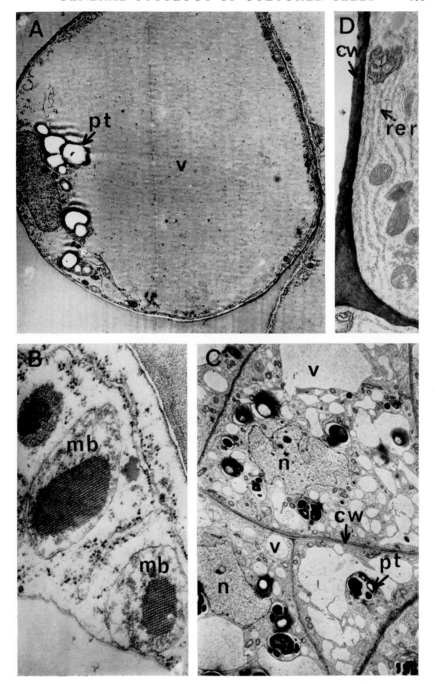

and dictyosomes per cell increase. The ribosomes occur free in the cytoplasm (many as circular or spherical polysomes) and attached to the E.R. membranes. Some of the mitochondria are dumbell-shaped, suggesting mitochondria in division. Their cristae are narrow and well defined. The dictyosomes are associated with clusters of vesicles. Starch is now less prominent, though amyloplasts still occur. Some cells show numerous proplastids. E.R. membranes often show parallel arrangement along the line of the cell wall (Fig. 6.18D).

Cells harvested at day 6 of incubation (these cells are in the short exponential growth phase observed in batch culture, see Chapter 11) have even denser cytoplasm and show a maximum density of rough E.R. profiles and ribosomes (Fig. 6.19A). It is at this stage that peak values for RNA and protein content per cell are attained. Many cells can be observed in mitosis and cytokinesis (Fig. 6.19, B and C). The initial stationary phase cells become segmented into two, four or six cells within the old cell wall; the daughter cells are each

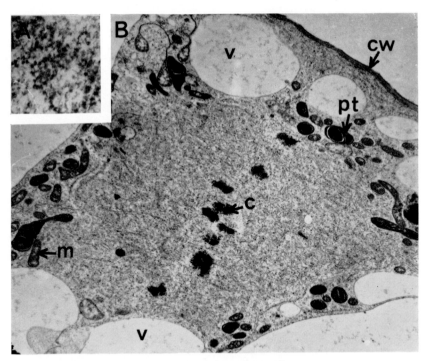

Fig. 6.19. *Electron micrographs of suspension culture cells.*
A. Spiral polyribosomes in the cytoplasm of a soybean (*Glycine max*) (× 60,000).
B. A sycamore (*Acer pseudoplatanus*) cell in mitosis. The metaphase chromosomes lie across the equator of the cell, and organelles such as mitochondria and plastids are confined to the periphery of the cytoplasm (× 5000).

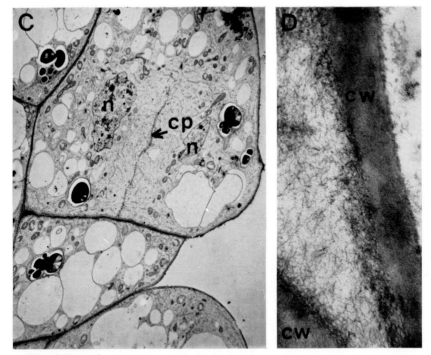

(FIG. 6.19 cont.)
C. A sycamore cell in cytokinesis. The new cell plate is developing in the cytoplasm between the daughter nuclei (× 6000).
D. Separation of cell walls during the breakdown of a cellular aggregate in a sycamore culture showing the microfibrillar structure (× 44,000).
(A from Short & Davey previously unpublished; B and C from Davey 1970; D from Sutton-Jones & Street 1968.) (Key as Fig. 6.20.)

bonded by a complete new wall within the wall of the original cell. These parent walls are ruptured and during the process often become teased out to show their microfibrillar structure (Fig. 6.19D).

Studies on dividing cells indicate that the cell plate formed between daughter cells arises by the fusion of membrane-bound vesicles. Roberts & Northcote (1970) consider that two distinct types of vesicle are involved. One of these (60–100 nm diameter) has an electron-dense core and corresponds closely to the cisternae of the dictyosomes which abound in the phragmosome region. The second type of vesicle is of unidentified origin and has a fairly electron-transparent centre. Microtubules, which increase in number during anaphase, are especially numerous at the growing edges of the cell plate. Smooth E.R. profiles are prominent near the advancing edges of the cell plate; rough parallel-stacked E.R. profiles prominent further back in the phragmosome where the cell plate is more developed. Numerous mitochondria, often accompanied by structures corresponding to the multivesicular bodies (Fig. 6.20A) of Jensen (1965) aggregate at the periphery of the phragmoplast area and are intermingled with the profiles of rough E.R. The cell plate usually fuses first with one wall and then later with the wall furthest from the daughter nuclei. The newly formed dividing wall often follows a zig-zag course.

With the progress of the phase of rapid cell division the average number of cells per cellular aggregate increases. Such aggregates begin to fragment as the rate of cell division begins to slow down and the culture approaches stationary phase. The cells of the aggregates are in protoplasmic continuity via plasmadesmata (Fig. 6.20B). Generally the cell walls of sycamore cells in suspension culture remain thin and uniform but the cell walls which delimit the surface of the cellular aggregates (walls at the aggregate-culture medium interface) often show characteristic internal wall thickenings, commonly ridge-like (Fig. 6.20C) but sometimes in the form of labyrinths of micro-fibrillar material connected to the original cell wall by arm-like processes.

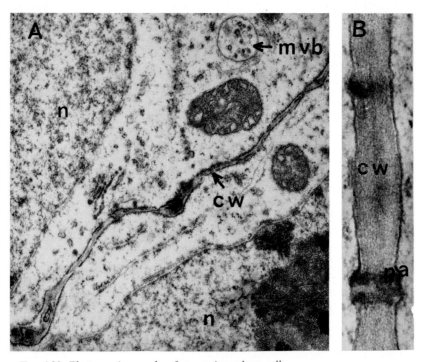

FIG. 6.20. *Electron micrographs of suspension culture cells.*
A. Multivesicular body in the cytoplasm adjacent to the newly formed zig-zag cell wall between two daughter nuclei in a recently divided sycamore cell (×27,000).
B. Plasmadesmata (pa) connecting adjacent cells of belladonna (*Atropa belladonna*) in a cell aggregate (×72,000).

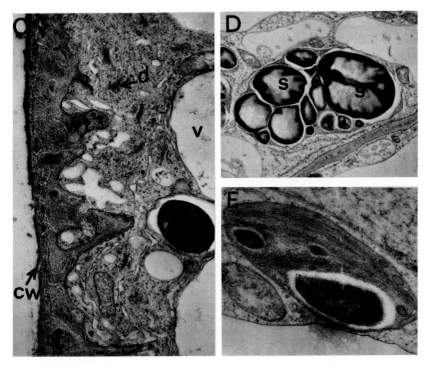

(FIG. 6.20 *cont.*)

C. Ridge-like thickenings lining the outer wall of one of the surface cells of a cell aggregate of sycamore cells (× 12,000).

D. Plastid containing numerous starch grains from a sycamore cell cultured in standard medium containing 2,4-D (× 20,000).

E. Chloroplast in a sycamore cell cultured in medium containing NAA instead of 2,4-D as the auxin addition (× 26,000).

(A from Sutton-Jones & Street 1968; B from Davey 1970; C from Davey & Street, 1971; D from Carceller, Davey, Fowler & Street, 1971; E from Davey, 1970.)

Key (for Figs. 6.18, 6.19, 6.20): c=chromosome cp=cell plate cw=cell wall d=dictyosome m=mitochondrion mb=microbody mvb=multivesicular body n=nucleus pa=plasmadesmata pt=plastid rer=rough endoplasmic reticulum s=starch v=vacuole wt=wall thickening.

The cytoplasm adjacent to these thickenings is rich in dictyosomes and electron-dense globules (dictyosme vesicles?). The significance of these wall thickenings is uncertain but they may serve to greatly increase the plasmalemma area over which solute transfer occurs. They clearly resemble the wall modifications observed by Bowes (1969) in callus of *Andrographis paniculata* and the wall invaginations of transfer cells (Gunning & Pate 1969).

As the aggregates break up and stationary phase is approached the mean cell size increases and the cells become more highly vacuolated and less rich in cytoplasm and cytoplasmic organelles. The only aspects of fine structure to increase at this stage are the number of microbodies and the degree of starch deposition in the amyloplasts. Another feature, not confined to stationary phase cells but most frequently observed as this stage is reached, are paramural bodies (resembling the plasmalemmasomes of Marchant & Robards 1968) associated with the internal surfaces of the cell walls. Structures transitional in form between these and the multivesicular bodies suggest that the latter are formed from the former. The functional significance of these structures is obscure.

The plastids of cultured sycamore cells are oval or elongate and occasionally amoeboid (Newcomb 1967). They are normally non-photosynthetic with poorly differentiated internal structures. They contain globules (larger and less osmiophilic than those of chloroplasts) and vesicles which appear to arise from the inner membrane. They enlarge enormously when swollen with starch grains (Fig. 6.20D). Occasional inclusions are strongly suggestive of phytoferritin (Hyde, Hodge, Kahn & Birnstiel 1963). The standard culture medium in which the plastids have the structure described above contains 2,4-D. The culture is white or very pale yellow. Sycamore suspensions grown in a medium in which the 2,4-D is replaced by NAA and illuminated show a limited degree of greening. In such suspensions *some* of the cells contain well-differentiated chloroplasts (Fig. 6.20E) along with undifferentiated plastids. Other cells remain hyaline. More detailed studies on chloroplast differentiation in cultured cells have, however, been made in callus cultures (see this chapter, p. 139).

Sycamore suspension cultures give rise to appreciable amounts of lignin when cultured in a medium of high sucrose content containing enhanced levels of 2,4-D (or a high level of kinetin). The cells, however, show no true secondary cell-wall formation or cell-wall lignification (Carceller, Davey, Fowler & Street 1971). The lignin is partly retained within the cell protoplasts, partly deposited between the cells associated in aggregates and partly released into the medium.

CHAPTER 7

NUCLEAR CYTOLOGY

N. SUNDERLAND

Introduction		161
Choice of material		161
Techniques		163
Chromosomal mutation		166
The problem of stability		186
In conclusion		189

INTRODUCTION

The perpetuation of a genotype during growth and development depends upon the regular and ordered processes of chromosomal replication and division: imperfections and occasional irregularities in these processes provide the basis for variation and evolution. Plant tissues and cells, like their animal counterparts, display more than the usual degree of nuclear irregularity when they are removed from the stabilizing environment of the intact organism and plunged into the alien environment of the culture vessel: variation and evolution in such cell populations can therefore be a serious problem. There are two kinds of nuclear irregularity, chromosomal mutations, many of which are detected simply by microscopical examination of the chromosomes, and gene mutations which normally can only be recognized by reference to the phenotype. Chromosomal mutations, with which this chapter is largely concerned, are front-line markers of genetic change. There is in the culture system a much less severe sieving process to eliminate unwanted variants than there is *in vivo*; abnormal cells therefore accumulate in culture and may ultimately replace the stemline or cell lineage which one is aiming to preserve. It follows that studies of cells and tissues in culture cannot be adequately carried out without constant checking of the chromosome complement. A skill in cytological techniques and a knowledge of the mechanisms underlying chromosomal irregularities are as essential as is a knowledge of culture media, asepsis and the like. Outlined below is a résumé of the principal kinds of chromosomal change that may be encountered.

CHOICE OF MATERIAL

The detection of chromosomal aberrations depends largely on the size and

number of chromosomes concerned, and the ease with which the chromosomes can be handled. This at once introduces a difficulty because what may
be regarded as the classical plant culture systems, tobacco, carrot, Jerusalem
artichoke and sycamore, possess such small or such large numbers of
chromosomes that karyotypic analyses are painstakingly tedious and accurate
interpretation of them well-nigh impossible: small changes in the length of a
chromosome can easily go unnoticed, as can the presence of rearrangements
between chromosomes. In recognizing the difficulty of manipulating the
carrot chromosomes, Steward and his collaborators turned to *Haplopappus
gracilis* ($2n = 4$), a species with few and relatively large chromosomes (Mitra,
Mapes & Steward 1960, Mitra & Steward 1961, Blakely & Steward 1964c).
Similarly, Melchers and Sacristán changed from the largely intractable
tobacco system ($2n = 48$) to *Crepis capillaris* ($2n = 6$), another Composite
with a small number of chromosomes (Sacristán 1967, Sacristán & Melchers
1969, Sacristán & Wendt-Gallitelli 1971, Sacristán 1971). *Haplopappus* and
Crepis have an additional advantage in possessing chromosomes with a
distinctive morphology; this simplifies the identification of a karyotype. Of
the two, *Crepis* is to be preferred because conditions have been established
for regenerating plants from its cultures (Sacristán 1971), and this is an
important criterion for the assessment of genetic change. *Crepis* is also a
nonpolysomatic species, that is, its cells do not undergo somatic endoreduplication during differentiation (see p. 181); thus explants taken from any
part of the plant consist entirely of cells of one ploidy level.

Pisum sativum ($2n = 14$) and *Vicia faba* ($2n = 12$) are both polysomatic
species, the latter in particular being highly favoured for *in vivo* cytological
studies on account of its large chromosomes. Cultures of both species have
proved valuable in *in vitro* studies, though conditions have still to be worked
out, as in *Haplopappus*, for the regeneration of plants. *Pisum* cultures have
been used extensively to study the effects of various components of the
culture medium such as kinetin, auxins and yeast extract on chromosomal
instability (Torrey 1959, 1961, 1965), and also in studies of the influence of
polyploidization on rhizogenesis (Torrey 1967a). *Vicia* cultures have also been
used for investigating mutational effects of the medium, in this instance, of
compounds such as nucleic acid derivatives that are likely to accumulate
during the culture process (Venketeswaran & Spiess 1963, 1964).

For large and distinctive chromosomes, many of the monocotyledons are
unsurpassed, but apart from *Triticum* species (Shimada, Sasakuma &
Tsunewaki 1969, Kao, Miller, Gamborg & Harvey 1970, Shimada 1971) and
Lolium perenne (Norstog, Wall & Howland 1969) cultures of monocotyledons
have so far been little exploited in cytological work. Hopefully, one envisages
diminishing use of the more traditional species in favour of low-chromosome-
numbered monocots.

Even with ideal species, interpretation of a karyotype is not straight-forward. In *Haplopappus*, the diploid complement consists of two chromosomes with subterminal centromeres and two with submedian centromeres, hereafter referred to as A and B chromosomes respectively (Fig. 7.1a, b, c). The A chromosome also carries a satellite, so that at metaphase two constrictions should be seen, one in the centromere region and the other at the point of attachment of the satellite. The detection of two constrictions at metaphase, however, depends largely on the quality of the preparation, and more often than not the secondary constriction is obscured (Fig. 7.1b). The satellite is best observed at anaphase when it tends to trail behind the short arm of the chromatid, but here again, the satellite may be obscured or even lost during preparation.

The presence or absence of the satellite was crucial in the interpretation of the karyotype of a strain of *Haplopappus* callus (strain A1) examined by Marks & Sunderland (1966). The strain had eight chromosomes, but the complement was not a true tetraploid (four A chromosomes with satellites and four B chromosomes); instead, only three chromosomes possessed submedian centromeres whereas five had subterminal centromeres (Fig. 7.1d, e). An exhaustive search failed to reveal conclusively the presence of a satellite on each of the five chromosomes with subterminal centromeres, and it was thus uncertain whether or not they were all true A chromosomes. Furthermore, only two of the chromosomes with submedian centromeres appeared to be normal in having arm-lengths equal to those of a typical B chromosome; in the third, the arm-lengths were shorter (Fig. 7.1d, e). Marks and Sunderland concluded that in all probability six of the chromosomes were normal, four A chromosomes, and two B chromosomes, while the remaining two were new chromosomes introduced into the complement by breakage and reunion involving two B chromosomes of the normal tetraploid complement (see also p. 181). One retained a submedian centromere, but in the other the break occurred so as to yield a chromosome with the centromere in a subterminal position and with arm-lengths very similar to those of a typical A chromosome.

TECHNIQUES

Before embarking on the nuclear cytology of a tissue or cell culture it is essential to determine the karyotype of the species concerned. For this purpose, root-tips are the easiest material to handle. Beginners are recommended to practise on root-tips of *Vicia faba*, but should not be surprised if in using *primary* roots they observe not only diploid but also occasionally tetraploid mitoses, and mitoses exhibiting chromosome breakage. For solely

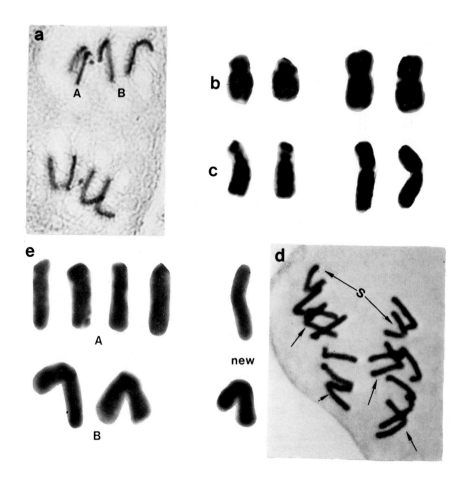

FIG. 7.1. *Metaphase and anaphase preparation of* Haplopappus gracilis *chromosomes.*
a: Root-tipa naphase showing two typical A chromatids with subterminal centromeres and two B chromatids with submedian centromeres. The satellite is out of
focus in one A chromatid, but can be clearly seen in the other. b–c: Ideograms of
root-tip metaphase chromosomes. Two constrictions can be seen on the A chromosomes in ideogram c, but the secondary constriction is obliterated in ideogram b
probably because of too-severe a colchicine treatment. d: Callus anaphase (strain
A1) showing five chromatids with subterminal centromeres and three with submedian. One of the chromatids with submedian centromeres has shorter arms (s→)
than the other two (→). The chromatids with subterminal centromeres all appear
morphologically alike. e: Ideogram of the metaphase chromosomes of strain A1,
showing four A chromosomes, two B chromosomes and two chromosomes new to
the complement.

diploid mitoses, lateral root-tips should be used; laterals arise from the pericycle, the cells of which are known to be wholly diploid. *Primary* root-tips of *Pisum sativum* may also show a mixed cell population. In establishing the karyotype, two series of observations should be made, one on untreated cells for examination of the chromosomes at anaphase, and another on cells pre-treated with a spindle inhibitor for examination of chromosome morphology at metaphase. The inhibitor-treatment provides an excess of metaphases; it shortens the chromosomes, cuts down overlap between them and generally facilitates counting. There is no short cut. DNA measurements by Feulgen photometry distinguish clearly between mitoses of different ploidy levels (see Patau & Das 1961) but give no information about the number or the morphology of the chromosomes concerned. Measurement of the diameter of the metaphase plate is also a useful guide to increasing ploidy levels (Cooper, Cooper, Hildebrandt & Riker 1964), but again gives no indication of the karyotype. Comparison of the chromosomes at both metaphase and anaphase also permits a distinction to be drawn between true aberrations which arise from the culture procedure and aberrations induced by the inhibitor-treatment.

Techniques of nuclear cytology are well documented (Darlington & La Cour 1970). Cells are fixed, hydrolysed to soften the cell wall, and then stained. Darlington and La Cour give a list of fixatives from which Carnoy's formula or acetic acid: ethanol, 1:3 v/v, may be recommended. Marks & Sunderland (1966) found a modified Carnoy consisting of methanol: chloroform: propionic acid, 6:2:1 v/v, advantageous for *Haplopappus* callus. Fixation is conveniently carried out overnight at 4°C, and if cells cannot be used immediately thereafter they may be transferred into 90% v/v ethanol and stored at 4°C. For hydrolysis, cells are taken down to water through a graded series of ethanol and treated for 1 hour at room temperature with 5 N HCl (concentrated HCl: water, 1:1 v/v) (Fox 1969), or by the more conventional method of heating the cells in 1 N HCl at 60°C for approximately 10 minutes. The former procedure appears to be generally applicable to cultured tissues, whereas the latter usually demands a precise determination of hydrolysis time which varies from one species to another. After removal of surplus HCl, the cells are transferred into Feulgen reagent for at least 2 hours. The cells may then be squashed in propionic orcein (4 g orcein refluxed in 50% v/v propionic acid for 24 hours). Slides may be made permanent by placing them on dry ice for 30–60 seconds; the cover slip is then levered off abruptly and the slide plunged immediately into absolute ethanol. Slides are subsequently taken up through a graded series of ethanol plus euparal, and finally mounted in euparal (see also Chapter 3, p. 53 and Chapter 4, p. 93).

For shortening the chromosomes, cells should be treated prior to fixation in a solution of colchicine ($0.05–0.1\%$ w/v) made up in liquid culture medium. The most effective concentration and the duration of exposure to the drug have to be determined for each species; times of 4–5 hours are usually adequate. 8-hydroxyquinoline (0.002 M) or a saturated aqueous solution of α-bromonaphthalene may also be used.

In order to achieve adequately flattened preparations only small groups of cells should be squashed. Non-dividing zones in a callus should if possible be separated from the dividing zones. With *Haplopappus* callus, the Feulgen treatment recommended above assists in the location of the dividing zones. In contrast with root-tip meristems, which stain densely in Feulgen reagent, the dividing zones in *Haplopappus* callus are little stained. Non-dividing zones, on the other hand, readily take up the stain probably because of their heavily thickened walls and the presence of tannin-like deposits upon them.

Cultures should be examined during the phase of maximum division. Samples of cell suspensions may be taken at appropriate intervals and centrifuged before treatment. Callus may be pricked out in small portions (less than 50 mg), put on fresh medium, and sampled from about 7–14 days. Samples should be taken at random from all parts of the callus. It is not sufficient to snip off the corner of a callus, count a few mitoses in it, and expect this to be representative of the culture as a whole (see, e.g. Demoise & Partanen 1969). Metaphases should be scored in which the chromosomes can be accurately counted; any cell in which the membrane is broken and the chromosomes dispersed should be rejected. Camera lucida drawings are helpful as an aid in counting large numbers of small chromosomes. Mixtures of pectinase and cellulase (0.5% w/v of each in 0.1 M sodium acetate buffer pH 4·5, for 1–2 hours) are a useful aid in the softening of cell walls before squashing (Kao *et al.* 1970). The numbers of cells in mitosis may also be increased by cold shocks. Kao *et al.* (1970) used a regime of 12–24 hours at 15°C in darkness followed by 11 hours at 27–29°C in light. Colchicine was applied during the final 5 hours of treatment.

CHROMOSOMAL MUTATION

The principal sources of chromosomal irregularity as observed *in vivo* and to which tissues and cells are subjected *in vitro* are represented diagrammatically in Figs. 7.2 and 7.3. These errors can be grouped into three types, those which arise (i) from spindle abnormalities, (ii) from chromosome breakage, or (iii) by repeated DNA replications without the intervention of mitosis. The first group leads to changes in chromosome number and involves either single chromosomes or sets of chromosomes—these errors are accompanied by a change in the absolute amount of DNA per nucleus but not in the amount per chromosome. The second group which involves chromosome breakage (in the transverse direction) with or without union of broken ends, leads to structural rearrangements between chromosomes, to loss of parts of chromosomes, and hence to a change in the DNA content of these chromosomes. Breakage may also entail a change in chromosome number. The third group leads to an increase in the girth of the chromosomes, and to an increase in their DNA content, but does not involve a change in chromosome number (somatic endoreduplication leading to polyteny).

In the account which follows, the number of sets of chromosomes in a complement is designated as n. A diploid cell thus has $2n$ chromosomes (one set contributed by one parent and one by the other), a tetraploid has $4n$ chromosomes, an octoploid $8n$ chromosomes, and so on. During the G1 phase of the cell cycle, each chromosome is present as a single chromatid, and the DNA content of one such set is designated as the C value. A diploid cell in G1 thus consists of $2n$ chromosomes with a DNA value of 2C, a tetraploid consists of $4n$ chromosomes with a DNA value of 4C, and so on. During the S phase of the cell cycle, the chromosomes are replicated and the DNA content of the nucleus is doubled. When a diploid nucleus enters prophase therefore the complement appears as two sets ($2n$) of two-chromatid chromosomes with a DNA value of 4C. Similarly, when a tetraploid nucleus enters prophase the complement appears as four sets ($4n$) of two-chromatid chromosomes with a DNA value of 8C. At anaphase, sister-chromatids separate, two identical nuclei are formed, and the G1 situation is restored.

1. *Increase in number without breakage*

The simplest change is a repeated doubling of the basic set of chromosomes, $2n$, $4n$, $8n$, $16n$ and so on (polyploidy). The phenomenon as observed in early culture studies has been discussed in detail by Partanen (1963b, 1965b). Doublings which give rise to exact replicas of the basic set take place mainly by the process of *endomitosis* (Geitler 1939). The term is an unfortunate one and is not liked by some cytologists (see, e.g. Patau & Das 1961), but it will be used here for the sake of continuity. Endomitosis consists of a separation of sister-chromatids within the nuclear membrane but without the formation of a spindle (see Figs. 7.2 and 7.9a). What was a diploid cell now comes to consist of four sets ($4n$) of one-chromatid chromosomes with a DNA level of 4C. In the absence of a spindle, the chromatids condense together and a restitution nucleus is formed within the original nuclear membrane. This nucleus subsequently undergoes a further DNA replication bringing the DNA level up to 8C; it then consists of four sets ($4n$) of two-chromatid chromosomes as in a true tetraploid cell in G2. The nucleus may then enter prophase and proceed through a normal mitosis giving rise to two

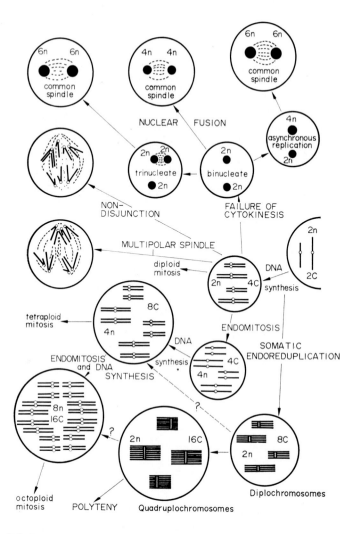

FIG. 7.2. *Diagram illustrating known nuclear aberrations as they might occur in a diploid* Haplopappus *cell*; nuclear fusion, non-disjunction, multipolar spindle formation, endomitosis and somatic endoreduplication.

identical tetraploid nuclei. Alternatively, sister-chromatids may again fall apart without the intervention of a spindle, and another restitution nucleus is produced consisting of eight sets ($8n$) of one-chromatid chromosomes. In this event, DNA replication followed by a normal mitosis results in the formation of two identical octoploid nuclei (Fig. 7.2).

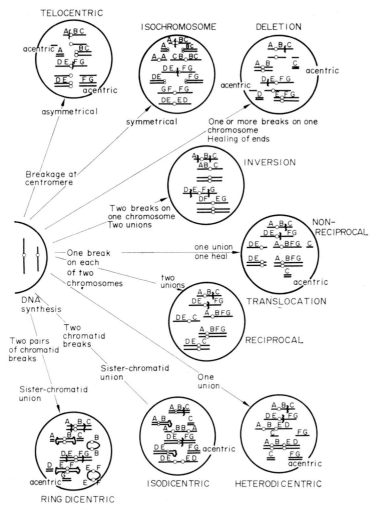

FIG. 7.3. *Diagram illustrating structural rearrangements between and within chromosomes as a result of breakage with or without union of broken ends.*

Survival of such polyploid cells in a culture depends largely on their rates of proliferation relative to that of the stem-line, and to the rate at which further polyploidization takes place. If the rates of proliferation of the polyploid cells do not markedly differ from that of the diploid they may be expected to survive from one subculture to another, and providing the rate of polyploidization is low such polyploid cells will remain a minor component of the cell population. Cooper *et al.* (1964), for instance, analysed chromosome numbers in two 8-year-old clones of tobacco callus, both of which originated from single cells of a crown-gall callus. Countable mitoses revealed, in both clones, numbers of 48 (2n), 96 (4n) and 192 (8n) indicating that both clones originated from cells having the diploid number of chromosomes (though not necessarily the true diploid complement). So far as could be judged from measurements made on the diameters of the equatorial plates, about 75% of mitoses in one clone had numbers around the diploid mode, 20% around the tetraploid and 5% around the octoploid mode. The diploid mode was even more predominant in the second clone. Little can be deduced about rates of change from a single measurement of this kind, but it seems safe to deduce that, in this instance, the polyploid cells showed no competitive advantage.

More often than not, however, polyploid cells predominate in established cultures. Sunderland (unpublished results) found that a suspension culture of *Haplopappus* cotyledon cells grown in a medium containing 2,4-D as the auxin component changed from wholly diploid to wholly tetraploid over a period of less than 6 months of subculture. Corresponding cultures given NAA instead of 2,4-D changed less rapidly, but they also grew less rapidly suggesting that 2,4-D favoured growth of the tetraploid line. The degree of cell dissociation was also considerably greater in the presence of 2,4-D. Shamina (1966) working on a strain of *Haplopappus* callus likewise observed a greater rate of change with 2,4-D than with NAA. In this instance, the resulting population consisted of several polyploid lines and it was concluded that 2,4-D operated by accelerating the rate of polyploidization.

Another factor contributing to an enhancement of polyploidization is the ploidy status of the explants. Sacristán (1971) compared the rate of polyploidization in a series of cultures derived from a haploid and a diploid plant of *Crepis capillaris* and found a much higher rate of change in the haploid series. About 40% of the replicate subcultures derived from haploid explants showed wholly diploid metaphases as early as the second transfer after isolation; the rate of change decreased thereafter with time reaching about 50% by the 7th transfer. A small proportion of the individual cultures examined showed wholly tetraploid metaphases at the 2nd transfer and this increased with time reaching about 20% by the 10th transfer. Relatively few cultures revealed mixed cell populations. In the replicate subcultures derived from different explants of the diploid plant, however, only 20% showed either wholly tetraploid metaphases or mixtures of diploid/tetraploid metaphases by the 16th transfer.

Plant tissue and cell cultures seldom keep to a rigid geometric series of chromosome doublings; cells with numbers intermediary in the series start to emerge. Shamina (1966) found about 13% of mitoses with a triploid number in *Haplopappus* after only 4 months of subculture, while Sidorenko & Kunakh (1970) examining the same strain several years later found a wide range of ploidies including $3n$, $5n$, $7n$, $9n$ and so on. Three lineages predominated possessing numbers of 4, 6 and 8 chromosomes; together they accounted for 60–70% of the cell population. The relative proportions of the three main lineages varied from one subculture to the next but either the lines with the triploid or the tetraploid numbers predominated. A similar wide range of odd ploidy values was found by Torrey (1967a) in long-established *Pisum* calluses, and this heterogeneity was correlated at least in part with a loss in rhizogenetic potential. Here again, 2,4-D was implicated in accelerating the rate of change. Plants possessing triploid and pentaploid chromosome numbers have been successfully regenerated from calluses derived by anther culture (see Chapter 9).

Another nuclear irregularity which leads to the formation of polyploid cells, both of the normal geometric series and of intermediate ploidies, is *nuclear fusion*. The idea that nuclei may coalesce together in cultured cells emerged early in the history of tissue culture from observations such as those described by Naylor, Sander & Skoog (1954). Tobacco pith cells were induced into prophase by an auxin-treatment, but they did not divide and instead became multinucleate. Some of the multinucleate cells contained large lobed nuclei suggestive of several nuclei coalescing together. The phenomenon was commented on later by other workers using different species (Torrey 1959, 1961, Partanen 1963b). Cooper *et al.* (1964) probably gave the most likely interpretation of this phenomenon in pointing to the known nuclear mishap of 'sticky' chromosomes which prevents daughter nuclei from separating completely at anaphase; mitosis fails and the partially separated nuclei coalesce together forming a restitution (polyploid) nucleus.

It is now known that nuclear fusion generally involves mitosis on a common spindle. In the classical case of fertilization, for instance, the two haploid nuclei do not simply coalesce but enter mitosis simultaneously on a common spindle so that two diploid nuclei arise and not one (see Lewis & John 1963). Common spindle formation can be induced by certain chemical treatments (Kihlman 1966) and is the accepted mechanism for fusion in animal cell hybrids (Harris 1970). For further information on synchronous mitoses in multinucleate plant cells the papers of Gonzales-Fernandez and his colleagues (1971) should be consulted.

Nuclear fusion is a complex process and requires a series of very ordered events. It requires, one, failure of cytokinesis leading to the formation of a binucleate cell; two, simultaneous DNA replication in both nuclei followed by entry into mitosis on a common spindle; three, a precise segregation of appropriate chromatids to the respective poles, and four, cytokinesis cutting off two identical cells. Asynchronous replication or fusion in a multinucleate cell can be seen as leading to cells with intermediary ploidy numbers (see Fig. 7.2). For the formation of a triploid cell by this process, a haploid nucleus must first be formed, though this does not apply in the case of genuine triploid formation in pollen grains. Mishaps might be expected to occur in such an ordered sequence so that nuclear fusion should also be regarded as a potential source of aneuploid cells.

2. *Decrease in number without breakage*

There is no known mechanism, other than the controversial one of somatic pairing, by which a reduction in whole sets of chromosomes can be achieved in one step. Somatic pairing though known to occur in certain animal systems is a rarity in plants, and most plant cytologists discount it. Mitra & Steward (1961) interpreted certain mitotic configurations in *Haplopappus* cultures as paired homologous chromosomes, and correlated them with the presence of occasional haploid metaphases. Partanen (1963b) has already commented on this interpretation. Haploid mitoses in *Haplopappus* cultures have also been reported by Shamina (1966) and Sidorenko & Kunakh (1970), though none of these authors attempted to account for them.

In a species with only four chromosomes, the formation of a haploid cell can be accounted for by several known mechanisms, of which multipolar spindle formation, non-disjunction or lagging of chromatids may be cited. These phenomena occur fairly frequently in *Haplopappus* cultures (Fig. 7.4). The formation of a four-pole spindle could result, for instance, in the distribution of two chromatids to each of the four poles, and conceivably a nucleus with an A and a B chromosome might subsequently be cut off. Alternatively, it needs only one irregular two-pole mitosis involving non-disjunction of an A and a B chromosome to produce a triploid nucleus at one pole and a haploid at the other, or else successive division involving the non-disjunction of one chromosome at a time (Fig. 7.5). For those interested in permutations and combinations a satisfying hour can be spent rearranging the diploid *Haplopappus* complement in this way. Not many changes are needed to produce the range of numbers reported by Sidorenko & Kunakh (1970). Lagging of individual chromatids leads to a similar situation providing that at cytokinesis the entire chromatid is incorporated into one of the daughter cells.

The greater the number of chromosomes in the complement the less chance there is of a cell with the haploid number being produced. In highly polyploid populations, non-disjunction, lagging or multipolar spindles can be seen as leading simply to a grossly hypo- or hyperploid condition. A long-established tobacco pith callus (KX-1) examined by Fox (1963) showed a range of chromosome numbers varying from about 130 to over 220 with more than 85% above the tetraploid level. Cells with numbers between 130 and 190 accounted for more than 50% of the population, that is, the strain appeared to be grossly hypo-octoploid. Compared with the tobacco clones of Cooper *et al.* (1964) which remained predominantly diploid this strain probably evolved by polyploidization to the octoploid level followed by a gradual break-up of the complement into aneuploidy.

Heinz, Mee & Nickell (1969) also found a highly aneuploid situation in

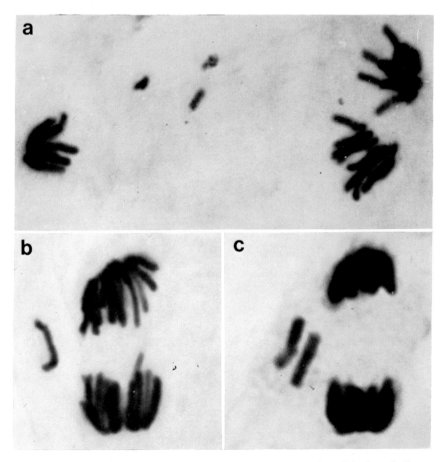

FIG. 7.4. *Nuclear aberrations in* Haplopappus *callus (strain A2)*. a: Tripolar spindle. b–c: Lagging chromatids.

suspension cultures of *Saccharum* cells. A 6-year-old culture initiated from a single callus cell showed numbers varying from 111–140, another culture numbers from 71–300 with a peak around 71–90, while a third culture had numbers varying from 51–185.

The initial numbers in each clone could not be ascertained because the strain from which the clones were isolated was itself highly aneuploid with

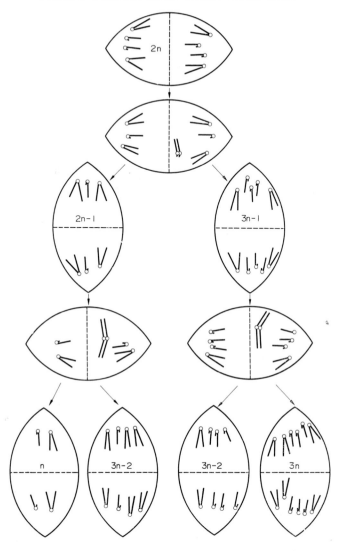

FIG. 7.5. *Diagram illustrating the effect of repeated non-disjunctions in the diploid* Haploppapus *complement.*

numbers varying from 71–300. Other strains isolated from different species of *Saccharum* (chromosome numbers of 114–122) all showed this heterogeneity after prolonged culture. In this instance polyploidization did not appear to have played a prominent role in the evolution of the cultures and this may have been because the *Saccharum* hybrids used were themselves of highly polyploid origin (Heinz *et al.* 1969).

The medium used for *Saccharum* lacked kinetin which, as a result of the work on *Pisum* cultures (Torrey 1959, 1961, 1965, 1967a), has become regarded as an effective agent for the acceleration of polyploidization. Heinz *et al.* (1969) tested the effect of kinetin on a *Saccharum* strain which consistently showed chromosome numbers lower than that of the donor plant, but kinetin failed to induce polyploidy. This failure suggests that polyploidization is mainly a problem with explants of initial low ploidy status; in other words the degree of polyploidization in a culture may be expected to decrease with increasing ploidy status of the explants (cf. Sacristán 1971). The composition of the medium is probably largely irrelevant, though this is not to deny that compounds such as 2,4-D and yeast extract, both of which were present in the *Saccharum* medium and which are known to accelerate chromosomal mutation (Torrey 1959, 1961), may have been partially responsible for speeding the break-up of the *Saccharum* karyotypes.

3.   *Chromosome breakage without numerical change*

Preoccupation with chromosome numbers has tended to divert attention from the far more important source of variation in tissue and cell cultures, namely that of structural change and loss of genetic material. That such abberrations occur in the classical systems such as tobacco has been admirably demonstrated by the work of Shimada & Tabata (1967). In a highly aneuploid mitosis from a 1-year-old pith callus, many of the chromosomes were found to have different lengths from those of the normal root-tip karyotype. The ratio in length of the longest chromosome to the shortest, for instance, was 4·3 in the callus compared with only 2·6 in the root-tip.

Changes in length of a chromosome arise principally by translocations and deletions (see Fig. 7.3). The break may occur at any time during the cell cycle. Non-reciprocal translocations are probably more deleterious in culture than reciprocal translocations since they incur loss of genetic material in the form of acentric fragments; the reciprocal form merely causes a rearrangement of the genes between chromosomes. Deletions also lead to loss of genes as does the formation of telocentrics and isochromosomes by breakage in the centromere region (Marks 1957).

Structural changes may begin very early in the life of a culture. In the work on *Crepis* already mentioned (Sacristán 1971), a number of abnormal karyotypes emerged during the first year. Several of these could be traced to random translocations and deletions (Fig. 7.6). Only one of the subcultures derived from the haploid plant and which retained the haploid number throughout, showed any structural changes, but changes occurred in about 5% of the subcultures derived from the diploid plant and in which the chromosome number remained constant. Structural changes were also

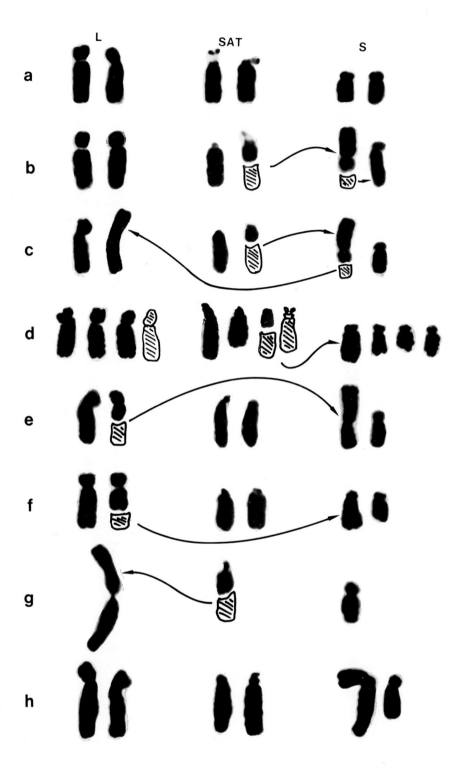

observed in subcultures which underwent polyploidization in culture. In both series, the degree of structural change increased with increase in ploidy level. Thus, in the subcultures derived from the haploid plant, 9% of those carrying the diploid number, showed structural changes, compared with 16% of those carrying the tetraploid number. Similarly, in the subcultures derived from the diploid plant, 8% of those in which the tetraploid line became predominant showed structural changes compared with the 5% which retained the initial diploid number. It may therefore be inferred that there was probably a high rate of structural change in the highly polyploid and aneuploid cultures of tobacco and *Saccharum* referred to above (Fox 1963, Heinz *et al.* 1969).

Structural changes at present can only be detected by reference to the morphology of the chromosomes. There is a real need for the perfecting of a differential staining procedure such as the fluorescent technique described by Caspersson, Zech, Modest, Foley, Wagh & Simonsson (1969) and Vosa (1970) which reveals characteristic banding patterns along the chromosomes. The ability to label plant chromosomes in this way would aid not only in the identification of structural changes but also in the location of those points at which chromosomes break.

4.   *Chromosome breakage with numerical change*

A change in chromosome number through breakage almost invariably results from the formation of ring, dicentric or tricentric chromosomes. The formation of such chromosomes in tissue and cell cultures is not a rare event and has been reported for several species (Torrey 1959, Mitra *et al.* 1960, Mitra & Steward 1961, and Norstog *et al.* 1969). Dicentric chromosomes are probably those most frequently encountered, and in two instances at least they have been found in the karyotype of the main cell-lineages (Marks & Sunderland 1966, Kao *et al.* 1970).

FIG. 7.6. *Ideograms showing seven abnormal karyotypes produced in callus cultures of* Crepis capillaris. a: Normal diploid karyotype—two long chromosomes with subterminal centromeres (L), two with satellites (SAT), and two short chromosomes with subterminal centromeres (S). b–d: Abnormal karyotypes produced in cultures derived from a diploid plant. b: Karyotype with translocations between the SAT and S chromosomes. c: Karyotype with rearrangements between the L, SAT and S chromosomes. d: Hypotetraploid karyotype—loss of one L and one SAT chromosome and a translocation between one SAT and one S chromsome. e–h: Abnormal karyotypes produced in cultures derived from a haploid plant. e–f: Diploid karyotypes each with a translocation between one L and one S chromosome. g: Haploid karyotype with a translocation between one L and one SAT chromosome. h: Diploid karyotype with rearrangements between one (extra) L chromosome and one S chromosome. This karyotype cannot be accounted for solely by translocations (adapted from the data of Sacristán (1971) and reproduced by courtesy of the author).

Norstog *et al.* (1969) examined mitoses in a 10-year-old callus of *Lolium perenne*. The callus originated from a triploid endosperm ($3n = 21$) and was found to have numbers ranging from 18 to about 50 with more than half of them still near to the triploid mode. None of the karyotypes analysed with the exact triploid number carried precisely three copies of each chromosome, and several mitoses examined with numbers of 22 and 25 possessed two dicentrics. In a suspension culture of *Triticum monococcum* cells ($2n = 14$) examined after 2 years of maintenance, and again after 3, Kao *et al.* (1970) found numbers varying from 27 to 30. The majority of cells had twenty-eight chromosomes, one or two of which were dicentrics. The authors calculated that the dicentric condition had persisted through at least sixty transfers equivalent to more than 200 cell generations. These results illustrate two further patterns of cell evolution. In the *Lolium* population, polyploidization played only a minor role, structural rearrangements between chromosomes and the formation of new chromosomes by dicentric breakage being the main sources of instability. In the *Triticum* cultures, one doubling of the diploid set accompanied by structural rearrangement and the formation of dicentric chromosomes conferred selective advantage, and the diploid cell-lineage was eliminated.

There are two types of dicentric chromosome which differ according to the number of chromosomes involved in their formation (see Fig. 7.3). When only one chromosome is involved an *isodicentric* is produced which has genetically identical arms of equal length, and an intercentric region carrying two mirror-image gene sequences. When two chromosomes are involved a *heterodicentric* is usually produced, and this has unequal arms with different gene sequences (one from each of the two chromosomes concerned), and an intercentric region, which combines genes from each of the donor chromosomes. Marks & Sunderland (1966) were fortunate in discovering the presence of a dicentric chromosome in a strain of *Haplopappus* callus (strain A2) immediately after its formation. The callus arose spontaneously from strain A1, the karyotype of which is shown in Fig. 7.1d, e. The karyotype of strain A2 consisted of seven chromosomes, four of which appeared to be A chromosomes, two were B chromosomes and the seventh was a heterodicentric. This latter chromosome was probably formed by breakage and union between the two abnormal chromosomes present in the A1 complement. Marks and Sunderland studied the behaviour of this chromosome, mostly at anaphase, through several transfers during which cells with fourteen chromosomes were produced among which were two dicentrics. These workers were able to observe in operation a breakage–fusion–bridge cycle similar to that originally described by McClintock (1951) in maize. This cycle is illustrated diagrammatically in Fig. 7.7. Dicentric chromatids separate in one of three ways; either by parallel separation (see Darlington & Wylie 1952), or by interlocked or criss-cross bridge-formation. Parallel separation, in which the two centro-meres of each chromatid pass to the same pole, restores the dicentric chromo-some and perpetuates it (Fig. 7.8a). In an interlocked bridge, the two cen-tromeres of each chromatid also pass to the same pole, but because the intercentric regions are interlocked the chromatids break. The original dicentric chromosome is, however, restored whenever the broken ends of the same chromatid fuse during the ensuing interphase (Fig. 7.7). In a criss-cross bridge, the centromeres of each chromatid pass to opposite poles and the chromatids again break. Fusion of the broken ends of the two chromatids which pass to the same pole again restores the dicentric condition, but because the break can occur at any point along the bridge, this event pro-duces new dicentrics with either longer or shorter intercentric regions than that of the original chromosome (Fig. 7.7).

About 98% of anaphases observed by Marks and Sunderland showed chromatid-bridges, and it was evident that this breakage–fusion–bridge cycle was largely responsible for perpetuation of the dicentric condition; parallel separation of chromatids as shown in Fig. 7.8a was a relatively rare event. However, most of the bridges which could be conclusively identified were of the criss-cross type. Hence it follows that the dicentrics present throughout the cell population were probably not identical, that they differed one from another in respect of their intercentric regions, and were probably constantly changing according to the point of breakage along the chromatid bridge.

At first, the cycle appeared consistantly to restore heterodicentric chromosomes. With time, however, as polyploid cells with two dicentrics emerged, an increasing proportion of the chromatid bridges showed the iso-type of dicentric (Fig. 7.8a, d). Since anaphases invariably showed one bridge at the lower mode, and two at the higher, it seemed unlikely that the isodicentrics were formed *de novo*, and that they must be further products of the breakage–fusion–bridge cycle. There are two ways in which the isodi-centrics could have arisen, both of which are illustrated in Fig. 7.7. One possibility is that broken ends of chromatids passing to opposite poles fused immediately after breakage, and that the entire chromosome was then incorporated in some way into one cell. The other is the process of sister-chromatid reunion discovered in maize by McClintock (1951). This process involves the fusion of sister-chromatid ends *after* DNA replication and an isodicentric chromosome is produced by opening out of the duplicated structure. In maize this event was confined to the reproductive phase and did not occur in somatic tissue, nor has this type of fusion been observed in other somatic tissues *in vivo*. Whatever the origin of the isodicentrics in strain A2, their appearance indicated yet a further source of variation.

The cycle is broken whenever healing of the broken ends occurs instead of fusion. Two new monocentric chromosomes are then introduced into the

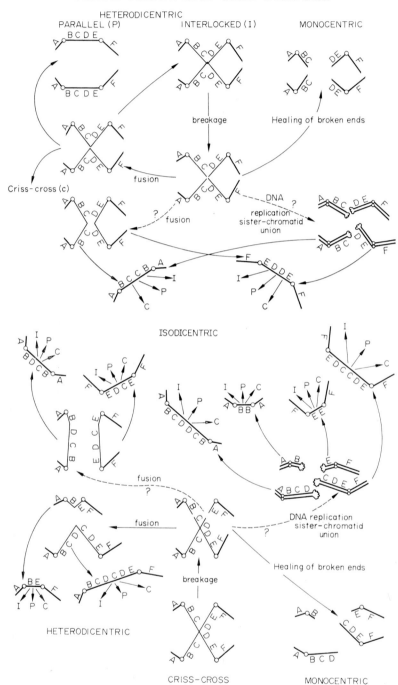

complement. In strain A2, cells with eight, fifteen and sixteen chromosomes emerged in addition to the main lines with seven and fourteen (compare Norstog *et al.* (1969) and Kao *et al.* (1970)). It may be stressed that the new chromosomes produced vary from cell to cell according to the point of breakage, and because the dicentrics from which they arise themselves differ from cell to cell, many of the new chromosomes though morphologically similar are in all probability genetically diverse. The number of permutations and combinations in such a system is very large, and the situation becomes progressively more complex when more than one dicentric is present; each one may be expected to behave independently of the others. Such cell populations can never be stable; change is inevitable. Marks and Sunderland thought that the two abnormal chromosomes in the karyotype of strain A1 (Fig. 7.1d) were produced by breakage of a dicentric bridge and subsequent healing of two ends, and that the chromosome concerned probably originated by breakage and union between two B chromosomes in a tetraploid cell (cf. p. 163).

Sunderland (unpublished results) re-examined the karyotype of strain A2 about 3 years after its isolation. The karyotype was then found to have lost the dicentric condition; cells with twelve chromosomes predominated though there were several variants present (Fig. 7.9). So far as could be judged from metaphases (a photographable anaphase could not be found), only two typical B chromosomes remained; they appeared in every karyotype examined. From eight to ten chromosomes with subterminal centromeres could be identified, but some karyotypes carried one or more chromosomes (possibly rings) the morphology of which could not be accurately assessed. Several of the chromosomes with subterminal centromeres carried satellites but the precise number of them could not be fixed. The manner in which these karyotypes evolved from the original is obscure.

5. *Somatic endoreduplication*

The third type of nuclear irregularity encountered in plant tissue and cell cultures is one which leads to the omission of mitosis altogether but yet permits DNA replication. It is therefore not concerned with structural changes in the chromosomes or changes in number *in vitro*. Endoreduplication

---

FIG. 7.7. *Diagram illustrating the breakage–fusion–bridge cycle.* The dicentric condition is perpetuated by parallel separation of chromatids (P) or fusion of chromatid ends following breakage of an interlocked bridge (I). Fusion of chromatid ends following breakage of a criss-cross heterodicentric bridge (C) results in the formation of new heterodicentric chromosomes. Possible routes for the generation of isodicentrics from a heterodicentric are indicated. The dicentric condition is lost whenever healing occurs.

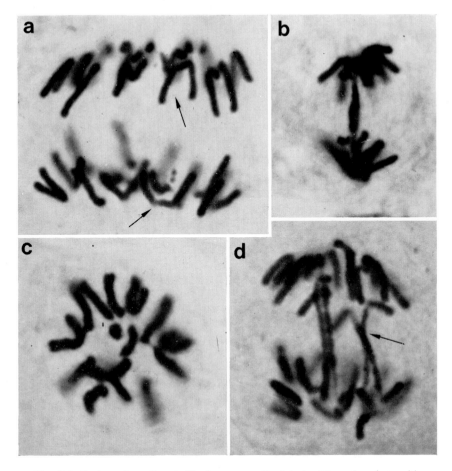

FIG. 7.8. *Nuclear aberrations in* Haploppapus *callus* (*strain A2*). a: Anaphase with fourteen chromatids showing parallel separation of an isodicentric chromosome (→). b: Anaphase with seven chromatids showing criss-cross bridge formation of a heterodicentric chromosome. c: Aberrant metaphase showing a ring chromosome. d: Anaphase with fourteen chromatids showing two criss-cross bridges; one bridge is of an isodicentric (→) and the other of a heterodicentric (reproduced by courtesy of G.E.Marks).

as described by D'Amato (1965) 'consists of a double, triple or multiple duplication of the interphase chromosomes, and might thus fall under the general definition of polyteny'. This geometrical increase in DNA causes an increase in the lateral multiplicity of DNA strands, and the process reaches giant proportions in the cells of Dipteran salivary glands, and in tapetal and suspensor-cells of higher plants. Non-dividing cells with nuclei possessing high DNA values of 16C and 32C have been reported in several plant tissue cultures and they are regarded as having arisen by endoreduplication (Patau, Das & Skoog 1957, Bennici, Buiatti & D'Amato 1968). The function of these cells, if any, is unknown. Polytene cells appear to have a secretory or storage function, and it is thus conceivable that endoreduplicated cells formed *in vitro* serve a similar purpose acting perhaps as a nurse to actively dividing cells.

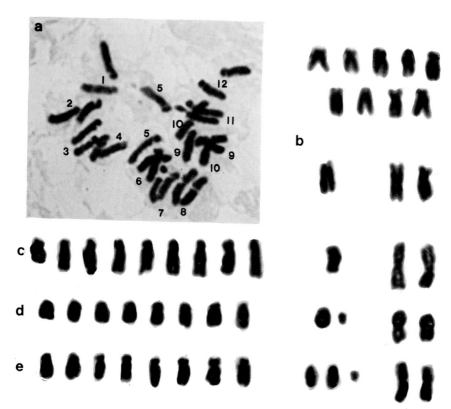

FIG. 7.9. *Karyotypes of* Haplopappus *strain A2 after 3 years of subculture*. a: A probable endomitotic figure showing chromatid separation. Only two pairs, 8 and 10, have submedian centromeres. b: Ideogram showing the predominant karyotype in which two chromosomes with submedian centromeres can be distinguished with certainty and at least nine with subterminal centromeres. c–e: Variants on b.

Polytene chromosomes do not split laterally or undergo mitosis. Endoreduplicated cells, on the other hand, which are common in differentiated regions of polysomatic plants, can be stimulated into prophase by suitable auxin-treatments. However, the chromosomes are structurally deficient in the sense that the association of chromatids at prophase is irregular. In a so-called tetraploid endoreduplicated cell the chromosomes appear as two sets (2n) of four-chromatid chromosomes (diplochromosomes) instead of the normal four sets (4n) of two-chromatid chromosomes. Likewise in a so-called octoploid endoreduplicated cell there are two sets (2n) of eight-chromatid chromosomes (quadruplochromosomes) instead of eight sets (8n) of two-chromatid chromosomes (D'Amato 1965) (Fig. 7.3). It may be inferred that replication of some component of the chromatids (centromere?) is incomplete and that this deficiency provides a barrier to normal chromatid separation.

Several investigators have argued, however, that endoreduplicated cells participate in callus formation and are in fact the chief source of polyploid cells in a culture. This may be so, but it is difficult for any reviewer in this field to make a reasoned appraisal of the claims on account of the lack of critical cytological data. It has not been demonstrated conclusively, for instance, that every cell in an explant which has a polyploid DNA value is indeed an endoreduplicated one, nor that an endoreduplicated cell, when stimulated into prophase by the cultural procedure, continues through a normal mitosis, undergoes cytokinesis and then continues to proliferate as a normal polyploid cell. Indeed, the evidence of Naylor et al. (1954) cited above indicates that auxin-induced mitoses in tobacco pith are highly irregular and give rise to multinucleate cells showing many chromosomal aberrations.

With the advent of kinetin and better cultural conditions, Patau & Das (1961) re-examined the cytological behaviour of cells in tobacco pith during callus initiation. The presence of nuclei with DNA levels of 2C, 4C, 8C and 16C was established by Feulgen photometry. Explants were then labelled with tritiated thymidine and DNA levels measured in both labelled and unlabelled mitoses at intervals over a culture period of 16 days. Only two 16C mitoses were in fact observed; both were labelled indicating that at least one DNA doubling had occurred during the culture period prior to the nuclei going into mitosis. Both labelled and unlabelled 8C mitoses were observed but these were all initial mitoses. Tetraploid mitoses were not seen in any new cell formed during the culture period, and continued division was found to be confined to diploid cells.

Thus the only tangible evidence in favour of the participation of endoreduplicated cells in these experiments comes from the unlabelled 8C mitoses, and their mitotic configuration at prophase si therefore crucial. Patau and Das did not record it 'suffice it to say that about a quarter of the polyploid *metaphases seemed* to contain diplochromosomes and that the three measured polyploid *anaphases seemed* to have normal chromosomes' (the italics are mine). If diplo- or quadruplochromosomes 'fall apart' as Patau and Das and other workers state, and the products of this disruption then continue through mitosis normally, how is one to distinguish such a mitosis (at metaphase or anaphase) from that of a cell doubled by endomitosis? Moreover, one would doubt whether in the tangled state of a tobacco prophase a distinction could be made between forty-eight four-chromatid chromosomes and ninety-six two-chromatid chromosomes. In the circumstances, the unlabelled 8C mitoses considered by Patau and Das to have arisen from endoreduplicated cells could equally have arisen from true tetraploid cells in G2 at the time of culture.

Partanen (1963a, 1965a) and Torrey (1965) using respectively root-explants of *Allium* and *Pisum* labelled with tritiated thymidine have also stressed the participation of endoreduplicated cells in callus initiation. Torrey, for instance found that, after only 3 days of culture in a kinetin-medium, unlabelled tetraploid mitoses could be observed and that *for the most part* these consisted of diplochromosomes, whereas, in an auxin-medium only diploid mitoses occurred. At 7 days, on the other hand, in the kinetin series, all tetraploid mitoses were labelled and consisted of normal chromosomes. It does not follow that the cells in mitosis at 7 days were derived from those in prophase at 3 days; they could equally have arisen *in vitro* by endomitosis from diploid cells. In any event, the present author thinks the argument is irrelevant in the case of *Pisum* because *primary* roots in this species do often possess a true tetraploid cell lineage. Incidentally Torrey demonstrated most elegantly that either the diploid line or the tetraploid could be triggered into mitosis by the choice of an appropriate hormonal treatment.

Later work carried out on the cytology of callus initiation in tobacco pith explants has done nothing to clarify the issue. Murashige & Nakano (1967), for instance, by means of the filter-paper raft technique, raised a number of clones from single cells of tobacco pith and compared them in respect of their chromosome numbers. The clones varied from diploid, tetraploid to highly aneuploid, and the authors concluded that these varied states were a reflection of a similar heterogeneous state in the pith cells themselves. This conclusion is untenable. In the first place, in order to isolate the single cells, the pith explants were pre-treated with auxins, the cytological effects of which have already been referred to above (Naylor *et al*. 1954). In the second, chromosome counts were not made on the clones until after about one month's growth on the filter-paper raft. During this period any manner of nuclear changes could have occurred. Since some of the clones consisted wholly of cells with the diploid number, all that can be claimed with certainty is that the technique selected diploid cells which subsequently varied in their behaviour in culture; some retained the diploid number, while others became tetraploid and even aneuploid (cf. Sacristán 1971).

Murashige and Nakano also claimed that explants taken from the pith at about 10 cm from the shoot-tip contained only 2*n* and 4*n* cells, whereas

G

explants taken further down the stem at about 20 cm were highly aneuploid. But how does aneuploidy develop in a non-dividing tissue? Aneuploidy in the cells of tobacco pith has also been stressed by Shimada & Tabata (1967). Contrary to Patau and Das, these workers found chromosome numbers ranging from 43–153 in the initial divisions in culture, and there was no definite modal distribution. The validity of these observations rests mainly on whether or not the mitoses counted were indeed the initial ones. There was a striking time-lag in these experiments before division commenced: about 6 days compared with 2 days in the experiments of Patau and Das. The present author considers it possible that the aneuploid condition could have developed rapidly through abnormal division of endoreduplicated cells, but until more precise information is obtained the interpretation of these varied results must remain obscure. The confusion simply underlines the point that polysomatic plants are not ideal for culture studies. and tobacco in particular for cytological observations.

THE PROBLEM OF STABILITY

There is no set pattern of variation and evolution in cultured populations of plant cells and tissues. Not only does the pattern change from one species to another, but the pattern differs between explants of the same species and subcultures prepared from them. While this inconsistency may be in part attributed to the many and varied nutrient media used by different workers, to different times and methods of subculture, and to mutagenic effects of metabolic products that accumulate in the medium, we are still largely ignorant of the fundamental causes of chromosomal mutation. The fact that, in some cells of a culture, there is breakdown solely in the mechanism controlling spindle formation, whereas in neighbouring cells breakdown extends to the entire mitotic phase but not to the S phase of the cell cycle, and that there is an enhanced rate of chromosomal breakage with breaks occurring at random throughout the population, strongly suggests the disruption of controlling forces between cells.

Despite this variable behaviour much progress has been achieved, and this has probably been largely due to the emergence of two types of cell population. The first is the type which becomes highly polyploid or which is initially derived from a highly polyploid species or a species in which the genome consists of a large number of small chromosomes. With such populations the rate of change is superficially slow—they reach and maintain a sort of equilibrium with their environment in the sense that the effects of a few mutations from one subculture to the next, or of loss of parts of chromosomes, are nullified by the activity of identical genes present on homologous chromosomes in the complement. The overall growth rate and metabolism of such populations may show little change with repeated subculture, and, within these limits, the population may be regarded as 'stable'. Furthermore, when such systems become highly aneuploid, they may still retain cells carrying copies of every gene, though in relatively different multiples, and for this reason may not entirely lose their morphogenetic potential. A striking example of this is provided by the work of Sacristán & Melchers (1969) who regenerated plants (morphologically abnormal and infertile) with grossly aneuploid chromosome complements from a number of long-established aneuploid tobacco calluses (Fig. 7.10).

The second is the type in which evolution is rapid and leads to the emergence of a cell lineage with a high selective advantage. Such events probably occur more frequently in cases where the genome consists of a few chromosomes. Providing there is no polyploidization, deletions and other forms of genetic loss may involve a relatively large proportion of the genome and could thus reduce the survival value of the cells; any advantageous mutant would soon take over. Under precise and constant cultural conditions such mutant lines also reach a form of 'stability' with their environment, the effects of chromosomal change being masked by the activity of the predominating cell-line.

Haplopappus strain A1 is a case in point. The strain arose spontaneously, as a purple-pigmented mutant, from a pale yellow callus and it continued thereafter to produce copious amounts of anthocyanins over many years of subculture (Stickland & Sunderland 1972a, b). Each periodic check of the chromosome complement revealed the presence of at least 90% of cells with the irregular karyotype shown in Fig. 7.1; the remainder possessed mainly twice the number though a few showed additional abberations. For studies of anthocyanin biosynthesis such a population is entirely acceptable, and within these limits, possession of an abnormal karyotype, or of a small proportion of other variants is largely irrelevant. Indeed, some of the polyploid and aberrant variants probably also continued to produce the pigments concerned.

Strain A1 is an auxin autotroph. Addition of auxin to the medium suppresses pigment formation, but as soon as the auxin is omitted pigment synthesis is resumed. Stocks of A1 kept in the presence of auxin showed a slightly higher proportion of aberrant mitoses but the karyotype of the stem-line remained unchanged. On one occasion, however, a culture that had been maintained in an auxin medium for several years reverted to pigment synthesis, and these coloured sectors continued thereafter to produce pigments even in the presence of augmented auxin levels. No significant change could be found in the karyotype of the reverted mutant and it seemed possible that this reversion may have been due to gene changes.

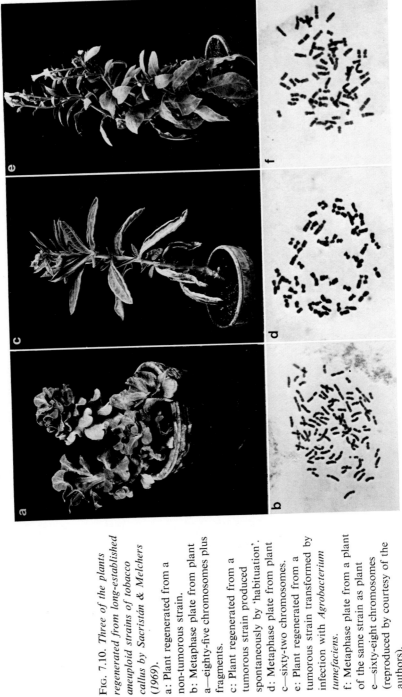

Fig. 7.10. *Three of the plants regenerated from long-established aneuploid strains of tobacco callus by Sacristán & Melchers (1969).*

a: Plant regenerated from a non-tumorous strain.

b: Metaphase plate from plant a—eighty-five chromosomes plus fragments.

c: Plant regenerated from a tumorous strain produced spontaneously by 'habituation'.

d: Metaphase plate from plant c—sixty-two chromosomes.

e: Plant regenerated from a tumorous strain transformed by infection with *Agrobacterium tumefaciens*.

f: Metaphase plate from a plant of the same strain as plant e—sixty-eight chromosomes (reproduced by courtesy of the authors).

Strain A2, which has evolved perhaps one of the most abnormal karyotypes yet described (Fig. 7.9) is an equally important mutant strain for the study of anthocyanin synthesis; the strain produces copious amounts of anthocyanin in darkness. There are other examples in the literature of spontaneous formation of chromosomal variants with such exploitable properties (see, e.g. Fox 1963, Blakely & Steward 1964c, Eriksson 1967b). The message spelled out for the future is clearly one of increased exploitation of chromosomal instability. The art of plant tissue culture can be seen as giving way to the science of plant cell breeding—the tailoring of cells to fit a specific role. Eriksson's isolation of a UV mutant in *Haplopappus* by selection for vigorous growth and high cell dissociation in suspension culture is perhaps one of the most successful applications in this area. The mutant possessed four chromosomes, though the karyotype was apparently slightly abnormal. The stemline has persisted over many years without polyploidization (Eriksson 1967b; Kao *et al.* 1970). As with strain A1, Eriksson's mutant consistently carries a residium of about 10% foreign cells.

IN CONCLUSION

The question inevitably arises whether or not preservation and perpetuation of valuable genotypes can be achieved by the culture techniques at present at our disposal. From past experience, the answer would appear to be in the negative; plant cell breeders must therefore seek alternative methods, as for instance, deep-freezing of cultures. Latta (1971) has shown that carrot cells grown in a conventional medium can survive freezing and short periods at temperatures as low as $-196°C$ in the presence of protecting agents such as glycerol. Upon thawing, vigorous growth is soon resumed. Cytological effects of such treatments have still to be assessed. However, considering the vast range of plant species which have not yet been examined seriously in culture an emphatic negative answer to the question would seem premature. Moreover, we know nothing as yet about the cytological effects of growing plant cells by some of the new procedures under development such as the continuous culture systems described in Chapter 3. Perhaps too much emphasis has been given in the past to the variable nature of cultured cells and the diverse behaviour of explants from the same plant. This preoccupation has tended to divert attention from the equally important fact that some cultures in a series do remain stable without any visible chromosomal change (though there may have been genetic change). In Sacristán's (1971) experiments on haploid and diploid plants of *Crepis*, 34% of the subcultures derived from the haploid isolates, and 79% from the diploid, maintained the original ploidy level for more than a year. Earlier, Reinert & Küster (1966) reported no change in chromosome number after 1 year in a culture derived from a diploid plant of the same species. Tommerup & Butcher (unpublished results) have also found a stable situation in cultures derived from the stem cortex of normal and crown-gall-infected plants of *Helianthus annuus*, another non-polysomatic species. Both series maintained the diploid number over a 9-month experimental period, and never more than 5% of the mitoses examined, in any isolate, showed the tetraploid configuration. Over relatively short-term maintenance periods, therefore, there is hope of achieving stability in certain species. Thus while chromosomal mutants are useful in many types of study, and the breeding of new cell-types a promising approach to the solution of others, there is also every reason for continuing the search for conditions whereby original genomes can be preserved and chromosomal instability minimized. Control is essential in certain problems, among which prevention of polyploidization in haploid cell systems is perhaps one of the most pressing.

CHAPTER 8

SINGLE-CELL CLONES

H. E. STREET

Introduction 191
The special growth requirements of isolated cells 191
Derivation of single-cell clones from plating 198
Differences observed between clones of single-cell origin 199
Techniques designed to assist selection of desired variants 202
Stability of clones 204

INTRODUCTION

The paper raft nurse technique (Chapter 1, p. 9), petri dish plating (Chapter 4, p. 95) and the growth of isolated cells in microchambers (Chapter 4, p. 97) have been used to obtain single cell clones. The isolation of such clones from established callus and suspension cultures has shown the heterogeneity of such cultures; the single-cell clones have differed in growth, texture, colour and morphogenetic potential from the parent culture and from one another. These studies have also drawn attention to the more exacting requirements for growth of isolated single cells compared with the requirements for growth of callus and cell suspension cultures.

Studies of the requirements for growth of isolated plant cells not only advance our knowledge of cell physiology but are essential if the full extent of variation in cultured cells is to be exposed. The recognition of the need for a minimum inoculum density for the successful initiation of both callus and suspension cultures and the exposure of the nurse effect of both callus tissue masses and high density cell suspensions on single cells (Chapter 4, p. 87) point to a mutually beneficial interaction between cultured cells; a dependence of each cell upon the cell population of which it is a constituent. The exact nature of this population dependence cannot at present be defined, although some progress has been made towards the identification of its components.

THE SPECIAL GROWTH REQUIREMENTS OF ISOLATED CELLS

The paper raft nurse technique shows that the nurse callus supplies by diffusion through the paper barrier not only all the essential nutrients of the medium capable of supporting the growth of the callus mass but those extra factors necessary to induce division in the cell beyond the intervening-paper barrier. This diffusion is also demonstrated when callus masses are implanted on petri dish plates seeded with isolated cells; cells begin division first in the region adjacent to the nurse callus and then the front of division initiation gradually moves outwards from the nurse (Fig. 8.1). The fact that a liquid medium which has supported the growth of a callus mass or cell suspension

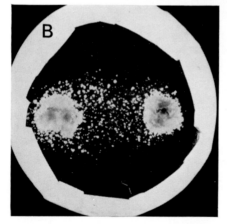 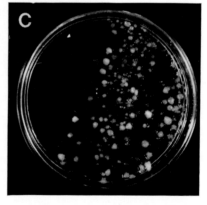

FIG. 8.1. *The nurse effect on the growth of isolated cells.*
A. Growth of single cells on upper surface of filter paper in contact with the 'nurse' callus (paper raft technique). From Muir (1953).
B. Bergmann type (1960) plate seeded with a suspension of *Acer pseudoplatanus* cells and two pieces of a callus culture of the same species. Note colonies arising from the suspension culture cells near to and in the zone between the two 'nurse' callus pieces.
C. Bergmann type (1960) plate seeded with a suspension of *A. pseudoplatanus* cells. Colony formation was first noticed on the right-hand side of the plate and then colony appearance spread to the left. The first formed colonies act as 'nurse' colonies to other cells.

for an appropriate time is then able to initiate division in a culture initiated below the critical initial cell density for new medium indicates that it has become enriched (conditioned) by the release into it of beneficial metabolites by the growing culture (Stuart & Street 1969). The success of the micro-chamber technique (Jones, Hildebrandt, Riker & Wu 1960) can similarly be interpreted as a consequence of the individual cell being able to effectively condition the very small volume of medium in which it is suspended.

The work of Duhamet (1957) and of Benbadis (1968) shows that medium conditioned by one tissue can be promotive to the growth of a low density culture of a different tissue. Such evidence of 'cross-conditioning' suggests that the metabolites whose endogenous levels are critical to the initiation of cell division may be the same in cells of different species.

The concept which emerges from these considerations is that the minimal medium for the growth of cells from a high population density is simpler than that for their growth from a lower population density and this again simpler than that for the growth of a single cell. Further, it can be postulated that this is not due to any division of labour within the cell population (reciprocal interchange of different essential metabolites) but due to the 'leaky' nature of cultured cells. Each cell is presumed capable, in the minimal medium, of synthesizing all the metabolites essential for its growth and division but the required endogenous concentrations of certain critical metabolites have to be established under conditions where they are continuously being lost to the bathing medium. Only, therefore, when the concentrations of these meta-bolites *external* to the cell reach appropriate levels which slow down the net efflux is the cell by its own biosynthetic activity able to establish the required intracellular concentrations. This situation comes about rapidly at a high population density (not only by 'leakage' but perhaps also by a proportion of cell death); it cannot ever occur if a single cell is in an infinite volume of medium. Therefore, at very low population densities growth will only be initiated if the minimal medium is either (i) supplemented by conditioning or by addition of the required metabolites at appropriate concentrations (Fig. 8.2) or (ii) altered in some way which either (a) enhances the rates of synthesis by the cells of these metabolites or (b) reduces the efflux of these metabolites from the cells (such modified media have been called 'Synthetic conditioned media'—Stuart & Street 1971).

Such a hypothesis is not at variance with the observations (Mehta 1963, Blakely 1964, Blakely & Steward 1964c, Earle & Torrey 1965) that division occurs first and with greater frequency in cell aggregates than in the free cells of a plated suspension. The aggregates can be regarded as sites of high cell density or as 'cellular units' with a lower ratio of surface area to volume than the separate cells.

Bearing in mind that it is possible to separate in time the conditioning process from the demonstration of enhanced growth-promoting activity of 'conditioned' medium (Stuart & Street 1969) some at least of the changes involved in conditioning must be stable changes. The observations that plating efficiency (Chapter 4, p. 96) can be increased by adjustment of the pH of culture medium and by raising the level of iron (Earles & Torrey 1965, Stuart & Street 1969) suggest that factors affecting cell permeability and the ionic status of the tissue may be important in conditioning. Attention has also been drawn to the possible importance of the ionic status of cells for the

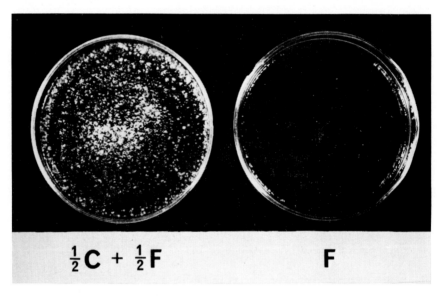

$\frac{1}{2}$C + $\frac{1}{2}$F   F

Fig. 8.2. *Demonstration of the growth-promoting activity of 'conditioned' medium upon colony formation from* Acer pseudoplatanus *cells*. The petri dish F contains new standard medium solidified with 0·6% agar. The petri dish $\frac{1}{2}$C+$\frac{1}{2}$F contains medium prepared by mixing equal volumes of sterile new medium (1·2% agar) and filter-sterilized 'conditioned' medium. Note absence, after 21 days' incubation, of visible colonies in F and in contrast numerous colonies on $\frac{1}{2}$C+$\frac{1}{2}$F. After Street (1968a).

expression of their biosynthetic potential by the work of Braun & Wood (1962) and Wood & Braun (1961) on normal and tumour tissues of *Catharanthus roseus* (*Vinca rosea*). The normal callus of this species requires for rapid growth a medium containing sugar, White's inorganic salts, White's vitamins, plus NAA, kinetin, asparagine, *myo*-inositol, cytidylic acid and guanylic acid. The fully transformed crown-gall tissue grows rapidly on White's basic medium without any of the six growth-factors listed above. The growth rate of the normal tissue can be raised to the growth rate of the fully-transformed tissue in a medium containing only three growth factors (NAA, kinetin and *myo*-inositol) provided that the concentrations of three of the salts (KCl, $NaNO_3$ and NaH_2PO_4) of White's medium are substantially increased; in this medium the normal tissue will grow slower but is capable of subculture without *myo*-inositol and can be shown to synthesize this growth factor. By adding an appropriate level of $(NH_4)_2SO_4$ in addition to the enhanced levels of the three salts listed the normal tissue can be grown in a medium containing only kinetin and *myo*-inositol (i.e. with the auxin omitted). These observations have been interpreted as showing the activation in the normal tissue of growth factor synthesizing systems by appropriate concentrations of inorganic ions (see Chapter 13, p. 377). The possibility that plating efficiencies could be improved by paying greater attention to the ionic composition of the culture medium should, therefore, be investigated in more detail.

As might be anticipated the highest plating efficiencies on Bergmann plates have been obtained with crown-gall tissues (Reinert, 1963a) or with normal tissues capable of growth in very simple media (Bergmann 1960, Gibbs & Dougall 1963, Reinert & von Ardenner 1964). Ability to form colonies when plated at low densities will depend both upon biosynthetic activity and efflux permeability to essential metabolites. Enhanced plating efficiency resulting from the use of minimal media supplemented with selected metabolites has been reported by a number of workers (Gibbs & Dougall 1963, Earle & Torrey 1965, Stuart & Street 1971). Thus for the growth of *Convolvulus* cells on plates inoculated at low density it is necessary to add a cytokinin and amino acids (neither being essential for normal propagation of the culture) and to adjust carefully the auxin level (Earle & Torrey 1965). Similarly for growth of sycamore cells at low plating density a medium containing a cytokinin, gibberellic acid and amino acids (corresponding to those detectable in conditioned medium) is required though none of these are included in the minimal medium satisfactory for normal propagation of the culture (Stuart & Street 1971).

When working with cell suspension cultures of sycamore the above-supplemented medium, carefully adjusted to pH 6·4, reduces the minimum effective initial cell density from $9-15 \times 10^3$ cells ml^{-1} (for the standard medium) to 2×10^3 cells ml^{-1}. The most active conditioned medium so far prepared has a minimum initial cell density of $1·0-1·5 \times 10^3$ cells ml^{-1}. Such media, although significantly more growth-promoting than minimal medium, clearly fall short of meeting the growth requirements of a single isolated sycamore cell. The failure of the conditioned medium (whether prepared by diffusion across a No. 3 sinter or a dialysis membrane—see Chapter 4, Fig. 4.18, p. 88) suggests that some unstable product(s) (a product either rapidly decomposing in aqueous solution or a volatile product) is required to initiate cell division in sycamore cells. Evidence for the involvement of a volatile factor (or factors) was obtained by Stuart & Street (1971). By using a special culture vessel (Fig. 4.3, p. 65) in which the low density culture in conditioned medium was exposed to the culture atmosphere of an actively growing high density culture, the minimum effective initial cell density could certainly be dropped to 600 cells ml^{-1} (Fig. 8.3). This beneficial effect

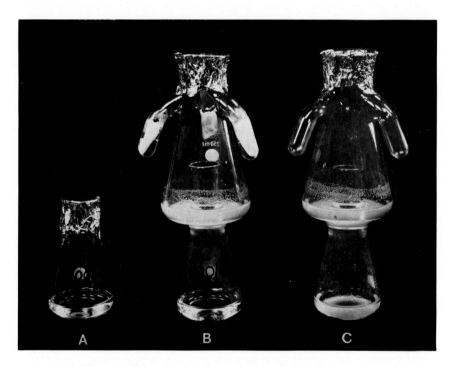

FIG. 8.3. *Evidence that CO_2 concentration is a factor in conditioning.* Cultures A and the lower tier flasks of B and C containing 20-ml conditioned medium and an initial density of 600 cells ml^{-1}. The upper tier flasks of B and C containing actively growing high density cultures. The side arms of B containing a CO_2 absorbent (40% KOH). All cultures after 17 days' incubation. Only in C is there visible growth of the culture at low density. After Street, King & Mansfield (1971).

was eliminated when a CO_2 absorbent was included in the culture atmosphere suggesting that a critical partial pressure of CO_2 was important for the initiation of growth in a low density culture. Subsequent work (unpublished work of K.J.Mansfield) has shown both in experiments with low density cell suspensions and low density platings that enhancing the CO_2 concentration up to 1% v/v promotes growth (concentrations above 2% being inhibitory) (Table 8.1). This growth-promotion can be further increased by simultaneous application of a low concentration (2·5 ppm) of ethylene. However, it has not been possible to fully reproduce the promotive effect of the volatile(s) from an actively growing culture by this adjustment of the culture atmosphere. Some other volatile may therefore also be involved.

Nearly all workers who have carried out extensive plating experiments have encountered high and inexplicable variability even when using cells harvested at a known point in the growth cycle of the parent culture, a synthetic medium and a carefully standardized plating technique (Earle & Torrey 1965). The recognition that the atmosphere above the culture must also be standardized may not only in future improve plating efficiency at low densities but reduce this frustrating occasion-to-occasion variation previously encountered in plating experiments. It is, however, at present not possible

TABLE 8.1. The plating efficiency in the presence and absence of carbon dioxide of *Acer pseudoplatanus* cells plated at low density (previously unpublished data by K.J.Mansfield)

Initial density cells ml^{-1}	No CO_2	1% CO_2
500	0	110%
1000	42%	95%
2000	60%	101%

to say whether further aspects of the special growth requirements of single cells have yet to be identified. The plating technique has so far been critically examined for only a very few plant cultures.

The possibility that significant differences in growth of colonies may be obtained with different samples of agar or by using alternative solidifying agents (biogel) cannot be ruled out but in our experience there seems no good evidence that bacteriological agar used at 0·6–0·8% has any deleterious effect on cell growth. Similarly the incorporation of the cells into the body of the thin film of agar in the Bergmann plates does not seem to be deleterious; we have obtained less satisfactory and more variable results from distributing the cells in a liquid film over the agar surface. It is important to prevent drying out of the agar surface and this has usually been achieved by sealing the dishes with sticking plaster or parafilm. Regulation of the gaseous atmosphere is, however, perhaps better achieved by enclosing the dishes in a larger container in which the gaseous composition can be controlled and maintained saturated with water vapour.

At present the most promising possibilities for raising plating efficiency at very low plating densities are further research into the sensitivity of low population densities to auxins and other known growth regulators, study of the influence of ionic salt concentrations, and further analysis of the atmosphere of actively growing cultures.

DERIVATION OF SINGLE-CELL CLONES FROM PLATING

The recognition that cell suspension cultures as at present available contain not only free cells but cell aggregates and that the aggregates have a higher growth expectation on plating than the free cells raises the problem that colonies developing on plates are not necessarily of single-cell origin. This problem is aggravated because highest plating efficiencies are obtained by plating from exponentially growing cultures and such cultures are more aggregated than at stationary phase (Henshaw, Jha, Mehta, Shakeshaft & Street 1966). Of course at least some colonies arising from plated aggregates may be single-cell clones since the aggregates in the suspension may contain only cells derived from a common mother cell. However, it would clearly be highly advantageous not only for studies on growth control and metabolism in cell suspension cultures but for single-cell cloning if growing suspension cultures could be obtained which consisted predominantly of single cells and if aggregation was limited to the transient formation of aggregates of very few cells (two to four cells). The solution of this problem would seem to be the isolation of variants differing from the normal in some specific metabolic lesion affecting middle lamella formation. Meanwhile significant improvement in the dispersion of cultured suspension can be obtained by incorporating low levels of cell-wall degrading enzymes into the culture medium. This approach borrows from observations made in the pioneer studies on the release of protoplasts following cell-wall lysis by these enzymes (see Chapter 5, p. 100). Studies with sycamore cell suspensions have shown that concentrations of macerozyme and cellulase (manufactured by All Japan Biochemicals) which do not significantly inhibit growth of the suspension can greatly increase the degree of cell separation. The most effective treatment involves incorporation of both enzymes, and the most effective non-toxic level of each enzyme must be experimentally determined for each batch of enzyme received. Although purified fractions obtained from these enzymes are effective at much lower concentrations than the crude enzymes, nevertheless there is no evidence that the crude enzymes are more toxic provided that at each subculture the enzyme-treated cells are washed free of 'spent' medium (which presumably contains degradation products of or impurities from the enzymes). As the concentration of the mixed enzymes is increased more effective cell separation occurs but so also does the proportion of cells undergoing destruction. It seems that this destruction is in part a result of cell bursting due to wall softening and enhanced separation without loss of cells can be obtained by incorporating 8% sorbitol into the medium to raise its osmotic potential (Fig. 8.4). The effectiveness of these treatments in increasing cell dispersion in sycamore suspensions is shown in Fig. 8.5. In the most successful treatments there are never aggregates of more than ten cells and by the beginning of stationary phase (when the suspension shows good plating efficiency) there are no aggregates of more than four cells and the culture consists predominantly of single cells and cell pairs. It is not certain that such suspensions can be repeatedly subcultured in presence of these enzymes without reduction of growth potential and increase in aggregation but

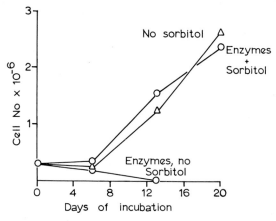

FIG. 8.4. *The protective action of sorbitol (8%) on* Acer pseudoplatanus *cells cultured in presence of the enzymes, macerozyme (0·05%) and cellulase (0·1%).* Progress of growth over 20 days' incubation followed by cell counting. Standard medium labelled 'no sorbitol'. Previously unpublished data of K.J.Mansfield.

cultures have been carried successfully through three passages of 28 days. The first passage cultures grown in presence of 0·05% macerozyme and 0·1% cellulase had unimpaired plating efficiency (plating efficiency of 14-day-old cells 78% at 5×10^3 cells ml^{-1}). Undoubtedly a much higher proportion of clones of single-cell origin will be obtained by using these more dispersed suspensions.

DIFFERENCES OBSERVED BETWEEN CLONES OF SINGLE-CELL ORIGIN

Since the pioneer studies of Muir (Muir, Hildebrandt & Riker 1958) with the paper raft technique, a number of workers using this method, or plating or growth in microchambers have isolated and described the characters of clones of presumed single-cell origin. The clones originally isolated by Muir, Hildebrandt & Riker (1958) from a crown-gall culture of marigold were described as differing in surface appearance, extent of friability and growth rate on different media. In 1961 Sievert, Hildebrandt, Burris and Riker

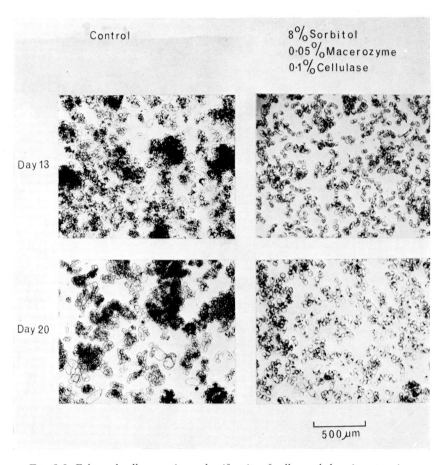

FIG. 8.5. *Enhanced cell separation and uniformity of cell morphology in suspensions of* Acer pseudoplatanus *cultured in presence of enzymes.* Previously unpublished photographs by K.J.Mansfield.

announced the isolation of single-cell clones and, after these had been several months in culture, of further secondary single-cell clones using as a source their primary single-cell clones. This work involved a culture of the hybrid *Nicotiana tabacum* ♀ × *N. glutinosa* ♂ and the paper raft technique. Their results were described in more detail by Sievert & Hildebrandt (1965). From two primary single-cell clones, fifty-four secondary single-cell clones were isolated. These secondary single-cell clones could be differentiated by using media from which particular growth factors (NAA, 2,4-D, or coconut milk) were omitted or where glucose was replaced by other sugars. For instance, only eight of the fifty-four clones showed any growth in galactose and only two of these showed good growth with this carbon source. These workers concluded that within several months the cells of the primary single-cell clones were no longer identical and considered that the variation arising so rapidly demanded more intensive cytological and nutritional study. The possibility of isolating clones differing in their ability to use different sugars and having different sugar concentration optima for growth was also indicated by studies on single-cell clones of both normal and *Phylloxera* gall tissues derived from grape (Arya, Hildebrandt & Riker 1962). Again from Hildebrandt's laboratory (Vasil & Hildebrandt 1965) it has been reported that single-cell clones derived from a recently initiated culture of the hybrid tobacco can differ in the ease with which they give rise to plantlets by shoot bud initiation; the plants obtained, however, have apparently not been scored for phenotypic variation. Similar variation in organ-forming potential of single-cell clones was reported by Murashige & Nakano (1965) in their studies with *Nicotiana tabacum* L. 'Wisconsin 38'. These workers also obtained evidence that single-cell clones initially totipotent lost their capacity for shoot and root initiation on continuing subculture. The totipotent single-cell clones when first isolated were either diploid or tetraploid; they became highly variable aneuploids as they lost their morphogenetic potential (Fox 1963). Partanen, Sussex & Steeves (1955) in their studies on cultured fern prothalli obtained at first a haploid prothallus showing normal organ differentiation. Later the callus became tumour-like, had little regenerative capacity and its chromosome number was increased to a state of variable aneuploidy in the 3–4n range. The studies of Earle & Torrey (1965) strongly indicate that cultures of *Convolvulus* root origin may contain a mixed cell population from the time of initiation. Recently initiated callus cultures were dispersed in a few minutes in a rotary shaker in liquid medium and the suspensions so obtained were immediately filtered (the filtered suspension containing 50% free cells and 91% as free cells plus aggregates containing less than six cells) and plated. Clones were isolated differing markedly in the extent to which they formed tracheids and in their ability to develop shoot buds.

Variants differing in definable metabolic differences have also been obtained by plating though in a number of these cases it is very unlikely that they were of single-cell origin. Strains of *Melilotus* and *Opuntia* calli differing in pigmentation and response to gibberellins were reported by Nickell & Tulecke (1959). Fox (1963) reported the isolation of strains of tobacco callus differing from 'wild' type; one strain had no requirement for exogenous auxin, a second strain required neither auxin nor cytokinin. Lescure (1970) treated a sycamore suspension with nitrosoguanidine (400 μg ml^{-1}) and by subsequent plating isolated two strains capable of growth in absence of 2,4-D; one strain was friable and developed very little chlorophyll, the other was compact and was much greener. Evidence was obtained that IAA oxidase was three times more active in the parent culture than in the friable strain with no auxin requirement. Blakely & Steward (1964c) isolated from carrot suspension cultures a friable strain and strains more resistant to acriflavin than the parent strain. Ericksson (1967b), following ultra-violet irradiation, of a suspension culture of *Haplopappus gracilis* isolated strains with very greatly increased anthocyanin content. Davey, Fowler & Street (1971) obtained three clones of *Atropa belladonna* var. *lutea* differing in growth rate, nutrient requirements, cellular fine structure and chloroplast pigment content when grown as cell suspensions.

Cloning techniques have, therefore, exposed a considerable range of variation and provided evidence that such variation arises spontaneously during culture. There is no convincing evidence so far that this variation is significantly increased by application of mutagenic treatments. No systematic work has been undertaken to isolate a particular class of variants and to characterize the biochemical differences between such variants. Many of the variants appear to differ in karyotype from one another suggesting that they differ in the activity of a number of genes.

There is an urgent need to examine how far cultural conditions can be modified to increase the genetic stability of cultured diploid cells.

TECHNIQUES DESIGNED TO ASSIST SELECTION OF DESIRED VARIANTS

Since the majority of mutations are likely to be deleterious and thus render the cells more exacting and less resistant to deviations from optimum, they may be expected to be at a selective disadvantage *vis-à-vis* the parent cells cultured in minimal medium. Such mutant cells will therefore be eliminated rapidly or only survive while the parent cells act as fully effective feeder-cells to them. Any systematic programme designed to isolate such mutants will need to include mutagenesis followed rapidly by an appropriate selection procedure. If the mutation renders the affected cell incapable of growth and DNA synthesis on minimal medium but does not prevent its survival under starvation conditions it could be placed under a selective advantage by a treatment lethal only to actively growing cells. Such a treatment, which has been successfully used in isolating deficiency mutants from animal cells, is treatment with bromodeoxyuridine (BUdR) followed immediately by exposure to light (Puck & Kao 1967, Puck 1971). Cells synthesizing DNA incorporate BUdR at sites where thymidine would normally be incorporated and light is lethal to such cells. Theoretically such a treatment should be capable of eliminating cells unimpaired and capable of growth on the minimal medium whereas if the mutants could not enter S-phase on this medium they would survive. When such a suspension, freed from added BUdR, is subsequently plated out on a more complex medium (now meeting the growth requirements of the mutant(s)) mutant colonies would develop and could be distinguished from any surviving normal colonies by subculture from each colony to minimal and complex medium. A particular class of mutant could be sought by selective supplementation of the minimal medium (e.g. amino-acid requiring mutants by using a medium supplemented by a balanced mixture of amino acids). The possibility that this approach might be effective with plant cells is supported by the successful isolation of auxotrophic mutants from spores of *Todea barbara* and from a haploid clone of *Nicotiana tabacum* using BUdR (Carlson 1969, 1970).

The account, given above, of some of the variants which have already been isolated from plant cultures indicates that another approach is to plate out cell suspensions (either with or without previous treatment with a mutagen) onto a medium incapable of supporting the growth of the parent culture because it lacks an essential nutrient or growth factor or contains an effective dose of a growth inhibitor (a chemical inhibitor or an inhibitory physical factor such as high temperature or an inhibitory light intensity). Further experimentation along these lines is urgently needed. A basically similar approach to this, and one which might enable natural variants arising over a period of time to be culled is the establishment of conditions unfavourable but not fully inhibitory or lethal to the normal cells. This implies growth over a long period in presence of a sub-optimal level of say an essential growth factor or a partially inhibitory concentration of an anti-metabolite. Any cell developing an enhanced capacity for endogenous synthesis or accumulation of the growth factor or an enhanced resistance to the anti-metabolite would then become at a selective advantage. A rise in growth rate of the culture under these conditions would signal the selective growth of the variant, and subsequent plating on media lacking the growth factor or containing the anti-metabolite at a level fully inhibitory to normal cells could lead to isolation of the variant uncontaminated with normal cells. This approach should now be tested using the continuous culture systems described in Chapter 4.

Haploid tissue cultures can be obtained via anther culture for a number of species (see Chapter 9), and cell suspensions have been prepared from such cultures and shown to be initially almost entirely haploid. Following mutagen treatment of such haploid suspensions, and selection of mutants, it should be possible to obtain homozygous mutants in the diploid condition by appropriate colchicine treatment.

STABILITY OF CLONES

Certain studies have shown that particular media may promote increase in the ploidy level of tissue cultures (Torrey 1959, 1961). However, all cultures even when propagated in a synthetic minimal medium appear to undergo change from an initially uniformly diploid condition to one in which a proportion of the cells are allopolyploid and aneuploid (see Chapter 7). Whether a more stable state of ploidy would occur in cultures derived from species where endopolyploidy does not occur during normal tissue differentiation (D'Amato 1965) has not been critically examined. However, even in such a case point mutations and chromosome mutations are likely to arise spontaneously during continuing growth in culture. Such mutants could quickly come to represent a significant element in the cell population particularly if the clone is maintained under a culture regime exerting a strong selection pressure towards some characteristic deviation from normal. Even under these conditions new mutations not affecting expression of the primary mutant character would survive.

With these considerations in mind great interest attaches to techniques of preserving clones in a non-growing condition. Caplin (1959) has succeeded in preserving cultures for long periods without loss of viability by covering them with a layer of mineral oil. More recently some success has been reported for freeze-preservation of cultured cells in presence of dimethyl sulphoxide (Quatrano 1968, Latta 1971). Further work along these lines is required to determine how widely applicable freeze-preservation is, how long cells can survive at very low temperature and to demonstrate that such preservation does not adversely affect the growth and morphogenetic potentiality of the clones.

CHAPTER 9

POLLEN AND ANTHER CULTURE

N. SUNDERLAND

Introduction 205
Microsporogenesis 206
Culture procedures 209
Culture media 213
Pollen age 219
Other parameters 221
Initial behaviour of the pollen 222
The vegetative cell 225
Ploidy status of pollen plants 234
Production of homozygous lines 235
Haploid mutants 236
Concluding remarks 238

INTRODUCTION

Unlike the somatic tissues of higher plants, pollen is produced in small discrete units. These are produced in relatively large numbers and in a readily accessible form, and consist of not more than three cells in the Angiosperms or five in the Gymnosperms. Each unit possesses a unique genome conferred upon it at meiosis by pairing and segregation of the chromosomes, and in the case of the pollen of a true diploid species every gene is present as a single copy. Exploitation of this unique genetic potential is the main purpose of pollen culture; to produce on the one hand, clones of cells, and, on the other, whole plants, carrying single-copy genetic information. Both these products have special relevance to mutation research, and haploid plants have many uses in plant breeding.

The first successful cultures were carried out in the 1950s by Tulecke who demonstrated that mature pollen grains of certain Gymnosperms could be switched from their normal role into callus formation by culturing them in an appropriate medium (*pollen culture*). Not all the grains responded in this way; some showed other forms of behaviour of relatively little significance, while many continued their normal determinate growth pattern and gave rise to pollen tubes and sperms. Attempts by Tulecke and others to stimulate Angiosperm pollens in a similar manner were unsuccessful; whenever growth occurred it was always in the direction of pollen tube formation. In 1964, Guha and Maheshwari discovered that the pollen of an Angiosperm, *Datura innoxia*, could be triggered into active growth by the simple expedient of culturing it undisturbed within the anther (*anther culture*), a discovery which was all the more remarkable because growth was organized and led to the formation of haploid plants (Guha & Maheshwari 1967). One pollen grain gave rise to one plant. Since then conditions have been established for the induction of growth in other Angiosperm pollens, and it is with these recent developments that this chapter is largely concerned. Reference is made, however, to those features of the early work on Gymnosperms, in which it parallels, or contrasts with, the more recent work on Angiosperms.

MICROSPOROGENESIS

Pollen grains (which represent the male gametophytic generation in higher plants) are products of the process called *microsporogenesis*. The term will be used here to cover the entire developmental sequence from spore mother cell to mature pollen grain, and, to facilitate the discussion, the sequence will be divided into three phases each characterized by a distinctive series of cellular events. Phase 1 is defined as that part of the sequence concerned with meiosis and the formation of spore-tetrads; phase 2 is concerned with the dissociation of the tetrads and the development of individual microspores, and phase 3 with the maturation of the microspores into pollen grains. Phase 2 ends and phase 3 begins with the first of the cell divisions that lead to the formation of the gametophyte (see Fig. 9.1). In phase 2 therefore pollen consists of *unicellular microspores*, and in phase 3 of *multicellular gametophytes* or pollen grains. Since much of what follows hinges on the cellular features of phases 2 and 3, a brief description at this point may be found useful.

In *Nicotiana tabacum*, which can be considered as a typical representative of the Angiosperms, the tetrads of spores formed at meiosis dissociate from the callose matrix surrounding them, and at the beginning of phase 2, they are thin-walled spherical cells with a high nuclear/cytoplasmic ratio. Their diameter is roughly 15 μm, and that of the nucleus about 10 μm. At the end of phase 3, when they are shed from the anther as bicellular pollen grains, their diameter is roughly 40 μm. This enlargement occurs mainly by vacuolation in phase 2, but after division of the microspore nucleus, the vacuole disappears and further enlargement is accompanied by active cytoplasmic synthesis. Specialized wall-formation is a prominent feature of phase 2. Sporopollenin deposition continues on the external surface of the spores as they enlarge, and each one becomes enclosed in a tough, resistant coat (exine) within which a typical cellulosic wall (intine) is laid down (Fig. 9.2a, b) (Dunwell, unpublished). The spores become spindle-shaped at this stage. The vacuole pushes the nucleus to one end of the spore and in this position the nucleus undergoes DNA replication.

The mitosis which follows, usually referred to as the *microspore* or *first pollen mitosis* is asymmetric and at cytokinesis two unequal cells are formed (Sax & Edmonds 1933, La Cour 1949). This type of grain is hereafter referred to as type A. The cell-plate curves around one of the daughter-nuclei (the

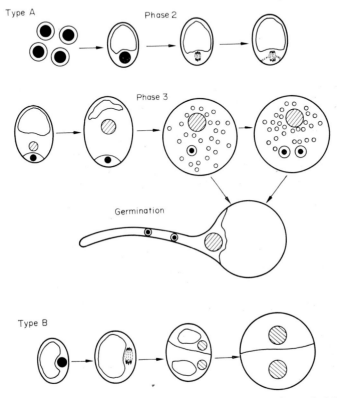

FIG. 9.1. *Developmental sequence in an Angiosperm pollen grain (type A) following its release from the tetrad as a uninucleate microspore.* Type B is an anomalous form.

future *generative* nucleus), and isolates it from the other (the future *vegetative* nucleus) with only a small portion of the microspore cytoplasm (Sanger & Jackson 1971, Vasart 1971a). The cell-plate fuses with the intine.

The generative and vegetative cells differentiate in different ways. The generative cell, which is destined to divide once again to produce the sperms, remains small and relatively inactive. Although the next division does not occur until after pollination, the nucleus appears to go rapidly into prophase and it remains suspended at this stage throughout the remainder of phase 3.

In this condensed form, the nucleus stains densely with the usual nucleic acid stains. Little if any cytoplasmic synthesis takes place in the generative cell; it is cut off with only a few mitochondria, an occasional plastid and golgi, and the organelle population remains sparse throughout the entire phase.

In contrast, the vegetative cell, which ultimately gives rise to the pollen tube, becomes metabolically highly active. Its nucleus enlarges into a diffuse, lightly staining body, which does not divide again. It contains a large well-defined nucleolus indicative of active transcription (Fig. 9.2b). As the nucleus enlarges, there is a rapid increase in the organelle population of the cell and the vacuole disappears. During this period of rapid cytoplasmic synthesis,

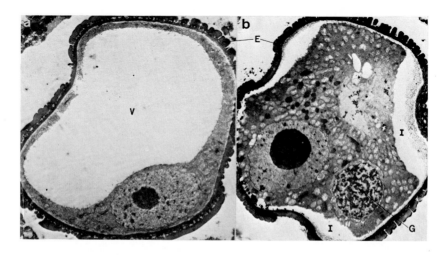

FIG. 9.2. *Electron micrographs of* (*a*) *a mid-phase 2 microspore, and* (*b*) *an early phase 3 pollen grain prior to starch-deposition in* Nicotiana tabacum *cv. White Burley.* E=exine; I=intine; V=vacuole; G=Generative cell (by courtesy of J.M.Dunwell).

the generative cell is detached from the intine and becomes suspended free inside the vegetative cell (Fig. 9.2b). Later, starch-deposition takes place in the plastids. In *Nicotiana* and many other Angiosperms the pollen grains are shed in the bicellular condition, though in some, the generative cell does complete its division before anthesis and the grains are then shed with two sperm cells suspended in the vegetative cytoplasm.

The differentiation of the vegetative nucleus into a diffuse, quiescent body is not a sign of degeneration as many of the earlier cytologists thought (Sax & Edmonds 1933, Brewbaker & Emery 1962). Its poor affinity for stains relative to that of the generative nucleus is due to a difference in DNA concentration brought about by condensation of the generative nucleus on the one hand, and enlargement of the vegetative nucleus on the other. DNA replication takes place in both nuclei, but at different rates. Species vary in the DNA content of the vegetative nucleus at anthesis. Woodard (1958) found that the DNA content of the vegetative nucleus in *Tradescantia paludosa* remained unchanged during phase 3, but previous to this Moses & Taylor (1955) reported levels varying from $1\cdot3$C to $1\cdot8$C (see Chapter 7) in the same species. In *Nicotiana tabacum*, both the generative and the vegetative nuclei contain the 2C level at anthesis and in *Hordeum vulgare*, which has a tri-cellular grain, the 2C level is attained in all three nuclei (D'Amato, Devreux & Scarascia Mugnozza 1965). Occasionally, the mechanism which causes asymmetry at the first pollen mitosis breaks down. The nucleus enters mitosis nearer to the centre of the microspore, and a normal symmetrical mitosis takes place (Fig. 9.1). This results in the formation of a pollen grain with two equal cells (La Cour 1949). Grains of this type, hereafter referred to as type B, are formed more frequently *in vitro* than *in vivo*.

In a typical Gymnosperm, such as *Ginkgo biloba* and *Ephedra foliata* more than one asymmetric division takes place during phase 3. The first two divisions give rise to small *prothallial* cells, the next to the generative and vegetative cells, and the generative cell may then undergo another division to form a *stalk* and a *body* cell. This latter division may occur either before or after the pollen is shed. Hence, there are four or five cells in the mature pollen grain. As in the Angiosperms, the bulk of the grain is occupied by the vegetative cell and it is rich in cytoplasmic organelles. The other cells remain small and protrude out into the cytoplasm of the vegetative cell. A detailed account of the ultrastructural aspects of microsporogenesis in the Gymnosperm, *Podocarpus macrophyllus*, will be found in the paper by Vasil & Aldrich (1970). Pollen of *Torreya nucifera* is atypical in being shed in the bicellular state; there are no prothallial cells (Tulecke & Sehgal 1963).

CULTURE PROCEDURES

Pollen culture

(a) *Gymnosperms*: Mature pollen grains of *Ginkgo biloba*, *Taxus brevifolia*(?) and *Torreya nucifera* can be triggered into active callus formation by simply distributing them on the surface of agar plates prepared with an appropriate medium (see Table 9.1). These pollens will also grow, as will that of *Ephedra foliata*, if undehisced microsporangia are plated on agar. The sporangia open and the pollen is shed on to the surface of the agar. Callus formation may occur from the shed pollen or from that remaining inside the sporangium. Stored pollen is more responsive than fresh (Tulecke 1960). To collect and store these pollens, undehisced sporangia are dissected from sterilized mature male cones, and placed in suitable sterilized containers. These are kept in a desiccator at about 4°C. The pollen is shed into the container and remains viable for several years if undisturbed (Tulecke 1954).

The resulting calluses consist of colourless parenchymatous cells which can be subcultured indefinitely. They are subject to polyploidization and other chromosomal aberrations as described in Chapter 7. They do not undergo morphogenesis. In both *Torreya* and *Taxus* the cells proliferate in an unusual manner. In the former, they proliferate by budding in a manner akin to that of yeasts, while in the latter the cells divide by asymmetric divisions similar to that of the microspore. The cells in *Taxus* are highly elongate. Prior to division the cytoplasm and nucleus accumulate at one end of the cell. Two unequal cells are produced and the smaller of the two continues growth and repeats the process. Branching also occurs and a callus is formed consisting of a dense reticulum of interwoven hyphal-like cells reminiscent of a fungal mycelium. Chlorophyllous calluses may arise occasionally in cultures exposed to light (Tulecke 1959).

(b) *Angiosperms*: Conditions for triggering Angiosperm pollens into active growth by direct plating procedures have not yet been established. However, mature pollen of *Brassica oleracea* and of the hybrid *B. oleracea* × *B. albo-glabra* may be induced to form 'cell clusters' in hanging drop cultures (Kameya & Hinata 1970). A drop of culture medium inoculated with pollen (50–80 grains) is placed on a cover slip which is then inverted over the well of a cavity slide and sealed. The pollen has to be inoculated at a low temperature and the cultures aerated. The seal is best applied to the cover slip beforehand as a circle of molten vaseline; at the same time a column of vaseline is placed in the centre of the circle, and is made of such a length that it touches the bottom of the well when the cover slip is placed in position. The culture drop is inoculated in the space between the seal and central column of vaseline. Aeration is afforded by rotating the slide such that the hanging drop flows around the central column.

A supply of uncontaminated pollen can be obtained from many Angiosperms by enclosing flower buds, just prior to opening, in a sterilized paper or polythene bag. When the flowers open, a hole is cut in the bag, and stamens are removed aseptically into a sterilized container. Kameya and Hinata recommend sterilizing the inflorescence beforehand but in the present author's experience (largely on glasshouse-grown plants) this is unnecessary.

Pollen of *Lycopersicon esculentum* (cv. Rutgers) can be stimulated into colony formation by nurse-culture (Sharp, Raskin & Sommer 1972). Anthers of this cultivar are placed on the surface of an agar medium and covered with a small disc of filter-paper. A drop of pollen-suspension (0·5 ml) containing about 10 grains is pipetted on to each disc. With incubation at 25°C in light, colonies of green parenchymatous cells appear on the discs within about 28 days. Plating efficiencies of up to 60% may be obtained. Colony formation does not occur in cultures lacking the nurse-anther. In this instance, immature pollen is used; flower buds 10 mm in length are surface-sterilized, anthers removed aseptically and the pollen released by cutting open the anthers in a small volume of liquid culture medium.

Anther culture

The species in which anther culture has been successful are listed in Table 9.1. Unopened flower buds are surface-sterilized, sepals and petals removed, and the anthers plated immediately on to an appropriate agar medium. They may also be floated on the surface of liquid media, or supported on filter-paper bridges over liquid media. There is no evidence as yet that liquid media are in any way advantageous. Contact between dissecting instruments and the anthers should be avoided as much as possible, and anthers badly damaged during dissection should be rejected. Direct contact between anthers and sterilants is also undesirable. In species with large flowers and anthers, contact can be largely avoided by removing whole stamens by the filament; the stamens are placed horizontally on the medium and the filament carefully broken off.

Dissection of anthers undamaged from species with minute flower buds (e.g. *Asparagus*, *Brassica*) is a laborious and painstaking procedure and can only be achieved successfully with the aid of a good dissecting microscope and fine forceps. In such cases, it is recommended that only the perianth is removed and the rest of the bud inoculated with the stamens intact. Experiments on *Nicotiana tabacum* have shown that the presence of other parts of the flower (whether free or attached) in the culture vessel does not affect the response of the pollen (Fig. 9.3c). However, the inoculum must be placed so that the anthers come into *direct* contact with the medium. The pollen will not grow if the filament lifts the anther clear of the medium. Evidently, the stimulus to growth cannot be transmitted via the filament (Sunderland & Dunwell 1971).

Pollens of *Atropa*, *Datura*, *Lolium*, *Nicotiana*, *Petunia* and *Oryza* give rise to embryoids which develop through stages not unlike those of true zygotic embryos to form plantlets. In *Nicotiana tabacum*, the anthers open after 4–5 weeks at 25°C and the embryoids are revealed at about the cotyledon stage (Fig. 9.3a). When large numbers of plantlets are produced in one anther it is necessary to separate them very soon after the anther opens (as in Fig. 9.3b). At this early stage, the plantlets can be readily teased apart and transplanted individually in fresh culture medium (see Fig. 9.3e). When the plantlets have developed an adequate root system they may be potted in compost or a mixture of peat and sand (Fig. 9.3f). Agar adhering to the roots need not be removed completely. Freshly potted plantlets should be kept preferably in a mist propagator for a few days until the roots get established.

If the plantlets are not separated at an early stage, roots and hypocotyls fuse inside the open anther and a callus is formed. This may give rise to additional shoots. These shoots have longer and narrower leaves than plantlets produced directly from the pollen (Fig. 9.3e, f). Callus may also develop on plantlets whose roots are damaged during the teasing process. This callus

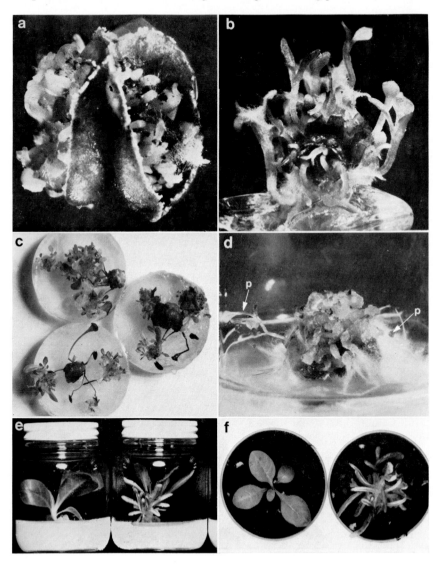

likewise undergoes organogenesis. At maturity, plants of callus origin are phenotypically indistinguishable from plants derived directly from the pollen.

Pollens of *Asparagus, Brassica, Hordeum, Lycopersicon* and *Solanum* give rise to callus before the anther opens. In all probability an embryoid is first formed but organized growth breaks down at an early stage. *Lolium* and *Oryza* also give rise to callus in certain media (see Table 9.1). Plants can be regenerated in all cases except *Lycopersicon*. In some instances, organogenesis takes place without a change of medium (as in *Nicotiana*), but in others, a change of medium is necessary.

CULTURE MEDIA

Culture media are made up of the same ingredients as are used in the culture of somatic tissues, and as with these it is the hormonal component of the medium which is critical for the initiation of growth. The composition of the basal medium (inorganic and organic constituents except hormones) is less critical. For culture of Angiosperm pollens the tonicity of the medium must be adjusted to prevent lysis of the pollen.

Hormones: For most species, both an auxin and a cytokinin is included in certain specific concentrations, though in a few instances, either one or the other suffices (see Table 9.1). In all cases the cytokinin can be replaced, usually more effectively, by coconut milk. There is no evidence to date that other plant hormones are involved (Nitsch 1970). *Nicotiana* species are exceptional in that growth occurs in the absence of exogenous hormones (Nitsch & Nitsch 1969, Nitsch 1970). Possibly, the endogenous levels in such anthers are present in a favourable balance for the initiation of growth.

By the use of the hormone components indicated in Table 9.1, it is possible

FIG. 9.3. *Anther culture of* Nicotiana tabacum *cv. White Burley*. a: Early phase 3 anther cultured without the filament for 28 days in H-medium at 25°C in continuous light; each embryoid has developed from one pollen grain. b: the same anther 7 days later; the plantlets can be readily teased apart at this stage and transplanted individually as in e. c: Stamens cultured intact with the ovary and style under the culture conditions of a; anthers in direct contact with the medium have produced plantlets as in b. Note callus formation from the ovary but no plantlet formation. Inclusion of the petals and sepals in the culture vessel does not affect the response of the anthers, but culture of intact buds is ineffective. d: Single anther cultured in N-medium + NAA 5·0 mg/litre at 25°C in darkness for 35 days, then in light for 14 days; anther tissues have formed callus from which numerous roots have developed. Two haploid plantlets (p) have grown out of the callus mass. e: Individual plantlets from b cultured for a further 3 weeks in H-medium lacking sucrose; callus producing shoots with unusually shaped leaves has developed from the hypocotyl region of the plantlet in the right-hand pot. f: the plantlets from e transplanted into a mixture of peat and sand, 1:1, and kept for 3 days in a mist propagator.

TABLE 9.1   Species successful in pollen and anther culture

Species	Basal medium	Growth component	Regeneration medium	Reference
		A. POLLEN CULTURE		
Brassica oleracea cv. Murasaki Kanran	N (mod)* 10–15% sucrose	CM 10%		Kameya & Hinata 1970
B. oleracea cv. Yoshin kanran × *B. alboglabra*				
Ephedra foliata	w (mod)	2,4-D 0·05		Konar 1963
Ginkgo biloba	w (mod)	CM 20%		Tulecke 1953, 1957
Lycopersicon esculentum cv. Rutgers	w (mod)	2,4-D 6·0		Sharp *et al.* 1972
Taxus brevifolia (?)	w (mod)	CM 15%+2,4-D 0·6		Tulecke 1959
Torreya nucifera	w (mod)	CM 18%+2,4-D 0·6		Tulecke & Sehgal 1963
		B. ANTHER CULTURE (embryoids)		
Atropa belladonna	LS	K 4·0+IAA 2·0		Zenkteler 1971
Datura innoxia	N	CM 15–30% or		Guha & Maheshwari 1964, 1966, 1967
D. stramonium		K 1×10^{-6} M		
D. metel	N (mod)	CM 15%		Narayanaswamy & Chandy 1971
D. meteloides	H minus vit.†			Nitsch 1972
D. muricata	H minus vit.			Nitsch 1972
Lolium multiflorum	LS 12% sucrose	CM 15%+NAA 1·0		Clapham 1971
Nicotiana alata 2n, 4n	H			Nitsch & Nitsch 1969
N. glutinosa 2n, 4n	H			Nitsch 1970, 1972

* (mod) indicates modified.   † vitamins omitted.

	Medium	Additions		Reference
N. rustica cv. Zlag and others	H			Nitsch 1970, 1972
N. sylvestris	H			Bourgin & Nitsch 1967
N. tabacum cv. Wisconsin 38, Maryland Mammoth and others	H			Bourgin & Nitsch 1967, Nitsch & Nitsch 1969
N. tabacum cv. Bright Yellow	Hildebrandt* C followed by LS	K 4·0+IAA 2·0		Nakata & Tanaka 1968
N. tabacum cv. White Burley and others	H			Sunderland & Wicks 1969, 1971, Noth & Abel 1971
N. tabacum cv. Java, Xanthi × Samsun and others	H			Melchers & Labib 1970
N. tabacum cv. Virginia Bright and others	H			Devreux et al 1971
N. tabacum cv. Hicks Broadleaf and others	H	IAA 0·1+K 0·1		Niizeki & Grant 1971
Oryza sativa indica	Blaydes (1966)	CM 15%+K 1·0+ IAA 2·0 and 2,4-D 2·0		Guha et al. 1970
Petunia hybrida / P. hybrida × P. axillaris	MS (Mod)	B 0·2–1·0+ NAA 0·1		Raquin & Pilet 1972

C. ANTHER CULTURE (callus)

	Medium	Additions		Reference
Asparagus officinalis	MS (mod)	B 1×10^{-6} M+ NAA $0·1 \times 10^{-6}$ M	omit NAA	Pelletier et al. 1972
Brassica oleracea cv. Murasaki Kanran and others / B. oleracea × B. alboglabra	N (mod)	CM 10% or K 1·6+2,4-D 1·0	no change or K 1·0+NAA 1·0	Kameya & Hinata 1970
Hordeum vulgare cv. Sabarlis	W (mod)+ sucrose 12%	CM 15%+NAA 1·0 or NAA 1·0		Clapham 1971

* Hildebrandt (1962).

TABLE 9.1 (*cont.*)

Species	Basal medium	Growth component	Regeneration medium	Reference
Lolium multiflorum		as for *Hordeum*		Clapham 1971
L. multiflorum ×		as for *Oryza* (Niizeki & Oono 1968)		Nitzsche 1970
Festuca arundinacea				
Lycopersicon esculentum cv. Yellow jubilee and others	w (mod) + sucrose up to 14%	K 0·13 + 2,4-D 0·32 or K 0·13 + NAA 1·0		Sharp & Dougall 1970 Sharp *et al.* 1971
Oryza sativa japonica cv. Norin 20 and others	MS	NAA 5×10^{-6} M or 2,4-D $1-5 \times 10^{-6}$ M	no change or K $1-2 \times 10^{-5}$ M + IAA 1×10^{-5} M	Niizeki & Oono 1971
O.s. japonica × *O.s. indica*	Blaydes (1966)	not known		Harn 1969, 1970
O.s. japonica Korean var.				
Solanum nigrum	MS (mod)	K 2·2 + 2,4-D 2·2 + NAA 1·9	omit auxins	Harn 1971, 1972

Quantities in mg/litre. CM=coconut milk; IAA=indol-3yl-acetic acid; NAA=naphthalene acetic acid; 2,4-D=2,4-dichlorophenoxyacetic acid; K=kinetin; B=6-benzylaminopurine. Formulae of N (Nitsch 1951) and H (Bourgin & Nitsch 1967) media given in text; those of MS (Murashige & Skoog 1962) and W (White 1963) on p. 39; LS=Linsmaier & Skoog (1965).

to trigger the growth of the pollen alone; the somatic tissues remain for the most part quiescent. Deviation from the recommended levels may result in the formation of callus (and possibly of organized growth) from the somatic tissues, and the pollen may then remain quiescent. This happens in *Datura innoxia* if auxins are included in the medium (Guha & Maheshwari 1967). Auxins have a similar effect in *Nicotiana tabacum* if added in high concentrations (NAA 5–10 mg/litre) though the development of the pollen may not be entirely suppressed (see Fig. 9.3d). In lower concentrations, the auxin has no effect on the somatic tissues but it tends to stimulate callus formation from the pollen either before or after the anther opens (Sunderland & Wicks 1971). Conditions have been found for triggering callus formation from somatic tissues in anthers of a wide range of species though the critical combinations for triggering growth solely of the pollen have still to be established (Konar & Nataraja 1965a, Sunderland, Wicks & Storey 1970, Fujii 1970, Sunderland, Dunwell & Lawes 1971, Niizeki & Grant 1971, Fowler, Hughes & Janick 1971, Abo el-Nil & Hildebrandt 1971a).

Basal Media: Three types of basal media are in use, all derived from formulae devised for the culture of somatic tissues. The original formulae have been modified in various ways presumably to meet the requirements of individual species, but it is not clear why the changes have been made; few are supported by quantitative data.

(1) Nitsch's (1951) medium (N) devised for the culture of zygotic embryos (*Datura, Brassica*). Formula: mg/litre Ca(NO$_3$).4H$_2$O 500, KNO$_3$ 125, MgSO$_4$.7H$_2$O 125, KH$_2$PO$_4$ 125, MnSO$_4$.4H$_2$O 3·0, ZnSO$_4$.7H$_2$O 0·5, H$_3$BO$_3$ 0·5, CuSO$_4$.5H$_2$O 0·025, Na$_2$MoO$_4$.2H$_2$O 0·025, Fe citrate 10, sucrose 20,000, agar 10,000 pH 6·0. It is used either alone (*Datura innoxia*) or with an organic supplement (each constituent in mg/litre): (a) thiamin HCl (T) 0·5, pyridoxin HCl (P) 0·5, nicotinic acid (NA) 5·0, glycine (G) 2·0 (*D. metel*), or (b) T 0·25, P 0·25, NA 1·25, G 7·5, yeast extract 2500 (*Brassica*). For pollen culture in *Brassica* the sucrose level must be raised to between 10 and 15%.

(2) White's (1963) general plant tissue culture medium (w) (*Ephedra, Ginkgo, Hordeum, Lolium, Lycopersicon, Taxus, Torreya*). Formula: see Tables 3.3, 3.4, p. 39. For *Ephedra*, the organic component is augmented by addition of a complex of amino acids (Reinert & White 1956), while for the other Gymnosperms the major change required is a tenfold increase in phosphate concentration. For *Lolium* and *Hordeum*, the minor minerals are replaced by those of Heller (1953) and the organic component increased to T 1·0, P 5·0, NA 5·0, G 20 and inositol 100 mg/litre. The sucrose concentration is also raised to 12% w/v though this is apparently optional (Clapham personal communication). For *Lycopersicon*, iron is supplied in the form and concentration recommended by Murashige and Skoog and the organic

H

component increased to T 1·0, P 1·0, NA 5·0, Ca pantothenate 0·1 and inositol 100 mg/litre. Concentrations of up to 14% of sucrose may be used in anther culture, but for nurse-cultures of individual grains 4% sucrose is recommended and a higher auxin concentration (Table 9.1).

(3) Murashige & Skoog's (1962) medium (MS) devised for the culture of somatic tissues of tobacco. Formula, see p. 39. It is used either unchanged (*Oryza, Solanum*) or in the Linsmaier & Skoog (1965) modification (LS) (*Atropa, Lolium*) or in the slightly less concentrated variants of Blaydes (1966) (*Oryza*) or Bourgin & Nitsch (1967) (*Nicotiana*). The latter variant (H) is highly successful for anther cultures of *Nicotiana* species. Formula: mg/litre KNO_3 950, NH_4NO_3 720, $MgSO_4.7H_2O$ 185, $CaCl_2$ 166, KH_2PO_4 68, $MnSO_4.4H_2O$ 25, H_3BO_3 10, $ZnSO_4.7H_2O$ 10, $Na_2MoO_4.2H_2O$ 0·25, $CuSO_4.5H_2O$ 0·025, $FeSO_4.7H_2O$ 27·8 mixed beforehand with Na_2EDTA 37·5, T 0·5, P 0·5, NA 5·0, G 2·0, folic acid 0·5, biotin 0·05, sucrose 20,000, agar 8000 pH 5·5. The organic component is optional (Nitsch 1972).

The bulk of the nutrients is supplied in order to sustain growth once it is under way. In *N. tabacum*, plantlets are not produced if iron is omitted from H-medium (Nitsch 1970). The iron concentration can be reduced fourfold without affecting the number of embryoids produced, but at this concentration growth of the embryoids is not maintained and few attain the plantlet stage (Sunderland 1971). Both N and W are low in iron relative to MS and H and this renders them less effective in anther culture of *N. tabacum*. With iron increased to the MS level, however, both N and W promote similar yields of plantlets to those promoted by either MS or H (Sunderland, unpublished). For certain cultivars of *N. tabacum*, iron-enriched W-medium is superior to H (Sharp *et al.* 1971). It has still to be demonstrated conclusively that high concentrations of many of the constituents as supplied under (3) are truly essential.

The following procedure is recommended for testing anthers of other species and cultivars. A range of cultivars and different genotypes should be tried wherever possible. A medium of relatively simple composition such as N should be used initially either alone or in combination with coconut milk. Cultures should be examined for dividing pollen or multicellular pollen grains after a relatively short period of say 14–21 days. If the pollen has not divided but is still healthy and stains well, and if callus is not developing from the somatic tissues (the cut end of the filament is especially prone to callus formation), another medium of a slightly more complex nature may be tried. By the use of this procedure, divisions have been observed in pollens of various species of *Paeonia*, cultivars of *Freesia* (Sunderland, Dunwell & Lawes 1971) and in *Capsicum annuum* and *Solanum tuberosum* (Sunderland & Dunwell, unpublished) though conditions have not yet been found to maintain growth beyond about the 8-celled stage.

Regeneration media: Media used for the regeneration of plants from pollen callus differ in composition from growth-initiating media usually in respect of the hormone component only; this is either omitted or its composition changed (Table 9.1). Guha *et al.* (1970) devised a medium for use with the *indica* varieties of *Oryza sativa* which permitted direct plantlet formation from the pollen without intervention of callus. Similar tests need to be made on the other species listed to discover whether callus formation can be avoided in these also. A simpler procedure might be to transfer the anthers from the growth-inducing medium to the regeneration medium at an early stage of culture probably just after growth has commenced.

POLLEN AGE

In addition to the hormone component of the medium the stage in microsporogenesis at which the pollen is inoculated may be critical. In testing different species therefore it is advisable to use sporangia of all ages. In Angiosperms with an indeterminate number of anthers per bud, e.g. *Anemone*, *Paeonia*, buds can be selected which contain several hundred anthers covering almost every stage in microsporogenesis. Different stages can probably be correlated with anther length. In species with a determinate number of anthers per bud, a series of buds is required to achieve the same range of stages. An age sequence can easily be assembled in species like *Nicotiana* by arranging buds in ascending order of petal length (not bud length), while in species like *Brassica oleracea*, which has a terminal inflorescence, the buds are already present in an age sequence starting with the youngest at the top and progressing downwards to the oldest at the base. The age sequence should ideally be taken from the same plant unless one is dealing with a clone.

In species having a determinate number of anthers per bud, the stage in microsporogenesis can be determined by removing one anther from each bud and examining the pollen after treatment with a suitable stain. Acetocarmine (4% w/v carmine refluxed for 24 hours in 50% v/v acetic acid and filtered) is a useful general stain, but with species possessing a low DNA content, it is recommended that the pollen should first be stained with Feulgen reagent. Whole anthers are fixed in acetic acid: ethanol, 1:3 v/v, for several hours at 4°C, taken down to water through a graded series of ethanol, hydrolysed in 5 N HCl for 1 hour at room temperature and stained in Feulgen reagent for at least 2 hours. The anthers are then squashed in acetocarmine. There is a slight asynchrony both between and within anthers of the same bud in *Nicotiana* (Sunderland & Wicks 1971).

A typical result obtained by culturing anthers of increasing age in *N. tabacum* is shown by the data of Fig. 9.4 (Sunderland & Dunwell 1971).

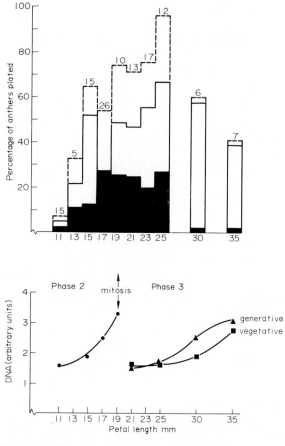

FIG. 9.4. *Change in anther response with increasing bud size in* Nicotiana tabacum *cv.* *White Burley.* H-Medium, 25°C, continuous light. Upper: solid histogram represents the proportion of anthers yielding abortive embryoids only; the open histogram represents the proportion of anthers yielding plantlets, and the dotted histogram the total proportion in which embryoid formation commenced. The numbers above the histogram represent the average number of plantlets per successful anther at each stage. Lower: average DNA values (50 grains per value) determined by Feulgen photometry.

When H-medium is used without hormones, anthers cultured during phase 1 do not respond. A few anthers give rise to plantlets when cultured at the beginning of phase 2, and the proportion increases thereafter with anther age. The maximum response is obtained with anthers cultured early in phase 3, that is to say, at the stage when the vegetative cell is in the process of rapid cytoplasmic synthesis. Thereafter the response declines. Such data are consistent with the view that the early stage of phase 3 is critical for embryogenesis—if this stage has been passed at the time of inoculation, embryogenesis can no longer take place. If this is so, the response obtained in older anthers must be due to grains retarded in their development, and which have not reached the critical point at the time of inoculation. The peak response in Fig. 9.4 coincides with the commencement of starch deposition in the pollen grains, and it has been suggested that plantlets cannot develop from the pollen when starch has been formed (Nitsch, Nitsch & Hamon, 1968). An alternative explanation is that the hormone balance in the anther tissues becomes unfavourable for growth as the anther matures, or that some other component essential for growth becomes depleted.

Other species have not been examined as intensively as *N. tabacum*. Results obtained so far suggest that the critical stage may vary from one species, and possibly cultivar, to another; it may also be influenced by the culture conditions and the environment under which the donor plants are grown. Nakata & Tanaka (1968) and Carlson (1970), for instance, using different cultivars of *N. tabacum* and different culture media, successfully raised plants from anthers cultured in late phase 1 when the pollen consisted mainly of tetrads. Nitsch & Nitsch (1969), on the other hand, found phase 2 anthers to be most favourable for a number of species of *Nicotiana*. Phase 2 anthers are also required in *Hordeum* (Clapham 1971) and *Oryza* (Niizeki & Oono 1971). In *Brassica oleracea*, only anthers taken from mature buds in late phase 3 give rise to pollen callus (Kameya & Hinata 1970). The latter findings are not consistent with the idea of a point of no return. *Brassica* pollen is shed in the tricellular condition (Brewbaker 1957), so the possibility cannot be ruled out that certain grains retain a potential for division even after the generative cell has divided to give the two sperm cells. The work on Gymnosperm pollens also indicates that the pollen can be triggered into growth when the grains are as far advanced in microsporogenesis as the 4- or 5-celled stage.

OTHER PARAMETERS

Effects of temperature and of light have not yet been critically examined; information is therefore fragmentary and restricted largely to *Nicotiana tabacum*. In general, high temperatures ranging from about 23° to 28° have been used. Growth-initiation is poor in cultures of *N. tabacum* incubated at 15°C; plantlet yields are accordingly poor. The response increases with increase in temperature, and reaches a maximum around 25°C (Sunderland 1971). Light is not essential for growth-initiation in *N. tabacum*, but it may be beneficial in post-initiation growth; slightly higher plantlet yields are ob-

tained in illuminated cultures (Sunderland 1971). Plant age is also an important parameter. In the cultivar White Burley, a succession of flowers is
produced over a period of about 4 weeks; anthers harvested and cultured at
the critical stage during the first week are generally more productive than
anthers treated later. At the end of the flowering period, anther response is
poor. Individual plants also vary considerably in anther response. Such
variability may be expected from a highly heterozygous species, but its
presence is an inbreeder like tobacco suggests that the factors concerned are
not genetical.

INITIAL BEHAVIOUR OF THE POLLEN

Angiosperms and Gymnosperms differ in the initial growth pattern of the
pollen. There are two basic forms which may be designated the Ginkgo and
Nicotiana types after the genera in which the patterns were first described.
These patterns are represented diagrammatically in Figs. 9.5 and 9.6.
Ginkgo type: Division commences in the vegetative cell. During the early stages

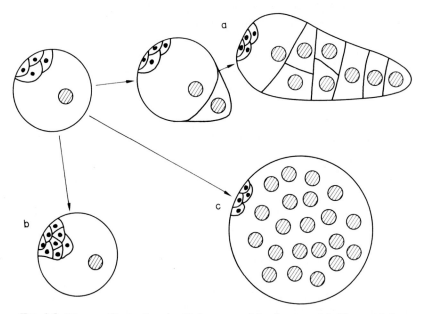

FIG. 9.5. *Diagram illustrating the Ginkgo type of development.* (a) The vegetative
cell divides to form a septate pollen tube. Also illustrated are two anomalous forms of
development; one in which there is a limited number of divisions in the remaining
gametophytic cells (b), and another in which the vegetative nucleus undergoes
repeated divisions without the formation of cell walls (c) (after Tulecke 1957).

of culture prior to division the exine is ruptured by swelling of a mucilaginous layer located between the exine and intine. The first division is accompanied by incipient tube formation (Fig. 9.5a). The next few divisions all take place in the transverse direction, and a uniseriate filament of cells is formed, or as Tulecke called it, a septate pollen tube. Divisions soon take place in different planes and the uniseriate state is lost. It is not clear whether the other cells

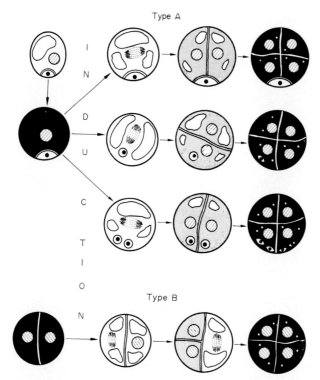

Fig. 9.6. *Diagram illustrating the Nicotiana type of development.* In type A grains the vegetative cell divides without tube formation. The generative cell may remain attached or become detached from the intine. At the first mitosis of the vegetative cell the grain is vacuolate and stains only lightly in acetocarmine. Rapid RNA and protein synthesis takes place immediately after the mitosis and two meristematic cells are formed. With succeeding divisions staining capacity of grains increases rapidly.

of the gametophyte may occasionally divide. Tulecke (1957) describes an anomalous form of development in which the vegetative cell remains quiescent, and the generative or body cells undergo a limited number of divisions (Fig. 9.5b). These grains do not form callus and presumably degenerate after the first few divisions.

Nicotiana type: The process of tube formation is suppressed. In type A grains, the embryoid is formed from the vegetative cell. The cell divides repeatedly within the framework of the exine to form a multicellular structure (Figs. 9.6 and 9.7d–f). During the early stages, divisions follow rapidly on each other with little or no cell enlargement, and as many as twenty to thirty cells come to occupy the volume of the mother cell (Fig. 9.7g). All derivatives possess

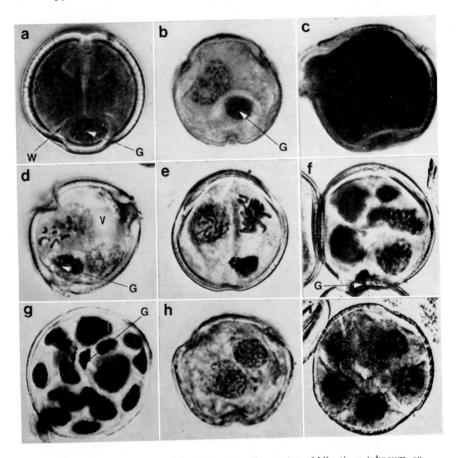

FIG. 9.7. *Stages in embryoid formation in pollen grains of* Nicotiana tabacum *cv. White Burley.* Culture conditions as in Fig. 9.3a. Grains a–d and h stained in aceto-carmine; grains e, f, g and i in Feulgen reagent. a: Early phase 3 grain, G = generative nucleus, W = cell-plate. b: Embryogenic grain after 9 days' culture; the vegetative nucleus is in prophase. c: Non-embryogenic grain from the same anther as (b). d: Vegetative nucleus in metaphase; note vacuole (V). e: Later stage: two vegetative derivatives in prophase, two generative derivatives, one in prophase. f–g: Multicellular embryoids each with one generative cell. h: Type B grain with both nuclei in prophase. i: Multicellular grain lacking generative derivatives and probably formed from a type B grain.

diffuse lightly staining nuclei characteristic of the mother cell. With continued division, the exine ruptures and the liberated embryoid continues its development free within the cavity of the sporangium (Sunderland & Wicks 1971). *Hordeum* follows a similar pattern except that, after rupture of the exine, organized growth breaks down and a callus is formed (Clapham 1971).

The generative cell may also divide, but the number of divisions is limited. Derivatives remain small and all possess highly condensed densely staining nuclei characteristic of the mother cell. The few cells formed do not contribute to the future embryoid or callus; they are probably lost when the exine ruptures. In *Hordeum* multicellular grains are formed with up to six generative daughter cells (Clapham 1971), and in *Nicotiana tabacum*, with up to four (Sunderland & Wicks 1971). Embryoids associated with one or two generative cells predominate in *N. tabacum*. The generative and vegetative cells divide independently though they may occasionally divide simultaneously. There is no constancy in the initial sequence; either cell may divide first, and one cell may divide more than once before the other begins (Sunderland & Wicks 1971).

In type B grains, either cell or both may contribute to the formation of the embryoid (Fig. 9.7h, i). These grains can be observed occasionally in anthers of *N. tabacum* cultured in early phase 3, when they are presumably formed by division of a retarded microspore. Type B grains are most frequently observed in anthers cultured at the beginning of phase 2.

The difference between the Ginkgo and Nicotiana types is probably due largely to the difference in pollen age at inoculation, and to the early rupture of the exine that takes place in the Gymnosperms. Konar (1963) describes two initial forms of behaviour in *Ephedra* pollen one of which results in the formation of a uniseriate filament of cells (Ginkgo type), whereas the other results in the formation of 'discoid multicellular masses'; this latter form may be analogous to the Nicotiana type. There has also been a suggestion of a Ginkgo type of development in *Datura* pollen (Narayanaswamy & Chandy 1971).

THE VEGETATIVE CELL

Cytochemical changes

The vegetative cell does not divide immediately. In *Nicotiana tabacum* there is a lag period which varies from about 6–12 days according to the temperature of incubation. This lag period is the important inductive period during which the switch mechanism comes into operation; it occurs whether the cell is formed in culture (phase 2 anthers) or before culture (phase 3 anthers). Induction is accompanied by a change in the staining properties of the cell

(Figs. 9.6 and 9.7) (Sunderland & Wicks 1969, 1971). In early phase 3 anthers (petal lengths, 20–25 mm, see Fig. 9.4), the vegetative cell stains uniformly and densely with acetocarmine because of the active cytoplasmic synthesis that is taking place. So far as can be judged by eye, the intensity of stain varies slightly between grains from the same anther but after about 24 hours in culture, all grains appear to be more or less equally-stained. In the light microscope, the grains appear to be non-vacuolate and the generative nucleus is the only structure that can be observed clearly (Fig. 9.7a). At the end of the inductive period, however, two classes of grain can be distinguished, one of which stains more densely than at inoculation and another which stains less densely (Fig. 9.7b, c). The densely staining class may develop starch and some may undergo pollen tube formation (see Fig. 9.11). This

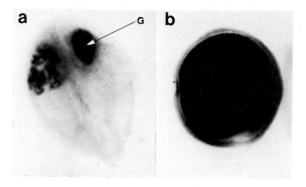

Fig. 9.8. *Embryogenic (a) and non-embryogenic (b) grains in* Nicotiana tabacum *stained by the gallocyanin-chrome alum technique for total nucleic acid.* G = generative nucleus. These grains are equivalent to those illustrated in Fig. 9.7b, c (by courtesy of S.S.Bhojwani).

class is non-embryogenic; division, if it occurs, is restricted to one division of the generative cell (second pollen mitosis). The lightly staining class is embryogenic. These grains are vacuolate, they lack starch and the vegetative nucleus can be clearly distinguished against the lightly stained background of the cytoplasm (Fig. 9.7b).

A similar picture emerges if the pollen is stained with specific stains and the intensity of stain estimated spectrophotometrically. With gallocyanin-chrome alum, for instance, which stains nucleic acids, two classes can again be distinguished towards the end of the inductive period; the cytoplasm is heavily and uniformly stained in one, but only lightly stained in the other (Fig. 9.8). These two classes are doubtless the same as those revealed by acetocarmine. Preliminary measurements indicate that there is about four times as much nucleic acid in the non-embryogenic class as there is in the embryogenic (Fig. 9.9) (Bho wani & Dunwell, unpublished). The total amount of nucleic acids in the embryogenic grains at the first mitosis of the vegetative nucleus is in fact about the same as in a microspore at the first pollen mitosis. In such microspores, the DNA level is 2C, but in the embryogenic grains since DNA replication occurs in both nuclei, the level may be as high as 4C. It follows therefore that there is less RNA in an embryogenic grain than in a late phase 2 microspore. An absolute stain value for RNA can be obtained by determining the DNA stain value separately on a series of RNAase-treated grains, and then substracting this value from that for total nucleic acid. When this is done, the RNA content of embryogenic grains at the first division of the vegetative cell is about three-quarters that of late phase 2 microspores.

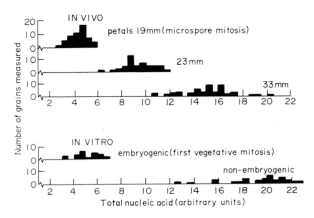

Fig. 9.9. *Total nucleic-acid contents of pollen grains in* Nicotiana tabacum *determined by gallocyanin photometry.* The average amount in embryogenic grains at the same stage as the one illustrated in Fig. 9.8a is lower than the average initial amount (bud with 23 mm petals). The amount in non-embryogenic grains from the same anther is about equal to that in mature grains at anthesis (by courtesy of S.S. Bhojwani).

The protein content of the grains can likewise be assessed by means of napthol yellow S, but in this instance, in order to ensure identification of the embryogenic class, the grains must also be stained with Feulgen reagent. With these double-stained grains, the protein value has to be measured off-peak to avoid interference by the Feulgen stain. This procedure again reveals two classes of grains, non-embryogenic with about fourfold the protein content of the embryogenic grains; again, the protein content of the embryogenic class at the first division of the vegetative cell is very similar to that of late phase 2 microspores.

These data indicate that, during the inductive period, predetermined biochemical pathways continue to operate in certain grains as in *vivo*. In others, these pathways are blocked and the normal accumulation of RNA and protein is checked. Thereafter loss of RNA and protein takes place, and only then does mitosis commence in the normally quiescent vegetative cell.

Ultrastructural features

While it is a relatively simple matter to locate embryogenic grains in the light microscope by the change in staining properties, it is less easy to locate them with certainty in the electron microscope. Since many grains degenerate in culture any ultrastructural changes observed during the inductive period may be the consequence of degenerative processes rather than of embryogenesis. Nevertheless, if there is a loss of RNA and protein during induction as the cytochemical data suggest, degradative processes must occur in grains destined to become embryoids. The problem is whether (a) distinctive ultra-structural changes are indeed associated with such controlled degradation, and (b) if so, how these changes can be distinguished from those caused by wholesale degeneration. In the first instance, reliable information is likely to be obtained from grains in which the vegetative cell is actually in process of division (as in Figs. 9.7d and 9.8a). Examination of the ultrastructure of many such grains will at least permit the salient features to be defined; it will then be easier to interpret changes observed during the inductive period.

An electron micrograph of a type A *Nicotiana* grain in which the first division of the vegetative cell has been completed is shown in Fig. 9.10c (Dunwell, unpublished). In this particular grain, the cell-plate has formed in the long axis of the grain and fused with the plate surrounding the generative cell. The generative cell is still attached to the intine, and bilateral symmetry has been established. Vasart (1971b) has also commented on this failure of the generative cell to become detached from the intine and has suggested that these grains are equivalent to the lightly staining class defined as embryogenic by Sunderland & Wicks (1969). However, division does take place in grains in which the generative cell has become detached from the intine, so that this condition cannot be regarded as a constant feature of the inductive process. Attachment of the generative cell to the intine could have some other morpho-genetic significance, as for instance, in the establishment of polarity.

There are other features of the embryogenic grain illustrated in Fig. 9.10c which may be relevant to induction. The first of these is the apparent lack of typical plastids. Plastids feature prominently in early phase 3 grains where they appear as one of the most electron-dense components of the cytoplasm (Fig. 9.2b). These plastids lack starch and they are distributed at random through-out the cytoplasm. After 2 days of culture, the plastids show the same marked contrast in electron density and the same random distribution (Fig. 9.10a).

At a later stage, however, there is much less contrast between the plastids and the rest of the cytoplasm. At 7 days, for instance, in what is judged to be an embryogenic grain on account of the presence of dense chromatic centres in the vegetative nucleus (Fig. 9.10b), the plastids have much the same electron density as the other components. They are, moreover, distributed in a more localized fashion, mainly around the vegetative nucleus. These plastids are highly amorphous in appearance; they contain no clearly defined ribosomal population or membranous inclusions, and some contain unusual electron-dense structures. After 14 days, when the vegetative cell has divided, plastids with neither the structure observed at 2 days nor that observed at 7 days can be discerned (Fig. 9.10c). In the grain illustrated there are several vesicular structures clustered around the two vegetative daughter nuclei. These vesicles are relatively large; they are the least electron-dense components of the cytoplasm, and at a glance might pass as vacuoles. Some of these however, have features more consistent with those of plastids. The vesicles concerned are bounded by a double membrane; they show occasional membranous inclusions and small opaque areas suggestive of the initial stages of starch-deposition. At a still later stage, when the embryoid consists of several cells, there is no difficulty in recognizing the plastids; they are mainly located around the nuclei and contain abundant starch (Fig. 9.10d).

The second feature of note in the embryogenic grain of Fig. 9.10c is the apparent lack of typical vacuoles. As has been seen the microspore vacuole is obliterated during early phase 3 of microsporogenesis (Fig. 9.2b). Vacuoles can be observed, however, in such grains after only 2 days of culture (Fig. 9.10a). These vacuoles, moreover, enlarge with time and after 7 days, the vacuolar/cytoplasmic ratio is high (Fig. 9.10b). Division commences in a vacuolate cell (Fig. 9.7d), yet when the division is complete this vacuolate condition has largely disappeared. It must be concluded that the first division is accompanied by rapid synthesis of new cellular components and this leads to the immediate formation of two meristematic-like cells. Soon afterwards, when these cells have undergone a few more divisions (Fig. 9.10d), typical vacuoles once more appear; these are localized near to the periphery of the cells.

A third feature of the embryogenic grain illustrated in Fig. 9.10c is the presence of lipid bodies in the cytoplasm. These are associated with a vesicular component. The lipids become more abundant as the embryoid develops (Fig. 9.10d). though they are not present to any extent in grains during the early stages of culture (Fig. 9.10b).

The electron micrographs of Fig. 9.10 strongly suggest that the changes observed in cytoplasmic RNA and protein levels during induction are indeed accompanied by changes in the ultrastructural features of the cell. Plastids may be one of the organelles most immediately affected.

Abnormal development

The vegetative cell may show anomalous forms of development other than those already mentioned. These lead to the formation of (1) non-embryogenic structures (2) abortive embryoids, and (3) abnormal plants.

(1) *Non-embryogenesis*: In Gymnosperms, the vegetative nucleus may divide

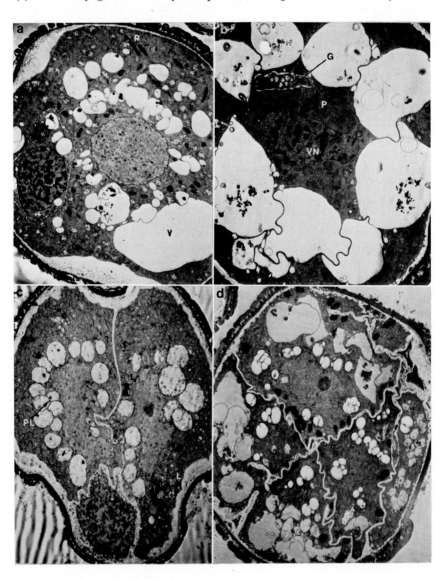

repeatedly without the formation of cell walls (Fig. 9.5c). The exineless grain enlarges to several times its original volume and as many as fifty nuclei may be formed (Tulecke 1957). Giant grains are also produced in Angiosperms especially in anthers cultured during phase 3. These giants have the normal complement of nuclei, either two or three. They are formed from the densely staining class of grains described above. The grains enlarge as *in vivo*, and become filled with starch, but instead of germinating they continue to enlarge and the exine is ruptured (Fig. 9.11). Expansion continues after rupture of the exine. The RNA and protein content of these grains may attain more than two or three times that of mature pollen grains.

In *Torreya*, the generative cell may be extruded from the exineless grain and continue to divide for a time after extrusion (Tulecke & Sehgal 1963). The vegetative cell also divides repeatedly after extrusion of the generative cell. (2) *Abortive embryoids*: In Angiosperms, the vegetative cell may stop dividing after a few divisions have taken place. Cell enlargement takes place, however, by vacuolation and this causes premature rupture of the exine (Fig. 9.12). The abortive embryoids may assume bizarre shapes. Cessation of division at an early stage in culture occurs on a large scale in *Nicotiana tabacum*. Division may commence in several thousand grains in one anther, but relatively few of these reach the plantlet stage (Sunderland & Wicks 1971). In certain anthers, all the embryoids may be abortive and plantlets are not produced (see Fig. 9.4). The reason for this is not clear. Embryoids are formed continuously over a period of days, so it is conceivable that a few of the first-formed embryoids compete successfully for available nutrients and continue their development at the expense of others. Competition does not satisfactorily account for cases in which all embryoids cease development at an early stage; nor does it account for the variable behaviour of anthers of the same plant. From one bud, each anther may on occasions give rise to more than fifty

---

FIG. 9.10. *Electron micrographs of cultured pollen grains of* Nicotiana tabacum *cv.* White Burley. Culture conditions as in Fig. 9.3a. a: After 2 days; note vacuolate (V) nature of the cytoplasm compared with that of the grain in Fig. 9.2b. The plastids (P) are the most electron-dense components of the cytoplasm. b: After 7 days; note dense chromatic centres in the vegetative nucleus (VN). A small portion of the generative cell can be seen. The plastids are relatively larger and less electron dense than at 2 days. c: After 12 days; the vegetative cell has divided once. Note lipid inclusions (L) associated with a vesicular component. The electron-opaque vesicles clustered around the vegetative nuclei have neither the appearance of plastids nor of typical vacuoles. Some (PL) are bounded by a double membrane and at higher magnification can be seen to contain occasional membranous structures. d: After about 21 days; multicellular grain. Note starch-filled plastids clustered around the nuclei, abundant lipid bodies, and typical vacuoles located near the periphery of the grain (by courtesy of J.M.Dunwell).

plantlets on average, whereas from another bud, of identical age and developmental stage treated under identical conditions, the anthers may be completely unproductive or yield only a few abortive embryoids.

Since type B grains occur sporadically and in relatively low frequency another possible explanation is that plantlets are formed only from type B grains; whereas type A grains give rise to abortive embryoids. It has not yet

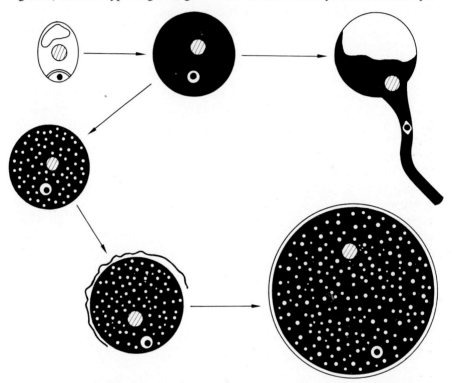

FIG. 9.11. *Diagram illustrating the behaviour of non-embryogenic grains in* Nicotiana tabacum. Some degenerate, others germinate inside the anther with or without starch-deposition. The generative cell may undergo the second pollen mitosis without germination. Starch-filled grains may enlarge and burst out of the exine forming giant grains with two or three nuclei (after Sunderland & Wicks 1971).

been possible, however, to establish a correlation between the numbers of type B embryoids formed and the numbers of plantlets produced. Moreover, type B grains arise most frequently in anthers cultured towards the beginning of phase 2, and these, as has been seen, are the least productive anthers in *N. tabacum* (Fig. 9.4). Also, generative cells can often be distinguished in viable embryoids consisting of many cells (see Fig. 9.7g). Sunderland & Wicks (1971) pointed out that perhaps one of the main criteria for ensuring continued growth is the early establishment of polarity within the embryoid, and it is in this sense, that attachment of the generative cell to the intine could have significance (see Fig. 9.10c). Errors in the initial sequence and pattern of divisions may lead to disorganized growth.

(3) *Abnormal plants*: In some Angiosperms, plants produced by anther culture are complete or partial albinos. The phenomenon has been observed in

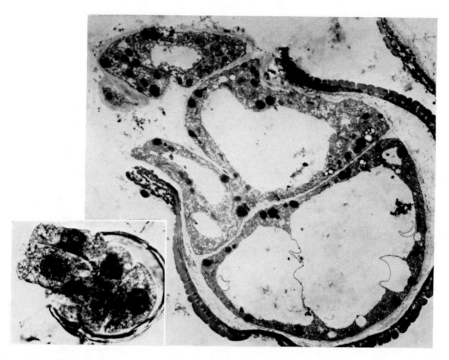

FIG. 9.12. *Abortive embryoids in* Nicotiana tabacum. Note the highly vacuolate nature of the cells (by courtesy of J.M.Dunwell).

N. tabacum (Devreux 1970), *Datura metel* (Narayanaswamy & Chandy 1971), *Oryza sativa* (Niizeki & Oono 1971), and is prevalent in *Hordeum vulgare* and *Lolium multiflorum* (Clapham, personal communication). On the assumption that the generative cell does not contain plastids, Devreux (1970) has suggested that albinos arise by repeated division of the generative cell. There is, however, no tangible evidence to support this view. If as is indicated by the electron micrographs of Fig. 9.10, the plastids undergo a fundamental reorganization process during the inductive period errors occurring at this stage could possibly lead to the appearance of albinos or plants with abnormal plastids.

PLOIDY STATUS OF POLLEN PLANTS

Populations of plants derived directly from pollen embryoids in *Nicotiana tabacum* are mostly haploid, as are plants derived from first-generation callus (as in Fig. 9.3e, f) (Sunderland 1970). The occasional plant will give rise to an inflorescence carrying haploid and diploid flowers and it is evident that endomitosis has occurred at some stage in the development of the plant to yield a diploid cell lineage. In other species, notably those in which plants have to be regenerated from a callus, plants of higher ploidies may be obtained. Narayanaswamy & Chandy (1971), for instance, found both diploids and triploids amongst plants derived from anthers of *Datura metel*; 70% were diploid, 23% triploid and only 7% haploid. Triploids have also been obtained in *Petunia* (Raquin & Pilet, 1972), triploids and aneuploids in *Solanum* (Harn 1971), and in *Oryza sativa*, plants showed ploidy levels ranging from haploid to pentaploid (Nishi & Mitsuoka 1969, Niizeki & Oono 1971).

Such plants, if they are indeed derived from the pollen, must arise by nuclear aberrations either in the pollen itself or in the callus derived from it. Endomitosis can account for the formation of diploids and tetraploids, and nuclear fusion for the triploids and pentaploids. It may be emphasized that the first step in nuclear fusion is the formation of a multinucleate cell (see p. 171). It is highly improbable that fusion can take place as suggested by Narayanaswamy & Chandy (1971) in a normal pollen grain in which the generative and vegetative cells are isolated from each other by a cell-plate.

The origin of these plants of different ploidies needs to be carefully checked. In the case of *Nicotiana tabacum*, there can be little doubt that the plants with mixed inflorescences are of pollen origin. Plants with no haploid cell lineage, however, may have developed from the anther wall (as, for instance, Konar & Nataraja 1965a). In a number of instances, anthers have been sectioned and examined for such development, but it may be doubted whether such an examination could be conclusive. Embryoids or callus may develop internally from the anther tissues simultaneously with the pollen. Tapetal cells (which have high DNA contents) and others may break away into the anther cavity during culture and become mixed with the pollen. Embryoids or callus derived from these free cells could be mistakenly interpreted as being of pollen origin. The most convincing evidence to date in favour of the pollen origin of certain diploid plants is that given by Niizeki & Oono (1971). These workers obtained both haploid and diploid plants from anthers of an F1 hybrid between the *japonica* and *indica* sub-species of *Oryza sativa*. The diploids showed segregation of several characters. This could not have occurred if the plants had come from callus derived from the somatic tissues.

PRODUCTION OF HOMOZYGOUS LINES

Doubling of the chromosome complement of a pollen plant leads directly to the formation of a fertile and completely homozygous plant. There are three routes by which this diploidization may be achieved. These are, (1) natural or spontaneous doubling in normal somatic tissues, (2) artificially induced doubling, and (3) regeneration from pollen callus. Whichever approach is adopted, the chromosome number of the plants or plantlets treated should first be checked.

(1) *Spontaneous doubling*: As indicated in the preceding section the frequency of spontaneous doubling in *Nicotiana tabacum* is low. About 1% of the plants examined by Sunderland (1970) gave rise to mixed inflorescences and about 2% of those examined by Kasperbauer & Collins (1972). Nevertheless in cases where large numbers of pollen plants can easily be produced a low frequency such as this will soon provide a working number of pure lines. Plants doubled spontaneously are preferable to those produced by either of the two following methods because they are less likely to show nuclear aberrations.

(2) *Artificially induced doubling*: The frequency of doubling may be increased by treating the haploids with a spindle inhibitor such as colchicine. The simplest procedure is probably that in which pollen plantlets are immersed in a solution of the drug as soon as they emerge from the anther (as in Fig. 9.3b). The chromosome number can be estimated from leaf squashes. With *N. tabacum* a 0·4% solution is recommended. Periods of immersion of up to 96 hours may be required, and the treatment may have to be repeated several times. Filter-sterilized colchicine solution should be used so that the plantlets can be transferred after treatment into fresh culture medium. Alternatively, the drug may be applied to mature plants in lanolin paste (0·4%) (Tanaka & Nakata 1969). The paste is applied to the axils of upper leaves, and the main axis decapitated in order to stimulate growth of the axillary branches. Details of the meiotic behaviour of several diploid plants derived in this way have been given by Nakata & Tanaka (1970); many of the plants showed nuclear aberrations.

(3) *Regeneration from callus*: This procedure used by Nitsch, Nitsch & Hamon (1969) to produce homozygotes in *Nicotiana*, involves the production of a callus from stem segments or other parts of the pollen plant, and then transferring the callus to a regeneration medium. The method exploits the known relatively high rate of endomitosis in callus tissues. A simple procedure has been reported by Kochbar, Sabharwal & Engelberg (1971) and another variation by Kasperbauer & Collins (1972). In the former, stem or petiole segments are inoculated into MS medium (see p. 39) supplemented with a mixture of NAA 0·2 mg/litre and IAA 0·2 mg/litre. The resulting callus is then placed on MS with the hormone component changed to a mixture of kinetin

2·0 and IAA 2·0 mg/litre. Roots appear after about 6 days and shoots after about 14. The plantlets can be transplanted into soil after an adequate root system has developed. Continued subculture in the callus-inducing medium may lead to the formation of tetraploid cells from which tetraploid plants can be regenerated. Homozygotes produced by this method need to be tested rigorously for gene and chromosome defects.

This rapid method of producing homozygotes is potentially of importance in plant breeding and in programmes of crop improvement (Sunderland 1970, 1971). For a more detailed account the reader is referred to papers by Melchers & Labib (1970) and Melchers (1972).

HAPLOID MUTANTS

Anther and pollen culture also offer considerable scope in the production of haploid mutants, and in the phenotypic expression of mutations. Tulecke (1960) isolated a number of naturally occurring arginine-requiring strains from *Ginkgo* pollen, by simply replacing the normal hormone component of the medium by relatively high levels (up to 1000 mg/l) of L-arginine. As in all the other work on Gymnosperm pollens, these calluses developed in areas of high pollen density. Several pollen grains may therefore have been involved in their formation, and the strains may not have been genetically uniform. In the Angiosperms, mutations have been induced by either irradiation or chemical mutagen treatments. A number of biochemical mutants have also been isolated by mutagen treatment of cell suspensions derived from pollen plants.

(1) *Irradiation*: In the method described by Nitsch, Nitsch & Péreau-Leroy (1969) plantlets of *Nicotiana* species are irradiated from a cobalt source just as they emerge from the anther (as in Fig. 9.3b). After irradiation the plantlets are transferred to soil. This treatment gives rise to a high proportion of aberrant phenotypes showing abnormal leaf shapes and leaf variegation, abnormal flower shapes and flower colour. The plants are chimeras. Nitsch (1972) describes a variegated mutant which after diploidization, produced seeds yielding either green or albino progenies. Further information on plastid inheritance in variegated pollen haploids of *N. tabacum* is given in the paper by Nilsson-Tillgren & Wettstein-Knowles (1970).

Devreux & Saccardo (1971) used X-rays for irradiation, and gave the treatment prior to culture of the anthers. In this method, flower buds of *N. tabacum* cv. Virginia bright are irradiated at a dosage of approximately 1000 R when the pollen is undergoing the microspore mitosis. The anthers are subsequently dissected from the buds and cultured in the usual way. The technique can yield up to 50% aberrant phenotypes differing in plant height, leaf shape and colour. Treatment of flower buds has the advantage that if the mutation is induced before the microspore division, the resulting plant is wholly aberrant. About 6% of the plantlets produced by Devreux and Saccardo showed chromosomal aberrations.

TABLE 9.2. Summary of the technique for inducing and isolating haploid auxotrophic mutants in *Nicotiana tabacum* cv. Wisconsin 38 (from Carlson 1970).

	Linsmaier & Skoog (1965) basal medium containing:			
	Agar %	Sucrose %	IAA	K
				mg/litre
1. Culture late phase 1 anthers	0·8	4·0	0·1	(a)
2. Culture stem segments of plants derived from 1	0·8	4·0	2·0	0·3 (b)
3. Maintain haploid callus obtained from 2. This callus is organogenic		4·0	2·0	0·3 (c)
4. Culture samples of callus from 3		4·0	2·0	0·03 (d)
5. Maintain in cell suspension culture the unorganized callus produced from 4	(c) medium			
6. Filter samples of cell suspension from 5 through four layers of cheese cloth				
7. Centrifuge filtrate which consists of single cells and small groups of cells and resuspend in (c) containing ethyl methane sulphonate, 0·25% for 1 hour Centrifuge, wash twice in (c), resuspend in (c) and incubate for about 96 hours to halt growth of mutant cells. Replace medium after 48 hours				
8. Add 5-bromodeoxyuridine to a final concentration of 10^{-5} M. Incubate in darkness for 36 hours				
9. Centrifuge, wash cells twice in (c). Mix cells with semisolid agar medium (e) and pour into petri dishes at a cell density of approximately 2500 cells/ml	0·5	2·0 plus T 10^{-5} M, CH 800 and YE 400 mg/litre	2·0	0·3 (e)
10. Test samples of surviving calluses in supplemented (b)	(b) plus amino acids 100 μg/litre, or nucleic acid bases 50 μg/litre, or vitamins 0·2 μg/litre			

IAA = indol-3yl-acetic acid; K = kinetin; T = thymidine; CH = casein hydrolysate; YE = yeast extract.

(2) *Chemical mutagens*: Attempts to induce mutations by direct chemical treatment of anthers prior to culture have so far met with little success. The present author has treated anthers of *N. tabacum* cv. White Burley during late phase 2 or early phase 3 with N-methyl-N′-nitro-nitrosoguanidine given in varying concentrations and for varying times. Anthers are dissected from

buds, floated on a solution of the mutagen, rinsed in sterile water and plated in the usual way. All plantlets so far obtained have shown normal phenotypes. Nitsch (1972) records a similar experience with this mutagen, but reports having obtained a white-flowered mutant by incorporating N-3-nitrophenyl-N'-phenylurea (10^{-6} M) into the culture medium.

Carlson (1970) isolated several biochemical mutants in *N. tabacum* by treating single haploid cells with ethyl methane sulphonate. The procedure is outlined in Table 9.2. Mutant cells are separated from wild-type cells by exposing the treated cell suspension to 5-bromodeoxyuridine. Cells which undergo DNA synthesis in the presence of the analogue incorporate it into their DNA. On subsequent plating and exposure to light these cells are killed. Mutant cells which do not synthesize DNA in the presence of the analogue survive the light-treatment. Auxotrophs are recovered by culturing survivor calluses on media supplemented with amino acids, vitamins, nucleic acid bases and so on. Carlson isolated 119 calluses in this way, among which six proved to be auxotrophic, one for each of the following compounds, hypoxanthine, biotin, p-amino-benzoic acid, arginine, lysine and proline. Plants were regenerated from four of these mutants.

An interesting feature of Carlson's experiments in relation to the data of Chapter 7 is that the haploid calluses initially isolated from the pollen plants remained chromosomally stable for about 6 months of subculture. They also retained a capacity for organogenesis. However, isolation of a haploid callus was a rare event; the bulk of the calluses produced were either diploid or tetraploid, and these lacked the capacity for organogenesis.

CONCLUDING REMARKS

The unit structure, unique genetic constitution, and partial synchrony of pollen, coupled with ease in handling, make it an ideal system for culture studies. The pollen grain is programmed for gamete and tube formation, but these processes are clearly not irrevocable and can be arrested in culture and the grain given an entirely new role. It has been suggested that culture effects a switch in the regulatory processes controlling DNA replication and mitosis in the vegetative cell (Sunderland 1971). This switch can be seen as leading to repression of the normal programme on the one hand, and depression of a latent programme on the other. The mechanism is not perfect and, as has been seen, errors manifest themselves in various anomalous forms of behaviour.

The cytochemical evidence, while still of a preliminary nature, suggests that replacement of one programme by another is only one facet of the switch mechanism. There is also probably a degradation of cytoplasmic components accompanied by a fundamental reorganization of the vegetative cell as it returns to the mitotic state. It is not unreasonable to suppose that the components degraded are those concerned in the execution of the pollen's normal role. If this is so, vacuoles formed during the inductive period could possibly be serving as reservoirs for breakdown products and a pool of nucleotide and peptide precursors is thus established which can be utilized in the synthesis of new cytoplasmic components when the latent set of instructions comes into operation. Indeed, this degradative process may be essential for derepression of the latent programme to occur.

Factors concerned in triggering the switch mechanism and in removing the block to mitosis in the vegetative cell are not fully understood. It is evident that the switch comes into operation as a result of the changed nutritional and physical environment of the pollen. At the beginning of phase 3, which appears to be critical in tobacco, the anther has completed its rapid phase of growth in terms of anther length, dry and wet weight; subsequent maturation is characterized by loss of water (Erickson 1948). Culture checks the dehydration process and presumably engulfs the pollen sac and its contents in an alien solution of nutrients and hormones. That this changed situation is essential for the switch to operate is indicated by the lack of response exhibited by stamens plated with only the filament in contact with the medium.

Other instances are known where an abrupt change in the environment of a cell brings about a change from the normal growth pattern. In fungi, for instance, the shape of the conidiophore, which is genetically controlled, appears to be linked closely to one particular set of environmental conditions. Exposure of hyphal tips to water for as little as 10 seconds is apparently sufficient to arrest growth. Hyphal branching and branch form can subsequently be manipulated by subtle changes in the tonicity of the bathing fluid (Robertson 1959). In anther culture of *N. tabacum* the arrest of one form of development and the switch to another operates if anthers are floated on water or a simple nutrient solution, but other factors are involved as is evidenced by the more complex hormonal requirements of other species, and by the highly variable response between anthers. These remarks are speculative and are put forward as a stimulus to further experimental work.

CHAPTER 10

GROWTH PATTERNS IN TISSUE (CALLUS) CULTURES

M. M. Yeoman and P. A. Aitchison

Establishment of the callus 240
Induction of growth 241
Metabolic consequences of induction 243
Importance of the wound reaction 243
Molecular events before division 246
Regressive change 247
Maintenance of division 254
Departure from division and the onset of differentiation 255
Growth of the established callus 257
Morphology of the established callus 259
Metabolic patterns in tissue cultures 260

ESTABLISHMENT OF THE CALLUS

Growth, regressive change, differentiation and pattern formation are the four major interacting processes which contribute to the development of a typical callus. Within this framework the course of callus development from a fragment of tissue may be divided into three stages, induction, division and differentiation. These three developmental stages may be characterized by changes in mean cell size of the population as well as in the structure and overall metabolic condition of the tissue. First there is an induction phase during which the cells prepare to divide, metabolism is activated and in which the cell size remains constant. The duration of this phase varies with the physiological state of the cells in the initial explant and the culture conditions employed. This is followed by a phase of active synthesis and decreasing cell size initiated by the occurrence of divisions in the peripheral layers of the explant. During this phase, regressive change, involving a progressive return to a meristematic or ground state (dedifferentiation) in the outer regions of the callus, results in the formation of a particular growth pattern. The distinctive feature of this phase is that division has proceeded throughout the outer regions of the callus leaving a small core of undivided cells within. Eventually the formation of differentiated structures (differentiation), begins to supersede the regressive changes and a new course of development is initiated. This new course is marked by increasing cellular differentiation through both the maturation of some cells and the expansion of other cells. However, this cell expansion is balanced by cell division continuing throughout this phase so that a more or less constant mean cell size is established. While each of the three stages is named by its predominant feature these are not exclusive terms, both division and differentiation proceed through the second and third phases. Growth of the tissue mass occurs throughout both the differentiation and division phases.

INDUCTION OF GROWTH

In order to understand how a quiescent cell may be activated and induced to divide it is instructive to compare the structure and metabolic state of such cells before and after induction. In addition it is vital to acquire an understanding of how the plant growth substances in question initiate cell division. Much information is available on the changes accompanying the addition of plant growth substances to plant tissues but the mechanism of action of these compounds is still not understood. The spectacular increase in RNA which follows soon after the addition of 2,4-dichlorophenoxyacetic acid (2,4-D) and related growth regulators to quiescent tissues (Steward, Mapes, Kent & Holsten (1964), Yeoman & Mitchell (1970) points to the importance of the nucleolus as a site of action of growth substances. Zwar & Brown (1968) have shown that ^{14}C labelled 2,4-D accumulates in the nucleoli of dividing cells of cultured explants from the Jerusalem artichoke tuber.

Other attempts to understand the process of growth induction have been made by studying the ultrastructure of cells before and after they have been stimulated to divide (see Chapter 6). Many of the cytological changes observed reflect the metabolic condition of the cell and a consideration of these helps to explain the transformation from quiescent to active cells. An important preliminary to such a consideration is an examination of the events which lead up to the establishment of quiescence in the cells of the explant. Discovery of what stopped the cells dividing might help to explain what is needed to initiate their division. Tissues commonly employed as a source of callus cultures are composed of differentiated cells which are the products of an earlier division in the intact plant. During the process of differentiation they have stopped dividing and acquired a specific function such as storage of reserves. Brown & Dyer (1972) consider that these cells still retain their potential for a meristematic state but it is repressed or inhibited. In order to return the differentiated cell to the dividing condition it is necessary only to remove this repression or inhibition. This concept of Brown & Dyer (1972) contrasts with the alternative and more popular theory that the cell loses the power to divide as differentiation proceeds and acquires it again as a result of the induction and activation processes.

Such a theory presupposes the synthesis of a new 'set of division machinery' in response to induction. The work of Harland (1971) on changes in the activity of enzymes during the induction of division in parenchyma cells supports the Brown and Dyer theory. Harland has shown that certain enzymes involved in the replication of DNA are present in fully differentiated non-dividing parenchyma cells and that when these cells are induced to divide synchronously by the addition of 2,4-D the levels of the enzymes remain more or less constant until after DNA replication has commenced. This suggests that in these differentiated cells the activity of some enzymes involved in division is preserved at a level sufficient to sustain DNA replication. It may be pointed out that synthesis and accumulation of these enzymes take place after DNA replication has started in preparation for the next wave of division. Another fact supporting the Brown and Dyer hypothesis is the remarkable degree of similarity that exists between the constituent cells from the storage parenchyma region of the Jerusalem artichoke tuber. These cells are similar with respect to DNA, RNA and protein content (Mitchell 1967, 1968, 1969). They are also alike with respect to their preparation for cell division, for, once induced, they proceed synchronously through the synthesis and accumulation of a variety of macromolecules and divide together. This would suggest that these cells have been brought to a halt at a similar stage of development. One way in which this state could have been achieved is by the presence of an inhibitor which stopped the cells dividing but preserved the necessary machinery for subsequent division. The removal of this inhibitor would then return the cell to the dividing state. A well-known characteristic of cell populations articificially held up by division inhibitors is that the removal of the inhibitor or its influence, gives rise to a population of cells which will divide synchronously. Indeed this procedure is a standard technique for obtaining synchronously dividing populations of micro-organisms (Zeuthen 1964, Padilla, Whitson & Cameron 1969). Cells can be stopped at many points in the division cycle by the use of inhibitors which are specific for the activities of a particular part of the cycle. For example fluorodeoxyuridine (FUdR) prevents DNA replication by restricting the supply of thymidine triphosphate (TTP). Cells treated with FUdR are therefore unable to synthesize DNA and divide until the effect of the inhibitor is overcome. Relief from inhibition by the addition of thymidine releases a population of cells all in late 'G_1' and they will replicate DNA immediately after the influence of the inhibitor has been removed. It is therefore not unreasonable to suppose the existence of a natural inhibitor present in cells which can also prevent division while keeping the division machinery intact. Little is known about the nature of this inhibitor except that it cannot be removed by washing and is therefore not freely soluble or volatile. It may be a protein. The nature or amount of such an inhibitor may well differ from plant to plant and may be related to the chemical substances necessary for induction. This could explain the differing responses of tissues to plant growth substances. The presence of an inhibitor in the nucleus in association with the nucleolus which can be removed by a growth regulator remains an attractive possibility.

METABOLIC CONSEQUENCES OF INDUCTION

The act of induction is followed by a succession of metabolic changes and syntheses which culminate in division. Not all the cells in a piece of excised tissue respond in this way to the new environment. Only the cells in the outer regions of the tissue are induced to divide (Yeoman, Dyer & Robertson 1965, Smithers & Sutcliffe 1967, Gautheret 1953, Nitsch & Nitsch 1956, Steward & Caplin 1954) and this results in an actively dividing periphery surrounding a non-dividing core. Recently Yeoman & Davidson (1971) have shown that in artichoke explants excised in the light, only some of the cells of the peripheral zone divide immediately after induction. Approximately half of the cells in this zone fail to divide until later in the development of the callus. However, if the explants are prepared in low intensity green light, all of the cells of the outer regions divide together soon after induction. The factor regulating the size of the dividing population appears to be a supply of reduced nitrogen and under light conditions a competition is set up between cell division and the synthesis of polyphenolic substances involved in secondary wall formation (Davidson 1971). This competition, which can be resolved by the addition of mixtures of amino acids, reduces the size of the dividing population substantially. The cells of the peripheral zone which fail to divide display a different pattern of synthesis to those cells in the core which also fail to divide. The non-dividing cells of the periphery accumulate substantial amounts of protein which is a preparation for a later division (Mitchell 1968). It is also important to note that under conditions of induction only cells which divide synthesize and accumulate RNA (Mitchell 1969; Yeoman & Mitchell 1970).

The initiation of division in the outer layers of the tissue is related to a number of interacting factors, including the wound response at the cut surface (Fosket & Roberts 1965, Yeoman, Naik & Robertson 1968), greater availability of oxygen, more rapid release of carbon dioxide, increased availability of nutrients. more rapid release of a volatile inhibitor (Laties 1962) and light (Yeoman & Davidson 1971).

IMPORTANCE OF THE WOUND REACTION

Very early work by Haberlandt (1902) showed the importance of substances liberated by wounded plant cells in the induction of cell division. He showed that the cut surface of a potato tuber soon becomes protected from the environment by the formation of new layers of cells and the deposition of polyphenolic materials. The divisions induced in the parenchyma of the potato (Fig. 10.1) are predominantly periclinal and give rise to radial rows of cells forming a wound cambium. Once an effective barrier is thus produced division stops. Haberlandt postulated that division is not simply due to the presence of a 'hormone' liberated by wounded cells (wound hormone) but is brought about by an interaction between the wound hormone and another 'hormone' present within the potato (leptohormone). The cessation of division once the wound has healed may presumably be due to the termination

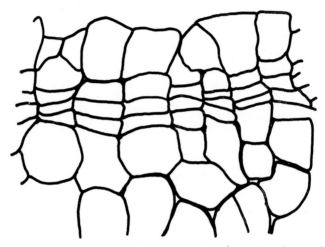

FIG. 10.1. *Regular radial rows of the wound cambium in cultured excised potato tissue.*

of the supply of either hormone. Had Haberlandt been able to add 2,4-D and coconut milk to his pieces of excised potato then division would have continued and a callus would have resulted (Steward & Caplin 1951). This would suggest that the wound hormone is important in the induction of division and perhaps during the early divisions but not later in the establishment of the callus. In an intact plant or with a cutting, damage can result in the formation of a massive callus without the intervention of applied growth substances (Fig. 10.2). Presumably the callus is nurtured from endogenous materials which are supplied from the other tissues of the plant. In cuttings the appearance of roots seems to stop further proliferation of the callus perhaps by restricting the supply of endogenous growth substances (Fig. 10.2). Clearly, an understanding of the wound reaction would help to answer many questions about the induction of cell division.

The act of removing a fragment of plant tissue damages cells at the surface of the explant. Two kinds of damage may be observed, (1) the breaking open of cells with the subsequent loss of contents and (2) the compression and bruising of cells underneath the broken layer; these cells remain intact. It is highly probable that the contents of the broken cells will be quickly lost to the environment or absorbed by the cell layers immediately below. The wound substances involved in the initiation of divisions are therefore presumably provided by the autolysis of the unbroken cell layers

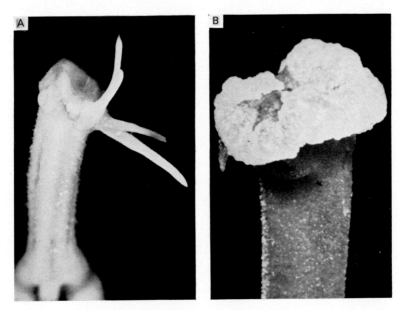

FIG. 10.2. *Callus formation and the subsequent development of roots on the petiole of a leaf detached from a French bean plant (A) and profuse callus formation at the cut surface of the node bearing the primary leaves of a young French bean plant (B).*

which lie above the dividing cells (Yeoman, Naik & Robertson 1968). This layering phenomenon also exists in cultured carrot explants (Fosket & Roberts 1965). This release of autolytic products begins soon after excision and continues until the outer cell layers have collapsed and become impregnated with ligno-suberin compounds. In the absence of added 2,4-D, explants from the artichoke tuber (Yeoman & Mitchell 1970) accumulate some protein, RNA and various polyphenolic substances but cell division occurs only in a few cells and no callus is formed. The presence of 2,4-D ensures the formation of a callus, probably the result of a successful interaction between the autolytic products and the added 2,4-D. In contrast, some explanted tissues may produce a substantial callus in response to a wound stimulus alone (see Fig. 3.2 in Chapter 3, p. 40). Two examples readily spring to mind, the cells of the immature lemon fruit (Kordan 1959) and isolated vascular cambia (Gautheret 1934, White 1963). It is perhaps significant that both of these tissues are meristematic at the time of excision and do not need to be induced to divide. Division can be maintained on a simple medium with sucrose. More usually the additional presence of added growth substances is essential to promote a callus.

MOLECULAR EVENTS BEFORE DIVISION

The induction of division in a tissue composed of like cells in a similar stage of development leads to a synchronous division in which all of the cells of the dividing population complete mitosis within a short space of time (Fig. 10.3).

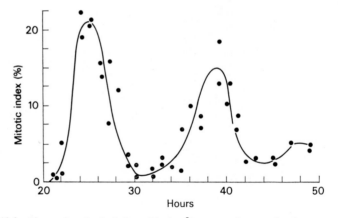

FIG. 10.3. *Change in mitotic index with time in a population of cultured explants from a Jerusalem artichoke tuber.* The tissue was excised in the light and grown in a stirred liquid medium containing 2,4-D and coconut milk (Evans 1967) approximately 35% of the cells divided synchronously at the first division.

The serendipity of synchrony has made it possible to examine in detail the whole process of cell division in developing callus cultures (Yeoman, Evans & Naik 1966; Yeoman & Evans 1967) and in particular to investigate the very earliest events which take place after excision. Two changes occur within minutes of excision and culture, a rapid increase in the rate of gaseous exchange (Yeoman 1970) and the appearance of polyribosomes. Some increase in the rate of gaseous exchange occurs in response to excision alone. However, the presence of growth substances ensures a much more spectacular increase. Polyribosomes too appear in response to excision alone but not in such large numbers. Tissues will take up various precursors of RNA after an hour or so and incorporate them into RNA. Incorporation of labelled amino acids into proteins takes a little longer. Within 4–6 hours a substantial accumulation of ribosomal RNA has taken place in potentially dividing cells. This increase of approximately 30–40% (Mitchell 1969) occurs quickly over about 1 hour and is followed by a period in which no further accumulation is detectable. However, subsequent to this, other periodic and more substantial increases in RNA take place as the cell moves towards division. The second increase of approximately 120% (ribosomal RNA) coincides with the commencement of DNA replication which also occurs in a step. The third and last increase in RNA (ribosomal and transfer) occurring just before mitosis brings the overall increase per cell up to 300%. The period of DNA synthesis lasts for 13–14 hours and is quickly followed by mitosis. Protein also accumulates periodically and the total increase per dividing cell is about 200%. The rate of oxygen uptake increases dramatically until the beginning of DNA replication and then stays constant (Yeoman 1970). After the 'S' period the increase in the rate of oxygen uptake is resumed.

During the period of preparation for division the protein pattern of the cell is also changing. This is shown in the electrophoretograms of Fig. 10.4. A marked change can be observed in the protein composition of quiescent cells and cells about to divide. Within the total protein of the cell there is a changing pattern of enzyme activities. Harland, Jackson & Yeoman (1973) have shown that enzymes involved in DNA replication increase in activity at a particular point in the cell cycle. Apart from DNA polymerase, which increases slightly before the start of DNA replication the other enzymes studied, TdR kinase and dTMP kinase showed no increase in activity until DNA synthesis had commenced (Fig. 10.5). Indeed dTMP Kinase exhibited a decrease in activity over cells in freshly excised tissue. All of the enzymes studied displayed large increases in activity after DNA replication had begun. Enzymes concerned with other facets of metabolism, hexokinase, glucose-6-phosphate dehydrogenase and malic dehydrogenase also increase in activity before division (Aitchison, 1972) (see also Chapter 11, p. 303 *et seq.* for discussion of synchronous division in cell suspension cultures).

REGRESSIVE CHANGE

Division once induced proceeds swiftly. The increase in cell number of an artichoke explant over the first 7 days of culture at 25°C is over 1000% (Yeoman, Dyer & Robertson 1965) and approximately 1000% in a carrot explant at the same temperature (Steward & Shantz 1956). However, the increase in fresh weight does not keep pace with the rise in cell number and

A   B

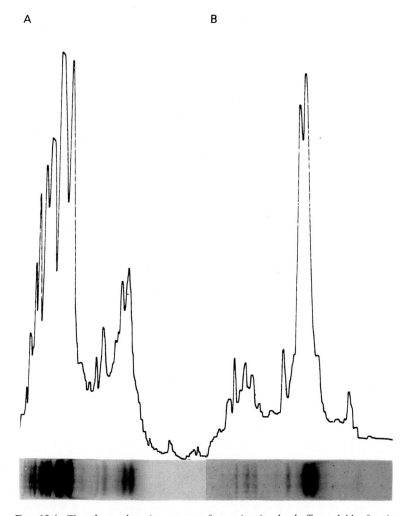

FIG. 10.4. *The electrophoretic pattern of proteins in the buffer soluble fraction* (*ribosomes removed*) *prepared from Jerusalem artichoke explants.*
A. Cultured explants in late 'S' a few hours before the first synchronous division.
B. Freshly excised tissue composed of quiescent cells. (Photograph provided by Mr P.Sealey.)

soon falls behind, leading to a spectacular reduction in average cell size (Fig. 10.6). If the biological equivalent of Newton's first law of motion is true, then presumably all that is required to maintain active cell division is that conditions remain the same. Despite extrinsic factors remaining constant, such a requirement can never of course be satisfied in this situation, because

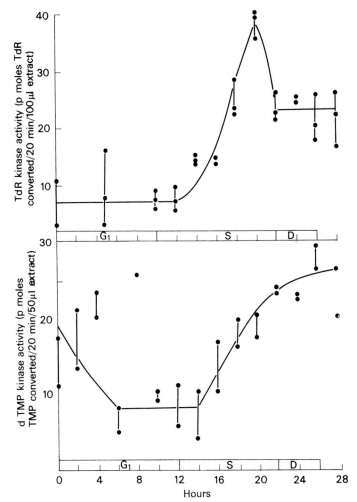

F𝐈𝐆. 10.5. *Change in the activity of TdR kinase and dTMP kinase during the first cell cycle in cultured artichoke explants (Harland 1971).* The timing of the 'S' period was determined by pulse labelling with ³H thymidine. About 60% of the cells divided at the first synchronous division. The approximate duration G_1, S and the time taken for the dividing population to complete G_2, mitosis, and cytokinesis (D) are shown on the graph.

an inevitable ancillary of growth is that novel intrinsic relationships of shape and size are continually being created.

It is during this period of dedifferentiation that a new situation is presented in which, apart from DNA and initially RNA, other constituents of the dividing cell diminish with succeeding divisions. Eventually a stage is reached

I

at which division and accumulation of the cellular constituents are in step and
the decrease in the size of the cell is arrested. At this point the dividing cell
is small, not highly vacuolate and more like a typical cell to be found at the
apex of the root or shoot, it has reverted to the meristematic state.

One constituent which increases per cell during induction and the initial
wave of division is RNA (Fig. 10.7). This reflects the colossal rate at which
RNA (mainly ribosomal) is being accumulated (Steward, Mapes, Kent &
Holsten 1964, Fraser 1968). Subsequently, RNA per cell declines and leaves

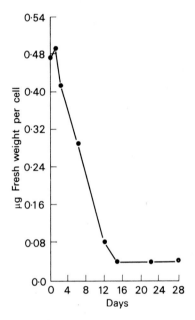

FIG. 10.6. *Change in average fresh weight per cell of cultured explants isolated from
the artichoke tuber.* (Robertson 1966.) Tissue was excised in the light and cultured
in a liquid medium with 2,4-D and coconut milk at 25°C.

only DNA to retain its constancy. The quantity of ribosomes present in the
cell at this time seems to be far in excess of requirements. Another most
interesting aspect to the phenomenon of decreasing cell size is that the
activity per cell of enzymes concerned with DNA replication remains constant at least during the initial period of division.

The marked decline in cell size observed during the first seven days
after induction is of course a result of cell division proceding much more
rapidly than cell growth, and treatments which interfere with the balance
between division and growth affect the final average cell size within the
developing callus. For example lowering the temperature of incubation not

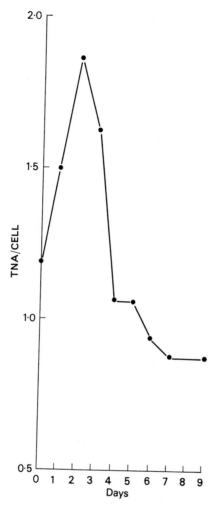

FIG. 10.7. *Change in total nucleic acid (TNA=approximately 90% RNA) per cell of cultured explants isolated from the artichoke tuber.* Tissue was excised in the dark and cultured in a medium with 2,4-D but without coconut milk at 25°C. (Davidson 1971).

only slows down the rate of division and growth but also leads to the establishment of a larger average cell size (Table 10.1). This is presumably related to the fact that division is more temperature-sensitive than expansion. Changes to the constituents of the culture medium may also affect the average cell size attained in Jerusalem artichoke cultures. In the presence of coconut milk and 2,4-D, the average cell size is much less than with explants cultured with 2,4-D alone. Adamson (1962) and Setterfield (1963) have shown that a

TABLE 10.1. Effect of temperature on the average fresh weight per cell (cell size) attained in the post-exponential phase of cell division in developing artichoke callus tissue when the average cell size is maintained at a fairly constant level (Naik 1965)

Temperature at which the tissue was cultured °C	Average fresh weight per cell (μg) in freshly excised tissue	Average fresh weight per cell (μg) in the post-exponential phase of division
10·0	0·44	0·20
12·5	0·52	0·15
15·0	0·42	0·12
25·0	0·44	0·07
30·0	0·42	0·06
32·5	0·52	0·03
*37·0	0·43	0·57

* No detectable increase in cell number.

balance exists between division and expansion in cultured discs of Jerusalem artichoke tuber which can be altered by pre-treatment of the tissue before the addition of growth substances.

Another feature of this regressive stage which has been explored is that the increase in fresh weight (Fig. 10.8) and cell number of carrot explants

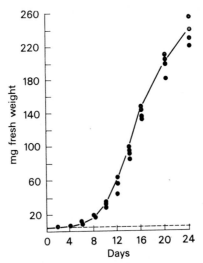

FIG. 10.8. *Growth curve of carrot phloem explants (4 mg) under standard conditions in a basal medium with coconut milk (solid line) compared with the initial weight (broken line).* (Steward & Caplin 1954.)

(Steward, Caplin & Millar 1952, Steward & Caplin 1954) and cell number of artichoke explants (Yeoman, Dyer & Robertson 1965, Yeoman 1970) show a definite exponential phase (Fig. 10.9). In addition, in the developing artichoke explant (Yeoman, Naik & Robertson 1968) changing the incubation temperature accelerates or slows down the rate of cell accumulation but does not eliminate the occurrence of a phase of exponential growth. The maintenance of such a constant relative rate of cell accumulation suggests

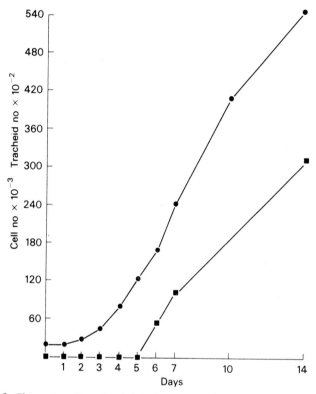

FIG. 10.9. *Change in cell number* (●) *and tracheid number* (■) *of cultured artichoke explants.* For conditions of culture see Fig. 10.6. (Robertson 1966.)

that division is being determined by the activity of all the cells in the system. It further suggests that the increment in cell number during any interval is influenced by all the cells at the beginning of that interval. If this is so, then the number of cells formed is determined by the quantity of synthetic product accumulated in the whole system. It may be supposed that each cell in the system synthesizes at a constant rate the materials required for the formation of other cells and the exponential character is maintained by the increasing number of cells available for synthesis. This situation is similar to that found in the growing apex of the root (Brown 1951).

MAINTENANCE OF DIVISION

The continued growth and division of the tissue culture is accompanied by massive increases in protein (Fig. 10.10), and nucleic-acid content. Synthesis

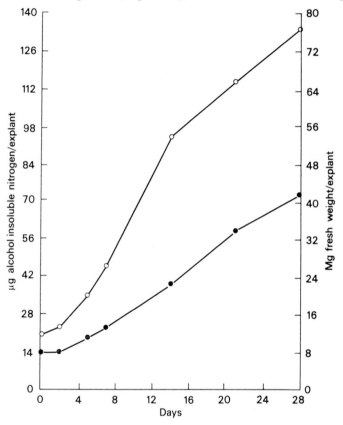

FIG. 10.10. *Change in alcohol insoluble nitrogen* (○) *and fresh weight* (●) *of cultured artichoke explants.* For conditions of culture see Fig. 10.6. (Robertson 1966.)

and accumulation of new cell walls and the laying down of secondary walls also proceed apace and contribute a significant part of the total dry matter. Energy requirements for these syntheses are revealed by the massive amount of oxygen taken up by the growing mass (Fig. 10.11). During this period of consolidation and increase, the average cell size of the developing callus remains constant, reflecting the achievement of a new balance within the system and it is during this time that a recognizable callus emerges and the differentiation of new structures commences. It is perhaps of interest to note that although the onset of differentiation is inevitable if the culture continues to grow on the original medium, such differentiation may be prevented and

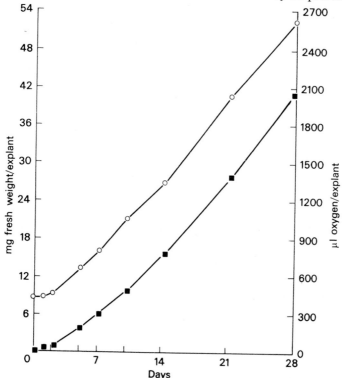

FIG. 10.11. *Comparison of the change in fresh weight* (○) *and net oxygen uptake since excision* (■) *of cultured artichoke explants.* For conditions of culture see Fig. 10.6. Oxygen uptake measured by the direct method of Warburg. (Robertson 1966.)

the callus maintained indefinitely in an undifferentiated but proliferating state by sub-culturing onto fresh medium (Skoog & Tsui 1948, Gautheret 1939, Nobécourt 1939a, and White 1939a).

DEPARTURE FROM DIVISION AND THE ONSET OF DIFFERENTIATION

As the end of the division phase is approached, another morphogenetic sequence appears. Previously, cell division has been located towards the periphery of the tissue and these divisions, which were mainly periclinal, contributed towards the formation of the radial rows of cells of the wound cambium. In most developing calluses these divisions in the surface layers slow down and may even stop (Sevenster & Karstens 1955, Yeoman, Dyer & Robertson 1965) and are replaced or supplemented by secondary growth formations. Eventually a new pattern may be superimposed upon the wound cambium region which is obliterated by the increase in size and number of secondary structures (Gautheret 1959, 1966). Frequently a feature of this departure from divisions at the surface to divisions deeper in the tissue is a change in orientation in the plane of division. This is reflected by the appearance of nodular structures resembling vascular bundles or apical meristems which contrast strikingly with the regular radial arrangement of the wound cambium.

The formation of meristematic nodules is a common feature in developing callus cultures (Gautheret 1959). Frequently they become growth centres which do not differentiate further and produce expanded parenchymatous cells from their periphery (Yeoman, Dyer & Robertson 1965). These cells form the 'frothy' proliferations characteristic of actively growing callus cultures. In addition, the products of the activity of the wound cambium expand considerably and contribute large inflated cells. White (1967) has pointed out that rapidly growing calluses are very much alike except in specific cellular details but as a callus slows down in growth rate it takes on characteristic forms and structures. Despite this all such calluses are similar in that they contain nodular or sheet meristems in groups or widely scattered throughout the tissue.

A major point of interest in the development of a callus is this transition from general division in the outermost regions of the callus to much more localized divisions deeper in the tissues. White (1967) has suggested that the appearance of localized regions of activity provides evidence of the existence of gradients of characteristic steepness which are important in morphogenesis and which can be suppressed, simplified, reoriented or modified experimentally. Morphogenetic effects of added growth substances could at least be partly explained on the basis of this hypothesis. Wetmore & Rier (1963) have applied auxin gradients to growing tissues, but other gradients were also introduced and the results, although suggestive, are not clear cut. Yeoman & Brown (1971) have shown that the plane of division in developing callus cultures of Jerusalem artichoke may be changed by the application of mechanical stress. They have erected a hypothesis based on these results to explain how pressures exerted on the plasmalemma may dictate the plane of cell division.

Comprehensive data on the changes which accompany the onset of differentiation are rare (Steward & Shantz 1956, Yeoman, Dyer & Robertson 1965, Yeoman, Naik & Robertson 1968, Robertson 1966) and concern a limited variety of tissues. A particularly marked change in the average cell size of the explant (Fig. 10.6) occurs at about the time the first signs of differentiation appear. Up to this point, the average cell size of the developing callus has been decreasing rapidly, but quite suddenly the decrease is arrested and throughout the phase that follows it is maintained at an approximately constant level. This sudden switch from predominantly division in the periphery of the tissue to more localized division and differentiation deeper in the explant suggests that a new state has been reached. This suggests that in a developing culture, once a particular average cell size has been reached regressive change is superseded by a phase of activity involving division, expansion and differentiation. The final pattern achieved depends on the nature of the callus and the conditions of growth employed.

GROWTH OF THE ESTABLISHED CALLUS

Callus tissue derived from the original explant can be maintained in an actively growing state by the transfer of fragments to a fresh medium. Alternatively a callus may be raised from single cells or groups of cells from a suspension culture (see Chapter 8). The established callus can then be propagated by sub-division and this procedure if repeated can provide a large population of cultures amenable for an investigation into the growth and metabolism of the tissue. Problems inherent in such an investigation are (1) the division of the callus into equal pieces, usually by weight, which must be conducted under conditions of strict asepsis, (2) to ensure that the pieces of tissue of similar mass also possess a similar growth potential and (3) to restrict damage incurred during subculture. Caplin (1947) in a study with tobacco callus has shown that the first problem can be solved by careful manipulation but the second is apparently insoluble, for in order to detect differences between pieces of callus which appear to be superficially similar, it is necessary to kill the tissue. Despite this difficulty several workers have managed to make a meaningful analysis of the growth of an established callus (Caplin 1947, Enderle 1951, Henderson, Durrell & Bonner 1952, Straus & La Rue 1954, and Lance 1957). Perhaps the most detailed study was that undertaken by Caplin (1947) with tobacco. He examined the change in external morphology of cubes of tobacco callus and this is shown in Fig. 10.12. The growth of the callus is spectacular and occurs at the surfaces not in contact with the agar medium. The proliferation rapidly leads to the deformation of the cube and the emergence of an irregular mass of tissue. It appears that transformation of the cube is due to the growth of knobs at or

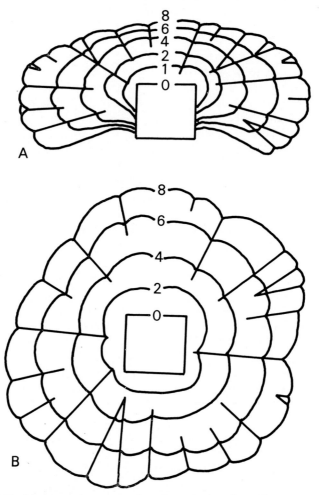

FIG. 10.12. *Diagrammatic sections through a tobacco culture illustrating morphological changes during 8 weeks' growth.* A, section perpendicular to nutrient agar
surface; B, section parallel to nutrient agar surface. Cubes change into somewhat
flattened hemispheres by growth of knobs in which proliferation of cells occurred in
and near the surface. Concentric lines represent shape of culture after indicated number of weeks' growth. Radial lines indicate surface of contact between adjacent
knobs. Points of radial lines nearest centre are points at which knobs have divided.
Where two lines radiate from such points, new knobs have been formed (Caplin
1947).

near to the surface of the callus. The rapid growth localized in the peripheral
regions of the callus is related to the wound damage at the surface of the
tissue as well as to the other factors already discussed (p. 243 *et seq.*). Increase
in fresh weight of the callus cubes is shown in Table 10.2. From these data

can be seen the variation that exists between individual cultures. This variability increases with the age of the culture and is always fairly considerable. Caplin (1947) suggests that the variability is due to the individual cultures growing at different rates and further points out that the differences in growth rate must be due to inherent differences in the tissue cubes at the time they were cut from the callus mass. It is not clear whether these differences are due to the growth rate of the culture from which the cube was obtained or related to the heterogeneity of the callus. It is a feature which has severely limited such

TABLE 10.2. Growth of uniformly cut cultures of tobacco callus (Caplin 1947)

No. of weeks' growth	No. of cultures removed	Weight of largest culture (mg)	Mean weight (mg)	Standard deviation	Coefficient of variability (%)
0	20	6·8	5·8	0·61	10·5
2	20	19·9	10·6	3·28	31·0
3	20	26·0	17·6	4·74	26·8
4	20	39·0	19·8	6·26	31·7
5	20	63·8	36·6	16·80	46·0
6	20	96·7	46·4	22·30	48·0
7	16	119·6	72·0	27·60	37·8
8	16	184·0	93·0	37·50	40·3
9	16	231·4	105·0	57·20	54·5
		× 34*	× 18*		× 5*

* Increase in 9 weeks.

investigations and is reflected in the scarcity of publications on the growth analysis of established callus cultures. It is clear from the data on fresh weight increase that growth of the callus occurs exponentially for most of the culture period. Towards the end of this period a diminution of the growth rate occurs. Caplin points out that the use of Blackman's formula for the growth of an annual plant, $W_1 = W_0 e^{rt}$, (Blackman 1919) is particulary applicable to the growth of this callus culture. The biological significance of this phase of exponential growth within the context of the developing callus has already been discussed earlier in this chapter.

MORPHOLOGY OF THE ESTABLISHED CALLUS

An obvious difference between established calluses is their texture and physical properties. Some are compact and quite hard and others are friable and disintegrate freely when handled or disturbed. The friable callus is a most suitable subject for growth as a suspension culture where mechanical agitation leads to a dispersion of the tissue. Non-friable tissues grow as lumps and do not give suspensions. It has been shown by Torrey & Shigemura (1957) working with pea callus and by Reinert & White (1956) with callus derived from *Picea glauca* that, with these tissues, compact cultures may give rise to the friable variety, but the reverse did not occur. Torrey & Shigemura (1957) have also shown that high concentrations of yeast extract relative to 2,4-D concentration induce friability in pea callus. Blakely & Steward (1961) have shown that friable and compact calluses derived from *Haplopappus gracilis* are interconvertible. The change from one form to another may be achieved by changing the level of coconut milk and napthalene acetic acid in the culture medium. Grant & Fuller (1968) have also shown that friable and compact forms of callus from the root of *Vicia faba* exist and may be interconverted. This may occur during the growth of the callus and be associated with the exhaustion of nutrients in the culture medium. Both Blakely & Steward (1961) and Grant & Fuller (1968) have investigated the basic anatomy of the different forms and have demonstrated quite spectacular differences. In bean callus the friable variety exhibits a large number of organized meristematic centres or nodules separated by large undifferentiated cells, whereas the non-friable type is much less differentiated and has a much larger proportion of large vacuolated cells. In *Haplopappus* cultures the major difference is in the packing of the cells, friable callus is composed of loosely arranged cells. In contrast the non-friable callus is made up of tightly packed cells with evidence of non-random divisions in localized areas, evidence for the onset of organization. Grant & Fuller (1968) have also compared the chemical composition of friable with non-friable callus. The differences show that the non-friable callus has a greater total amount of cell-wall polysaccharides, more of each particular cell-wall fraction per unit dry weight, but a decreased percentage of cellulose compared with pectic substances and hemicelluloses. The greater amount of cellulose will no doubt increase the rigidity of the cells and the increased amount of pectic substances ensures that the cells are held together more firmly and resist fragmentation. Attempts to quantitate friability and show how it changes during callus development and in response to specific chemical treatments are currently in progress in the authors' laboratory.

METABOLIC PATTERNS IN TISSUE CULTURES

Differences between callus and source material

Some of the early metabolic events observed in tissue cultures after removal from the plant are related to the damage caused by excision. Other events reflect normal cell cycle patterns amplified by division synchrony. These changes occurring during dedifferentiation have been described above. The metabolic patterns of tissue cultures maintained and subcultured for long periods well illustrate the balance between genotype and environment.

Many tissues in culture have been shown to retain some of the characteristics of the plant and organ of their origin. Slepyan, Vollosovich & Butenko (1968) pointed out that tissues of various medicinal herbs synthesize traces of biologically active products, such as alkaloids, steroids, vitamins and antibiotics which are identical or very closely related in structure and function to those isolated from intact plants in very much larger amounts. Calluses from selenium-tolerant species of *Astragalus* retained their tolerance in culture (Zeibur & Shrift, 1971). The use of tissue culture material for investigation of biosyntheses and physiological studies carries the tacit assumption that the results can be usefully extrapolated to the whole plant.

On the other hand tissue cultures share a form of growth, which is very similar irrespective of origin with its relatively rapid cell division, loose structure and lack of organization, and which differs from the organs from which cultures are derived (White 1967). The esterase isoenzyme pattern of calluses from different somatic organs of *Oryza sativa* was very similar whereas the enzyme isolated from whole organs displayed characteristic differences in numbers and density of bands in gel electrophoretograms (Wu & Li 1970). Immunoelectrophoresis has revealed changes in antigens during the cycle from excision to regeneration of tobacco plants. Some proteins disappeared during dedifferentiation while others appeared and seemed characteristic of the undifferentiated phase, as they were also detected in normal meristematic tissue (Butenko & Volodarsky 1968). Tissue grown from castor bean endosperm lost the ability to accumulate lipid, and no glyoxylate cycle activity could be detected (Brown, Canvin & Zilkey 1970). Tissue derived from Jerusalem artichoke lacked inulin, the characteristic storage carbohydrate of that plant (Kaneko 1967).

Frequently tissue cultures differ from the parent material, probably as a result of the different nutritional environment to which they are subjected. Thus kinetin promotes nicotine production in callus from the stem pith of tobacco, although nicotine is normally a product of the roots of *Nicotiana* species (Tabata, Yamamoto, Hiraoka, Marumoto & Konoshima 1971). Atropine occurs in cultured *Atropa belladonna* roots and roots derived from callus, but in the presence of auxin none could be detected in tissue cultures. When transferred to auxin-free medium atropine was produced. The investigators (Bhandary, Collin, Thomas & Street 1969) suggested that alkaloid synthesis only occurred in association with meristem organization, or at least in conditions which usually led to the formation of organized structures.

Nicotine and anatabine were not formed when IAA was replaced by 2,4-D in the medium of tobacco cultures (Furuya, Kojima & Syono 1971) whereas the sterols, β-sitosterol, stigmasterol, campesterol and cholesterol were produced in the presence of 2,4-D but not with IAA. Scopolin and scopoletin occurred in both conditions. Auxins also affect levels of anthocyanins (Klein & Hagan 1961) and coumarins (Sargent & Skoog 1960) in *Impatiens*. *Catharanthus roseus* cultures lost the ability of the whole plant to synthesize certain alkaloids, but cavincine and akuammicine previously detected only in the roots of whole plants were found in tissue cultured from leaves, and cathalanceine, yohimbine and lanceine were also identified. These latter alkaloids are known from related species but not whole plants of *C. roseus* (Patterson & Carew 1969). In this case organ, and even species specificity has been broken down in tissue culture.

The protein complement of tissue cultures has also been shown to vary under different environmental conditions. There were fewer isoenzymes of acid phosphatase, and more of esterase and peroxidase than occurred in the source (stem) material of *Dianthus* (McCown, McCown, Beck & Hall 1970). Many of the isoenzymes detected, corresponded with those directly extracted from stem tissue, but all three enzymes showed some different components, and the patterns varied under environmental combinations of high (25°C) and low (0–5°C) temperature, light and dark.

Green cultures of *Kalanchoë* could fix CO_2 into sugars, sugar phosphates and organic acids, and contained phospho-enol pyruvate carboxylase. However, no diurnal fluctuation in acid levels occurred, and there was no net CO_2 fixation. The tissue did not display a typical pattern of Crassulacean Acid Metabolism (McLaren & Thomas 1967). Although there are many instances of callus cultures producing chlorophyll and photosynthetic products when grown in the light (Venketeswaran 1965, Fukumi & Hildebrandt 1967, Sunderland & Wells 1968, Hanson & Edelman 1972) it remains doubtful whether these cultures are close to being photosynthetically self-sufficient. This is presumably due to a lack of organization, rather than a lack of expression of full synthetic potential. Corduan (1970), by growing calluses on a minimum of organic substrates, followed by a CO_2-enriched atmosphere, produced a callus of *Ruta graveolens* which when transferred to medium containing only mineral salts, grew autotrophically for a period of 2 years.

These examples demonstrate that the exact biochemical expression of cells in tissue culture may be considerably modified by conditions imposed by the medium. If the events of early callus development can be accurately regarded as a deprogramming process, it is not surprising to find that in its new state the cell in tissue cultures may not have the same access to genetic information that it had before. Excision and the various factors required for growth and maintenance of tissue cultures presumably act by altering the accessibility of genetic information (see also discussion in Chapter 11, p. 293 *et seq.*).

Some of the metabolic patterns observed in apparently undifferentiated tissue cultures are associated with a subsequent inception of morphogenesis, and such patterns have been studied with the hope of understanding the molecular links between exposure to morphogenic stimuli (often appropriate auxin/kinin balance in the medium, Skoog & Miller 1957), and the appearance of differentiated structures. In studies of the intact plant it may be difficult to distinguish between initiation of specialized meristematic regions and outgrowth from pre-existing primordia. In tissue cultures this problem need not arise. RNA and protein are increased in shoot-forming regions of tobacco (Thorpe & Murashige 1970), and some qualitative changes in the proteins of carrot were observed prior to visible morphological onset of root formation (Werner & Gogolin 1970). It is not known whether the exogenous auxin/kinin balance acts by causing intercellular variations in the concentrations of these substances which are themselves directly responsible for switching the pattern of synthesis, or whether they act through other regulatory systems. The effect of auxin/kinin interaction on levels of scopoletin and scopolin (Skoog & Montaldi 1961), and the correlation between scopoletin content of tobacco callus and subsequent shoot development (Tryon 1956) suggested the possibility of a regulatory system centred on the effect of phenolics on IAA oxidation. Grant & Fuller (1968) suggested that part of the control mechanism could be the result of a competition between starch and cell-wall synthesis for a common intermediate. Thorpe & Murashige (1968) detected starch accumulation in cells, at loci which ultimately give rise to primordia, before any visible morphological change, and suggested starch accumulation was a prerequisite rather than a consequence of organ initiation, as added gibberellin prevented both processes. This does not preclude starch formation from being incidental if gibberellin blocks an early stage in the sequence. Thorpe & Murashige (1968) suggested that starch acted as an energy reserve material which was utilized during organogenesis.

Studies on tissue culture metabolism

Whilst admitting that metabolic patterns observed in excised material are not bound to be identical with those in intact plants, many workers have seen in tissue cultures, an attractive system in which to investigate metabolic inter-relationships and pathways of biosynthesis. Tissue cultures offer a source of relatively homogeneous material which can be easily manipulated and obtained in large amounts in a controlled environment. It offers many of the technical advantages of work with micro-organisms to the plant investigator. Some of these studies are mentioned below.

(a) *Auxin metabolism*

The widespread dependence of normal cultures on an exogenous supply of growth regulators, particularly auxins, led to the early investigation of auxin metabolism in tissue cultures. Naturally occurring plant auxins are either IAA or are closely related to IAA. Different plant material varies in its sensitivity to added auxins and this could be due to a real difference in optimum endogenous levels for growth response, or could be due to differences in other factors affecting endogenous levels. IAA synthesis can apparently occur even in tissues which require supplementation with exogenous auxin for growth, and the pathway of synthesis appears to be similar to that of intact plants. IAA is formed from tryptophane via indolylpyruvic acid and indolylacetaldehyde in tissue slices of water melon (Dannenberg & Liverman 1957). The similar activity of enzymes of the IAA synthetic pathway in auxin sufficient (crown-gall) tissue and normal sunflower tissue requiring exogenous auxin suggests that different endogenous auxin levels do not reflect differing rates of synthesis (Henderson & Bonner 1952).

The variable binding of auxin to protein was not considered a likely regulatory mechanism (Butenko 1964), and a control acting via differential rates of auxin inactivation was favoured. Photo-oxidation and inactivation by pH changes have been demonstrated in cell-free systems but the most likely regulatory mechanism is the action of auxin oxidases. It is possible that the synthetic auxins such as NAA and 2,4-D act by competing for these enzymes, thus protecting endogenous IAA (Straus 1962). IAA oxidase activity has been detected in many cultured plant tissues, and depends on availability of cofactors, e.g. manganese, phenols (Goldacre, Galston & Weintraub 1953). IAA oxidase is sensitive to light in the presence of riboflavin (Galston 1959) and phytochrome in red light (Hillman & Galston 1957). Galston & Hillman (1961) postulated a two-stage oxidation involving formation of a peroxide in the presence of flavoprotein, followed by oxidation in the presence of peroxidase, on evidence that IAA degradation had an action spectrum similar to the absorption spectrum of riboflavin, and inhibition of oxidation by cyanide and catalase. Inhibitors have been isolated from carrot tissue which inhibited auxin oxidase activity in other tissues, e.g. *Partheno-cissus* (Butenko 1964). Scopoletin-inhibited IAA degradation in tobacco callus, and the level of this substance in the tissue was itself affected by exogenous IAA and kinetin (Skoog & Montaldi, 1961). Finally phytohor-mones may induce IAA oxidase synthesis. Two isoenzymes of IAA oxidase in particular were stimulated by kinetin at concentrations above $0.2 \mu M$ but not in the presence of Actinomycin D or cycloheximide, supporting *de novo* synthesis of the isoenzymes in response to kinetin (Lee 1971). Obviously there are numerous possible mechanisms by which auxin levels might be controlled, and these may assume different importance in different tissues and conditions. The primary target of auxin action has not, however, been elucidated by any of the work with tissue cultures.

(b) *Tumour tissue* (see also Chapter 13)

Tissue cultures have been used to investigate the difference between plant tumour cells and normal cells. Plant tumours, which are characteristically a mass of unorganized tissue of exceptionally rapid growth which rarely produce differentiated structures, can be induced by a variety of agents. Once established, however, the continued presence of the agent is not required. Tumours induced by bacteria and viruses can be freed of the tumorogenic agent by sterilization or temperature treatment (White & Braun 1942) and retain the characteristic growth form of tumours indefinitely in culture. Such permanent tumour cells are able to synthesize sufficient quantities of auxins and cytokinins for rapid growth, and unlike normal cells, require no supplementary supply (Braun 1962). Transformation is accompanied by increases in RNA (Srivastava 1968), total protein (Braun & Wood 1961, Scott, Smillie & Krotkov 1962) and the activity of certain enzymes (Reddi 1966). Tumour cells were also found to be self-sufficient for myo-inositol, glutamine and asparagine, which are required by normal tissue, and to exhibit enhanced rates of purine and pyrimidine synthesis. Braun & Lipetz (1966) view the tumour cell as being permanently trapped in a division-oriented pattern of metabolism, by constitutive synthesis of hormones, which in turn determine permanent excessive production of compounds that normal tissue cultures require from the medium. These are used for nucleic acid synthesis, mitotic proteins and membrane synthesis. The change in hormone synthesizing ability is probably quantitative rather than qualitative, as normal tissues do produce a certain amount of auxin (Dannenberg & Liverman 1957), and perhaps the difference lies in the mechanism by which auxin levels are regulated.

The ability to initiate metabolic sequences which lead to the formation of differentiated structures is not irretrievably lost in tumour cells. Some plant tumours do produce organized, albeit morphologically abnormal, leaves and buds in culture. These can be maintained indefinitely, but when consecutively grafted onto a series of healthy plants are reported to eventually revert to normal growth, and yield tissue material which is normal (i.e. non-tumourous) in culture (Braun 1959). In tumour cells, it appears that the aspects of cellular metabolism concerned with cell division are permanently activated rather than that the potential to initiate other patterns is permanently lost. Tumour cells either do not generate, or do not usually respond to conditions which cause a switch in the pattern of metabolism in normal cells.

(c) *Response to light*

Tissue cultures have recently been used to follow the development of chloroplasts and photosynthetic capacity under the influence of light. Sunderland & Wells (1968) followed greening in cells derived from *Oxalis dispar* endosperm grown on 2% sucrose. In the dark, proplastids formed amyloplasts and extensive starch deposition occurred. When these cultures were transferred to the light, chloroplasts developed. Thylakoids formed between starch grains and starch disappeared. Pigments accumulated as grana developed. The whole process was accompanied by a rapid decrease in the rate of cell division and expansion, which, it was suggested, was due to a diversion of raw materials into specific syntheses accompanying chloroplast maturation. The incompatability of general rapid cell proliferation with the intercellular differentiation process was indicated by the fact that auxin, at concentrations optimal for growth, completely suppressed chlorophyll formation. A similar suppression of chlorophyll synthesis by a high level of auxin and antagonism between most active cell division and chloroplast differentiation are also reported for *Atropa belladonna* (Davey, Fowler & Street 1971). The observation that starch deposition and formation of lamellar structures from the inner plastid membrane occurring in the dark in the *Oxalis* culture was not affected by the presence of auxins, in contrast to thylakoid development, was tentative evidence that thylakoid membranes had a different ultrastructural origin from the membranes of amyloplasts. However, auxin is not consistently involved in chloroplast development. At NAA concentrations optimal for growth, the development of pigments was unaffected in *Hypochaeris* cultures, stimulated in *Haplopappus gracilis* but partially inhibited in *Acer pseudoplatanus* as well as *Oxalis*. 2,4-D acted similarly but completely inhibited pigmentation in *Oxalis* (Sunderland 1966).

The transfer of *Kalanchoë* cultures to the light was followed by an increase in carotenoids (Stobart, McLaren & Thomas 1967) and chlorophyllase activity (Stobart & Thomas 1968). The implied involvement of chlorophyllase in the synthesis of chlorophyll *in vivo* was supported by the demonstration of a transient rise in the level of phytol, a substrate of chlorophyllase, preceding the detection of chlorophyll *a* (Stobart, Weir & Thomas 1968). The involvement of a mevalonate-activating system in pigment production was implicated by the increased activity of this system in the tissue (Thomas 1970). Another enzyme, aminolaevulinate dehydrase, showed specific activity increases in proportion to chlorophyll content in greening tobacco callus, and Schneider (1970) suggested this enzyme was intimately connected with plastid development, despite being recovered largely from supernatants in general fractionations.

Kasperbauer & Reinert (1967) found that stem pith cultures of *Nicotiana tabacum* showed a growth response to red light which was reversed by far-red. Phytochrome was particularly detectable in material grown in the dark or with short exposure to red light (5 minutes/day). They pointed out that the system could be useful in studying the effect of phytochrome in bringing about a growth response, as the tissue showed a light induced growth response but could grow heterotrophically in the dark.

(d) *Amino-acid metabolism*

Delmer & Mills (1968) used an amino-acid sufficient cell-clump culture of *Nicotiana* to study the biosynthesis of tryptophane by isotope competition and direct labelling, and found shikimic acid, anthranillic acid, indoleglycerol phosphate and indole could serve as precursors, which suggested a similar pathway to that found in micro-organisms. By monitoring the growth and amino-acid pools in rice callus tissues supplemented with various amino acids Furuhashi & Yatazawa (1970) investigated the interrelationships of amino acids in a manner similar to that used with colonies of micro-organisms. The characteristic poor growth in the absence of methionine only occurred when the medium contained both threonine and lysine, possibly because the biosynthesis of methionine is competitively inhibited by these two amino acids. On isoleucine deficient medium the threonine content decreased suggesting that *in vivo* isoleucine inhibited the decomposition of its precursor.

(e) *Secondary plant products*

Much recent attention has been paid to the synthesis of so-called secondary plant products in tissue cultures. Examples include steroids, e.g. diosgenin, β-sitasterol in *Solanum* (Heble, Narayanaswami & Chadha 1968, Kaul, Stohs & Staba 1969), β-sitasterol, stigmasterol, campesterol, cholesterol in *Nicotiana* (Furuya, Kojima & Syono 1971, Nickell & Tulecke 1961), cardiac glycosides in *Digitalis* (Buchner & Staba 1964); pyridine alkaloids, e.g. nicotine from *Nicotiana* (Furuya *et al.* 1971, Tabata, Yamamoto, Hiraoka, Marumoto & Konoshima 1971); tropane alkaloids in *Datura* (Chan & Staba 1965); indole alkaloids in *Catharanthus* (Patterson & Carew 1969), *Ipomoea* (Staba & Laursen 1966); caffeine in tea (Ogutuga & Northcote 1970a); coumarins, e.g. scopoletin and scopolin in *Nicotiana* (Fritig, Hirth & Ourisson 1970); tannins in *Juniperus communis* (Constabel 1968); lignins (Brown 1966); deoxyisoflavones in *Glycine* (Miller 1969); vitamins, e.g. thiamine (Dravnieks, Skoog & Burris 1969). These examples are incomplete and merely intended as entry points into the literature.

In some cases the aim has been that of understanding the mode of biosynthesis and the relevance of these products to cellular metabolism. West

& Mika (1957) used cultures of *Atropa belladonna* in studies of atropine synthesis. Ogutuga & Northcote (1970a) investigated the biosynthesis of caffeine in tea callus tissue and found evidence that caffeine arises from the breakdown of nucleic acids rather than direct from purine pools. In tobacco tissue cultures Fritig (Fritig *et al.* 1970), by feeding labelled compounds considered to be precursors, found evidence that scopolin, the glycoside of scopoletin which is often found in association with the coumarin, was probably not involved in the latter's synthesis, but either arose from scopoletin or independently, and behaved like a storage product.

Some workers have, more ambitiously, drawn attention to the possibility of applying the method of tissue culture to the controlled production of such compounds as alkaloids, steroids, vitamins, antibodies and enzymes on a commercial scale, an ambition not so far realized. Butenko (1964) cites some attempts to establish cultures of pharmacologically useful plants. Most of the work to date has been limited, however, to determination of the optimal conditions for growth and synthesis of the relevant products. In exceptional cases substances may be accumulated in tissue cultures in excess of the amounts present in the intact plant (Heble, Narayanaswami & Chadha 1968). The stated advantages of tissue cultures in this respect—rapid growth, sterility, determinable and controllable environmental conditions, ease of penetration and extraction, apply even more to cell suspensions, grown in chemostats (Wilson, King & Street 1971) and such systems would seem particularly suitable for exploitation in production of plant substances (see Chapter 11).

CHAPTER 11

GROWTH PATTERNS IN CELL CULTURES

P. J. KING and H. E. STREET

'exciting prospects lie in the possibility that (plant) cells, freed from the restraints of being part of a multicellular, multifunctional organism will, in effect, be a new group of micro-organisms with all the capabilities which this encompasses'
(Nickell & Tulecke 1959)

Introduction 269
Aggregation and heterogeneity of cell suspension cultures . . . 270
Growth patterns in batch cultures 274
Synchronous cultures 297
Semi-continuous and continuous cultures 309
Glossary of technical terms 334

INTRODUCTION

The earliest suspensions of cultured plant cells capable of repeated sub-culture (those of *Nicotiana tabacum* and *Tagetes erecta*) by Muir (1953) were developed as a source of single viable cells from which single-cell clones could be established (Muir, Hildebrandt & Riker 1958—see Chapter 8). Similar suspensions composed of free cells and aggregates were also obtained from cultured carrot-root explants (Steward & Shantz 1956), from normal and tumour tissues of *Picea glauca* (Reinert 1956) and from haploid tissue of *Antirrhinum majus* (Melchers & Bergmann 1959). Nickell (1956) first demonstrated the feasibility of growing such a cell suspension as a culture of 'micro-organisms' using a highly dispersed suspension of cells derived from the hypocotyl of *Phaseolus vulgaris*. Subsequent studies from his laboratory (Nickell & Tulecke 1960, Tulecke & Nickell 1959) described the use of microbial fermentation techniques for studies on growth kinetics, biochemical composition and production of particular metabolites by such plant suspension cultures.

The present chapter attempts to evaluate critically the contributions subsequently made by use of such cell suspension cultures to our knowledge of the control of cell growth and metabolism. Much of this work is based upon studies with **batch cultures*** but more recently **synchronized cultures**

---

* Terms shown, when first introduced, in bold type are defined in the 'Glossary of Technical Terms' which concludes this chapter.

and **continuous cultures** have been developed and the new possibilities thereby opened up will also be reviewed.

AGGREGATION AND HETEROGENEITY OF CELL SUSPENSION CULTURES

Heterogeneity of cell cultures

Suspension cultures are usually initiated by placing pieces of a callus culture into an agitated liquid medium. Many factors influence the friability of the tissue piece and its dispersion in the liquid medium (see Chapter 4, p. 83), and clearly different cultures all described by the blanket term 'suspension cultures' have varied very greatly in their degree of cell aggregation. For example, the suspension cultures of *Antirrhinum* studied by Melchers & Bergmann (1959) contained large spherical aggregates (up to 4 mm diam.) and very few free cells, whereas the cultures of *Rosa* sp.—Paul's Scarlet (Tulecke 1966)—are very highly dispersed and 'represent the closest approximation to bacterial cultures yet obtained with plant cells' (Liau & Boll 1971).

The morphological heterogeneity of cell cultures is well known (Street 1966b, Liau & Boll 1971, Nash & Davies 1972). Furthermore, there are several reports indicating a correlation between biochemical heterogeneity and aggregation. Mixtures of pigmented and non-pigmented cell groups have been observed in cultures of *Haplopappus gracilis* (Steward, Israel & Mapes 1968) and *Acer pseudoplatanus* (Wilson 1971). Non-pigmented suspensions of *Haplopappus gracilis* have given rise to pigmented and non-pigmented cell groups in the presence of a high level of naphthalene acetic acid (NAA) (10^{-5} M) (Constabel, Shyluk & Gamborg 1971). In cell cultures of *Nicotiana tabacum*, variation was observed in the activity of cytochrome oxidase between cell groups (de Jong, Jansen & Olsen 1967) and the activity of several enzymes was localized in particular cells within individual aggregates. Quantitative and qualitative differences in peroxidase and catalase activity between cell groups of different sizes occurred in cell cultures of *Arachis hypogoea* (Verma & van Huystee 1970a). Rates of protein synthesis and concentrations of free amino acids have also been shown to vary with aggregate size (Verma & van Huystee 1970b).

The ideal requirement for a cell suspension culture is that of morphological and biochemical homogeneity. Therefore, detailed study of aggregation, of the selection of easily dispersed cell types and of cultural procedures favouring cell separation is essential. However, it must be emphasized that all long-established cell cultures appear to show genetic diversity in their cell populations (see Chapter 7), usually extending to differences in chromosome number or chromosome morphology. Such heterogeneity will not be eliminated by obtaining from such a mixed cell population a truly free-cell culture; only free-cell cultures derived by single-cell cloning and maintained under conditions conferring nuclear stability meet the stringent (ideal) requirements for future work.

Factors affecting the degree of cell separation in cell suspension cultures

Not only the specific origin of the tissue culture but the composition of the culture medium may determine the degree of cell separation (Street, 1966b). For instance, better separation of cells of *Daucus* and *Convolvulus* was obtained in fully defined medium than in medium containing coconut milk or yeast extract and the vitamin requirements for high growth rates of the suspensions differed from those for the parent callus cultures (Torrey & Reinert 1961). Again the medium for optimum cell separation depends upon the species (Tulecke, Taggart & Colavito 1965); media suited to the culture of a monocotyledon (*Zea mays*) and a gymnosperm (*Ginkgo biloba*) were very different from those for the two dicotyledons tested (*Rosa* sp. and *Lycopersicon esculentum*). Often cell separation is critically affected by the levels of plant growth hormones in the medium. Relatively high auxin levels have been reported to increase and low auxin levels to decrease cell separation (Torrey, Reinert & Merkel 1962). The absolute and relative concentrations of 2,4-D and kinetin very significantly influenced cell separation in *Acer pseudoplatanus* cell suspensions (Simpkins, Collin & Street 1970). To establish well-dispersed liquid cultures of *Atropa belladonna* it was necessary, in presence of 2 mg/litre NAA, to reduce the kinetin level from 0·5 mg/litre (used for callus propagation) to 0·1 mg/litre (Davey, Fowler & Street 1971).

Selection pressure in favour of increased cell separation

Some limited improvement in cell separation can be obtained by techniques which preferentially select for subculture the free cells and smaller cell aggregates. Veliky & Martin (1970) subjected initial suspensions derived from callus cultures of *Phaseolus vulgaris*, *Pisum sativum*, *Glycine max* and *Ipomoea* sp. to a 'conditioning process' in a **V-fermenter** (for description see Chapter 4, p. 69). Periodically (*c.* 7 days) they allowed the crude suspension to settle and drew off the lower 80% of the culture volume (containing all the larger aggregates) and replaced it with new medium. After 40 days the cultures consisted of isolated cells and very small aggregates and their rate of biomass production was significantly higher than that of the primary suspension. Such a 'drain-and-refill' technique is the basis of a **semi-continuous culture** and it was shown that by its use the cultures could be maintained actively growing for many months. A similar 'selection' technique has been applied to *Acer pseudoplatanus* suspensions grown in small batch cultures (Street, King & Mansfield, 1971). At each subculture (21-day intervals) the culture was allowed to settle for 30 seconds and then cells for subculture withdrawn from the upper part of the suspension. After six passages the proportion of the cell population in aggregates of ten or less cells had increased to 30% as against 6% in the control cultures, but continuing the 'selection' for a further thirty passages effected no further increase in cell separation.

The changing degree of cell separation through the growth cycle

The degree of cell aggregation alters in a predictable way during growth in a batch cell culture. (Sussex 1965, Henshaw, Jha, Mehta, Shakeshaft & Street 1966—Fig. 11.1). Aggregation increases during the period of maximal cell

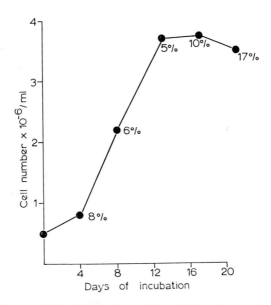

Fig. 11.1. *Changes in the degree of aggregation during the growth cycle of* Acer pseudoplatanus *cells in batch culture.* The percentage of the total cell population found in aggregates composed of ten cells or less are recorded beside the growth curve at each sampling time (previously unpublished data of K.J.Mansfield).

division and the incidence of mitotic figures (mitotic index) is greater in the cell aggregates than in the free cells (Torrey, Reinert & Merkel 1962, Henshaw, Jha, Mehta, Shakeshaft & Street 1966, Mehta, Henshaw & Street 1967). Thus the frequency of subculture may affect the degree of aggregation;

frequent transfers designed to maintain the cells in active division may result in increased cell aggregation (Henshaw *et al.* 1966, Liau & Boll 1971). However, this may not always be the case. Torrey & Reinert (1961) in their studies on suspensions of *Daucus carota* and *Convolvulus arvensis* noted that the proportion of the total cell fresh weight represented by aggregates <100 µm diameter was highest early in the growth cycle and that this proportion decreased markedly with the progress of growth despite the simultaneous increase in number of free cells per unit volume of culture. They therefore concluded, though did not convincingly demonstrate, that frequent subculture favoured maximum cell dispersion. A careful study of aggregation through to **stationary phase** in cultures initiated from various inoculum densities is necessary to establish the subculture regime most conducive to cell separation.

Criteria which precisely describe the aggregation in cell suspension cultures are not easy to determine experimentally. Various workers have used filtration techniques to separate 'fractions' falling within diameter limits and have termed their small fraction the 'free-cell fraction' (Torrey & Reinert 1961, Henshaw, Jha, Mehta, Shakeshaft & Street 1966). Alternatively the 'free-cell fraction' has been defined as composed of free cells and of aggregates containing less than a fixed number of cells (e.g. ten cells) and this expressed as a percentage of the total cell population (Fig. 11.1). Such criteria are probably less descriptive than photographic records (Street, Henshaw & Buiatti 1965). This aspect of culture growth is in need of more detailed study. However, the existing data serve to emphasize that even the most dispersed cell suspension cultures fall significantly short of being true free-cell cultures.

Free-cell cultures obtained by use of enzymes

The frequent reports that cell divisions occur most frequently in cell aggregates has raised the possibility that high growth rate in suspension cultures may be dependent upon the formation of aggregates; that within such aggregates there is either a differentiation into 'feeder' cells and meristematic cells or that aggregates achieve a volume to surface ratio compatible with the critical accumulation of growth-limiting factors. There are, however, lines of evidence contrary to this hypothesis. A culture of *Acer pseudoplatanus* cells in stationary phase contains a high number of free cells but when such a culture is used to initiate a low-density culture which subsequently shows division synchrony, the first division may achieve an almost exact doubling of the cell population. The stationary-phase culture does not contain cells of low or zero division potential and the free cells are not delayed in their division compared with the cells in the aggregates. Again, a more than normal degree of cell separation can be induced in the *A. pseudoplatanus* cell suspensions by incorporating into the culture medium low concentrations of cell-wall degrading enzymes together with sorbitol to increase the osmotic potential of the medium (see Street, King & Mansfield 1971, Street, 1973 and Chapter 8, p. 198). The presence of these enzymes and the consequent increased cell separation does not adversely affect growth rate and this growth rate and increased cell separation is maintained on subculture into the same medium. These highly dispersed suspensions were primarily developed to assist single-cell cloning but the results indicate that aggregation is not essential to high division potential. Reports of the division of single, isolated, cultured cells of various species in synthetic media support this view (Reinert 1965, Earle 1965, Hildebrandt 1965). Thus true free-cell suspension cultures now seem to be in sight of achievement. The enzyme-dispersed suspensions also show very uniform cell morphology (see Fig. 8.5, p. 200) and hence presumably are composed of cells more uniform in metabolic activity.

GROWTH PATTERNS IN BATCH CULTURES

Introduction

Most of the published physiological and biochemical investigations of growth and cell division in higher plant cell cultures have been carried out on batch cultures. This batch culture technique involves the isolation of an inoculum of cells in a finite volume of nutrient medium in a system which is *closed* except for exchange of gases and volatile metabolites with the outside air.

The change with time in the number of cells present in a batch culture follows the course shown in Fig. 4.16, p. 85 (see Glossary of Technical Terms, p. 335, for more detailed description of the separate phases labelled). This pattern of growth over the span from initiation of the culture to the next subculture of stationary phase cells is usually referred to as the **growth cycle** of the culture. The rate of increase in cell number during this growth cycle changes almost continuously from the initiation to the cessation of cell division. The occurrence and duration of each phase shown in the model curve depends very much upon the cell type (species and strain), the frequency of subculture, the initial density and the culture medium used. Fig. 4.16 is derived from observations upon a strain of *A. pseudoplatanus* cells maintained under standard conditions of culture and incubation in our laboratory. Cells of *Nicotiana tabacum* var. Xanthii (line XD) cultured by Filner (1965) in a minimal medium with nitrate as sole nitrogen source did not show a **lag phase** (Fig. 11.2) but cell number and cell dry weight per unit volume increased exponentially for 10 days (a straight line relationship between the logarithm of the growth parameter and time; Eqn (11), p. 335), before rapidly entering a stationary phase. Cell cultures of *Rosa* sp. (Paul's Scarlet), however, when grown in a chemically defined medium, more complex and concentrated than that of Filner (1965), showed a distinct lag phase, a shorter **exponential growth phase** (the duration of which was dependent upon the initial cell density) and a gradual decline in the rate of cell division before reaching stationary phase (Nash & Davies 1972) (Fig. 11.3).

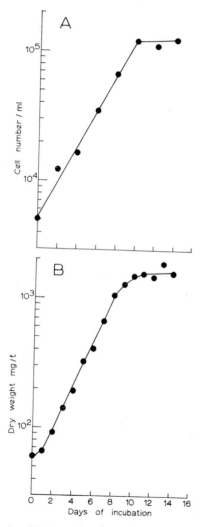

FIG. 11.2. *The growth of* Nicotiana tabacum *cells in batch suspension culture,* measured by cell number (A) and dry weight (B). Semi-logarithmic plots from Filner (1965).

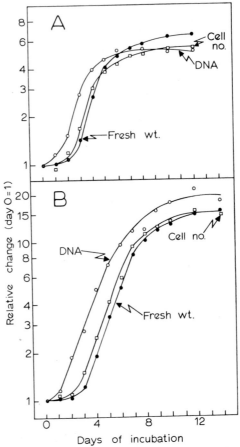

FIG. 11.3. *The growth of* Rosa *sp.* (*Paul's Scarlet*) *cells in batch suspension culture.* Semi-logarithmic plot of the relative changes in fresh weight, cell number and DNA during the growth cycle of a high inoculum density culture (A) and a low inoculum density culture (B). Absolute values for the day 0 cultures were:

	High ID	Low ID
Fresh weight	2·16 g	0·74 g
DNA	144 μg	31 μg
Cell number	44·1 × 10⁶	12·8 × 10⁶
Culture volume	72 ml	64 ml

(From Nash & Davies 1972).

Changes in the rate of cell division during the growth cycle

The changes which occur in the percentage of cells in mitosis during a batch culture (Fig. 11.4) also reveal the pattern of change in the rate of cell division. When divisions are occurring randomly in a population of cells, it can be

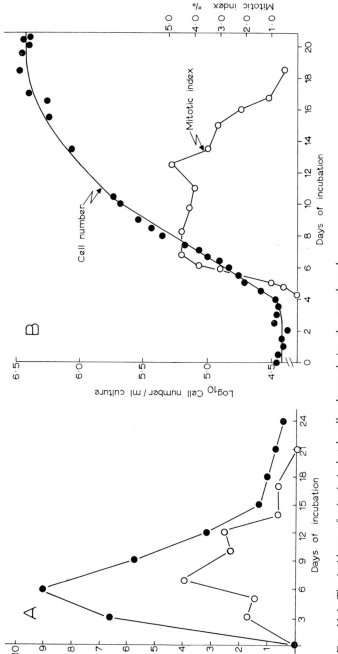

FIG. 11.4. *The incidence of mitosis in batch cell cultures during the growth cycle.*
A. Changes in the mitotic index of cultures of: *Rubus fruticosus* (●) (Henshaw *et al.* 1966) and *Convolvulus arvensis* (○) (Torrey *et al.* 1962). Data for all aggregate-size fractions of the cultures have been combined.
B. Changes in the mitotic index and cell number of *Acer pseudoplatanus* cells in a 4-litre batch culture initiated at low density (previously unpublished data of A.Gould).

shown that the number of cells in any particular phase of the **cell cycle** is proportional to the duration of that phase. Although to calculate *real* values for the duration of any phase a correction must be applied to allow for the form of age distribution in the population (see this chapter, p. 299), we may simply write:

$$\text{Mitotic index} \propto \frac{\text{duration of mitosis}}{\text{duration of cell cycle}}$$

Because in plant cell cultures the duration of mitosis is usually short relative to the total cycle time ($< 10\%$) and is relatively constant over a wide range of growth rates (Table 11.4), the mitotic index is inversely proportional to the mean cell cycle time of the population.

Therefore, in randomly dividing cultures, the peaks of mitotic index in Fig. 11.4A indicate an initial, rapid shortening in the mean cycle time (**mean generation time**) of the cultures during the transition from lag phase to the exponential growth phase. This, in turn, is followed by a longer period of gradual lengthening of mean generation time.* The period of minimum mean generation time (exponential growth phase) was, in these cultures (Fig. 11.4A), of very short duration. However, as already indicated, the exponential growth phase of a batch culture is extended when cultures are initiated at low cell densities and a constancy of mitotic index over 7 days is shown in Fig. 11.4B for a culture of this kind.

Various workers have reported minimal generation times of plant cells in batch culture, i.e. population **doubling times** (*td*) calculated from cell counts during the exponential growth phase: 48 hours for *Nicotiana tabacum* var. Xanthii (Filner 1965), 36 hours for *Rosa* sp. (Paul's Scarlet) (Nash & Davies 1972); 24 hours for *Phaseolus vulgaris* cv. Contender (Liau & Boll 1971) and 22 hours for *Haplopappus gracilis* (Eriksson 1967a). These doubling times are, in general, longer than cell cycle times reported from studies on whole plant meristems (8–20 hours: Evans & Rees 1971, Phillips & Torrey 1972) and are therefore likely to be shortened as more nearly optimal cultural conditions are established. Studies with *Acer pseudoplatanus* cell suspensions in our laboratory have shown that *td* is dependent upon the composition of the culture medium (Fig. 11.7B). While in our basal synthetic culture medium *td* = 70 hours, values as low as 20 hours have been recorded in cultures initiated at low density and embarking on exponential increase in cell number following a period of **synchronous division** (Fig. 11.29).

Unbalanced growth during the growth cycle

When parameters other than cell number are followed through the growth

* See footnote on p. 283.

cycle of batch cultures, similar patterns of change are obtained to that in Fig. 4.17; for instance: cell dry weight (Figs. 11.2 and 11.5), cell fresh weight (Figs. 11.3 and 11.5), total cellular protein, RNA or DNA (Figs. 11.3 and 11.5), oxygen demand, the total activities of some enzymes (e.g. peroxidase—Veliky, Sandkvist & Martin 1969, de Jong, Jansen & Olson 1967, pyruvate-kinase and glucose-6-phosphate dehydrogenase—Fowler 1971), production of total phenolics (Nash & Davies 1972) (Fig. 11.18). However, these separate patterns differ from one another in two important ways:

(i) During the exponential growth phase a number of parameters may show exponential increase at different rates. Thus, during the three generations of *A. pseudoplatanus* cells shown in Fig. 11.6A the **specific growth rate** of the cells was constant whether measured as increase in cell number, cell dry weight or cell protein. However, the rate of accumulation of dry weight or protein was slower than the increase in cell number and, therefore, the mean composition of the cells was undergoing continuous change (Fig. 11.6B). Just prior to the onset of stationary phase the rate of increase of cell fresh

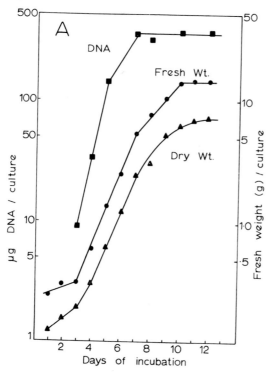

FIG. 11.5. *Use of various parameters to monitor growth of batch cell cultures.* Data presented as semi-logarithmic plots.
A. *Rosa* sp. (Paul's Scarlet); from Fletcher & Beevers (1970).

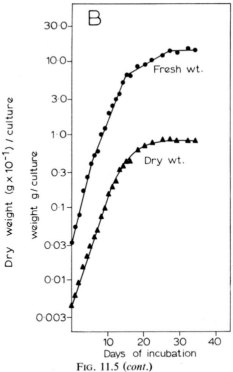

FIG. 11.5 (*cont.*)

B. *Carnation*; from Engvild (1972).

weight was greater than the increase in cell number; the trend towards small cells occurring during the exponential phase was reversed and cell expansion continued into stationary phase. Such **unbalanced growth** is characteristic of batch cultures. This uncoupling between biosynthesis and cell division may be modified by the composition of the culture medium. Thus, the rate of cell division in cultures of *A. pseudoplatanus* can be altered by varying the initial concentration of 2,4-D in the basic synthetic medium (Fig. 11.7A). Raising the concentration of 2,4-D from 4·5 to $9·0 \times 10^{-6}$ M increased the rate of cell division (*td* was reduced by 20 hours, i.e. by 33%) but there was no corresponding increase in the rate of synthesis of cell material (Fig. 11.7B).

(ii) The initiation or suppression of synthesis of particular cell metabolites may be quite uncoupled from the initiation or cessation of cell division. Thus the synthesis of RNA by *A. pseudoplatanus* (Short, Brown & Street 1969b) or by *Rosa* sp. cells (Nash & Davies 1972) was initiated prior to cell division, occurred at a greater rate than cell number increase and ceased before cell division (Fig. 11.8A). Hence the amount of RNA per cell rose to a peak value and then decayed (Fig. 11.8B) during a period when cell division rate remained constant. Again net synthesis of some metabolites ceases before or very shortly after the onset of division so that the concentration of the metabolite per cell progressively declines as division continues. This is exemplified by the free nucleotide content of *Acer pseudoplatanus* cells (Brown & Short 1969, Fig. 11.9A), or by their total soluble nitrogen pool (Simpkins & Street, 1970, Fig. 11.9B).

Stationary phase and the associated lag phase following subculture are phases peculiar to batch cultures of cells. During lag phase, the expanded stationary phase cells of low metabolic activity are transformed into cells capable of embarking upon active division. This transformation involves

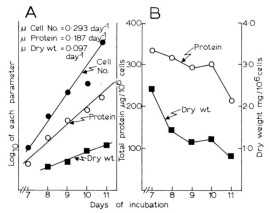

FIG. 11.6. *Unbalanced growth of* Acer pseudoplatanus *cells during the exponential growth phase of a batch culture.*
A. Semi-logarithmic plot showing rates of change of cell number, total protein and cell dry weight per unit volume of culture. The slope of the line of best fit (calculated by linear regression analysis, $P < 0.01$) was used to determine the specific growth rate (μ) of each parameter (see Eqn 11).
B. Changes in total protein content and dry weight of cells with time calculated from data in A.

major changes in cell structure (see Chapter 6) and metabolism. The marked increase in free nucleotides (mainly due to increase in levels of UDP-glucose and ATP) noted above has been interpreted as an essential preparation for subsequent synthesis of large amounts of cell-wall polysaccharides and for the provision of the appropriate energy source for the endergonic processes involved in cell division (Brown & Short 1969). Transient high activity in carbohydrate oxidation by the pentose phosphate pathway (relative to the EMP pathway) during lag phase has been interpreted as providing the necessary NADPH for biosynthesis (Fowler, 1971). Givan & Collin (1967) noted a transient peak of QO_2 (per cell) as lag phase reached its close. The rate of DNA synthesis might be expected to be closely coupled to the rate of cell division. However, DNA synthesis was initiated before cell division

K

occurred and cell division continued after DNA synthesis had virtually ceased in cultures of *Rosa* sp. (Nash & Davies 1972) and *Acer pseudoplatanus* (Short, Brown & Street 1969b, Phillips 1970). The period during which mean DNA content per cell was constant (during which cytokinesis and DNA synthesis continued uninterrupted and at the same rate) was of short duration —*c.* 72 hours in *Rosa* sp. cultures initiated at low density (Fig. 11.3B and

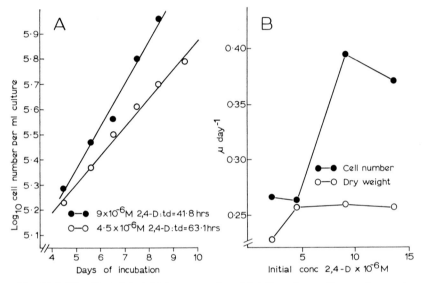

FIG. 11.7. *The effect of the initial concentration of 2,4-D on the specific growth rate of* Acer pseudoplatanus *cells in batch suspension culture in a synthetic medium* (King 1973).

A. Rate of change of cell number (semi-logarithmic plot) at two concentrations of 2,4-D.

B. The influence of the initial concentration of 2,4-D on specific growth rate during the exponential growth phase.

Specific growth rates (μ) were calculated from the slope of the line of best fit (linear regression analysis, $P < 0.01$) for both cell number and dry weight data (see Eqn 11). Population doubling times (*td*) were calculated from the equation: $\mu = \log_e 2/td$ (see Eqn 9).

Fig. 11.10). This period is not clearly defined by the data of Short, Brown & Street (1969b) because the intervals between their samples was too long (Fig. 11.10). The separation in time between the initiation of DNA synthesis and the initiation of cell division in these randomly dividing cultures (Fig. 11.3B) and the increase to a relative DNA content per cell of 2·0 units relative to the initial value (Fig. 11.10) suggests: (i) that cells of both species accumulate in G1 during stationary phase; (ii) that during lag phase each cell goes through a DNA replication step before dividing; (iii) that during

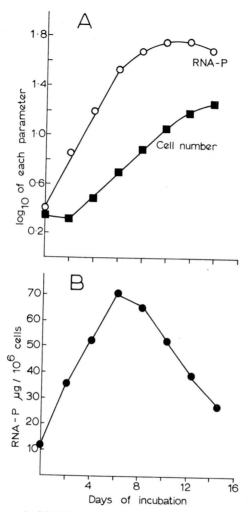

Fig. 11.8. *Changes in RNA-P during the growth of* Acer pseudoplatanus *cells in batch suspension culture.*
A. Rate of change of cell number and total RNA-P per ml of culture (semi-logarithmic plot of the original data of Short, Brown & Street 1969b).
B. Change in RNA-P content of the cells, calculated from the data in A.

the exponential growth phase, G2 is the predominant phase of the cell cycle; and (iv) that the mean generation time lengthens prior to stationary phase due to an extension of G1.* Densitometric studies of nuclei of cultured *Acer pseudoplatanus* cells support these conclusion (Fig. 11.11).

* It is difficult to distinguish between cells actually halted in any phase of the cell cycle and cells progressing through that phase at a decreased rate. The mean generation time of a culture would appear to progressively increase if cells at random were blocked at a specific point in the cycle.

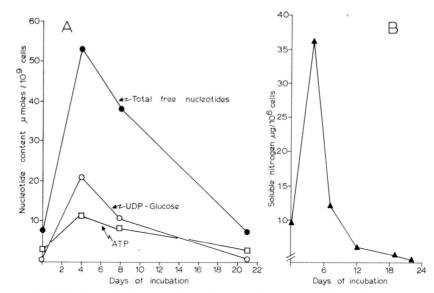

FIG. 11.9. *Changes in the composition of* Acer pseudoplatanus *cells in batch suspension culture.*
A. Free nucleotide contents (from Brown & Short 1969).
B. Total soluble nitrogen content (from Simpkins & Street 1970).

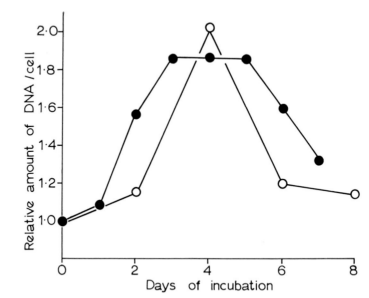

FIG. 11.10. *Relative change in DNA content of cells in batch suspension culture.* Rosa sp. (Paul's Scarlet) (●) calculated from data of Nash & Davies (1972) (Fig. 11.3B). *Acer pseudoplatanus* (○), from Short, Brown & Street (1969b).

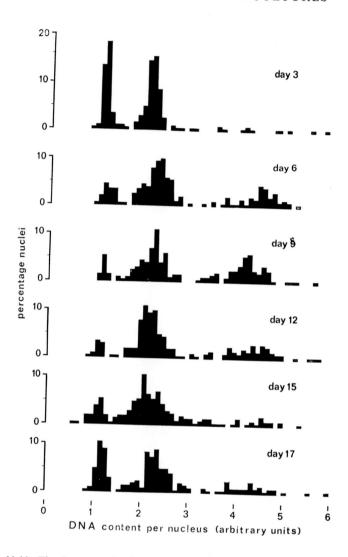

FIG. 11.11. *The frequency distribution of values for the nuclear DNA content of* Acer pseudoplatanus *cells at intervals during their growth cycle in a 4-litre batch culture.* DNA content was estimated by microspectrophotometry of Feulgen-stained nuclei. 200 nuclei were examined at each interval. In these cultures two modes of chromosome number, at 75 and 130, predominate (see also legend to Fig. 11.22). Thus the peaks from left to right represent interphase nuclei of mode 75 in G1 (1·0 to 1·5), mode 75 in G2 together with mode 130 in G1 (2·0 to 3·0) and mode 130 in G2 (4·0 to 5·0) (previously unpublished data of M.W.Bayliss).

Changing metabolic activity during the growth cycle

In discussing changing cell composition some indications have already been quoted of large changes in the metabolic activity of the cells as a batch culture proceeds through its growth cycle. These changes have been the subject of a number of recent papers. Continuous changes in metabolic activity of cell cultures of *Ipomoea* sp. are described by Rose, Martin &

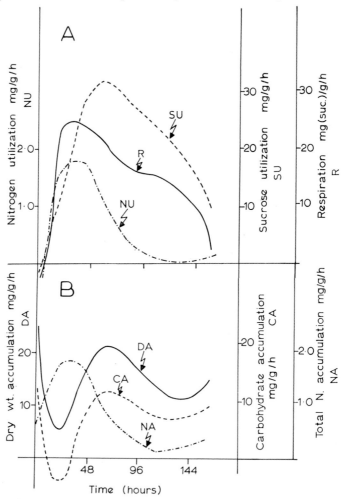

FIG. 11.12. *Metabolic rates of* Ipomoea *sp. cells during growth in 14-litre batch suspension cultures.* Polynomial functions of time fitted to chemical analysis data and the metabolic rates calculated by dividing the first derivative of these functions by the dry weight of cells per unit volume of culture at each sampling time (from Rose, Martin & Clay 1972).

Clay (1972) from their studies on rates of utilization of nitrogen and sucrose and of the accumulation of cellular nitrogen and carbohydrate (Fig. 11.12). Significant changes in acetate metabolism during the growth cycle of *Rosa* sp. (Paul's Scarlet) and in cellular composition (amino acids and organic acids) are described by Fletcher & Beevers (1970) (Table 11.1). When acetate-^{14}C was fed to the cells throughout the exponential growth phase, ^{14}C was effectively channelled through relatively small, rapidly turned-over pools of amino acids and organic acids to protein and CO_2 (Fig. 11.13A). Later in the growth cycle those pools expanded or were in rapid equilibrium with the same constituents elsewhere in the cell (Fig. 11.13B).

Evidence for the operation of mechanisms which rapidly initiate or

TABLE 11.1. Changes in composition of *Rosa* sp. cells with time in batch culture (from Fletcher & Beevers 1970; see also Fig. 11.13)

Time after inoculation	Protein	Soluble amino acids	Malate	Other organic acids
days	μ moles α-amino N/g fresh wt.		μ moles/g fresh wt.	
4	91·5	9·6	7·5	<1
12	21·9	8·4	23·8	<1

repress metabolic pathways comes from studies on ethylene production by plant cell suspension cultures. A very sharp peak of ethylene production occurred late in the exponential growth phase of batch cultures of *Acer pseudoplatanus* (Mackenzie & Street 1970). At this stage the culture was dividing and was close to the point of maximum cell aggregation. Almost immediately after this there was rapid repression of ethylene production and the culture entered the phase where the cells expanded and the aggregates broke up. An essentially similar pattern of ethylene production has been reported for cell cultures of *Rosa* sp. (Fig. 11.14), *Glycine max*, *Triticum monococcum*, *Melilotus alba*, *Haplopappus gracilis* and *Ruta graveolens* (La Rue & Gamborg 1971). Peak ethylene production occurred while the cultures were still dividing and no evidence was obtained that the ethylene accumulating within the culture itself exerted any inhibitory effect on culture growth.

There are a number of reports of changes in the total and specific *in vitro* activities of many regulatory enzymes during the progress of batch cultures. Large changes in activity of phenylalanine ammonia-lyase (PAL) and in *p*-coumarate:CoA ligase occurred prior to stationary phase in cultures of

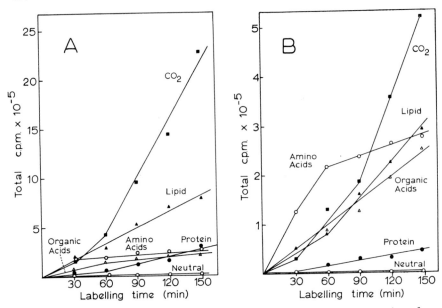

FIG. 11.13. *Acetate metabolism in batch cell cultures of* Rosa sp. Incorporation of ^{14}C into various fractions during continuous feeding of acetate-1-^{14}C to cells at two points in the growth cycle: A, after 4 days' incubation and B, after 12 days. Labelled substrate (2 μC) was added to 1 g fresh weight of cells in 5 ml of growth medium. Samples of cells were removed at 30-minute intervals for fractionation (from Fletcher & Beevers 1970).

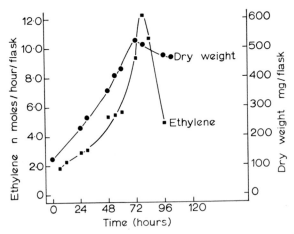

FIG. 11.14. *Ethylene production during the growth cycle of* Rosa sp. *cells in batch culture.* The cells were cultured in 40 ml of a basal medium (B5) supplemented with 0·2% casein hydrolysate and 1 mg l^{-1} NAA (from LaRue & Gamborg 1971).

Glycine max (Hahlbrock, Kuhlen & Lindl 1971, Fig. 11.15A). These changes have been interpreted as indicative of the opening up of aspects of secondary product metabolism. In contrast the activity of an enzyme (acetate:CoA ligase) involved in primary metabolic pathways did not change significantly during this period. A similar peak of PAL activity has been reported in cell cultures of *Rosa* sp. (Fig. 11.15B); it coincides with the period of maximum production of total phenols (Davies 1971). In this latter study a pronounced

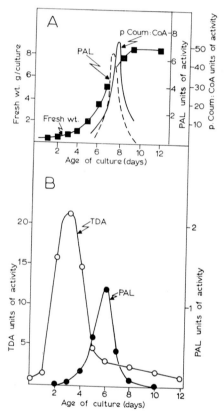

FIG. 11.15. *Changes in enzyme activity during the growth cycle of batch cell cultures.*
A. Phenylalanine ammonia-lyase (PAL) and *p*-coumarate: CoA ligase (*p*. Coum: CoA) activity in cultures of *Glycine max* grown in the dark. Units of activity: PAL, μ moles cinnamic acid mg protein^{-1} × minutes ; *p*. Coum:CoA, n moles hydroxamic acid mg protein^{-1} × minutes (from Hahlbrock, Kuhlen & Lindl 1971).
B. Phenylalanine ammonia-lyase (PAL) and threonine dreaminase (TDA) activity in cultures of *Rosa* sp. Acetone powder preparations of cells harvested at intervals during the growth cycle were accumulated and assayed simultaneously at the end of the culture period. One unit of enzyme activity is that required to produce 1 μ mole product hour^{-1} g fresh weight^{-1} (from Davies 1971).

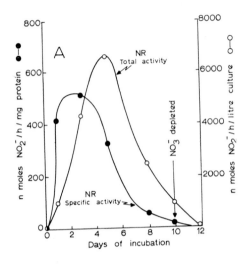

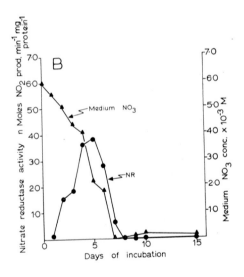

FIG. 11.16. *Regulation of nitrate reductase in cultured cells.*
A. Total and specific activities of nitrate reductase (NR) during the growth cycle of *Nicotiana tabacum* cells. The NO_3^- concentration in the medium on day $0 = 2 \cdot 5 \times 10^{-3}$ M (from Filner 1966).
B. Levels of nitrate in the culture medium and the specific activities of nitrate reductase recorded in *Acer pseudoplatanus* cells at different stages of growth in batch culture. Nitrate reductase was assayed by monitoring nitrite formation under anaerobic conditions in the presence of NADH and benzylviologen (previously unpublished data of M.Young).

peak of threonine deaminase activity occurred at a quite different point in the growth cycle.

Filner (1966) has studied the induction of nitrate reductase (NR) which occurs when cultures of *Nicotiana tabacum* are supplied with nitrate as sole nitrogen source; the activity of the enzyme increased rapidly during the first 24 hours of culture, remained highly active for only a very limited period and then declined rapidly as nitrate was depleted from the medium (Fig. 11.16A). An essentially similar pattern of NR activity was observed in *Acer pseudoplatanus* cell cultures transferred from nitrate-omitted to nitrate-containing medium (M. Young—previously unpublished data—Fig. 11.16B). Induction of NR by nitrate in cell cultures of *Glycine max* is reported to depend upon the presence of ammonia or substances synthesized by the cells in ammonium-containing medium (Bayley, King & Gamborg 1972). A further interesting example of enzyme induction from cell cultures is the report of fructose-1-6-diphosphatase in *Acer pseudoplatanus* suspensions supplied with glycerol as their sole source of carbon (Scala & Semersky 1971).

In a number of cases changes in enzyme activity associated with the growth cycle pose difficult problems of interpretation. The activities of acid and neutral invertases both in the soluble and cell wall fractions of cultured *A. pseudoplatanus* cells increased to peaks towards the end of the exponential growth phase and earlier showed evidence of movement between the cell wall and the cytoplasm (Copping & Street, 1972). The patterns of changing activity of these enzymes were similar whether the cells were supplied with sucrose, glucose or sucrose plus glucose. Whilst these enzymes may yet prove to be directly involved in the growth process it could be that their changing activities result from the association of their structural genes with others of the same operon whose products are regulating growth. Situations where enzymes appear extracellularly and hence are released from normal intracellular feed-back mechanisms also pose problems of interpretation. Such enzymes include glycosidases (β-glucosidase and α- and β-galactosidases by *A. pseudoplatanus* cells—Keegstra & Albersheim 1970), amylases (*Saccharum officinarum*—Maretzki, dela Cruz & Nickell 1971) and peroxidases (Veliky, Sandkvist & Martin 1969). Significant changes with age of the culture are reported for the concentration, composition and viscosity of extracellular, non-dialysable macromolecules released by cell cultures of *Nicotiana tabacum* (Olson 1971, Fig. 11.17). The greatest changes in these extracellular macromolecules occurred during the period when the culture showed most rapid increase in cell dry weight. Similarities in the changes which occurred in both cell-wall preparations and in these extracellular products suggest that the latter were cell-wall precursors which failed to be retained in the cell walls under the culture environment. Carceller, Davey, Fowler & Street (1971) have advanced a similar interpretation to explain

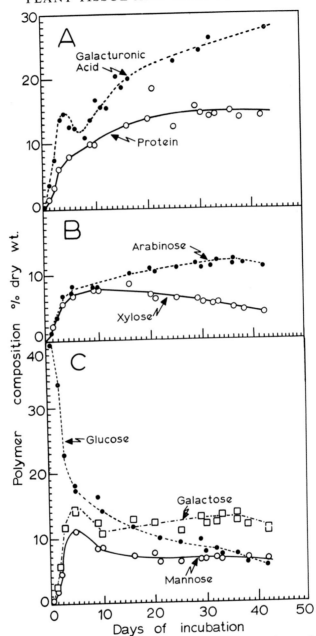

FIG. 11.17. *Polysaccharides and proteins secreted by suspension cultures of* Nicotiana tabacum. Changes in levels of analytical components of the non-dialysable polymer secreted into the culture medium during the growth of a batch culture (from Olson 1971—see also Olson, Evans, Frederick & Jansen 1969).

the intercellular and extracellular appearance of lignin in *Acer pseudoplatanus* suspension cultures.

Synthesis of secondary plant products

A large number of callus cultures, usually harvested after their rapid growth phase has been completed, have been examined (albeit usually with disappointing results) for the presence of secondary products characteristic of the whole plant or some particular organ of the plant (for reviews see Kaul & Staba 1967, Puhan & Martin 1971, Krikorian & Steward 1969, Turner

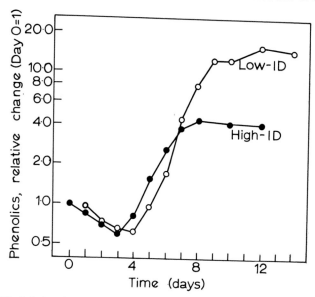

Fig. 11.18. *Relative change in total phenolics during the growth of batch cell cultures of* Rosa sp. (*Paul's Scarlet*). Initial values (day 0) were: High-ID (high inoculum density), 4·0 mg/culture; Low-ID, 1·25 mg/culture. For initial cell densities see Fig. 11.3 (from Nash & Davies 1972).

1971). Very few systematic studies have been made of secondary product metabolism *during the growth cycle* of cell or tissue cultures.

Nash & Davies (1972) have reported that the accumulation of polyphenols by cell cultures of *Rosa* sp. was restricted to the late exponential growth phase (Fig. 11.18). This fraction contained some fourteen separate compounds each of which changed in a characteristic way during the growth cycle. The duration of the accumulation of polyphenols was controlled by carbohydrate availability but the time at which their synthesis was initiated and their initial rates of synthesis were controlled by the level of 2,4-D and the light intensity (Davies 1972). Studies on anthocyanin accumulation by cultures of *Haplopappus gracilis* (Constabel, Shyluk & Gamborg 1971, Fig. 11.19) also illustrated how the timing of accumulation of a secondary product can be controlled by the level of a plant growth regulator (in this case by NAA). Similarly the 2,4-D and kinetin levels and the level of sucrose were all shown to significantly influence in time and quantity the accumulation of Klason lignin by *Acer pseudoplatanus* cell cultures (Carceller, Davey, Fowler & Street 1971).

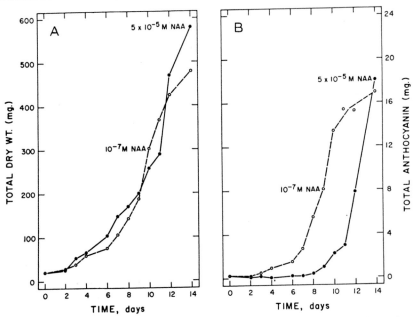

FIG. 11.19. *The effect of the concentration of NAA on growth* (A) *and anthocyanin production* (B) *in batch cell cultures of* Haplopappus gracilis.
A. Change in the total dry weight of cells in 200 ml of culture.
B. Change in the anthocyanin extracted from the cells harvested from 200 ml of culture. Anthocyanin was estimated by measuring the optical density of *n*-propanol/HCl extracts of cells at 525 nm (from Constabel, Shyluk & Gamborg 1971).

Several secondary products of pharmaceutical importance have been detected in cell cultures showing only limited morphological differentiation (Table 11.2) and in a few cases yields from such cultures (as percentage of culture dry weight) are higher than those from the normal whole plant source. For instance, visnagin (a physiologically active furanochromone) is normally present in the seeds of *Ammi visnaga* to the extent of 0·04–0·06% of their dry weight. At least during a transient period in the growth cycle, batch cultures of *A. visnaga* contained the active compound at 0·3% of their

TABLE 11.2. Compounds with pharmaceutical uses detected in plant cell suspension cultures

Compound	Species	Yield as % dry weight	Reference
Alkaloids	Ruta graveolens	*	Steck, Bailey, Shyluk & Gamborg (1971)
Alkaloids (Tropane)	Datura innoxia	0·016	Tabata, Yamamoto & Hiraoka (1971)
Alkaloids (Tropane)	Scopolia japonica	0·008	Tabata, Yamamoto & Hiraoka (1971)
Antibiotics	'Many species'	*	Khanna & Staba (1967)
Cardenolides	Digitalis purpurea	0·02–0·002	Büchner & Staba (1964)
Coumarins	Ruta graveolens	*	Steck, Bailey, Shyluk & Gamborg (1971)
Diosgenin	Dioscorea deltoidea	1·0	Kaul & Staba (1968)
Diosgenin	Solanum xanthocarpum	0·008	Heble, Narayaraswami & Chadha (1968)
Nicotine	Nicotiana tabacum	0·7	Speake, McCloskey, Smith, Scott & Hussey (1964)
β-sitoserol	Solanum xanthocarpum	0·04	Heble, Narayaraswami & Chadha (1968)
Stigmasterol	Dioscorea deltoidea	*	Kaul & Staba (1968)
Vindolin	Catharanthus roseus	0·01	Boden, Gorman, Johnson & Simpson (1964)
Vindolinum	C. roseus	0·01	Boden, Gorman, Johnson & Simpson (1964)
Visnagin	Ammi visnaga	0·31–0·13	Kaul & Staba (1967)

* No quantitative estimations made.

cell dry weight (Kaul & Staba, 1967, Fig. 11.20). Multi-litre cultures of this species were shown to produce 62·7 mg visnagin 1^{-1} day^{-1}.

The possibility that plant cell cultures could be used to effect important biotransformations has been examined in only a very preliminary way (Stohs & Staba 1965, Graves & Smith 1967, Furuya, Hirotani & Shinohana 1970). Thus, esterification of an exogenously supplied steroid (progesterone) was shown to occur in cultures of *Nicotiana tabacum* and *Sophora angustifolia*

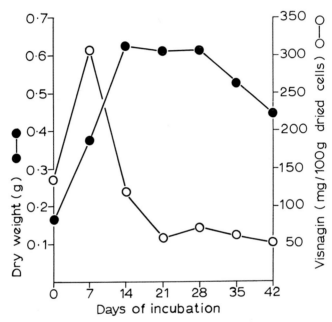

FIG. 11.20. *Growth and visnagin production in 50-ml batch cell cultures of* Ammi visnaga. Visnagin was extracted from dried cells with ethylene dichloride and purified by repeated thin-layer chromatography on silica-gel plates. To estimate the visnagin content of the extracts, ethylene dichloride eluates of the visnagin zones were evaporated to dryness and the residues redissolved in methanol. The optical densities of the methanolic solutions were determined at 242 nm (from Kaul & Staba 1967).

(Furuya, Hirotani & Kawaguchi 1971), but no attempt was made to plot the activity of this transformation during the growth cycle.

Clearly, more detailed studies of the growth kinetics and metabolism of cultures of appropriate plant species are essential for any assessment of the possible economic exploitation of plant suspension cultures as synthetic systems (see for example: Ogutuga & Northcote 1970b, Tabata, Yamomoto & Hiraoka 1971). Such studies should be linked to mutant selection as outlined in Chapter 8.

Short-term studies of metabolism with suspension cultures

The absence of balanced growth in batch cell suspension cultures does not preclude their use for intensive study of particular aspects of cell metabolism. Already many important questions regarding metabolic regulation have been opened up by study of the transient changes occurring during batch culture and by short-term studies centred upon a particular stage of the growth cycle. Examples of the latter include studies on glucose transport (Maretzi & Thom 1972), on shikimic-acid metabolism (Gamborg 1966), on amino-acid synthesis (Dougall 1971), on arginine catabolism (Maretzi, Thom & Nickell 1969), on DNA replication (Filner, 1965), on the kinetics of ^{14}C incorporation into RNA from 2-^{14}C-uridine (Cox, Turnock & Street 1973) and of the incorporation of adenine into adenosine nucleotides (Dorée, Leguay, Terrine, Sadorge, Trapy & Guern 1971).

The study of unbalanced growth has its own intrinsic interest; unbalanced growth is characteristic of primordia (cell division without growth) and of cytodifferentiation (growth without cell division) (Steward & Mohan Ram 1961).

Limiting factors operating to control growth in batch culture

The ever-changing environment of cells propagated by batch culture represents a complex situation in which it is difficult to identify the factor(s) which determine the growth pattern and the metabolic fluctuations.

Several studies have focused attention on nutrient depletion at the end of the growth cycle to try to identify factors which limit the duration of the exponential growth phase (Henshaw, Jha, Mehta, Shakeshaft & Street 1966, Street, King & Mansfield 1971, Wilson 1971, Rose, Martin & Clay 1972). However, the continuous changes in cell composition characteristic of the exponential growth phase in batch cultures is clearly a situation which could not continue indefinitely and raises the question of whether the retardation of cell division prior to stationary phase may be determined by the factors which cause these compositional changes and which operate early in the growth cycle before any nutrient limitation develops. A systematic study of limiting factors can, however, now be made with synchronized cell populations (see this Chapter, p. 303), or **steady-state** cultures (see this Chapter, p. 317) which permit the effects of single, defined perturbations of the culture environment to be investigated.

SYNCHRONOUS CULTURES

Potential of synchronous cultures

Since 1953 methods have been developed for the production of highly synchronous cultures (Zeuthen 1964, Cameron & Padilla 1966, Padilla, Whitson & Cameron 1969) and work with such cultures of bacteria (Donachie & Masters 1969), algae (Schmidt 1969), yeast (Mitchison 1971), protozoa (Zeuthen 1964) and mammalian cells (Robbins & Scharff 1966, Petersen, Tobey & Anderson 1969) has already made significant contributions to our understanding of gene expression, replication and the triggering of cell division. Synchronous cultures have also been used in studies on development (Cameron, Padilla & Zimmerman 1971).

Synchronous cultures of higher plant cells, if available, would not only enable a detailed description of the events of the cell cycle but would make possible identification of the factors operating to control the orderly sequence of biochemical events which separate one generation of cells from its daughter generation (Street 1973). Knowledge of the factors controlling the duration of the sequential phases of the cell cycle and identification of those phases most susceptible to change in extrinsic factors would improve our understanding of the differential rates of growth and cell division which occur in organized meristems (Steward & Mohan Ram 1961, Phillips & Torrey 1972). The evidence that cytodifferentiation may be determined during an earlier critical cell cycle (Torrey 1971) suggests that synchronous cultures will be necessary to achieve high levels of cytodifferentiation in tissue and cell cultures (Fosket & Torrey 1969, Fosket 1970). The successful development of large scale plant cell cultures (see Chapter 4) means that once such cultures can be obtained in synchrony it will be possible to withdraw samples frequently and in sufficient amount for detailed biochemical and cytological study without disturbing the progression of the growth of the remaining cell population.

Naturally occurring synchrony in higher plants

Chamberlain, as early as 1935, reported a high degree of synchrony of mitosis in the female gametophyte of the cycad, Dioon (8 successive divisions). Similar synchronized mitoses were observed during female gametophyte and pro-embryo development in a number of other cycads and conifers. Maheshwari, in 1950, described the occurrence of synchronous mitoses during embryo-sac development in flowering plants; in the Polygonum-type of embryo-sac the megaspore nucleus gave rise by three synchronous mitoses to an 8-nucleate syncytial embryo sac. The triploid endosperms of many flowering plants show synchrony of mitoses—sometimes the synchrony is uniform throughout the endosperm, in others a gradient of mitotic stages can be traced from the antipodal to the micropylar end of the ovule (Erickson 1964).

The sporogenous tissue of the angiosperm anther shows synchrony of mitoses and of the meiotic division of the spore mother cells. These divisions are often very extended (in *Trillium erectum* meiosis takes up to 100 days, microspore interphase being up to 40 days and microspore mitosis up to 14 days). Some studies have been undertaken with the microspores of *Trillium* and *Lilium* on changes in sulphydryl concentration, in respiration, in the activities of respiratory enzymes and in nucleic acids during the cell cycle (Erickson 1964). However, the challenge very much remains of identifying the factors inducing synchronous division *in vivo*, the factors which can uncouple nuclear division from cytokinesis and those which can induce simultaneous cytokinesis without any accompanying nuclear division. The problem is that of working experimentally with these *in vivo* systems and hence great interest attaches to the possibility of studying induced synchrony of mitosis and/or cytokinesis in plant cell suspension cultures.

Study of the cell cycle in asynchronous cell populations

Since the classic study of the cell cycle in cells in root-tips of *Vicia faba* by Howard & Pelc (1951), techniques have been developed which have enabled the duration of the separate phases of the cell cycle to be determined for asynchronous cells of a number of tissues (Table 11.3). These techniques should be applied to asynchronous cell cultures as a prelude to and in parallel with attempts at their synchronization.

Determination of the duration of all phases of the cell cycle can be made by a combination of mitotic figure scoring, ^{3}H-thymidine labelling and autoradiography (Cleaver 1967). The duration of mitosis and S-phase may be estimated directly from the mitotic index (see Chapters 3, p. 51, and 4, p. 91) and the labelling index (the fraction of the population labelled after a brief exposure to ^{3}H-thymidine) (Table 11.4). The relationship between the index and the phase duration depends upon the age distribution of the cells in the population. The two age distributions most commonly encountered in plant cell cultures are those in steady-state populations, in which the total cell number per unit volume is constant and those in exponential growth. The equations applicable to these two states are:

Duration of mitosis

Steady-state population:
$$MI = \frac{tm}{T} \qquad (1)$$

Exponential population:
$$MI = tm.\frac{ln2}{T} \qquad (2)$$

Duration of S-phase

Steady-state population:
$$LI = \frac{ts}{T} \qquad (3)$$

TABLE 11.3. Duration of cell cycle phases in root tip meristems and asynchronous cell cultures

Species	Duration of phases (hours)					Reference
	G1	S	G2	M	Total	
Tradescanthia sp.	1·0	10·5	2·5	3·0	17·0	Wimber & Quastler (1963)
Allium sp.	4·0	10·8	2·7	2·5	20·0	Wimber (1960)
Vicia faba	4·0	7·5	4·9	2·0	19·0	Evans & Scott (1964)
Haplopappus gracilis	1·4	6·8	2·6	1·2	11·9	Ames & Mitra (1966)
*H. gracilis**	9·34	6·4	4·86	1·4	22·0	Eriksson (1967a)

* Cell culture.

TABLE 11.4. The duration of mitosis (tm) and S-phase (ts) of *Acer pseudoplatanus* cells in suspension culture, estimated from the mitotic index (MI) and labelling index (LI)

td‡ hours	Mitosis*			S-phase†		
	MI %	tm hours	td hours	LI %	ts hours	
1. Batch Culture: exponential growth phase						
26	7·9	2·98				
35	6·5	3·45	35	17·0	8·63	
36	4·4	2·29	36	16·8	8·54	
36	6·9	3·32	36	15·5	8·10	
42	5·7	3·45				
50	4·5	3·24	50	13·0	9·43	
2. Chemostat: steady-state population						
66·5	3·2	2·12	66·5	11·3	7·55	
85	4·9	4·19	85	10·9	9·26	
103	3·1	3·18				

* tm calculated using equations (1) and (2), p. 299.
† ts calculated for the steady-state population using Eqn (3), p. 299. The value of ts for the exponential population was calculated using Eqn (2), p. 299, by assuming that the duration of ts was short relative to the cell cycle time, T.
‡ It has been assumed that the population doubling time, td, is equal to the average cycle time, T, of the cells.
(Values of tm and ts for the exponential populations are calculated from previously unpublished data of A. Gould.)

Exponential population:

$$LI = (\exp.tsln2/T - 1)\exp.t_2ln2/T \qquad (4)$$

Where MI = mitotic index; LI = labelling index; tm = duration of mitosis; ts = duration of S-phase; t_2 = duration of G_2 plus one-half of mitosis (tm); and T = duration of cell cycle.

Calculations of the durations of G1, G2 and the total cycle plus mitosis and S-phase can be made simultaneously by the 'pulse-chase' technique used by Howard & Pelc (1953). Cultures are exposed to ^{3}H-thymidine during a brief period (the pulse); only cells in S-phase during this period are labelled.

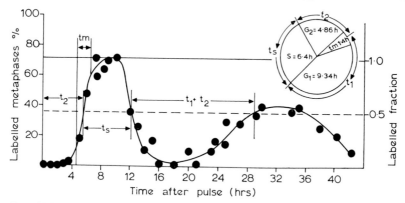

FIG. 11.21. *Distribution of labelled metaphases plotted against the time after a 30-minute pulse-labelling with* ^{3}H-*thymidine of a batch culture of* Haplopappus gracilis *cells.* The duration of the phases of the cell cycle, estimated from changes in the fraction of labelled metaphases, are shown in the inset (top right). The two waves of labelled mitoses represent successive divisions of the fraction of the population in S-phase during the pulse. The fraction of labelled mitoses remained at zero until a period of time equal to the duration of G2, then rose to 1·0 in a further period equal to the duration of mitosis (tm). The labelled fraction remained at 1·0 for a period ($S - tm$) and then fell to zero. The second wave appeared after a further interval equal to the duration of G1 + G2. All calculations (t_1, t_2 and ts) were made from the mid-points of the curves as indicated (from Ericksson 1967a).

Immediately following the pulse, unlabelled thymidine is added to the growth medium to dilute rapidly the pools of DNA-precursor molecules and the cultures allowed to continue growth. The percentage of labelled metaphases are followed through two successive divisions (Wimber 1960, Cleaver 1967). Fig. 11.21 shows the results of such a pulse labelling using a suspension culture of *Haplopappus gracilis* (Ericksson 1967a) in exponential growth.

An alternative technique which has been applied to suspension cultures is that of combined microspectrophotometry and autoradiography. The *relative* duration of each phase of the cycle is calculated by a combination of

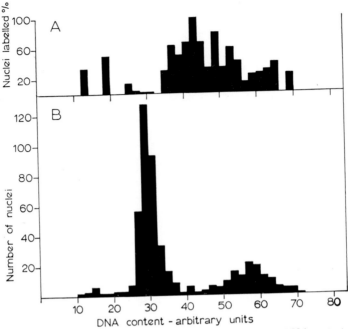

FIG. 11.22. *The frequency distribution of values for the nuclear DNA content of an exponentially increasing population of* Acer pseudoplatanus *cells. Distribution of 498 nuclei among different DNA-content classes.*

A. Labelled nuclei as percentage of those in each class after a 10-minute pulse with ^{3}H-thymidine (0·25 μCi ml^{-1}).

B. Total nuclei in each class.

Samples of the cell population were pulsed with ^{3}H-thymidine, fixed and mounted on microscope slides. After Feulgen staining, autoradiographs were prepared on the slides by the stripping-film technique, and incubated at 4°C for 4 days. After development of the film, the DNA content of randomly selected interphase nuclei was estimated by microspectrophotometry through the stripping-film. The presence of label in each nucleus examined was noted. (The presence of label did not interfere significantly with the estimation of DNA content.) Calculation of the relative duration of cell cycle phases from these data is complicated by the presence in the culture of cells with two modes of chromosome number at 75 (75% of population) and 130 (25%) (see legend Fig. 11.11). However, assuming similar cycle times for both modes ($td = 36$ hours), the duration of cell cycle phases have been calculated (Cleaver 1967) and are as follows:

	Mode	G1	S	G2
Fraction of cycle	75	0·051	0·062	0·886
	130	0·023	0·326	0·651
Duration in hours	75	1·8	2·2	31·9
	130	0·8	11·7	23·4

These data disregard nuclei in mitosis (3%) and assume a uniform age distribution for the population. The latter should result in an overestimation of G1 and S-phases (previously unpublished data of A.Gould).

Feulgen staining (see Chapters 3, p. 53, and 4, p. 93) and ^{3}H-thymidine labelling. The *absolute* duration of the cell cycle is calculated from the exponential growth curve. Data from a cell suspension culture of *Acer pseudoplatanus* examined by this technique is presented in Fig. 11.22.

Synchronization of plant cell cultures

A number of techniques are now available for the induction and selection of synchrony in both procaryote and eucaryote cells (James 1966). Assessment of the success of these techniques is based upon the persistence and the degree of synchrony induced. The degree of synchrony is usually expressed as a percentage synchrony (Engelberg 1961, 1964) based upon such determinations as (i) the percentage of cells at a specific point in the cell cycle (e.g. metaphase index) at one moment of time; (ii) the percentage of cells passing a specific point in the cycle (e.g. entering mitosis) during a brief specified period; or (iii) the percentage of the total cycle time needed for all the cells to pass a specific point in the cycle (e.g. the anaphase of mitosis). In comparing the degree of synchrony obtained by different workers it is important to examine precisely the basis upon which the percentage synchrony has been calculated.

Synchrony has been induced in mammalian cell cultures by the selective inhibition of DNA synthesis by addition of 5-aminouracil, hydroxyurea, 5-fluorodeoxyuridine or excess of thymidine (Puck 1964, Petersen, Tobey & Anderson 1969). The cells accumulate in G1 until the inhibition is removed and then enter the next division in synchrony, the percentage synchrony depending both upon the inhibitor used and the cell line involved. Neither excess thymidine nor a cold treatment was effective in synchronizing cell cultures of *A. pseudoplatanus* (Roberts & Northcote 1970). Eriksson (1966), however, induced synchrony in suspension cultures of *Haplopappus gracilis* by treating dividing cells ($MI = 5\%$) for 12–24 hours with four different DNA synthesis inhibitors (Table 11.5); peaks of mitosis occurred 10–16 hours after removal of the inhibitor. Hydroxyurea at an optimum concentration (3 mM) gave a peak mitotic index of 30% over 4·5 hours (0·2 of the total cycle time of 22 hours) and accounted for 97% of the cell population (assuming a duration of mitosis of 1·4 hours—Eriksson 1967a) (Fig. 11.23). The duration of 4·5 hours for the mitotic peak set against the figure for the duration of mitosis of 1·4 hours led Eriksson to use the term 'partial synchrony'. Only a single mitotic peak was reported after all treatments, suggesting that the synchrony was limited to a single division cycle. This may be related to the observation that both hydroxyurea and 5-aminouracil caused chromosome damage (up to 15% abnormal metaphases were observed). Confirmation of the effective synchronization of the population (even for a single cycle) would require either further peaks of mitotic activity or a rapid reversion to the mitotic index characteristic of the randomly dividing population before treatment.

Cell cultures of *Nicotiana tabacum* have recently been synchronized by transfer to cytokinin-omitted medium followed by the later addition of the cytokinin essential for growth in culture (Péaud-Lenöel & Jouanneau 1971; Jouanneau 1971). When cultures were initiated from stationary phase cells and the additions of 2,4-D and kinetin delayed for 24–72 hours, a transient mitotic index of up to 7% was obtained (as compared with 2–4% for a randomly dividing exponential culture) (Fig. 11.24A). Despite the relatively

TABLE 11.5. Peaks of mitotic index induced in cell cultures of *Haplopappus gracilis* by treatment with DNA-synthesis inhibitors (from Eriksson 1966)

Inhibitor	Optimum concentration M	Peak mitotic index %	Peak duration*		Proportion of* population in mitosis during peak
			Hours	Fraction of cycle	
5-Aminouracil	6×10^{-3}	25	4·4	0·20	79
Hydroxyurea	3×10^{-3}	30	4·5	0·20	97
5-Fluorodeoxyuridine	2×10^{-7}	15	6·9	0·31	75
Thymidine	6×10^{-3}	22	5·2	0·24	82

* Values calculated from the published data.

low value of this peak it was calculated that 70–85% of the cells subsequently divided. It was further shown that the temporary absence of the cytokinin and not of the auxin was the factor inducing synchrony (Fig. 11.24B). N-6-(isopentenyl) adenosine was as effective as kinetin. The degree of synchrony could be further slightly increased by witholding 2,4-D for 24 hours and the cytokinin for a further 10 hours.

Additions of cytokinins (kinetin or 6-benzylaminopurine) also increased the initial peak of mitosis which occurred in cultures of *Acer pseudoplatanus* between 40–72 hours after subculture (timing of this peak being dependent upon age of the stock culture, initial cell density and temperature). Roberts & Northcote (1970) used this approach (Fig. 11.25) to obtain a high proportion of dividing cells for microscopical study. However, it is not clear whether there occurred in their cultures any synchronization, i.e. that division was other than random. A single peak of mitotic index occurs in randomly dividing batch cultures of a number of species (see Fig. 11.4) when the transition from lag phase to the exponential growth phase is followed closely by a period of declining rate of cell division caused by nutrient limitation; when there is a transient decrease in the mean cell cycle time. Any enhancement in this mitotic peak by cytokinin could be due to either a decrease in the mean cell cycle time or to an increase in the duration of mitosis (Eqn (2), p. 299).

However, a high and persistent level of *cell division* synchrony has been

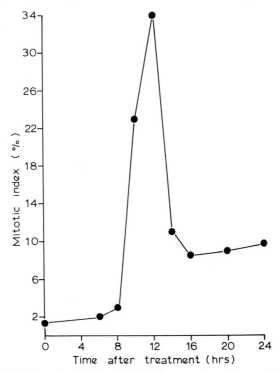

Fig. 11.23. *Partial synchronization of mitosis in suspension cultures of* Haplopappus gracilis. One-day-old cultures, from a strain maintained by subculture every second day, were treated for 12–24 hours with 3 mM hydroxyurea, washed twice in medium (minus hydroxyurea) and then incubated in fresh medium (minus hydroxyurea). A peak of mitotic index appeared 10–12 hours after treatment (from Eriksson 1966).

demonstrated in suspension cultures of *A. pseudoplatanus* (King, Mansfield & Street, 1973). This was achieved by establishing cultures at low initial density ($c\ 20 \times 10^3$ cells ml^{-1}) from cells well into stationary phase (24–28-day-old stock cultures) (for the culture system used see Chapter 4, Figs. 4.9 and 4.10). The induced synchrony persisted through five to six generations, the increase in cell number at each step normally falling within the range 70–95% (Fig. 11.26, p. 308). There was no progressive decline in the percentage increase in cell number at each successive cytokinesis. However, there was some variation in the duration of the intervening plateaus.

A correlation between cell division synchrony and mitosis is indicated by peaks of mitotic index recorded *c*. 10 hours before cytokinesis steps (Fig. 11.27). The mitotic peaks (*c*. 7·5%) persisted for less than 0·2 of the cell cycle and were followed by a 'step-up' in cell number of similar duration when 72% of the cells divided. However, a puzzling feature of these (and subsequent) data is the relatively high 'background' mitotic activity recorded

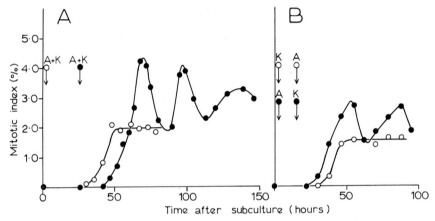

FIG. 11.24. *Synchronization of mitosis in cell cultures of* Nicotiana tabacum. The cells were subcultured by transfer of 25 ml of stationary phase suspension to 400 ml of new medium.
A. 2,4-D (A) and kinetin (K) (to final concentrations of $3·75 \times 10^{-7}$ M and $2·5 \times 10^{-7}$ M respectively) were added to the cultures together, either at the time of subculture (○) or after a delay of 25 hours (●).
B. 2,4-D and kinetin (final concentrations as in legend for A) were added at separate times. 2,4-D was added at the time of subculture and kinetin 15 hours later (●) or vice versa (○), (from Peaud-Lenoël & Jouanneau 1971).

during the 'interphase' period. The peak of mitosis would account for the subsequent division of 72% of the cells if the duration of mitosis in these cells was *c*. 1·5 hours (a relatively low value—see Table 11.4). If the mitotic activity recorded during 'interphase' represents the random division of the remaining 28% of the cells (a possibility not excluded by the cell number data) then the duration of mitosis in these latter cells must be much longer (*c*. 5 hours). The mitotic figures scored during 'interphase' are predominantly metaphases (prophase : metaphase ratio is 1 : 64). Therefore, the possibility arises that a proportion of cells become arrested in metaphase and do not contribute to cell number increase but dominate the interphase mitotic index. Alternatively, the degree of synchrony of cytokinesis in these cultures may be much higher than that of other events in the cell cycle (including mitosis).

However, two enzymes which may be involved in nucleic-acid biosynthesis showed marked changes during the 'interphase' periods in these cultures (Fig. 11.28, p. 310). The specific activity of aspartate transcarbamoylase rose late in interphase. This may be related to the formation of pyrimidine precursors for nucleic-acid synthesis early in the succeeding cell cycle. The peak of thymidine kinase activity at or immediately following mitosis and a

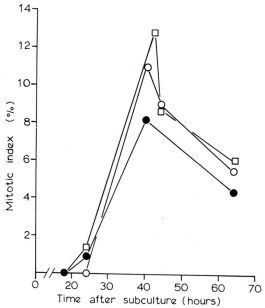

FIG. 11.25. *The incidence of mitosis in batch cell cultures of* A. pseudoplatanus. Cultures were set up from stock suspensions (maintained in absence of an added cytokinin) by subculturing 2 ml of suspension into 10 ml fresh medium in 50 ml roller-tubes. No additions were made to the control cultures (●), but, at the time of subculture, kinetin (○) or 6-benzylaminopurine (□), already dissolved in sterile medium, was added to some of the tubes to give a final concentration of 5.6μ mole 1^{-1}. Mitotic indices were measured using an aceto-orcein squash method (from Roberts & Northcote 1970).

transient peak of ^{3}H-thymidine incorporation at the beginning of 'interphase' (Cox 1972) may further indicate that the initiation of DNA synthesis occurred shortly after cytokinesis (see also p. 283).

A very sharp transition from synchronous to asynchronous cell division may occur in synchronized cultures of *A. pseudoplatanus* (Fig. 11.29, p. 311). Such transitions are characterized by a rapid decrease in mean cell cycle time and an uncoupling of the synthesis of DNA, RNA and protein from cell division (Table 11.6) resulting in DNA levels of ×0·5 and ×0·25 that observed

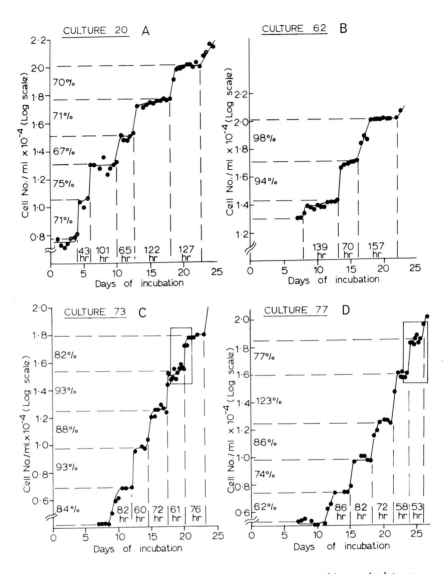

Fig. 11.26. *Cell division synchrony in 4-litre batch cell cultures of* A. pseudoplatanus. The cell number data were obtained from samples (*c.* 5 ml) taken by an automatic device at intervals of 3–6 hours. The percentage increase in cell number at each step is shown against the ordinate. The duration of each step (the cell cycle time) is shown against the abscissa. The rectangles superimposed on cultures 73(C) and 77(D) indicate cell cycles during which mitotic indices were scored (Fig. 11.27) and biochemical analyses undertaken (Fig. 11.28).

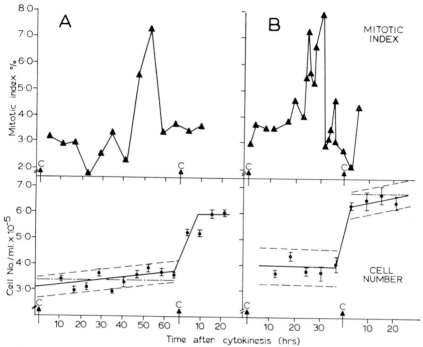

Fig. 11.27. *Changes in the cell density and mitotic index of synchronized batch cultures of* A. pseudoplatanus. *Mitotic indices were obtained by examining 1000 nuclei after staining with lacto-propionic orcein. Lines of best fit to the cell number data were calculated by linear regression analysis. A mean line of zero slope (— . —) falls within the 95% confidence limits (- - -) of the best fit line in both cultures. C = cytokinesis.*

The two cell cycles illustrated (A and B) occurred in the intervals marked by the rectangles superimposed on growth data for cultures 73 and 77 (Figs. 11.26C and D) respectively.

during the period of synchronous division. This could indicate that the clone has developed polyteny by endoreduplication and that this process is reversible (see Chapter 7, p. 181).

SEMI-CONTINUOUS AND CONTINUOUS CULTURES

Introduction

At no time during a batch culture growth cycle are cells in a state of **balanced growth.** Cell division and the synthesis of metabolites are initiated or cease at different points in time. During the transient exponential growth phase, rates of cell division and of synthesis of many cell materials may be constant with time but quite different (see Fig. 11.6). Consequently the mean composition and metabolic activity of the cells changes continuously. Perhaps causally related to these changes in growth pattern are the changes which occur in the nutrient environment of the cells.

Batch cultures, particularly in so far as balanced growth does not occur, are of limited value for studies on metabolic regulation in actively dividing cell cultures. Presumably the changing cell composition observed during their transient exponential growth phase would not continue indefinitely if

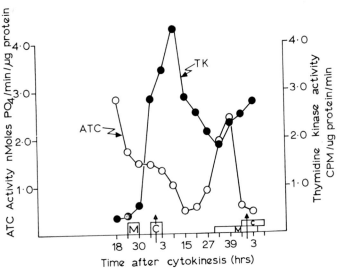

FIG. 11.28. *Changes in the activities of thymidine kinase (TK) and aspartate trans-carbamoylase (ATC) during a cell cycle* in a synchronized culture of* Acer pseudo-platanus *cells.* C = duration of cytokinesis from cell count data; M = duration of mitosis from mitotic index data. For references to enzyme assay techniques see Street (1973).

    * The cell cycle enclosed in a rectangle in Fig. 11.26D. (Previously unpublished data of M.W.Fowler.)

this phase could be extended by preventing the development of growth limiting conditions. In this way, balanced growth might be achieved.

Great interest therefore attaches to the various techniques which have been used to extend the exponential growth phase and to data obtained in such experiments indicative of the development of cells of constant cell composition and metabolic activity.

Frequent serial subculture

Attempts have been made to maintain stock cultures in exponential growth

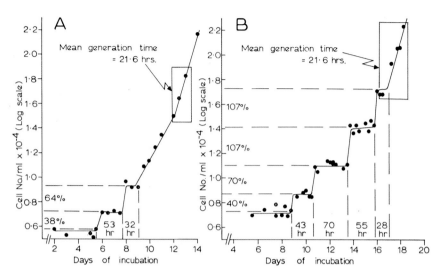

FIG. 11.29. *Rapid transition between synchronous and asynchronous growth in* Acer pseudoplatanus *cell cultures*. The transitions to exponential growth occurred after two steps-up in cell number (A) or four steps (B) and were complete within 30 hours of the preceding step. The mean generation times after the transitions (calculated from the slopes of the $\log_{10}$ plots) are much shorter than the duration of the preceding cell cycles and shorter than generation times (70 hours) normally associated with exponential growth in batch cultures initiated in the same culture medium at $25–50 \times 10^4$ cells per millilitre ($\log_{10}$ *c*. 1·5) (see also Table 11.6).

TABLE 11.6. Changes in DNA, RNA and protein per cell in *Acer pseudoplatanus* cultures after transition from synchronous to asynchronous growth

| Sample times (days) | *Culture A | | | | *Culture B | | | |
| | Cells per ml $\times 10^{-6}$ | μg per 10^6 cells | | | Cells per ml $\times 10^{-6}$ | μg per 10^6 cells | | |
		DNA	RNA	Protein		DNA	RNA	Protein
0	0·32	8·5	99·1	664	0·65	5·3	79·0	460
1	0·67	3·8	61·3	417	1·03	3·1	48·0	249
2	1·46	1·9	27·5	230	1·85	2·1	37·4	170

* See Fig. 11.29.

These data were obtained during the periods of culture enclosed in rectangles in Fig. 11.29. The zero-time samples were taken on day 12 (A) and day 16·5 (B).

by preventing nutrient limitation by sufficiently frequent subculture. Thus Eriksson maintain cultures of *Haplopappus gracilis* (doubling time $(td) = 22$ hours) by subculture every 48 hours and used 24-hour cultures for studies on growth requirements (1965), effects of radiation (1967b), synchronization of cell division (1966) and the duration of the cell cycle (1967a). In their studies on adenine utilization by cultures of *Acer pseudoplatanus* ($td = 48$ hours), Dorée, Leguay, Terrine, Sadorge, Trapy & Guern (1971) subcultured every 7 days, transferring 80 ml of culture to 200 ml new medium (giving an initial density of 2×10^5 cells ml^{-1}).

It is not clear whether these techniques have led to a state of balanced growth; certainly the cultures produced do not appear to correspond to a growth state encountered during standard batch culture propogation. Thus cells of *Phaseolus vulgaris* ($td = 24$ hours), when subcultured every fifth day through four passages, were smaller than cells on day 5 of the first passage and more aggregated (Liau & Boll 1971). Similarly Henshaw, Jha, Mehta, Shakeshaft & Street (1966) found that cultures of *Parthenocissus tricuspidata* ($td = 90$ hours) when subcultured every 7 days for six passages had a much lower mean cell fresh weight than day 7 cultures of the first passage and contained a higher proportion of aggregates greater than 200 μm in diameter.

Generation-interval subculture

In order to restrict changes in cell density to within more precisely defined limits, Wilson (1971) adopted the technique of sub-culturing cells of *Acer pseudoplatanus* at intervals equal to the cell population doubling time. The limits of the cell density chosen (2–5×10^5 cells ml^{-1}) were those at which the rate of cell division was not restricted by changes in nutrient concentrations. A 400-ml culture was established at an initial density of *c.* 2×10^5 cells ml^{-1}. The time for the population to double was determined by cell counting, after which 200 ml of the cell suspension was transferred to 200 ml of fresh medium. The process was then repeated at similar time-intervals. The excess 200 ml of culture remaining at each subculture were used to determine the composition of the cells. The cell density of the cultures oscillated about a fixed mean (Fig. 11.30A) but the cumulative growth of the culture was exponential for at least eight generations ($td = 33$ hours) (Fig. 11.30B).

The composition of cells maintained in this way resembled that of cells late in the exponential growth phase of normal batch cultures of *A. pseudoplatanus* (Table 11.7). Furthermore, data for cell dry weight, fresh weight (Fig. 11.31) and insoluble nitrogen content suggests that the state of the cultures approximated to one of balanced growth.

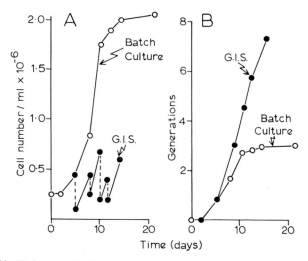

FIG. 11.30. *High rates of cell division maintained in 400-ml cultures of* Acer *pseudo-platanus cells by generation-interval subculture (GIS).*
A. The cell numbers recorded (●) at the beginning and end of subcultures of *c.* 40 hours duration. Each 'peak' of cell number represents the termination of an individual subculture, when the culture was diluted × 2 with fresh medium. Also shown (○) is the growth of a 400-ml batch culture not subjected to dilution.
B. The cumulative growth by cell division (expressed in terms of number of cell generations with time) in a batch culture (○) and a culture subjected to GIS (●). With these axes a straight line represents an exponential increase in cell number. Thus the overall rate of cell division in the diluted culture was constant for *c.* 240 hours ($td = 33$ hours) (from Wilson 1971).

TABLE 11.7. Composition of *Acer pseudoplatanus* cells maintained* by generation-interval subculture (GIS) compared with composition of cells in batch cultures (from Wilson 1971)

		GIS	Batch culture cells	
			Exponential phase	Stationary phase
Doubling time, td	Hours	33	60	∞
Cell fresh weight	$\mu g \times 10^{-3}$	$36 \pm 2 \cdot 1$	38	50
Cell dry weight	$\mu g \times 10^{-4}$	48 ± 4	45	25
TCA insoluble/N	$\mu g \times 10^{-5}$	$16 \cdot 3 \pm 2 \cdot 7$	15	$4 \cdot 5$
Insoluble/N per g fresh weight	mg N/g	$4 \cdot 3 \pm 0 \cdot 8$	4	$1 \cdot 0$

* For up to ten generations.

L

Large-scale semi-continuous culture

Large-scale semi-continuous culture systems have been devised primarily with a view to producing large volumes of reproducibly uniform cell cultures for biochemical study.

As a source of material for their studies of protein biosynthesis in cell-free systems, Graebe & Novelli (1966) used cell cultures of *Zea mays* endosperm tissue maintained in exponential growth over at least 160 days by a semi-continuous system (Fig. 11.32A). Two 4-litre cultures were established in

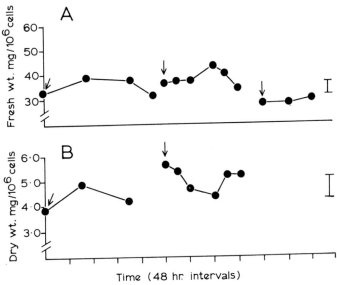

FIG. 11.31. *Balanced growth of* Acer pseudoplatanus *cells during generation-interval subculture.* Each group of points (starting at arrows) represents the sequential values obtained during continuous periods of maintained cell division in separate experiments under the same conditions. The vertical lines (right of figure) denote × 4 the standard error of the means positioned with the means as midpoints (from Wilson 1971).

two linked 6-litre flasks. At intervals approximately corresponding to the population doubling time ($td = 300$ hours) one culture was harvested, the other divided between the two flasks and fresh medium added to bring both cultures back to 4 litres. Although growth was only measured by settled volume, a constant yield of cells (3·6–4·5 g fresh weight per litre of culture per day) was achieved.

Veliky & Martin (1970) have established semi-continuous cultures from several species using a V-fermenter (see Chapter 4, Fig. 4.6). These cultures were maintained for several months as sources of uniform inocula either for small-scale batch culture experiments or to initiate other fermenters. Data for one such culture of *Phaseolus vulgaris* (td = 63 hours) which was diluted to twice its volume every 7–8 days suggest that balanced growth was *not* achieved. The rate of biomass production was still increasing after 41 days and the cell dry weight still decreasing.

Large, homogeneous cell suspensions of *Arachis hypogaea* (td = 120 hours) were maintained in exponential growth for about five generations by Verma & van Huystee (1971) (Fig. 11.32B). A series of 100 ml samples,

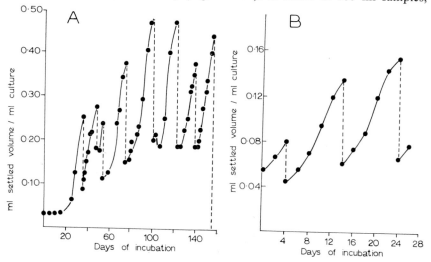

FIG. 11.32. *Prolongation of a high growth rate by the semi-continuous culture of cell suspensions.* At each peak (dashed lines) a volume of culture was harvested and fresh medium added. Growth followed by increase in volume of cell material (allowed to settle in measuring cylinder) per unit volume of total culture.
A. *Zea mays;* from Graebe & Novelli (1966).
B. *Arachis hypogaea;* from Verma & van Huystee (1971).

removed from a 2·5-litre culture every 2 days, were replaced after 10 days by 1·0 litre of fresh medium. Again the data for these cultures do *not* clearly indicate that balanced growth was achieved.

In these cultures although the number of cells increased exponentially the cell density was maintained within fixed limits by the periodic replacement of harvested *culture* by fresh medium. Such a culture is termed *open* (Ricica 1966). However, during the intervals between additions of new medium both the cell density and the nutrient environment were in continuous change.

Closed continuous culture

A *closed* culture is one in which cells are retained and hence there is progressive increase in cell density whilst growth continues (Ricica, 1966). A **closed continuous system** is one in which nutrients in excess of the culture requirements are supplied by continuous inflow of fresh medium and this is balanced by continuous harvesting of *spent medium*. In a culture system (4 litres) of this type tested with a culture of *Acer pseudoplatanus* (Wilson, King & Street, 1971—see also Chapter 4, Figs. 4.11 and 4.12) a constant volume of culture was maintained by the continuous outflow of spent medium via a siphon. The cells were separated from the spent medium by gravity

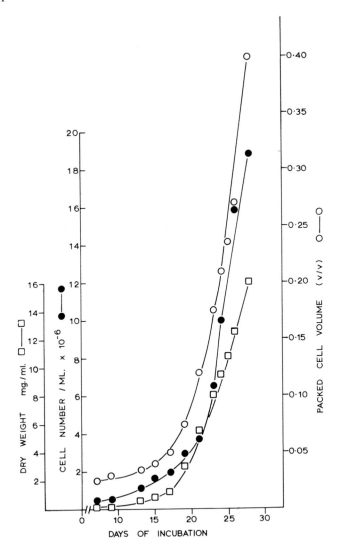

sedimentation in a wide-bore stilling tube. In this system cells accumulated exponentially in the culture vessel through six generations ($td = 72$ hours) before oxygen availability limited further growth (Fig. 11.33). During the period of constant specific growth rate, cell composition varied between relatively narrow limits, i.e. a state close to balanced growth was achieved (Fig. 11.34). The mean values for cell dry weight, cell volume and protein were lower than for cells late in the limited exponential growth phase of a batch culture (see Fig. 11.6). Such cells might well repay careful comparison with the cells of plant meristems.

Since the demands of an exponentially increasing population can only be met for a limited time in such a closed system, its potential value is rather in the direction of studies in cytodifferentiation where it may be important to grow cells under a particular regulated environment and then maintain them for a considerable period in a non-dividing but viable state.

Open continuous culture—turbidostat

In order to maintain non-limiting conditions indefinitely and to establish a constant culture environment it is necessary to employ automated control of culture volume and inflow of fresh medium. In such a system the cell population density is monitored in the culture vessel by a physical (e.g. light transmission or scattering) or a physico-chemical method (e.g. pH determination) (see Málek & Fencl 1966, Norris & Ribbons 1971, and Wilson, King & Street 1971, for constructional details). This information is fed back to a fresh medium-input control system linked to a system for a balancing harvest of culture. The medium input system operates each time the population density exceeds a pre-selected value and dilution of the culture reduces the cell density to just below this value (the fresh medium-pulse volume is small compared to the culture volume). The effect is to produce a culture of

---

FIG. 11.33. *Growth of* Acer pseudoplatanus *cells in a 4-litre closed continuous culture.* Fresh medium was continuously supplied from day 11. The rates of supply of air and fresh medium were as follows:

Day	Medium flow rate litre day^{-1}	Air flow rate ml min^{-1} (G3 sinter)
0	0·0	500
11	0·2	500
18	0·8	500
21	0·8	600
23	0·8	1000
24	1·8	1000
25	2·0	2000

constant volume and constant population density growing at a constant rate. Thereby balanced growth of plant cells in an unchanging environment, a steady state, can be rapidly achieved (Fig. 11.34).

In systems of this kind developed for micro-organisms, the intensity of the light transmitted through the culture is measured by a system of photo-cells. Such a system, called a 'turbidostat', was first successfully used in studies of microbial selection by Bryson (1952) and for the examination of steady-state cultures of bacteria by Northrop (1954). A turbidostat for the culture of plant cell suspensions was described by Wilson, King & Street in 1971 (see Chapter 4, Figs. 4.15 and 4.16).

There have been several reports of the *turbidometric measurement* of the

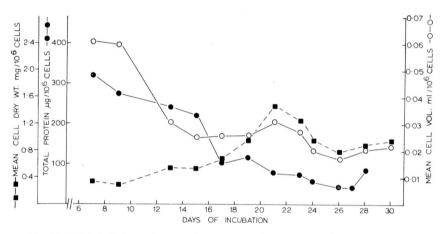

FIG. 11.34. *Limited change in size and composition of* Acer pseudoplatanus *cells during prolonged exponential growth in a closed continuous culture system.* Dilution began on day 11 (see Fig. 11.33 for further experimental details).

growth of plant cell cultures. Eriksson (1965), using a rapidly growing culture of *Haplopappus gracilis*, demonstrated a relationship between optical density (at 610 nm) and culture dry weight (Fig. 11.35C). The growth of cell suspension cultures was also monitored by colorimetry by Dougall (1965) working with *Rosa* sp. (Paul's Scarlet) and by Bellamy & Bieleski (1966) working with *Nicotiana tabacum*. A useful, working relationship has been found between the opacity of *Acer pseudoplatanus* cell suspensions and both their cell density and dry weight (Fig. 11.35A, B). Cells from **chemostat** cultures, in which steady states at various growth rates had been established, were pumped continuously through a sterile glass cuvette held between two cadmium sulphide light-sensitive resistors which formed the arms of a Wheatstone bridge circuit (Wilson, King & Street 1971). These relationships, not vitiated by the degree of cellular aggregation in *A. pseudoplatanus*

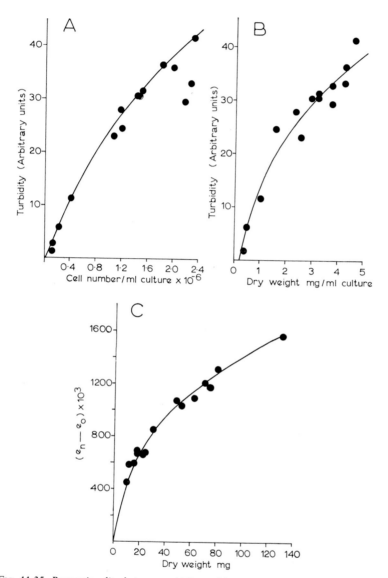

FIG. 11.35. *Proportionality between turbidity and biomass in cell suspension cultures.*
A and B. The relationship between cell number and dry weight per millilitre of
Acer pseudoplatanus suspensions and their turbidity. The electrical output of a
bridge circuit, which included the density detectors, was used as a measure of
turbidity (see text).
C. The relationship between the extinction at 610 nm of growing suspensions of
Haplopappus gracilis and their dry weight, recorded using a specially constructed
apparatus (Eriksson 1965). e_o=initial extinction; e_n=extinction at sample time.

suspensions, not only provide a convenient method for monitoring the cell density of large-scale batch or continuous cultures but are also sufficiently sensitive to permit the turbidometric *control* of cell density and hence the operation of steady state turbidostat cultures.

In any open continuous culture system the following equation holds:

$$\frac{dx}{dt} = \mu x - Dx \tag{5}$$

(see Eqn (16) Glossary of technical terms, p. 337, where μ = specific growth rate; x = cell density (cells per unit volume); D = **dilution rate** and t = time. In a turbidostat where an upper limit to cell density is imposed electronically, μ and D are tightly linked by feed-back control of the inflow of fresh medium. Thus as $\mu = D$, $dx/dt = 0$, i.e. the cell density remains constant with time and the specific growth rate can be calculated from the rate at which fresh medium enters the culture. In practice the input of fresh medium into the turbidostat is intermittent (occurring each time the cell density exceeds the imposed value) and hence the cell density is not absolutely constant but oscillates around a mean. However, the sensitivity of the control system can restrict this oscillation. For example, in a 4-litre turbidostat culture of *A. pseudoplatanus* cells, each pulse of fresh medium only dilutes the cell population by 2%; the oscillation of the cell population is less than can be detected by the available cell counting technique (Chapter 4, p. 90).

A basic feature of a turbidostat culture is that its growth rate is independent of cell density within defined limits and is at a maximum value determined only by the chemical and physical *quality* of the environment and by the associated velocity of metabolic processes in the cell and not by the operation of any substrate limitation (for further discussion see Bryson 1959). Fig. 11.36A shows a steady state established in a turbidostat culture of *A. pseudoplatanus* following a restriction on cell density operating from zero time. Stabilization of culture biomass was expressed by four parameters. At the indicated transition point, the limit on cell density was 'stepped-up'. A second steady state (Fig. 11.36B) became rapidly established at about twice the previous cell density but, as predicted, at the same specific growth rate. Preliminary examination of the steady-state levels of a number of major nutrients in these two cultures suggests that nutrient availability remained non-limiting. However, at higher cell densities, when one or more nutrients is depleted to below a limiting level, the culture system becomes insensitive and small changes in cell density produce large changes in growth rate. Fig. 11.37A shows a 'step-up' transition to a cell density higher than the defined limit; the population, initially slow growing, ceased growth after 170 hours under these conditions. However, a 'step-down' from such a density re-established a stable population growing at close to the maximum specific growth rate (Fig. 11.37B). The maximum specific growth rate of suspension culture cells of *A. pseudoplatanus* as determined in the turbidostat was 0.225 day^{-1}, agreeing well with values for chemostat cultures at 'wash-out' (Figs. 11.42, 11.43) and for batch cultures during the exponential growth phase (Fig. 11.7B—growth at 4.5×10^{-6} M, 2,4-D) using the same medium and the same incubation temperature.

The particular value of turbidostat cultures for studies on the regulation of cell metabolism is that it enables the operation of environmental factors (e.g. temperature, light), of specific metabolites and antimetabolites and of

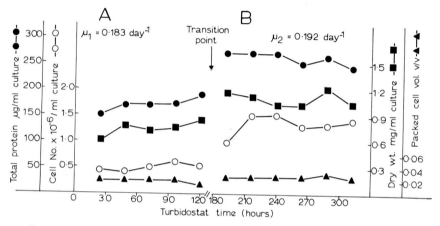

FIG. 11.36. *Steady states at high growth rates in a turbidostat culture of* Acer pseudoplatanus *cells.* After 120 hours operation in state A, the balance-point of a density-detection Wheatstone bridge circuit was raised by a potentiometer adjustment; fresh medium additions temporarily ceased. Forty-eight hours later the culture density exceeded the new balance-point and this initiated a further series of medium additions. A new steady state (B) rapidly stabilized at a growth rate (μ) not significantly different to that of state (A).

intrinsic (genetic) factors to be studied under conditions where growth is not limited by availability of essential nutrients.

Open continuous culture—chemostat

The work of Monod (1950) and Novick & Szilard (1950) first demonstrated that steady states of growth could be achieved in cultures of micro-organisms by continuous dilution of a fixed volume of culture with fresh medium at a *chosen* rate. Any chosen dilution rate (within certain limits) produced a steady-state population with a characteristic cell density. Expressed in terms of the equation for an open continuous culture system (see Eqn (16), p. 337) this implies that any perturbation of either μ, D or x produces compensatory

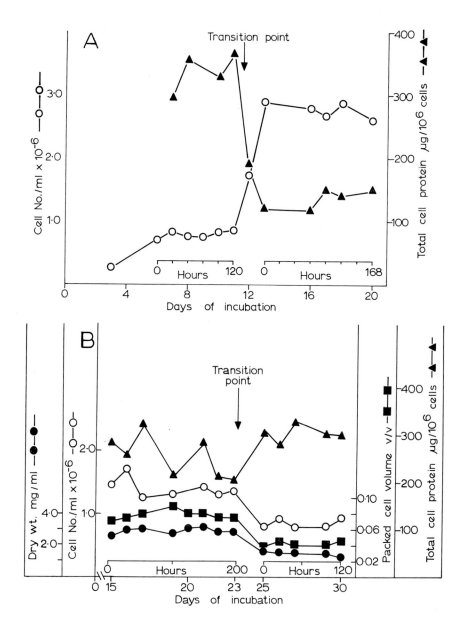

FIG. 11.37. *Transitions between different biomass levels in turbidostat cultures of Acer pseudoplatanus cells.* Specific growth rates recorded were:

A. 0·237 day^{-1} and 0·110 day^{-1}.

B. 0·132 day^{-1} and 0·195 day^{-1}.

Adjustments made as described in Fig. 11.36 except that lowering the balance-point in (B) resulted in a continual inflow of fresh medium until the cell density had been diluted to below the threshold level.

changes in one or both of the other two terms, thereby maintaining $dx/dt = 0$. During each of these steady states characteristic levels (s) of individual nutrients are established in the culture medium. The value of each equilibrium substrate (nutrient) concentration follows the general relationship:

Equilibrium
substrate   = Input of   − Output of   − Consumption
concentration   substrate   substrate   of substrate
(s)

which can be expressed mathematically:

$$\frac{ds}{dt} = DS_R - Ds - \frac{\mu x}{Y} = 0 \tag{6}$$

where S_R = concentration of nutrient in the input medium; s = equilibrium nutrient level in the culture and Y = the yield coefficient (cells formed/ nutrient used). Monod (1950) further observed that the specific growth rate (μ) was dependent upon the equilibrium concentration of one growth-limiting nutrient (s), the relationship being a type of saturation curve obeying the equation:

$$\mu = \mu\text{max} \frac{s}{K_s + s} \tag{7}$$

where K_s = a 'saturation constant' numerically equal to the nutrient concentration at which the specific growth rate is one-half of its maximum value. Hence in a chemostat it follows that a steady state only occurs when the dilution rate chosen is less* than that equivalent to μ_{max} and if the dilution rate exceeds this value, the culture becomes unstable and ultimately all the cells will be washed out of the culture vessel. For a more detailed discussion of chemostat theory the reader is referred to Herbert, Elsworth & Telling (1956), Herbert (1959), Málek & Fencl (1966), Kubitschek (1970), and Norris & Ribbons (1971).

Until recently there had been no attempt to examine in detail to what extent the kinetics of growth of higher plant cells in chemostat cultures conform to the 'Monod equations'. The emphasis of the early work with plant cell chemostat cultures was solely on biomass production. Thus Tulecke, Taggart & Colavito (1965) described the dilution of 8-litre 'phyto-stat' cultures of Rosa sp. (Paul's Scarlet) by replacing 1–2 litres of culture with the same volume of fresh medium at approximately daily intervals

---

* There is also a practical limit to the *lowest* dilution rate which will produce a stable culture.

$(D = 0.1–0.2$ day^{-1}; $td = 88–250$ hours*). The yield values quoted (Fig. 11.38) suggest that some steady states were achieved. A chemostat culture of *Glycine max* produced by Miller, Shyluk, Gamborg & Kirkpatrick (1968) by the continuous dosing of a 2-litre culture with fresh medium and the removal of 10 ml portions of culture every 20 minutes ($D = 0.36$ day^{-1}, $td = 46$ hours*) gave a yield of 1.34 mg ml^{-1} cell dry weight and was demonstrated to be stable in output for a short period. A constant yield of cells from a culture of *Haplopappus gracilis* was reported by Constabel, Shyluk & Gamborg (1971) using a similar system. Kurz (1971) has cultured suspensions of *Glycine max*

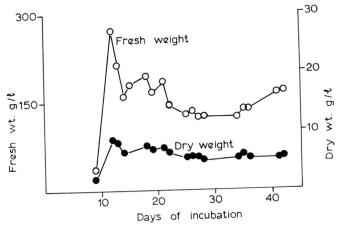

FIG. 11.38. *The production of* Rosa *sp. cells by semi-continuous culture in a phytostat.* Harvests of biomass from an 8-litre culture when 1 litre of culture was collected at approximately daily intervals and replaced manually by 1 litre of fresh medium. The culture was maintained for 43 days during which time 33 litres of culture were harvested. The average daily harvest was: 6.0 g/litre dry weight (146 g/litre fresh weight); the total tissue harvested was 169 g dry weight (4835 g fresh weight) (from Tulecke, Taggart & Colavito 1965).

and *Triticum monococcum* in a continuous-flow chemostat (see Chapter 4, Fig. 4.5) and reported, for the *Glycine max* culture, the achievement of a steady state after 4–6 days' dilution with a surprisingly high specific growth rate ($td = 30$ hours) and a constant yield of 1.3 mg ml^{-1} cell dry weight.

The 4-litre chemostat culture system developed by Wilson, King & Street (1971) (see also Chapter 4, Fig. 4.13) has been used to establish steady-state populations of *Acer pseudoplatanus* cells over a wide range of growth rates. With this system individual experiments have extended up to 200 days before being terminated by contamination or mechanical failure. Data for one steady state, which was maintained for 314 hours before being perturbed in

* These values are calculated from the published data.

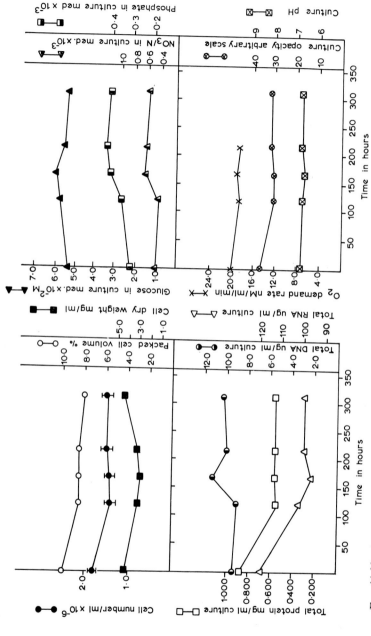

FIG. 11.39. *A steady state established in a 4-litre chemostat culture of* Acer pseudoplatanus *cells.* The culture was diluted for *c.* 400 hours at a rate of 0·194 day⁻¹. Samples (50 ml) were withdrawn at intervals for biomass measurements, nutrient analysis and respiration rate determinations. Culture opacity and pH were monitored continuously in the culture vessel.

the course of an experiment, are shown in Fig. 11.39. The limiting nutrient for this culture was nitrate-nitrogen (see below). Similar steady states of growth of *A. pseudoplatanus* suspension cultures have been obtained by Wilson (1971) under conditions where phosphorus was the limiting nutrient (Fig. 11.40). These results indicate that such steady states were characterized by steady levels of cell morphology and composition and by steady levels of medium constituents and pH. Studies on the nitrogen metabolism of *A.*

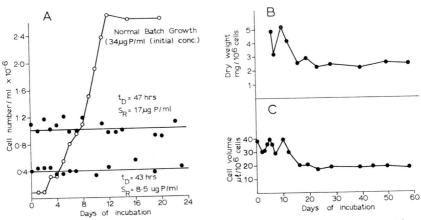

FIG. 11.40. *Steady-state growth of* Acer pseudoplatanus *cells in a chemostat with phosphorus as the limiting nutrient.*
A. Random, short-term fluctuations in the number of cells in two chemostat cultures (●) with similar doubling times (*td*), diluted with media containing different concentrations (S_R) of phosphorus. The plots show best fit lines of zero slope. For comparison, the change in cell number normally recorded during the growth of a batch culture in phosphorus-containing medium is superimposed (○).
B and C. Steady states of cell dry weight and cell volume arising in a chemostat diluted at a rate of 0·0161 hour^{-1} (*td* = 43 hours) with medium containing phosphorus at a concentration of 8·5 μg ml^{-1} (from Wilson 1971).

pseudoplatanus cells using chemostat cultures also point to the establishment of steady levels of enzyme activity. Fig. 11.41 gives data for a steady state in which the equilibrium and limiting level of nitrogen was 0·03 μmoles/ml (input level of nitrate/N plus urea/N = 13·0 μmoles/ml). In this culture the rate of nitrate consumption (0·46 nanomoles ml^{-1} min^{-1}) corresponded closely to the potential nitrate reductase activity as assessed by *in vitro* assay (0·57 nanomoles ml^{-1} min^{-1}).

Differing equilibrium densities of *A. pseudoplatanus* cells obtained at different dilution rates in chemostat cultures are shown in Fig. 11.42. These results not only point to the limitation of the growth rate of the cells by a single nutrient but, in so far as each increase in dilution rate leads to a

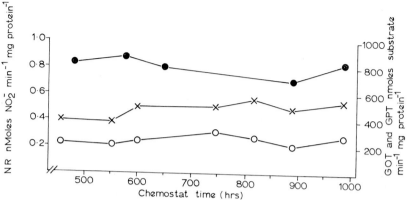

FIG. 11.41. *Steady states of nitrate reductase* (NR – ● –), *glutamate-oxaloacetate tran-saminase* (GOT –×–) *and glutamate-pyruvate transaminase* (GPT –○–) *activity in a chemostat culture of* Acer pseudoplatanus *cells.* The culture was N-limited and was diluted at the rate of 0·094 day^{-1} (td=178 hours). The steady-state biomass was characterized by a cell density of $2·32 \times 10^6$ cells ml^{-1} and a total-protein content of 705 μg ml^{-1}. GOT and GPT were measured by a coupled assay technique with α-oxoglutarate and aspartate (GOT) or alanine (GPT) as substrates. The oxalo-glutarate (GOT) or pyruvate (GPT) generated was determined by monitoring the change in extinction at 340 nm in the presence of malic dehydrogenase (GOT) or lactic dehydrogenase (GPT) (previously unpublished data of M.Young).

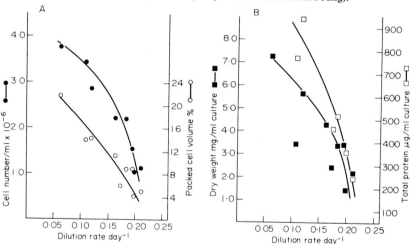

FIG. 11.42. *Relationship between steady-state biomass and dilution rate in chemostat cultures of* Acer pseudoplatanus *cells.* Each point in the curve for any one parameter in A and B represents the mean of a series of values obtained at intervals during a steady state established at the dilution rate indicated. The standard errors of these means were all less than 10% of the mean values. The data were obtained during nine different steady states (all in excess of 400 hours) established in five separate chemostat cultures.

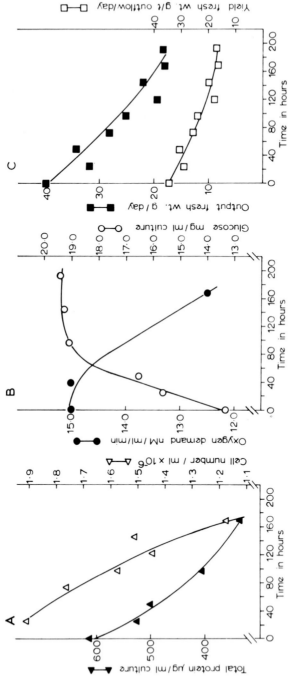

FIG. 11.43. *'Wash-out' of* Acer pseudoplatanus *cells from a chemostat culture diluted at a rate in excess of the critical dilution rate.* The dilution rate was 'stepped-up' from 0·182 day⁻¹ (steady state) to 0·274 day⁻¹ at time = 0 hours and cells were washed-out at a rate of 0·038 day⁻¹ indicating a critical dilution rate of 0·236 day⁻¹.

A. Decay in cell number and protein per millilitre of culture.

B. Decline in oxygen demand and increase in the concentration of glucose per ml of culture (input glucose concentration, $S_R = 20$ mg/ml).

C. Decline in the total output of biomass per day and in the daily yield per litre of outflow (the latter is equal to the biomass concentration in the culture vessel).

decline in the steady-state population density, suggest that the cells have a relatively low affinity for this limiting nutrient (that K_s of the Monod equation —see Eqn (7), p. 323—relating s to μ, has a relatively high value). The trends shown in Fig. 11.42 also indicate an upper limit of dilution rate. The population density did not decline linearly but progressively more steeply as this critical dilution rate was approached. When the critical dilution rate was exceeded (Fig. 11.43 where $D = 0\cdot274$ day^{-1}) the cells were washed out at a steady rate ($0\cdot038$ day^{-1}) even though availability of the potentially limiting nutrients was high and the population density low. This critical dilution rate corresponds to a maximum specific growth rate (μ_{max}) of $0\cdot236$ day^{-1}

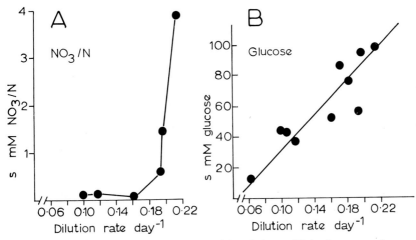

FIG. 11.44. *Steady-state levels (s) of nitrate (A) and glucose (B) in chemostat cultures of* Acer pseudoplatanus *at different dilution rates.*

($td = 70$ hours), which is in agreement with the value noted in turbidostat cultures (Fig. 11.37).

Study of the steady-state levels of nutrients at different dilution rates in these cultures showed that the relationship for NO_3/N differed from that of other nutrients (e.g. glucose) (Fig. 11.44). When the data for nitrate are plotted as in Fig. 11.45A, a curve is obtained of the kind expected from the Monod equation (Eqn (7) if nitrate is the limiting nutrient. The same data submitted to a Lineweaver/Burk plot (Fig. 11.45B) gives values for μ_{max} of $0\cdot225$ day^{-1} and for K_s (NO_3/N) of $0\cdot13$ mM. This latter value is three orders of magnitude greater than K_s values for amino nitrogen as a limiting nutrient for bacterial cultures; the plant cells have a relatively low affinity for nitrate. Wilson (1971) reported that K_s (phosphorus) was $0\cdot032$ mM for *A. pseudoplatanus* cells; this contrasts with $1\cdot56$ μM for bacterial cultures.

The predictable response of *A. pseudoplatanus* cells to a change in the concentration of the limiting nutrient in the inflowing medium at a constant dilution rate further illustrates the conformity of higher plant cell suspension cultures to microbial theory. Since in a steady state $\mu = D$, Eqn (6) may be simplified to:

$$x = Y(S_R - s) \tag{8}$$

and hence, if Y is a constant and s is very small relative to S_R then:

$$x \propto S_R$$

In the experiment shown in Fig. 11.46, when the NO_3/N concentration in the

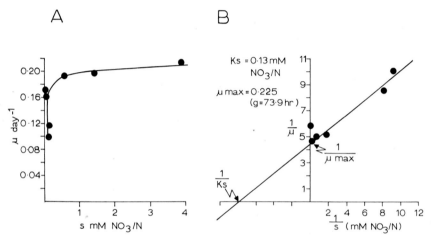

FIG. 11.45. *Relationship between specific growth rate and the steady-state level of nitrate/N in chemostat cultures of* Acer pseudoplatanus. *The specific growth rates* (μ) *are calculated from the dilution rates applied (see Eqn 16, p. 337 and text, p. 320).*

input medium (S_R) was raised from 7·0 mM to 10·5 mM, an increase in cell density was observed which corresponded well with the prediction made from Eqn (8).

A similar response to enhanced S_R (phosphorus) has been reported by Wilson (1971) (see Fig. 11.40A), although he observed considerable variation in Y at different growth rates, possibly related to the large changes in cell nucleic acid content (relative to changes in total cell protein) which occur with changes in the growth rate of *A. pseudoplatanus* cells in chemostat culture (Fig. 11.47B).

A. pseudoplatanus cells in steady states at different specific growth rates differ significantly in mean size, composition and physiology (Fig. 11.47). Dry weight per cell declines as growth rate increases; at the higher growth rates this is associated with a decline in cell volume. QO_2 and RNA values

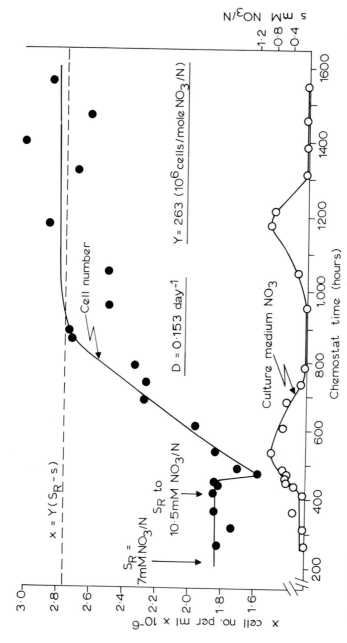

Fig. 11.46. *Change in the biomass of a 4-litre chemostat culture of Acer pseudoplatanus cells following a change in the concentration of the limiting nutrient (NO_3/N) in the inflowing medium (S_R). The culture was diluted at a constant rate of 0·153 day^{-1} (td=108 hours). The yield coefficient, Y, for the first steady state (ended after 420 hours) = 263×10^6 cells moleN^{-1}. s = steady-state concentration of NO_3/N in the culture vessel.*

rise progressively with growth rate; protein content remains steady over a range of growth rates but increases as μ_{max} is approached. Glucose consumption reaches a peak at an intermediate growth rate and at this time the percentage of the glucose respired is at a minimum. Wilson (1971) has shown

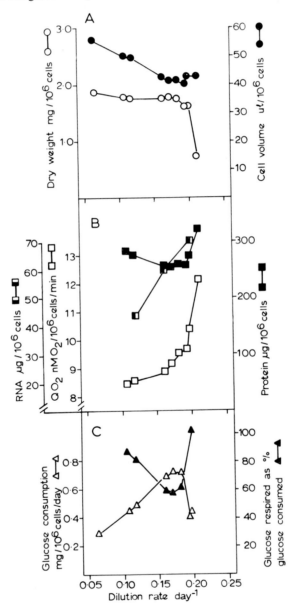

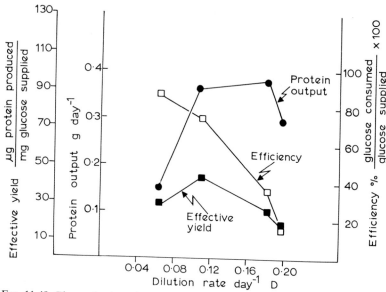

FIG. 11.48. *The production of protein by a 4-litre chemostat culture of* Acer pseudo-platanus *cells at different dilution rates.* Protein output (*Fx*) calculated from the flow rate *(F in ml/day⁻¹)* and the protein concentration in the culture vessel (*x* in g ml⁻¹).

that significant synthesis of leucoanthocyanins and catechins occurs at low but not at high growth rates.

These observations suggest that cell populations differing in their metabolism can be selected simply by altering the potentiometer on the metering pump delivering fresh medium to a chemostat. It can therefore be predicted that such chemostat cultures will, in due course, lead to the discovery of the special cultural environments needed to elicit the activity of particular metabolic sequences in cultured plant cells (Street 1973). The concept is illustrated by changes in the protein yield of cultures of *A. pseudoplatanus* (Fig. 11.48). If total protein is the desired product, an intermediate growth rate is economically most productive; a different growth rate might be needed for maximum yield of a *particular* protein.

FIG. 11.47. *Relationships between composition, physiological activities and growth rate of* Acer pseudoplatanus *cells in suspension culture.* Steady states of cells at different growth rates were established in 4-litre chemostat cultures with nitrate/N as the limiting nutrient. Points in A and B were calculated from the trend lines drawn through data relating the mean cell number, packed cell volume, dry weight, total protein and total RNA per millilitre of cultures in steady states at the dilution rates indicated (see Fig. 11.42). Glucose consumption was determined by analysis of the culture medium. The proportion of glucose respired was calculated from data of oxygen uptake and respiratory quotients.

The growth of *A. pseudoplatanus* cells in chemostats is not entirely predictable (Street 1973) and it would be surprising to find that cell cultures of all plant species were completely amenable to chemostat operation; this is not the case even for bacterial species (Fencl 1966). Because of the relatively low metabolic rates and long generation times of plant cells, long periods of operation are necessary to achieve steady states initially or following perturbations (see Fig. 11.46). Nevertheless continuous cultures are a significant advance in the techniques of plant cell culture and possess high potential for studies in metabolic regulation, the maximized production of secondary metabolites, cytodifferentiation and organogenesis.

GLOSSARY OF TECHNICAL TERMS

The development of synchronous and continuous plant cell cultures introduces a number of new technical terms into plant tissue and cell culture from the field of microbial methodology. The terms and definitions are mainly based upon Herbert, Elsworth & Telling (1956) and Ricica (1966) (see also Málek & Fencl 1966).

Batch culture: cells growing in a finite volume of nutrient medium. *At its simplest growth ceases when an essential nutrient is depleted.*

Continuous culture: a culture continuously supplied with nutrients by the inflow of fresh medium. *The culture volume is normally constant.*

Open continuous culture: a continuous culture in which inflow of fresh medium is balanced by outflow of corresponding volume of CULTURE. *Cells are constantly washed out with the outflowing liquid. In a steady state, the rate of cell wash-out equals the rate of formation of new cells in the system* (see Chemostat and Turbidostat).

Closed continuous culture: a continuous culture in which inflow of fresh medium is balanced by outflow of corresponding volumes of SPENT MEDIUM. *Cells are separated mechanically from outflowing medium and accumulate in the system.*

Semi-continuous culture: a continuous culture in which the inflow of fresh medium is controlled manually at infrequent intervals by a 'drain-and-refill' operation. *Equivalent to rapid serial transfer of batch cultures.*

Chemostat: an open continuous culture in which growth rate and cell density are held constant by a fixed rate of input of a growth-limiting nutrient.

Turbidostat: an open continuous culture into which fresh medium flows in response to an increase in the turbidity of the culture. *A preselected biomass density is uniformly maintained by wash-out of excess cells.*

Phytostat: the name adopted by Tulecke (1966) for an apparatus designed for the semi-continuous chemostat culture of plant cells. Also used by Miller, Shyluk, Gamborg & Kirkpatrick (1968).

Fermenter: a large-scale vessel ($c.$ 1 litre and upwards) used for production of cell material or a specific biological compound by a batch or continuous culture process. *Normally used where micro-organisms are the synthetic agent. A term now in use for eucaryote cell cultures.*

Growth cycle: the sequence of events occurring in a batch culture made apparent by changing rates of production of cells and cell material. *The cycle repeats itself when cells at the end of one batch culture are used to initiate another.*

Lag phase: the initial period of a batch culture when no cell division is apparent. *May also be used with reference to the synthesis of a specific metabolite or the rate of a physiological activity. Some cultures show no lag phase.*

Exponential growth phase: (=*logarithmic phase*) a finite period of time early in a batch culture during which the rate of increase of biomass per unit of biomass concentration (*specific growth rate*) is constant and measurable. *Biomass is usually referred to in terms of the number of cells per ml of culture.*
    Mathematically:

$$\frac{1}{x}\cdot\frac{dx}{dt} = \frac{d(\log_e x)}{dt} = \frac{\log_e 2}{td} = \mu \qquad (9)$$

where x = *number of cells per unit volume at time,* t; μ = *the specific growth rate* (*time*$^{-1}$) *and td is the doubling time, i.e. the time required for the cell density to double.*
    Eqn (9) may be rewritten to express the increase in number of cells within a certain time, t:

$$\mu = \frac{d(\log_e x)}{dt} = \frac{\log_e x + \log_e x_0}{t} \qquad (10)$$

where x_0 = *cell density initially and* x = *cell density after time,* t.
    Rearranging (10) gives an equation of a straight line:

$$\log_e x = \mu t + \log_e x_0 \qquad (11)$$

Plotting values of t (*abscissa*) *and of* $\log_e x$ (*ordinate*), *a straight line is obtained for the exponential growth phase, with slope* $= \mu$. (*Plotting* $\log_{10} x$ *against* t *gives slope* $= \mu/2\cdot303$.)

Specific growth rate: μ: (=*relative growth rate*) the rate of increase of biomass of a cell population per unit of biomass concentration (see Exponential growth phase).

Maximum specific growth rate: μ_{max}: the specific growth rate of cells in a non-restrictive medium when all nutrients are present at non-limiting concentrations, e.g. *a low density of cells in a turbidostat culture. The specific growth rate may be maximal during the exponential growth phase of a batch culture* (but see Wilson 1971). *The actual* μ_{max} *recorded in a non-restrictive medium may depend upon environmental factors such as pH and temperature or upon the concentration of growth hormones. The absolute maximum specific growth rate will depend upon an intracellular rate-limiting step.*

Doubling time: *td:* the time required for the concentration of biomass of a population of cells to double.

Mean generation time: g: usually regarded as synonymous with doubling time. *But doubling time refers to total population of cells in a culture and is not strictly identical to the mean of the individual cell generation times* (Herbert 1961).

Linear phase: Progressive deceleration phase: transient phases of the batch culture growth cycle following the exponential growth phase when the specific growth rate declines uniformly with time (linear phase) and then decays more and more rapidly (progressive deceleration phase). *The duration of these phases varies between cultures of different species but probably depends greatly upon the complexity of the culture medium.*

Stationary phase: the terminal phase of the batch culture growth cycle where no net synthesis of biomass or increase in cell number is apparent. *Cells may remain viable for many days by utilizing intracellular reserves or recycling metabolites released by lysed cells.*

Cell cycle (=*mitotic cycle*): the repetitive sequence of events normally exhibited by a dividing cell consisting of a biosynthetic phase (interphase), nuclear division (mitosis) and cell division (cytokinesis). *The interphase may be further subdivided into a presynthetic phase (G1) following cytokinesis, a phase of DNA replication (S) and a postsynthetic phase (G2) preceding mitosis. The duration, or indeed existence, of each phase depends upon both the cell type and the environment.*

Synchronous culture: a culture in which the cell cycles (or a specific phase of the cycles) of a proportion of the cells (often a majority) are synchronous. *Thus changes in the culture as a whole are an amplification of events of the cell cycle of an individual cell.*

Balanced growth: 'growth is balanced over a time interval if, during that interval, every extensive property of the growing system increases by the same factor' (Campbell 1957). *Thus if the rates of biosynthesis and cell division in a random population of cells are constant and equal and therefore the mean composition of the cells of the culture does not vary with time, a culture is in a state of balanced growth. Such a state is probably never realised in a batch culture but persists in continuous cultures.*

Dilution rate: the ratio of the flow rate (f) (volume per unit time) to culture volume (V) in a continuous culture:

$$D = \frac{f}{V} \, [\text{time}^{-1}] \tag{12}$$

Thus D = the number of culture volumes displaced per unit of time. In the steady state, $D = \mu$, where μ = specific growth rate. Thus D is related to the doubling time of the cell population:

$$D = \mu = \frac{\log_e 2}{td} \tag{13}$$

Steady state: the situation in an open continuous culture where balanced growth proceeds in an unchanging environment. *Sometimes used rather loosely to describe short periods of balanced growth in closed systems when the environment is changing with time.*

The specific growth rate, μ of cells in culture may be written as:

$$\frac{dx}{dt} = \mu x \tag{14}$$

by rearranging Eqn (9), where x = the concentration of biomass and t = time. In an open *continuous culture subjected to constant dilution (at a rate = D), the rate of washing out of biomass is given by:*

$$-\frac{dx}{dt} = Dx \tag{15}$$

Net change in biomass in the vessel is the resultant of (14) and (15)

$$\frac{dx}{dt} = (\mu - D)x \tag{16}$$

Hence if $\mu > D$, the concentration of biomass will increase, while if $\mu < D$, wash-out will occur, When $\mu = D$, $dx/dt = 0$, which by definition is the case for the steady state.

CHAPTER 12

ASPECTS OF ORGANIZATION— ORGANOGENESIS AND EMBRYOGENESIS

J. REINERT

Introduction 338
Organogenesis 339
Embryogenesis 348
Conclusions 354

INTRODUCTION

The organization and morphogenesis of multicellular plants depends upon the integration and mutual interaction of various organs, tissues and cells which are separated from one another in space. Together they form a complex system for the analysis of which we need to know not only about the substances primarily involved in morphogenesis, for instance hormones, but also about the correlations which result from the differential capability of different parts of the plant to synthesize certain compounds and the transport of these substances. This system and its complex control mechanisms for organ or embryo formation can be simplified by isolating cells, tissues and organs and their subsequent cultivation *in vitro*. This and the demonstrated persistence of the potency of the isolated plant cells and tissues, as manifested, for instance, in the regeneration of leaves, flowers and embryos, are the primary reasons for the increasing use of cell and tissue cultures in morphogenetic studies. The use of such cultures implies the assumption that the processes involved in the formation of cells, meristems, organs and of whole plants are the same *in vitro* as they are in the intact plant.

In addition cell and tissue cultures offer the advantage of an almost perfect control of exogenous factors. This applies, for example, to the exclusion of biological contaminants (bacteria, fungi) which is attained by the use of aseptic methods and by the control of the uptake of nutrients, when the cultures are grown on chemically defined nutrient media without the addition of complex components (e.g. coconut milk, yeast extract, etc.).

Morphogenesis of higher plants can be considered from a number of quite different aspects. In the context of this book it is appropriate that we should deal here mainly with the initiation and the development of organs and of embryos *in vitro* and with the cultural factors and—as far as they are known—

with cytological and metabolic changes involved. In investigations concerning morphogenesis cultures of dicotyledonous plants have almost always been used. Considerable difficulties have been encountered in the culture of tissues derived from monocotyledonous and other groups of plants.

ORGANOGENESIS

Cell and tissue cultures are able to form a variety of organs, e.g. roots, shoots, leaves and flowers. As a rule the formation of these organs does not depend on the presence of pre-existing initials. The fact that the rapidly growing primordia cannot be demonstrated for at least 1 or 2 weeks after the isolation of the explants indicates that they are formed *de novo*. The formation of these new meristems involves two distinct growth phases. The first is the dedifferentiation of the original explant. It begins shortly after the isolation of the tissue with an acceleration of cell division and a consequent formation of a mass of undifferentiated cells. In this process all living cells, even such specialized types as collenchyma cells or already partially lignified fibres can revert to an embryonal status. In the second phase different types of specialized cells will again differentiate. This demonstrates clearly that tissue cultures are excellent examples for the persistence of the potencies and of the plasticity of form in plant cells. On the other hand by far the greatest proportion of the newly differentiated cells are vacuolated parenchyma cells, indicating that the cells of plant tissue cultures do not have the same capacity for variation as observed in the differentiation of cells of intact plants (Gautheret 1966, Reinert 1968).

The primordia of organs are formed as a rule from single or small groups of differentiated cells (Fig. 12.1) which give rise to small meristems with cells densely filled with protoplasm and with strikingly large nuclei (Buvat 1945). Some time after the formation of these meristems, the polarity of the longitudinal axis of the organizing growing points of the organs can be seen; this is coincidental with the mitoses.

The most frequent type of regeneration is root formation; this is true for callus as well as cell suspensions. Roots were first observed by Nobécourt (1939b) in cultures from carrots and have since been reported for many other tissues including woody plants (Jaquiot 1951), parenchyma from tobacco shoots (Skoog & Miller 1957), pea roots (Torrey 1959) as well as virus-induced tumours (Nickell 1955). These are just a few examples from a vast number of publications; however, it is not certain in all of these cases whether the root meristems in the culture were newly-formed or pre-existing. In this connection it is of interest that the frequently quoted observations of White (1939b) on root formation in cultures from *Nicotiana glauca* × *N. langsdorffii*

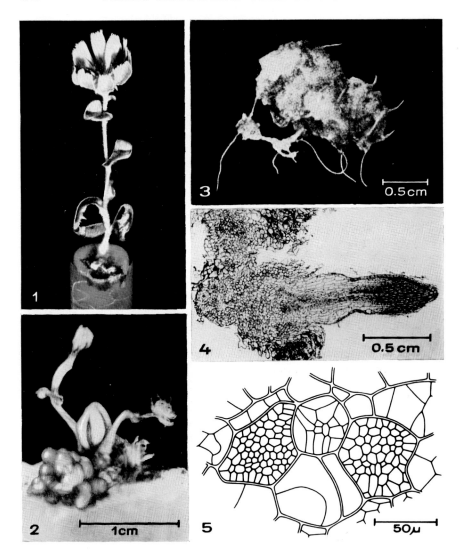

Fig. 12.1

12.1.1 Flower formation *in vitro* by a short-term culture from the root of *Cichorium intybus* (from Dr Colette Nitsch).

12.1.2. Flower formation *in vitro* by 2-year-old cultures derived from leaves of *Crepis capillaris*.

12.1.3. Root formation by 1-year-old carrot culture.

12.1.4. Median section through a young elongating root in tissue culture from pea roots. (from Dr J.G.Torrey).

12.1.5. Cell division in parenchyma cells of a carrot culture. These limited meristematic zones develop later into root primordia.

hybrids concern only adventitious roots from shoots and not from new meristems in the callus.

Roots will form, from whatever part of the plants the explants are taken; in other words, tissue from shoots and other organs will also form roots. Normally they are distributed irregularly over the surface of the callus and in suspensions they can grow on various components. They are monopolar organs and their structure may be entirely normal (see Fig. 12.1). The expression 'normal' here applies only to the root meristem and the growth and differentiating zone behind it, the base ends blind in the parenchyma. When isolated and grown in suitable nutrient solutions these roots scarcely differ from those of intact plants.

Shoot buds are formed nearly as frequently as roots; mostly they occur on separate cultures but sometimes they can be formed on the same tissue. Much less frequently observed in tissue cultures are leaves, singly or in groups. Many objects, including short-term cultures of chicory roots (Camus 1949), cuttings from young *Sequoia* stems (Ball 1950) and carrots (Levine 1951), have long been known to produce typical monopolar shoots having a perfectly normal organization. These findings have been confirmed later on by a large number of similar observations on, among others, explants of carrots (Steward, Mapes & Mears 1958), potatoes (Fellenberg 1963) and *Convolvulus* roots (Bonnett & Torrey 1966). The normal development is the rule; however, there have also been cases of extremely anomalous structures that can only be described as leaf-like or bud-like. Such teratomas were observed on cultures derived from tumours of tobacco shoots (Braun 1959), but also on normal tissues of *Taraxacum* roots (Bowes 1971).

The anatomical events during the induction and development of shoots and leaves are identical with the early phases of the initiation of roots, i.e. there must first be dedifferentiation before the primordia can be laid down.

In short-term cultures the localization of the newly initiating meristems may still bear some relation to the original organization of the explant. Thus, shoot primordia arise in segments of tobacco stalk mostly from elements of the external phloem near the surface (Sterling 1951). In chicory cultures it is the cambium (Camus 1949) and in explants from *Convolvulus* roots it is the cells near the protoxylem from which buds or shoots develop (Bonnett & Torrey 1966). Similar results have been reported for short-term suspensions of carrot cells; here root primordia appeared in association with endogenous protoxylem strands which later—after the formation of a root and transfer to agar medium—developed into complete plants with leaf-bearing shoots (Steward *et al.* 1958). However, this does not apply to older cultures, for callus from *Convolvulus* roots, cultivated for a longer period, produces meristems of shoot buds and roots, also from the parenchymatous parts of the culture.

In long-term cultures the primordia for shoots and buds lie near the surface, i.e. they arise exogenously, as in the intact plant (Reinert 1962). However, this is not always so, since some cultures may produce shoot initials both endo- and exogenously; this pertains, for instance, for *Convolvulus* callus (Bonnett & Torrey 1966).

Some of the facts concerning root and shoot formation *in vitro* have been known for quite a long time; flower formation in tissue cultures, on the contrary, has a relatively short history. Flower primordia *in vitro* and their subsequent development to complete flowers were first observed about 10 years ago in explants of stalks of a non-photoperiodically reacting tobacco variety (Aghion-Prat 1965). Segments from different parts of the stalks of flowering plants behaved differently; at the older basal internodes only vegetative buds were produced, while flower buds formed on the young upper parts, particularly on explants from the inflorescence. These findings were later confirmed by experiments with other photoperiodic as well as with cold-requiring plants. Thus, the formation of flowers could be induced by vernalization in *Cichorium intybus* and *Lunaria annua* cultures (Nitsch & Nitsch 1964, Pierik 1967). The same effect has been achieved by photoperiodic induction of cultures derived from various organs, including roots, of the short-day plant *Plumbago indica* (Nitsch 1968). It was these studies which helped to carlify, at least to a certain extent, some of the fundamental questions of flower formation *in vitro*. See Fig. 12.1.

Factors controlling organogenesis

Despite numerous studies and observations on organogenesis it is difficult to make extensive generalizations about these processes. In particular this is true for our knowledge about the endogenous factors involved. However, it is certain that the inception of organs occurs most frequently in recently isolated tissues and this ability decreases with increasing duration of culture, and eventually disappears completely. However, there are exceptions to this generalization. While many cultures remain indefinitely without any potential for regeneration others appear to have an unlimited capacity for organ formation. The best known example for the latter type is *Amorphophallus* (Morel & Wetmore 1951), but also carrot cultures are able to produce roots for years (Reinert, Backs-Husemann & Zerban 1971) (Fig. 12.2). These alternatives do not exhaust by any means the possibilities of organogenic expression. Thus, cultures of crown-gall tissue of *Scorzonera hispanica* did not form roots and shoots until after some years without any organization (Démétriades 1954). Similar observations have been made also with virus-induced tumour tissues (Nickell 1955) and other pathologically altered cultures (Sacristán & Melchers 1969).

This variability in tissues cultured *in vitro*, which for the present cannot be satisfactorily explained, is matched by the instability of the genome (see also Chapter 7). Early cytological studies revealed a variable degree of ploidy in cultured normal and tumour cells. This instability has now been particularly clearly shown by recent investigations of diploid callus tissues from the meristem of pea roots. From this work it is evident that only a few weeks after the isolation of these tissues the number of diploid cells starts to decline and the callus comes to consist mainly of tetraploid and sometimes even octoploid cells (Torrey 1967a). The instability of the genome of these cells is not an isolated phenomenon for similar changes have also been observed with cultures derived from *Haplopappus* shoots (Bennici *et al.* 1971), pollen grain cultures of the *Gingko* tree (Tulecke 1957) and cultures of the medullary parenchyma of tobacco stalks (Fox 1963), to mention just a few examples. Some idea of the frequency with which such polysomy occurs can be gathered from the fact that very few objects, namely cultures from tubers of *Helianthus tuberosus*, *Crepis capillaris* leaves and *Medicago sativa* (Street 1966b) are known to contain only the normal diploid set of chromosomes.

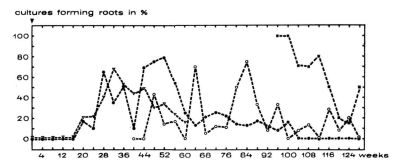

Fig. 12.2. *Pattern of root formation in freshly isolated* (● – – – ● ; ▽ – – – ▽) *and in older carrot cultures* (○ – – – ○ ; ■ – – – ■) growing on an agar medium containing a low nitrogen level (3·2 mM) and auxin. Abscissa: Weeks after isolation of the cultures (Reinert *et al.* 1971).

In some cases correlations have been found between the degree of ploidy and the capacity for organ formation. Pea-root callus, which had formed diploid root meristems after a considerable period of culture, lost this ability after all its cells turned polyploid (Torrey 1959). There has been a similar report for carrot-root cells (Mitra *et al.* 1960), although the data are not completely convincing in this case. However, any generalized assumption that there is a causal connection between the disappearance of diploid cells and the loss of capacity for organogenesis cannot be upheld. Contrary to this assumption, callus cultures will, under suitable culture conditions, produce organs with polyploid meristems (Torrey 1967a). Tissue cultures thus behave exactly like intact plants, i.e. endomitosis and the resulting polyploidy do not necessarily prevent the induction and continuation of morphogenetic processes (d'Amato 1952). This does not, however, exclude a blocking of morphogenesis by other genetic alterations, for example, by a high degree of aneuploidy (Murashige & Nakano 1967) or even by somatic mutations. But in this case there are clearly exceptions, e.g. callus from normal and tumerous tobacco plants can form shoots despite aneuploidy (Sacristán & Melchers 1969).

In connection with the evaluation of morphogenetic systems involving the cells of higher plants it is important to recognize that the loss of capacity for organogenesis in cultures can be reversible. Apart from a few more or less random observations (Partanen 1963b) this has been demonstrated clearly only for embryogenesis *in vitro* (see this chapter, section 'Embryogenesis', p. 348). In this example obviously the conditions of culture and other non-genetical factors are important because they may facilitate or prevent the manifestation of genetical information.

The labile properties of cells growing *in vitro* create difficulties in studies on the mechanism of organogenesis for it is seldom possible to work for long periods with the same system. This is one of the principal reasons why freshly isolated or short-term cultures of tissues are mainly used for experimental work. The search for cultures having properties which will remain stable for months or years is thus one of the priorities in this field.

As regards exogenous factors, there is so far little known about the influence of physical factors upon organogenesis. But from the work on flowering we know that light is definitely necessary for the induction and development of flowers in cultures from photoperiodic species (Nitsch 1968). Similarly, a period at a lower temperature is necessary for flower induction in cold-requiring plants (Pierik 1967). The drastic consequences of a change in the consistency of the nutrient medium have been known for a long time from studies with cultures of genetical tumours from tobacco hybrids. On simple synthetic agar medium these cultures grow as undifferentiated callus, whereas when immersed in a liquid medium of similar composition they produce buds and shoots (White 1939b). Low-intensity white light and temperatures below 25°C will enhance bud formation in tobacco cultures; higher intensity light and higher temperatures, however, inhibit it (Skoog 1944). It has also been demonstrated that light can exert a positive influence on rhizogenesis in artichoke cultures. These investigations showed that low-intensity white light acted synergistically to the rhizogenic effect of auxin (Gautheret 1971).

We know much more about the effect of chemical factors on organogenesis, especially those of phytohormones. Gautheret (1945) was the first to show that auxin at optimal concentrations could induce root primordia in carrot explants, but in a great number of subsequent investigations it turned out that a particular inductive treatment developed for a particular culture does not guarantee its positive effect on other cultures. Auxin, for instance can inhibit root formation in carrot and pea cultures and it can be without effect in tissues where it is not a limiting factor. The entirety of these observations leaves no doubt that organogenesis *in vitro* depends upon a complex system involving a number of limiting and interacting factors. It is this situation which has led to different theories about the regulation of organogenesis *in vitro*. One of these postulates that minimal quantitative changes in the ratio of certain components of the nutrient media are decisive and determine whether organs are initiated or not. The classical example for such a regulation is seen in tissues derived from the pith parenchyma of tobacco shoots (Skoog & Miller 1957). When this tissue is isolated and grown *in vitro* cell division and true growth occur on a synthetic medium only if it contains auxin and kinetin (6-furfurylaminopurine) or some other effective cytokinin. These two substances determine not only whether the tissue grows, but also how it grows. On an agar medium containing 2 mg/litre of auxin and 0·1 mg/litre of kinetin only an undifferentiated callus results. The two substances in this case work additively. However, if the kinetin concentration is lowered to 0·02 mg without altering the auxin level, roots will arise in the cultures. Higher concentrations of kinetin (0·5 mg/litre) lead, conversely, to the initiation of shoots; root formation is then suppressed. Other compounds interfere in this interaction. Raising the phosphate concentration of the medium can reinforce the shoot formation and suppress or weaken the root-promoting effect of auxin. Adenine and amino acids, such as tyrosine, act similarly for in suitable concentrations they augment the kinetin effect and reduce that of auxin.

So far there are only a few observations with other tissues supporting the principle that organogenesis is regulated by quantitative shifts in the ratio of hormones or other substances. One of these is chicory roots in which the zone which forms shoots contains a relatively large amount of native cytokinin whereas auxin predominates in the root-forming region (Vardjan & Nitsch 1961). Experiments with tissues from cyclamen tubers have also shown interactions between auxin and purine derivatives (adenine, guanine) in the regulation of root and shoot formation. But the decisive factor for root or shoot formation in these cultures apparently was not the ratio between these substances, but rather the absolute concentration of the auxin in the medium (Stichel 1959). In cultures of a variety of *Armoracea rusticana* it was even possible to induce shoot formation by raising the auxin concentration and to inhibit it by adding kinetin (Sastri 1963). Kinetin in physiological concentrations can also block morphogenesis in carrot cultures (Reinert 1959). Thus it could be that cultures from the medullary parenchyma of tobacco shoots are a somewhat special case so that the principle of regulation

M

of organogenesis by quantitative changes of the ratio between certain specific compounds cannot be generalized to cover cultured tissues in general.

It is equally impossible to deduce from the limited number of experimental data currently available on flower formation in tissue cultures as to whether this process is controlled by quantitative or qualitative changes. The factors that have been found to be essential for the initiation of flower primordia in explants of tobacco (Aghion-Prat 1965), *Plumbago indica* (Nitsch 1968) and *Lunaria annua* (Pierik 1967) are, apart from the light involved in photoperiodic induction and the low temperatures in vernalization, a relatively high level of nitrogen in the medium, plus cytokinin and various constituents of nucleic acids (adenine and orotic acid). Auxins, gibberellins and various organic nitrogen compounds (with the exception of urea) had an inhibitory effect. A quantitative relationship between these factors, similar to those between auxin and kinetin in root and shoot formation in tobacco parenchyma, has been observed only recently in experiments on flower formation. In this work the inhibiting action of IAA on flower formation in tobacco callus could be counteracted by adding different nucleic acid-base analogues (Skoog 1971). Although these results are partly contradictory they do not exclude the possibility that the reaction mechanism for organogenesis in the cells of higher plants are alike, differing only in that different limiting factors are decisive for different species or families.

There are other important questions which are still unsolved. Thus it is not clear whether inorganic nitrogen compounds or hormones like auxin and cytokinin induce organogenesis in cell cultures *de novo* or whether they enhance merely a process which begins already with the isolation of the cells. We know too almost nothing about the mechanism of action of these substances. If this background is taken into account it is not surprising that recently the rhizocaline theory had a revival. This theory implies that not quantitative but rather qualitative changes, i.e. the synthesis of the hypothetical rhizocaline determines the initiation and the development of roots. The basis of this concept is experimental data on the root formation of

---

FIG. 12.3

12.3.1–3. One-, two-, four-, and six-celled stages formed in the parenchyma of carrot cultures. These stages resemble closely those which occur during embryogenesis in seeds.

12.3.4–6. Globular and heart-shaped embryo stages formed in parenchymatic regions of carrot cultures.

12.3.7. Perfect torpedo-stage embryo isolated by immersion of a carrot culture in liquid nutrient. Note the suspensor-like cells at the radicular pole.

12.3.8. Young carrot plantlet with main root, cotyledons and first leaves removed from a carrot culture.

12.3.9. Late development stage of a carrot plant from a culture which was illuminated.

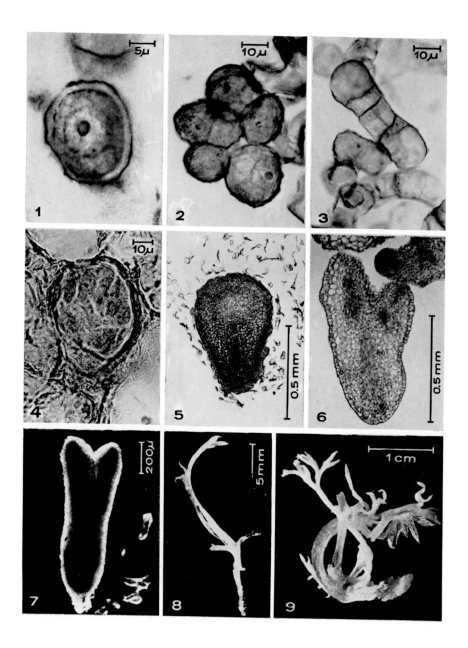

cultures from *Helianthus tuberosus* indicating that in addition to auxin and sugar, temperature and light are essential factors and play a decisive role in the formation of the rhizocaline in cultures (Gautheret 1966). According to Gautheret his data show that sugar is a precursor in the biogenesis of the rhizocaline but this has to be stabilized by auxin and light before it induces root formation. This concept can be extended to include the synthesis of other organ-specific substances in shoot and leaf production, i.e. substances like caulocaline and phyllocaline.

EMBRYOGENESIS

The most spectacular demonstration of the persistence of the potencies in the cells of higher plants—that is of totipotency—is the regeneration of whole plants following the induction of embryos in somatic cell cultures and recently also in generative cells, i.e. in pollen grains. The formation of bipolar adventive embryos begins just as in organogenesis with the dedifferentiation of the isolated tissue or cells but proceeds subsequently in a different way from the development of monopolar organs like shoots or roots. The process is best known for cells from tissue cultures of carrot roots which produce embryos by a developmental sequence through pro-embryonal, globular and torpedo stages, which are strikingly like those occurring after the fertilization of egg cells (Reinert 1958, 1959) (see Fig. 12.3). At first these findings were regarded as exceptional; the main reasons for this were other studies on the organogenesis in carrot tissue which were derived from cell suspension cultures (Steward, Mapes & Mears 1958). According to this work the regeneration of whole plants starts with the development of internally located vascular tissue in cell aggregates of the suspension which is followed by the formation of root primordia. After transfer to an agar medium an organized shoot and thereby a whole plant was formed by the root. This developmental sequences has still a highly hypothetical character because no detailed anatomical studies were made during the phase of shoot initiation (Steward *et al.* 1964). However, the events of embryo formation in cells or callus cultures have been confirmed later with both cultivated and wild varieties of carrot (Kato & Takeuchi 1963, Steward *et al.* 1964, Halperin 1966a). Embryogenesis *in vitro* has, moreover, been proved to occur not only in cells of carrots and other Umbelliferae but also in species from widely differing families of Angiosperms, including *Solanum menongema* (Yamada, Nakagawa & Sinobo 1967), *Atropa belladonna* (Konar, Thomas & Street 1972a), *Bromus inermis* (Constabel, Miller & Gamborg 1971) and various tissues of Cycads (Norstog & Rhamstine 1967). Therefore it is clear that the capacity for embryogenesis is widely distributed and possibly it is a fundamental property of somatic plant cells.

These results quite naturally raise the question as to the criteria for adventive embryos, in particular how they are to be clearly distinguished from developing buds. The best definition has been given very recently (Haccius 1971) after a thorough study of all facets of the problem. This is that the decisive feature for categorizing a plant structure as an embryo, besides other morphological properties, is its bipolarity and the fact that at the earliest developmental stage it has at opposite ends a shoot and a radicular pole. Furthermore, this system must not be connected with the vascular tissue of the mother plant or the explant during its initiation and development. With monopolar buds and roots, on the other hand, it is always possible to show their connection with the vascular elements of the mother plant or in the callus.

Under appropriate experimental conditions embryo formation *in vitro* is a continuous process and various embryonal stages occur side by side. On the basis of the developmental history that can be deduced by combining these stages, it has been postulated that embryos in tissue cultures are formed by single cells, usually at the upper surface, of the callus. This concept is based on the almost identical pattern of cell division during embryogenesis *in vitro* and in seeds (Reinert 1964). The situation is similar for the embryo formation in callus of *Ranunculus sceleratus*. In these cultures the majority of embryos originate from single cells at the surface of the callus (Konar *et al.* 1972b). Similar observations on suspensions containing single cells and cell aggregates have later on led to other conclusions (Steward, Kent & Mapes 1966), i.e. that isolated single cells develop directly by segmentation without callus formation to pro-embryos and these then go on to form embryos and normal plantlets. This implies a direct development of a plant from a single isolated cell without the participation of other cells. In addition it has been assumed that the physiological isolation of single cells in suspension is a necessary precondition for this embryogenesis. Furthermore, it has been assumed by the same group that coconut milk (COM) is essential for the initiation and development of such embryos. These postulates were criticized first on the basis of theoretical reasons (Gautheret 1966). Later on it was also shown experimentally by several workers that Steward's hypothesis cannot be generalized and may perhaps be wrong, for cells from carrot cultures are able to form embryos on synthetic, chemically defined nutrients without COM; indeed, COM and also cytokinins can even inhibit embryogenesis *in vitro*. COM and cytokinins obviously stimulate only after late embryo stages have been formed (Reinert 1963b, Halperin & Wetherall 1964). Moreover, by direct microscopic observations it could be proved that isolated parenchymatous single cells from carrot cultures must first produce a multicellular aggregate before embryos can be initiated (Reinert *et al.* 1971) (see Fig. 12.4). As a consequence of careful electron-microscope studies on callus from *R*.

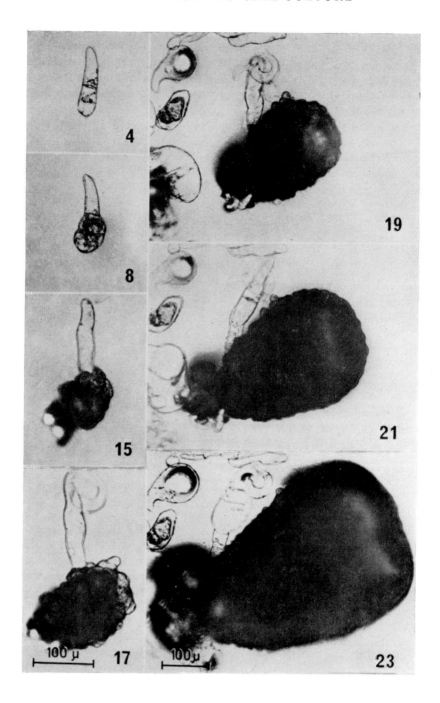

sceleratus, it is now also clear that the embryogenic cells in these tissues possess protoplasmic strands linking them with neighbouring cells during the initiation of embryos. Therefore, physiological isolation is not a necessary precondition for embryogenesis *in vitro* (Thomas, Konar & Street 1972). On the contrary, it appears that 'synergistic' cells are necessary for this process.

Factors controlling embryogenesis in vitro

In connection with the discussion of the effect of COM and of physiological isolation on embryogenesis *in vitro* we have already touched upon some of the factors involved. The situation resembles closely that for organogenesis in tissue cultures and here again we know much more about exogenous factors than about endogenous ones. The first clear-cut proof of the induction and development of embryos in carrot tissues was achieved by a succession of changes in the nutrient media. The decisive step in these experiments was the transfer from auxin-containing to auxin-free media and the addition of nitrogen in the form of amino acids (Reinert 1959). These observations were subsequently confirmed and it was discovered that embryogenesis *in vitro* can be induced by both inorganic (KNO_3, NH_4NO_3) and organic nitrogen compounds, such as amino acids and amides (Tazawa & Reinert 1969). It is now also clear that in order to initiate embryogenesis most salts must be present in high concentrations (Butenko, Strogonor & Babaeva 1967) whereas compounds like ammonium (Halperin 1966b) and glutamine are effective at very low levels. However these substances are not specific for embryogenesis *in vitro*; in other words, they are interchangeable. Thus the non-inductive medium prepared according to White's formula (1954) and which contains only 3·2 mM nitrogen, can be converted into an inductive form by the addition of only 5 mM glutamine. If, instead of the amide, KNO_3 is added to the same medium a considerable higher amount (40 mM) is required to produce the same effect. It is not the absolute quantity of the nitrogen which determines the triggering of embryogenesis *in vitro* although the ratio of nitrogen to auxin is important; embryo formation can be triggered on White's medium, in spite of the low nitrogen content (3·2 mM), by omitting auxin.

The significance of the successive changes in the nutrient media mentioned above for the process of embryo formation *in vitro* has been confirmed with tissues other than carrot (Steward, Kent & Mapes 1967), and not only for

---

FIG. 12.4. *Embryo formation by a single somatic cell* isolated from a callus culture derived from a carrot root and growing on an agar medium containing auxin and a high nitrogen level (60 mM). Embryogenesis starts with an unequal division (4) and the formation of a complex of small cells (8–17) developing into the early stage of a heart-shaped embryo (23). Numbers in the lower corner: days after isolation of the cell. Age of the 'mother culture': 16 weeks (Backs-Hüsemann & Reinert 1970).

embryogenesis but also for organogenesis (Torrey 1966) and cytodifferentiation (Fosket & Torrey 1969). The developmental response, which occurs particularly following switching from auxin-containing to auxin-free media, is
readily understandable in the light of the continual changes to which the
'system embryo' is submitted during the course of its development. It is plain
that changes in the nutrient substrate are necessary, not only for the development of the embryo, but also for the callus cells. During cultivation *in vitro*
carrot callus also goes through various metabolic phases, characterized by
different morphogenetic capacities (Syono 1965).

The parallel to organogenesis extends to the endogenous factors, in
particular as far as changes in embryogenic expression. As a rule the inception
of embryos occurs most frequently in freshly isolated or short-term cultures.
With increasing duration of culture this ability decreases and eventually

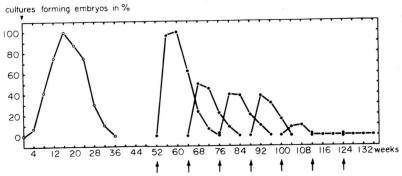

Fig. 12.5 *Loss and maintenance of the ability for embryo formation in carrot
cultures* isolated and subcultured on an 'inductive' agar medium containing auxin
and a high (60 mM) nitrogen level (○ —— ○) and of cultures isolated on a 'non-
inductive' low nitrogen (3·2 mM) medium but transferred after various periods
(indicated by arrows) to the 'inductive' high nitrogen medium (● —— ●). The
control was the low nitrogen substrate with no embryo formation (Reinert *et al.* 1971).

disappears. Here again, however, there are examples of embryogenesis in
cultures which have been cultivated for years. The cause of this variability is
probably the same or similar to that of change in capacity for organogenesis.
At present we are not able to explain this phenomenon satisfactorily; in fact
we have no exact knowledge about the mechanism involved. An intensive and
exact characterization of the entire process is called for. A beginning may have
been made with observations on carrot cultures (Reinert *et al.* 1971). Embryo
formation in these cultures begins 4–6 weeks after isolation of the tissues and
attains a maximum after about 15 weeks (Fig. 12.5). This is the rule and
exceptions to this seldom occur. The same series of investigations yielded two
additional interesting findings. First, the capacity of carrot cells for morpho-

genesis can be sustained over prolonged periods (more than 20 weeks) on non-inductive media, after which it slowly declines. Secondly, after the capacity for embryogenesis has apparently been lost it can be reinduced by an appropriate inductive medium with a high nitrogen level. These results prove unambiguously that cell cultures *in vitro* are pliable systems. They also demonstrate that the frequently observed loss of the characteristic properties of cultures need not always depend upon alterations in the genome (with an associated loss of genetic information) but may be traced to other endogenous physiological factors. Among these factors are presumably substances which are present immediately following isolation but are either not synthesized or synthesized only in insufficient quantity during *in vitro* culture. Successful work with cell and tissue cultures over extended periods is possible only on the basis of an exact knowledge of the properties of each particular system used.

Embryogenesis in generative (haploid) cells

In recent years the success achieved with somatic cells has been supplemented by extremely important results on the formation of embryos and even entire plants from haploid cells, that is from pollen grains. By using such cultures it is possible for the first time to produce large numbers of haploid plantlets within a short period. Certain aspects of this androgenesis, e.g. the influence of the age of the mother plant, the stage of development of the pollen, etc., and hence the part played in this process by endogenous factors, has already been discussed (see Chapter 9). Therefore this section can be restricted to a consideration of endogenous factors. There are two possibilities which have to be considered here: firstly, the direct initiation of embryogenesis in pollen, for instance in Solanaceae, leading to the production of haploid plantlets; or secondly, their indirect initiation with preceding formation of a haploid callus from anthers, which is necessary when Brassicaceae and Gramineae are used.

Anthers contain both haploid and diploid cells and this creates certain difficulties in selectively inducing the growth of haploid cells. In the simplest case, that of uninuclear pollen stages from *Nicotiana tabacum*, this aim can be achieved by the use of a minimal medium containing only sucrose (Nitsch 1970). Hormones have no positive effect in these experiments or even inhibit the development of haploid cells. However, bearing in mind factors promotive of embryogenesis from somatic cells, it is interesting that certain nitrogen compounds (arginine, asparagine and glutamine) also stimulate embryogenesis in the *Nicotiana* pollen (Nitsch 1970). Another species, *Atropa belladonna*, in contrast requires the use of more elaborate nutrient media for the initiation of androgenesis; White's medium (1954) for example was ineffective (Zenkteler 1971). In quite a number of investigations, skilful manipulation of phytohormones has selectively stimulated growth of the haploid cells. This pertains to the first successful experiments with anthers of *Datura innoxia* (Guha & Maheshwari 1964, 1967) where the presence of kinetin in the nutrient medium was necessary for the induction of embryogenesis from the pollen. In contrast the addition of auxin resulted in the formation of callus from the diploid connective tissue. Yeast extract and casein hydrolysate had a similar effect, i.e. they enhanced only the growth of the diploid cells. In other species, e.g. Brassicaceae, the haploid cells react best to the addition of kinetin *plus* auxin (Kameya & Hinata 1970). In *Oryza sativa* the addition of auxin alone induces the production of haploid callus, followed by the regeneration of plantlets (Niizeki & Oono 1971).

Much less is known about the effect of physical factors, but we know that androgenesis in *Nicotiana* pollen is highly sensitive to temperature, the optimal range being 23–28°C; below 15°C the process is blocked. Periodic changes of day and night temperatures have apparently no effect at all. The effect of light is important only to prevent the etiolation of the developing plantlets but does not influence the induction process in *Nicotiana* pollen (Devreux 1970, Sunderland 1971).

Studies on androgenesis are still very limited but they already indicate a similar situation to that existing for embryogenesis from somatic cells. It is quite clear from the available data that a particular inductive treatment successful in one species may be ineffective in another. Nevertheless, the fact that in organogenesis as well as in embryogenesis the same substances, like auxin, cytokinin and certain nitrogen compounds, are, in many cases, effective probably means that they are generally involved in morphogenetic processes although they are not, in every case, the limiting factors.

CONCLUSIONS

In conclusion it is necessary to consider some matters which have implications beyond those concerning morphogenesis in cell and tissue cultures. A problem which is most important for those engaged in basic research is that of the mechanism and site of action of endogenous and exogenous factors, particularly of the phytohormones and various nitrogen compounds promotive of morphogenesis in cell and tissue cultures. It is still not known whether these substances merely enhance the rate of synthesis of certain compounds not normally produced in vacuolated cells and hence indirectly create the conditions necessary for the following through of a predetermined 'developmental programme' in which other metabolites become active, or whether these substances are in fact more specific components of the apparatus of morphogenesis, possibly determining the pattern of gene activation and/or repression. In order to answer these questions we will need 'better' cultures and methods than those used hitherto. Here the most urgent question is that of the control of endomitosis and hence of the genetic stability of cell cultures from higher plants. We will not have ideal cultures for further research until we have means to stabilize their genomes.

Tissue cultures, particularly those derived from anthers, can now be used to produce and to propagate practically unlimited numbers of haploid or isogenic cells and plants. For the geneticist such materials offer unique opportunities for investigation of problems of gene mutation and of plasmatic inheritance.

Cell and tissue cultures, quite apart from their increasing importance for basic research, will also come to play an increasing role in practical science. The results now available from work on fundamental problems constitute a broad basis from which to investigate questions in applied botany which previously could not or could only be tackled inadequately. This applies both to cultures of somatic cells and from pollen. In both cases we are only at the start of a critical evaluation of their future role in the breeding of plants with new desirable attributes for agriculture and for the preservation of our environment. This is equally true in regard to the future use of cell cultures as biochemical tools for the synthesis or partial synthesis of secondary plant products such as alkaloids and glycosides (see also Chapters 10 and 11). Cell and tissue cultures are therefore likely to prove an excellent example of the essentiality of fundamental research for the solution of economic and ecological problems.

CHAPTER 13

THE ORIGINS, CHARACTERISTICS AND CULTURE OF PLANT TUMOUR CELLS

D. N. BUTCHER

Introduction 356
Crown-gall tumours 356
Virus tumours 382
Genetic tumours 384
Habituated tissue cultures 388
Conclusions 390

INTRODUCTION

Plant tumour diseases have attracted the attention of many investigators. One reason for this interest is that they are in many ways comparable to animal cancers and therefore present an alternative experimental system for investigating the fundamental processes of tumorigenesis. Another more practical reason is that it is possible with tissue culture methods to analyse critically the process of transformation of a normal intact cell to a tumour cell. In this chapter the term tumour will be restricted to non-self limiting neoplasms or overgrowths which once they have been induced exhibit continuous autonomous disorganized growth in the absence of the inciting agent.

The tumour diseases which have been studied in most detail are the crown-gall disease caused by *Agrobacterium tumefaciens* (Smith and Townsend) Conn., the wound tumour disease induced by the virus *Aureogenus magnivena* Black and the genetic tumour disease which occurs on certain interspecific hybrids, e.g. *Nicotiana glauca* × *N. langsdorffii*. This text will be confined to a consideration of these three diseases and to the phenomenon known as habituation which sometimes occurs spontaneously in established lines of callus cultures. The latter is relevant since it results in tissues having tumour-like properties.

CROWN-GALL TUMOURS

Introduction

The popularity of crown-gall tumours in studies of tumorigenesis is due primarily to the relative ease with which the tumours can be induced experimentally with *Agrobacterium tumefaciens*. This bacterium, a member of the

Rhizobiaceae, is a facultative anaerobe which is commonly found in the soil. It forms gram-negative rods $0.7-0.8$ μ by $2.5-3.0$ μ and may have from one to six flagella. Many strains of bacteria have been isolated and used in experimental work. A list of the most commonly used strains and their characteristics is given in Table 13.1.

The growth form of tumours depends both on the strain of *Agrobacterium tumefaciens* and the nature of the susceptible host. For example strain B_6 induces large undifferentiated tumours on a wide variety of plants. On the other hand strain T_{37} is only moderately virulent, causing either slowly growing tumours or highly complex tumours known as teratomas. The latter are formed on plants with relatively high regenerative capacities such as *Nicotiana tabacum* and *Kalanchoë daigremontiana*. *Agrobacterium radiobacter* a species closely related to *A. tumefaciens*, is non-tumorigenic and often used as a control in studies of tumour initiation.

In experimental conditions crown-gall tumours can be evoked on a large number of species. Elliot (1951) listed about 142 genera belonging to sixty-one families and Gadgil & Roy (1961a) have since added other susceptible genera. Many dicotyledonous plants and some gymnosperms (Smith 1943) are susceptible. In contrast there are only a few unsubstantiated reports of crown-gall tumours on monocotyledonous plants (Braun & Stonier 1958). The fact that the bacterium can incite tumours on such a wide variety of plants suggests that the transformation of a normal cell to a tumour cell involves an interference of fundamental growth regulating mechanisms. Hence the great interest in the problems of crown-gall.

Experimental techniques for acquiring bacteria-free tumour cultures
Induction of tumours

The cultures of *A. tumefaciens* may be maintained on a nutrient agar medium consisting of 17.2 g nutrient agar (Oxoid CM3), 1 g yeast extract (Oxoid No. L21), 5 g sucrose dissolved in 1000 ml of distilled water (Lippincott & Heberlein 1965b). Stock cultures may be kept as agar slants and stored at $4°C$. Prior to tumour initiation the bacteria are transferred to a liquid medium and incubated at $25°C$. After about 10 hours the bacteria are in a logarithmic growth phase and suitable for inducing tumours.

The age of the prospective host is not critical, but it is usual to use well-established seedlings or cuttings with thick stems (*c.* 5 mm diameter) when the eventual objective is to obtain bacteria-free tumour cultures. Tumours are most commonly induced on the stem, but they can also be induced on leaves and roots. The following procedure has been successful for a large number of species. The youngest fully elongated internode is selected as the site for tumour induction and surface sterilized with cotton wool soaked in ethanol.

TABLE 13.1 The origin and characteristics of *Agrobacterium tumefaciens* strains commonly used in crown-gall studies

Strain	Origin	Author and date	Guanidine bases of tumours	Phages isolated from bacteria	Remarks
A_6	*Rubus occidentalis*	Riker (1923)	Octopine	PA_6 Roussaux et al. (1968)	Virulent. Cloned by Wright et al. (1930)
A_{66}	*R. occidentalis*	Hendrickson et al. (1934)	Octopine and nopaline	—	Attenuated strain, but tumours produced when host is treated with auxin. Mutant from strain A_6
B_2	*Malus sp.*	Kent (1937)	Octopine	PB_2 Stonier et al. (1967)	Virulent. Originally isolated from apple and re-isolated after three passages in tomato
B_6	*Malus sp.*	Kent (1937)	Octopine	Omega Beardsley (1955) PB_6 Roussaux et al. (1968)	Highly virulent. Most frequently used strain
B_6 806	*Malus sp.*	Kent (1937)	—	—	Virulent. Phage-sensitive strain obtained after strong UV irradiation. Beardsley (1955)
B_6 S	*Malus sp.*	Kent (1937)	—	—	Virulent. Spontaneous mutant sensitive to phage PB_6. Isolated by Beardsley
T_{37}	*Juglans regia*	Riker (1926) Riker et al. (1930)	Nopaline	LT_{37} Zimmerer et al. (1966)	Moderately virulent. Induces teratomas on *Nicotiana tabacum* and *Kalanchoë daigremontiana*
$11BN_7$	*Chrysanthemum frutescens*	Stapp (1927)	Nopaline	$L11BN_7$ Zimmerer et al. (1966)	Moderately virulent. Induces teratomas on *Nicotiana tabacum* and *Kalanchoë daigremontiana*. Isolated from Stapp IIB by Braun
$11BNV_6$	*C. frutescens*	Stapp (1927)	Octopine	$L11BN_6$ Zimmerer et al. (1966)	Avirulent. Tumours produced when auxin applied at the same time as inoculation
V1	*Zinnia sp.*	Zimmerer et al. (1966)	—	LV1 Zimmerer et al. (1966)	Virulent
CV1	*Zinnia sp.*	Zimmerer et al. (1966)	—	—	Sensitive to phage LV1. Obtained by heat treatment of strain VI
Agrobacterium radiobacter	—	—	—	—	Avirulent. Used as control for virulent strains of *Agrobacterium tumefaciens*

Next a wound is made by piercing with a sterilized needle and immediately covered with moist cotton wool bandage to reduce contamination and prevent drying. (This wounding is a prerequisite for tumour induction and will be discussed in a later section.) After 48 hours the wound is temporarily uncovered and inoculated directly by inserting a needle previously dipped in a suspension of logarithmic phase *A. tumefaciens* cells. The bandage is finally removed 3 days later and small tumours are usually evident after 2 weeks. When the tumours are required for initiating bacteria-free tissue cultures they are allowed to develop until they are 20–30 mm in diameter (4–5 weeks old). The tumours most suitable for this purpose are those which are compact and smooth with a minimum of senescing tissues.

Elimination of bacteria from tumour tissues

In order to exploit tissue culture techniques it is necessary to eliminate *A. tumefaciens* from the tumour tissues. This is important for two reasons. Firstly the causal organism once it has initiated the tumour is no longer needed for the continuance of the tumorous properties and its presence would unduly complicate experiments. Secondly there is the practical problem that the bacteria would overgrow the tumour tissues.

There are several ways of obtaining bacteria-free crown-gall tumours and the choice of method depends on the morphological characteristics of the tumours. The easiest way is to initiate cultures from the so-called secondary tumours which sometimes develop at a distance from the primary tumours. Such tumours are similar to the primary tumours in all respects except for the absence of *A. tumefaciens*. Cultures can be initiated by taking small explants from surface sterilized secondary tumours and placing them on a simple medium (White's basic medium 1943) without auxins or cytokinins. Unfortunately secondary tumours occur only on a few plants such as sunflower, Paris daisy and chrysanthemum (White & Braun 1941).

Bacteria-free tumour cultures may also be obtained directly from the primary tumours of many plants where the bacterium is confined to certain regions of the tumour, e.g. sunflower, artichoke and tobacco. In such tumours explants are taken from the sterile parts of the tumour (usually the inner regions) after surface sterilization with 5% calcium hypochlorite or 0·1% mercuric chloride and placed on a simple medium without growth hormones. As would be expected the success rate is less than that for secondary tumours, none the less many bacteria-free tumour cell lines have been started in this way (De Ropp 1947a). The chances of success with this method are increased somewhat if very small explants are taken. Not all tumours respond to this procedure as some such as those on carrot and tomato appear to have bacteria dispersed throughout the tissues.

Another method for ridding tumour tissues of the inciting bacterium is based on the observation that cells of *A. tumefaciens* are killed by prolonged exposure to temperatures of 46–47°C. Braun (1943), White (1945) and De Ropp (1947b) succeeded in killing the bacteria in infected tissues of *Vinca rosea* without killing the tumour tissues by giving the plants a heat treatment. Tissue cultures were then obtained from the surviving tumour tissues. The procedure was as follows. *V. rosea* plants with woody stems were inoculated with bacteria and placed for 5 days in an incubator at 25°C to allow tumour initiation. The plants were then transferred to an oven kept at 46–47°C and left for a further 5 days. This treatment caused severe damage to the leaves and young stem tissues, but the thicker woody stems and tumour tissues survived when the plants were returned to 25°C. Tissue cultures were initiated from the tumours after further development. Theis, Theis, Riker & Allen (1950) have shown that success rates with this method are improved if the relative humidity during the heat treatment is kept above 65%. Unfortunately this method has only been successful for *V. rosea*. Most other species tested have been unable to tolerate the high temperature treatment (Gadgil & Roy 1961b). Recently Manasse & Lipetz (1971) have described a more convenient method for giving a heat treatment to tumorous *V. rosea* tissues. Surface sterilized stem segments bearing tumours were placed in test-tubes containing modified White's agar medium (10 × normal salt concentration) so that the tumour sites were above the agar. The tube plus segments were then put in a waterbath at 41°C and kept in total darkness. After 7 days they were transferred to a room at 25°C and 40–50% relative humidity. Six weeks later the tumour proliferations were subcultured and tested for the presence of bacteria. Fifty to 75% of the tumours were found to be free of viable bacteria.

Gadgil & Roy (1961b) investigated the possibility of using antibiotics for freeing crown galls of *Agrobacterium tumefaciens*. They immersed explants from 40-day-old tumours of hollyhock in a solution of streptomycin (80 µg per ml) for 2 minutes before transferring them to a medium containing up to 40 µg streptomycin per ml. After incubating for 28 days at 25°C they were tested for the presence of bacteria. Out of forty explants fifteen were found to be bacteria-free when placed on a streptomycin-free medium. Morel (see Gautheret 1959) has found that aureomycin is useful for obtaining bacteria-free cultures from tomato and virginia creeper crown-gall tumours.

It is clear that none of the methods available for obtaining bacteria-free tumour cultures are ideal. Complete success is rarely achieved and even partial success depends very much on the type of tumour. Fortunately for many purposes this is not important since the objective is to obtain one or a few tumour isolates. Once these have been acquired they can be multiplied by normal callus culture procedures.

It is not usually necessary to follow elaborate procedures to demonstrate the absence of the bacteria since all the known strains of *A. tumefaciens* grow well on tissue culture media. However, if tests are required to confirm their absence the tumour tissues can be macerated and plated out onto a nutrient agar medium which is highly suitable for bacterial growth. Alternatively the tumour tissues can be examined by electron microscopy.

Studies of tumorigenesis are often concerned with comparisons between tumour and normal callus isolates. The aim of these studies is to detect physiological and biochemical differences which result directly from the transformation. Such comparisons are valid only if the representative isolates are identical in all respects except for the factors involved in tumorigenesis. Thus the selection of isolates is very important in comparative studies. Ideally they should have been isolated from plants of the same genotype at the same time, and if possible from similar tissues, e.g. stem cortex. The maintenance media for tumour tissues and normal callus tissues should be identical except for the additional growth factors required by the latter (see Table 13.2). Another safeguard against studying differences which have arisen subsequent to the isolations is to make periodic examinations of the nuclear cytology. A further problem is that isolates initiated under apparently identical conditions often differ from one another in growth characteristics. It is therefore not wise to depend on a single pair of isolates (one tumour and one normal callus). If these pitfalls in comparative studies are ignored they are likely to lead to erroneous interpretations.

Studies of tumour transformation

The earlier physiological studies of tumour induction by *A. tumefaciens* led Braun to propose that tumour formation takes place in two phases. The inception phase during which the bacteria transform normal cells into tumour cells and the development phase in which the resulting transformed cells continue abnormal autonomous proliferation.

Inception phase and the nature of the tumour-inducing principle (TIP)

An essential prerequisite for the inception is that the host cells in the vicinity of the infection must be conditioned by causing mechanical damage to the tissues. The significance of this requirement is not understood, but it has been suggested that juices released from the ruptured cells may activate the wound healing cycle (Braun 1947; Klein 1955). However, the more recent work of Lange & Rosenstock (1970) and Therman & Kupila-Ahvenniemi (1971) indicates that the situation is more complex than this.

Braun (1952) utilized the knowledge that tumour induction in *Vinca rosea* takes place at 25°C but is prevented at temperatures above 32°C to characterize the conditioning period. The inciting bacterium was allowed to act on the

TABLE 13.2 The nutritional requirements of crown-gall and comparable normal callus tissue cultures

Species	Origin	Inorganic medium	Carbon source	Auxin	Cytokinin	Vitamins and other growth factors	Authors
Datura stramonium L.	Stem	Murashige & Skoog (1962)	Glucose	2,4-D	—	Meso-inositol, nicotinamide, choline, B_2, Bc, B_1, PYR, B_5	Chan & Staba (1965)
	Stem tumour (B_6)	Skoog*	Glucose	—	—	Meso-inositol, nicotinic acid, B_1, H, B_5	Morel (1967)*
Daucus carota L.	Root	Braun & Wood (1962)	Sucrose	2,4-D	Kinetin	Meso-inositol, nicotinic acid, B_1, PYR	Butcher & Sogeke (1972) (unpublished)
	Root tumour (B_6)	Braun & Wood (1962)	Sucrose	—	—	Meso-inositol, nicotinic acid, B_1, PYR	Butcher & Sogeke (1973) (unpublished)
Helianthus annuus L. var. Russian Giant	Stem cortex	White (1943)	Sucrose	2,4-D	Kinetin†	Meso-inositol, nicotinic acid, B_1, PYR	
	Stem tumour (B_6)	White (1943)	Sucrose	—	—	Meso-inositol, nicotinic acid, B_1, PYR	
H. tuberosus L.	Rhizome	Knop [Gautheret 1942]	Glucose	NAA	—	—	Gautheret (1941)
	Rhizome tumour (A_6)	Heller (1953)	Glucose	NAA	Kinetin	—	Gautheret (1947)* Morel (1948)*
Nicotiana tabacum L. var. White Burley	Stem	Skoog	Glucose	NAA	Kinetin	Meso-inositol	Morel (1946)*
	Stem tumour (A_6)	Skoog	Glucose	—	—	—	Gautheret (1941)
Scorzonera hispanica L.	Root	Heller (1953)	Glucose	IAA	—	—	Gautheret (1948)*
	Root tumour (A_6)	Heller (1953)	Sucrose	NAA	Kinetin	Meso-inositol, nicotinic acid, B_1, PYR, asparagine, glutamine, cytidylic acid, guanylic acid	Braun & Wood (1962)
Vinca rosea L.	Stem	Braun & Wood (1962)	Sucrose	—	Kinetin	Meso-inositol, nicotinic acid, B_1, PYR	Braun & Wood (1962)
	Stem tumour (B_6)	White (1943)	Sucrose	—	—	Meso-inositol, nicotinic acid, B_1, PYR	Braun & Wood (1962)

* From Goldman–Ménagé (1971). † Stimulatory but not essential. B_1 = thiamine hydrochloride, B_2 = riboflavine, B_5 = pantothenic acid, PYR = pyridoxine hydrochloride, Bc = folic acid, H = biotin.

host tissues for 24 hour periods at 25°C at different times after wounding and then incubated at 32°C. It was shown that rapidly growing tumours resulted when the bacteria were added 48 hours after wounding. while no tumours were formed when the bacteria were added at the time of the wounding and the plants held at 25°C for 24 hours before transfer to 32°C. Conditioning of the host cells was shown to take place gradually reaching a maximum between the second and third day after wounding and then declining as wound healing progressed towards completion. Braun concluded that conditioning is related to the wound healing cycle and has postulated that cells preparing to divide or in the state of division may be vulnerable to the action of *Agrobacterium tumefaciens*. Moreover, he suggested that the pattern of metabolism found in dividing cells must be established prior to transformation so that it can be perpetuated (Braun 1962). The studies of Therman & Kupila-Ahvenniemi (1971) also support the view that transformation occurs at a particular stage in the wound healing cycle.

The earlier work on crown gall indicated that although the presence of viable cells of *A. tumefaciens* in the wounded tissues is essential for tumour induction the bacteria are only observed in intercellular spaces and damaged cells. Hence it was concluded that the bacteria remain extracellular within the host tissues (Beardsley 1972). From these observations Braun & Mandle (1948) proposed that the bacterium releases a tumour inducing principle (TIP) which accomplishes the change of a normal cell to a tumour cell. As would be expected many studies have been concerned with the identification of this hypothetical factor.

Braun (1947, 1951) attempted to analyse the transformation by TIP by exposing *Vinca rosea* plants to *Agrobacterium tumefaciens* for specified periods (1·5 to 5 days). Tumorigenesis was terminated by subjecting the plants to temperatures of 46–47°C. This treatment arrested tumorigenesis and killed the bacteria, but not the tumour tissues (see p. 360). An induction period of 36 hours at 25°C gave small slowly growing tumours at the sites of infection, whereas when the induction period was 3 days tumours of moderate size and growth rate developed. Tumours of maximum size and growth rate were produced when the induction period was 4 days or more. Braun isolated bacteria-free tissue cultures from these tumours and demonstrated that the capacity of converted cells for autonomous growth was reflected in differences in the rate at which cells grow on White's basic medium. Such differences in the growth rate of cultures were found to persist indefinitely. Furthermore it was shown that the various levels of growth autonomy exhibited *in vivo* relate to differences in capacity to grow on media lacking certain critical growth factors (Braun 1958). From these observations it was suggested that the small and moderate-sized tumours represent partially transformed cells and that the changes from a normal to a fully transformed autonomous cell take place gradually and progressively over a 3–4 day period due to the accumulative action of TIP. Studies with *Kalanchoë daigremontiana* also indicated that the size of tumours developed increased with the increasing duration of the induction period (Braun 1950). However, in this case tumour tissue cultures were not obtained to demonstrate differences in degree of autonomy. Braun & Mandle (1948) alternated periods during which infected *K. daigremontiana* were maintained at 25°C with periods of exposure to 32°C. From their results it was concluded that TIP accumulated during periods at 25°C but was inactivated at 32°C.

These interpretations are based on the assumption that the conditioning of the host cells in the vicinity of the wound, the wound-healing process and the bacteria are unaffected by the higher temperatures. The earlier work by Braun (1947) indicated that this assumption is valid with *Vinca rosea*, but Lipetz (1966) showed clearly that both wound healing and conditioning of *Kalanchoë daigremontiana* tissues are accelerated when the temperature is raised from 25 to 32°C. Since it has been established that the stage in the wound-healing cycle determines the growth characteristics of the tumour it is possible that the so called partially transformed tissues result from the physiological state of the host and not the time of exposure to the bacterium. Thus the evidence that TIP is inactivated at 32°C and transformation is a gradual process is inconclusive. Beardsley (1972) has recently published a comprehensive review of this aspect of tumorigenesis.

The early attempts to isolate TIP have been reviewed by Braun & Stonier (1958), Klein (1965), Braun (1962) and Schilperoort (1969) and for the most part are only of historical interest. At the present time there appear to be two chief candidates for the title of TIP. First there is considerable evidence suggesting that it is a nucleic-acid fraction released from the bacterium which is capable of transforming the host cells. Secondly there is also evidence which indicates that a virus (phage) is transmitted from the bacterium to the host cells during transformation. In addition there is the third possibility that TIP is the bacterium itself which enters the wounded cells and becomes altered in its morphology and physiology in such a way that it remains undetected in the host cells.

The fact that the transformation of a normal cell to a tumour cell involves changes in properties and behaviour which are perpetuated in subsequent cell generations suggests that nucleic acids and particularly DNA may be involved. Several investigations have been designed to establish that the tumorous properties result from the presence of foreign DNA in the host cells. One approach has been to test the effect of various inhibitors and other factors which would be expected to interfere with the metabolism and functions of DNA. Braun (1962) and Schilperoort (1969) have discussed the earlier work which was inconclusive, but not inconsistent with the view that DNA is involved in the induction of tumours. Recently Beiderbeck (1970, 1971) has shown that rifampicin, an inhibitor of DNA-dependent RNA polymerase, and polyornithine (PO) inhibit tumour induction. PO was effective at concentrations (10 μg/ml) which do not influence the growth of *Agrobacterium tumefaciens* and inhibited tumour induction only when added at the same time as the bacteria. *In vitro* it was shown that PO forms complexes with DNA and RNA (low MW) and *in vivo* that added RNA can antagonize the inhibition of tumour induction by PO. It was concluded that PO influences the induction of tumours by complexing with a nucleic acid which is necessary for transformation. Braun & Wood (1966) have also reported that RNAase A, but not DNAase significantly inhibits tumour induction on *Kalanchoë daigremontiana* at concentrations which did not affect the growth rate of the bacterium or the wound-healing response. This was taken as evidence that RNA may be involved in the inception of the tumours. In general these studies on the effects of inhibitors and other factors on tumour induction remain inconclusive, which is not surprising in view of the difficulty of distinguishing between effects on tumour induction and effects on the metabolism of the bacteria and host cells.

A more direct approach towards the identification of TIP has been to attempt to induce tumours with cell-free preparations from the bacteria, particularly with preparations of their DNA and RNA. The results from such studies have been controversial because it has proved difficult to establish unequivocally that the preparations were completely free of viable bacteria and that the induced overgrowths are tumorous. In order to establish the absence of the bacteria in extracts it is necessary to demonstrate their sterility by plating out onto a culture medium suitable for *Agrobacterium tumefaciens*, e.g. nutrient agar.

There are several methods in current use for testing the tumour-inducing ability of preparations. One is to treat a previously wounded susceptible plant directly with the extract in a similar way to that described for inducing tumours with *A. tumefaciens*. Plants of *Helianthus annuus*, *Nicotiana tabacum*, *Kalanchoë daigremontiana* and *Lycopersicum esculentum* have given consistent results and are suitable for this purpose. The advantages of this method are that tumour development can be observed for extended periods and the tumorous properties of the overgrowths can often be confirmed by isolating tissue cultures and demonstrating that the cells are capable of continuous growth in the absence of auxin and cytokinin. The disadvantages are that the experiments take up to 2 months to complete and require considerable glasshouse space.

A quicker method for assessing the tumour-inducing properties of extracts has been described by Klein & Tenenbaum (1955). Radial cylinders (13 mm diameter, 2·5 mm thick) are taken aseptically from the young phloem tissues of carrot roots and placed cambium side uppermost on a moistened filter paper in petri dishes. The cambial tissues are then treated directly with the test solutions and incubated at 20°C. Treatment with virulent *Agrobacterium tumefaciens* induces tumours within 7 days and quantitative estimates of the amount of tumour tissue may be obtained after 15 days by removing the tumours with a scalpel and weighing. This assay is relatively quick and can be carried out under aseptic conditions. The induced overgrowths can be tested for continuous autonomous growth by transferring to an auxin and cytokinin omitted medium. Root tissues of *Beta vulgaris* and *Pastinaca sativa* can be substituted for carrot and give similar results.

Leff & Beardsley (1970) have used tobacco stem segments cultured on Murashige & Skoog's (1962) medium to test the tumour-inducing activity of DNA preparations from phage PS 8 which was isolated from bacteria-free crown-gall tissues and propagated on *Agrobacterium tumefaciens* strain B6 806. The tobacco segments were treated directly with a drop of DNA solution 3 days after they were placed on the medium.

Lippincott & Heberlein (1965a, b) have described a method using the primary leaves of Pinto beans. This method developed primarily for studies of the initiation and growth of tumours by *A. tumefaciens*, has also been used for testing cell-free preparations. Primary leaves, selected 7 days after sowing, are lightly dusted with carborundum powder before treating with 0·1 ml of test solution. The mixture is then spread over the leaf surface with a glass rod. The numbers of tumours initiated on each leaf is determined after 7 days using a dissecting microscope. This assay is quick and allows a measurement of a large number of localized responses. However, it suffers from variability due to leaf damage and weather conditions. Another disadvantage is that the overgrowths cannot be used for initiating tissue cultures. It is therefore not possible to confirm their tumorous properties. A similar method using plants of *Ricinus communis* has been described by El Khalifa & El Nur (1970).

Beiderbeck (1970) has developed a quantitative method using leaves of *Kalanchoë daigremontiana*. A holder with thirty needles is used to make a large number of uniform wounds at regular intervals before treatment with test solution. Suspensions of *Agrobacterium tumefaciens* produce small tumours after 7 days and a quantitative evaluation is possible after 14 days. The number of tumours increased linearly with the logarithm of the concentration of bacteria.

Stroun, Anker, Charles & Ledoux (1967) have used shoot cuttings of tomato for their studies of the uptake and properties of bacterial nucleic acids by plant cells. The cut ends of the shoots are placed in aqueous DNA preparations or bacterial suspensions for 21 hours. Wounds are made on the stems by piercing with a needle. The shoots are then examined periodically for the presence of overgrowths. Leff & Beardsley (1970) have employed sunflower shoots in a similar way in their experiments with DNA preparations from phage PS 8. After removing the roots the cut ends are placed in the test solutions for 5 hours at 27°C under continuous illumination. After the solutions have been absorbed the shoots are planted in vermiculite, wounded and left to allow tumours to develop.

It would seem from the above that the best methods for confirming the tumorous properties of induced overgrowths are those which allow tissue cultures to be initiated. If such cultures grow indefinitely in the absence of auxin and cytokinin it is good evidence that the tissues are tumorous. This evidence is strengthened if the cultured tissues can be grafted onto healthy plants to produce large tumours (Fig. 13.1). Even if the results are positive it does not necessarily mean that the induced tumorous tissues are equivalent to crown-gall tumours. It is possible that the tumour-like properties result from phenomena such as habituation and chemical induction which may be unrelated to crown-gall transformations. However, the recent work of Morel and co-workers (Morel 1971) may provide a useful marker for crown-gall transformations. They have discovered that crown-gall tissues, but not normal callus, habituated or genetic tumour tissues, contain abnormal amino acids (Fig. 13.2). They contain either octopine or nopaline depending on the strain of bacterium which induced the original tumour (Table 13.1). A second possibility is to establish that *A. tumefaciens* phages are present in the tumour tissues, since Tourneur & Morel (1970) have shown that all the crown-gall cultures they examined contained phages which cause lysis in a phage sensitive strain (B_6 806) of the bacterium.

The earlier attempts to induce tumours by nucleic acids and other bacterial preparations have been discussed by Braun (1962). At that time he considered that the evidence for implicating DNA was inconclusive. Since then there have been several reports of DNA preparations inducing tumour-like tissues. Kovoor (1967) has claimed to have transformed callus cultures of *Scorzonera hispanica* with relatively large amounts of DNA extracted from *Agrobacterium tumefaciens* strain B_6. The transformed tissues appeared to be autonomous in culture and transplantable to intact plants. Similarly Beltrá & Rodriques (1971) have reported that tumours have been induced on the stems of *Phaseolus vulgaris* and aseptic carrot root discs with *Agrobacterium tumefaciens* DNA preparations. The carrot tumours produced proliferated on a medium with no auxin and cytokinin. Recently Kado, Heskett & Langley (1972) observed that a small proportion of their bacterial high MW DNA preparations caused crown-gall-like overgrowths on sunflower. However, the overgrowths could not be induced at will. Preparations of circular satellite DNA did not induce tumours. Perhaps the most interesting report so far has been that of Leff & Beardsley (1970) where it is claimed that DNA of the phage PS 8, but not the phage itself, induced tumour-like proliferations on Pinto bean leaves, sun-

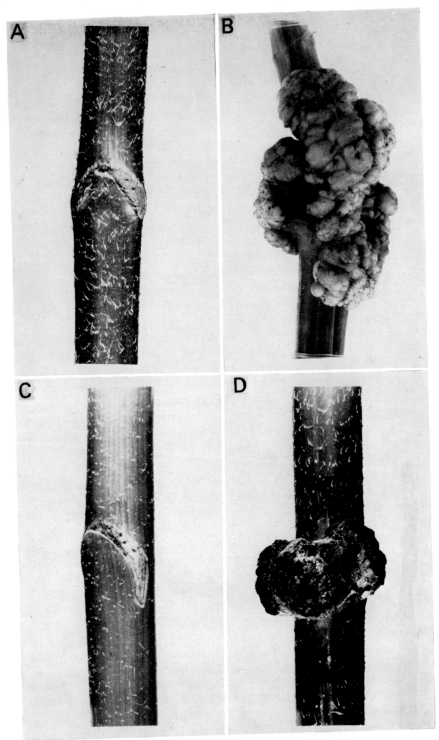

flower stems and stem explants of tobacco. The phage PS 8 was originally isolated from established crown-gall tumour tissue cultures and then propagated on the phage-sensitive strain of *A. tumefaciens* B_6 806. The tumours induced on sunflower were transplantable and the cellular proliferations on the tobacco explants appeared to be capable of growth without auxin and cytokinin. If these results can be confirmed they could be highly significant, but are not necessarily incompatible with the positive results with bacterial DNA since it is possible that the latter include small amounts of phage DNA. It is interesting that Beardsley himself has been very cautious about the significance of these results (Beardsley 1972). However, in spite of these positive indications that TIP is a DNA molecule it should be recognized that there have

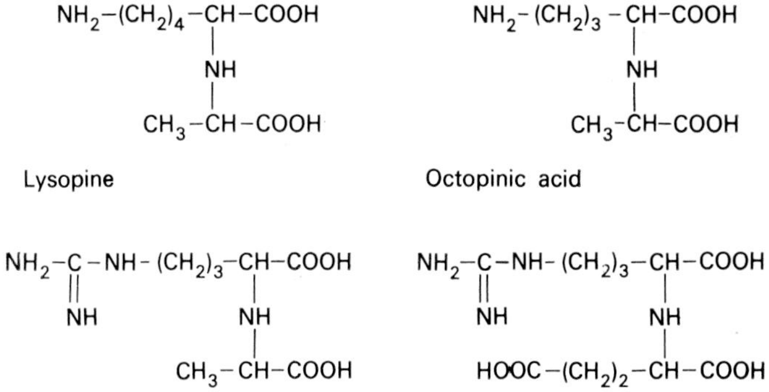

$$NH_2-(CH_2)_4-\underset{\underset{CH_3-CH-COOH}{|}}{\underset{NH}{|}}{CH-COOH}$$

Lysopine

$$NH_2-(CH_2)_3-\underset{\underset{CH_3-CH-COOH}{|}}{\underset{NH}{|}}{CH-COOH}$$

Octopinic acid

$$NH_2-\underset{NH}{\overset{||}{C}}-NH-(CH_2)_3-\underset{\underset{CH_3-CH-COOH}{|}}{\underset{NH}{|}}{CH-COOH}$$

Octopine

$$NH_2-\underset{NH}{\overset{||}{C}}-NH-(CH_2)_3-\underset{\underset{HOOC-(CH_2)_2-CH-COOH}{|}}{\underset{NH}{|}}{CH-COOH}$$

Nopaline

FIG. 13.2. *Abnormal amino acids found in crown-gall tissue cultures.*

been many reports of unsuccessful attempts to induce tumours with bacterial DNA (Gribnau & Veldstra 1969, Bieber & Sarfert 1968, and Stroun, Anker, Gahan, Rossier & Greppin 1971).

There have been fewer attempts to initiate tumours with RNA preparations, but Kado *et al.* (1972) have recently reported that 70s ribosomes, messenger RNA, ribosomal RNA and transfer RNA do not induce tumours on sunflower plants.

If in fact transformation is a result of the transfer of bacterial or phage DNA from the bacterium to the host cell it should be possible to detect and locate it within the host cells. Several investigations motivated by the detection of viral DNA in nuclei of polyoma (mice) and SV40 (monkey) transformed cells have attempted to demonstrate the presence of *A. tumefaciens* DNA in

FIG. 13.1. *Crown-galls induced experimentally on sunflower stems.* A. Normal wound response; B. Tumour induced by *Agrobacterium tumefaciens* strain B_6; C. Stem grafted with normal callus culture after 1 year in culture; D. Stem grafted with bacteria-free tumour tissues after 1 year in culture. Appearance of stems 5 weeks after treatment.

bacteria-free crown-gall cultures with nucleic-acid hybridization techniques. Schilperoort, Veldstra, Warnaar, Mulder & Cohen (1967) looked for complementary sequences between bacterial RNA and DNA from tobacco tumour cells. They synthesized ^{3}H-RNA using *Agrobacterium* (Strains B_6, A_6 and *A. radiobacter*) DNAs as templates and a DNA-dependent RNA polymerase from *Escherichia coli*. It was found that all three ^{3}H-RNAs complexed with DNA from bacteria-free crown-gall tissues, but not with DNA from tobacco leaves. This was taken as evidence that common sequences of nucleotides occur in the bacteria and in the tumour cells and further that bacterial DNA may be present in the bacteria-free crown-gall cultures. The observation that complexing occurred between the ^{3}H-RNA synthesized from the DNA of the avirulent *Agrobacterium radiobacter* as well as with the RNAs prepared from DNAs of the virulent strains B_6 and A_6 was considered to be evidence for a close relationship between these bacterial species. Milo & Srivastava (1969c) performed the reciprocal experiments in which 14–18s RNA from crown-gall tissues and normal callus were hybridized with DNA from *A. tumefaciens* strains 4–32, B_6 (virulent) and 11BNV$_6$ (avirulent). Significant complexing occurred between the tumour RNA and the DNA from the bacterial strains, but not between the normal callus RNA and bacterial DNA. They concluded that bacterial specific RNAs are present in bacteria-free tumour cells.

At about the same time Quetier, Huguet & Guillé (1969) obtained evidence for partial homology between *A. tumefaciens* DNA and tumour tissue DNA of *Scorzonera hispanica* and *Nicotiana tabacum*. They suggested that a fraction of bacterial DNA is TIP and that it is incorporated into the plant genome and responsible for the increased synthesis of growth hormones and the abnormal nitrogen metabolism. They also reported a partial homology between *Agrobacterium tumefaciens* DNA and nuclear heavy satellite DNA (Nh, G-C rich) which appears in stressed (low temperatures) or damaged tobacco tissues. They speculated that this DNA is amplified during wounding and may provide the active sites necessary for the incorporation of bacterial DNA. Subsequently Guillé & Grisvard (1971) reported a similar satellite DNA to be present in crown-gall, but not in healthy tissues of *Scorzonera hispanica* and *Nicotiana tabacum*. They proposed that bacterial DNA or host sequences homologous to it are integrated into certain sites on the amplified DNA where it modifies a quantitative control mechanism. However, Srivastava (1970), in contrast to Quetier *et al.* (1969), detected a significant amount of homology between bacterial DNA and normal callus tissue DNA. The extent of this homology was about half that between bacterial DNA and tumour tissue DNA. It was suggested that during transformation the bacterial DNA integrates with the host genome by complexing with the DNA with homologous sequences. Unlike Quetier *et al.* (1969), Srivastava did not detect satellite DNA in his preparations and suggested that the wound requirement for tumour induction is necessary to trigger off cell division. He considered that this could lead to an unfolding of the DNA chains and provide conditions necessary for the integration of foreign DNA. In support of these ideas Srivastava & Chadha (1970) showed that sheared tumour tissue DNA preparations contained two bands, one corresponding to host DNA and another smaller one corresponding to *Agrobacterium tumefaciens* DNA.

Clearly this evidence from nucleic acid hybridization studies is in line with the view that bacterial DNA is incorporated into the host genome during transformation. However, at this stage the results should not be taken as conclusive since the observed DNA-DNA and DNA-RNA complexes could have resulted from non-specific reactions. Also the reported presence of satellite DNAs in stressed tissues and tumour tissues needs to be confirmed. It is interesting to note that Pearson & Ingle (1972) have recently suggested that the satellite DNA induced in stressed plants comes from contaminating bacteria.

If bacterial DNA is present in the host cells it is pertinent to ask whether or not it transcribes bacterial proteins in the tumour cells. Schilperoort, Meijs, Pippel & Veldstra (1969) using immuno-diffusion techniques reported the presence of *A. tumefaciens* cross-reacting antigens in bacteria-free crown-gall tissues of tobacco which were originally induced by strain A_6. The detected antigens were not found in comparable normal callus tissues of tobacco. Four precipitation lines were formed when soluble antigens from virulent bacterial strains were tested against tumour tissue antiserum. Only two lines were formed with antigens from the avirulent *A. radiobacter*. It was concluded that the presence of these tumour-specific antigens may be part of the expression of bacterial genes in the crown-gall tissues. Chadha & Srivastava (1971) have extended these results working with another tobacco variety. They were able to detect cross-reacting antigens in bacteria-free crown-gall tissues, but not normal callus tissues tested with serum prepared against *A. tumefaciens*. Furthermore they detected cross-reacting antigens in the bacteria after the anti-tumour serum had been absorbed with a callus antigen preparation. They also compared the soluble proteins from tumour tissues, normal callus tissues and *A. tumefaciens* after separation on DEAE-sephadex (A-50) columns. They observed three protein peaks in the tumour tissues which were absent in the normal callus tissue extracts. Column fractions corresponding to these peaks contained cross-reacting antigens when tested with serum prepared against sterile tumour tissue.

Other evidence for the presence of bacterial proteins in bacteria-free crown-gall tissues comes from the work of Morel and co-workers (Petit, Delhaye, Tempé & Morel 1970, Morel 1971). They have made a detailed study of the nitrogen metabolism of crown-gall tissues in culture and have observed abnormalities which appear to be specific for the strain of inciting bacterium used. In all the tumour cultures examined the arginase activity is low and 'new' enzymes occur which lead to the production of abnormal amino acids (octopine and nopaline) which are never found in healthy plants or normal callus tissues. Two kinds of tumour are distinguishable, one of which contains octopine and the other nopaline. The tumour tissues containing octopine and the *A. tumefaciens* strains which induce them, e.g. B_6, A_6 possess octopine dehydrogenase, while the tumour tissues containing nopaline and the bacterial strains which produce them, e.g. T_{37} possess nopaline dehydrogenase. This unique relationship between the inciting strain of bacterium and the nitrogen metabolism of tumour tissues provides indirect evidence that bacterial information is transferred to the host cells during transformation.

The possibility that viruses or bacteriophages may be involved in the transformation of normal cells to tumour cells by *A. tumefaciens* has been considered by many investigators. The subject has recently been reviewed by Tourneur & Morel (1971). Interest in the phage hypothesis was renewed following reports by Beardsley (1955, 1960) that lysogeny occurred in strain B_6, an isolate commonly used in crown-gall studies. It was found that small doses of UV irradiation caused lysis and release of mature phage from the bacterium. The isolated phage called omega lysed a sensitive strain B_6 806 which had been obtained previously after strong UV irradiation of B_6. At that time there was no evidence that the omega phage was involved in tumorigenesis since the B_6 806 strain which had been cured of omega was as virulent as the original B_6. Furthermore attempts to induce tumours with bacteria-free lysates containing omega phage failed (Klein & Beardsley 1957). More recent attempts to induce tumours with phage preparations from *A. tumefaciens* have also been unsuccessful (Stonier, McSharry & Speital 1967; Roussaux, Kurkdjian & Beardsley 1968; Brunner & Pootjes 1969). Zimmerer, Hamilton & Pootjes (1966) tested 130 strains of *A. tumefaciens* and found that lysogeny was quite common. Five of the isolated phages were examined by electron microscopy and found to be morphologically similar. The phage LV1 from bacterial strain V1 from Zinnia had a polyhedral shaped head (71×63 nm) and a flexible tail (211 nm $\times 9·5$ nm) composed of helical units. Korant & Pootjes (1970) showed that it contained a linear DNA molecule 14 nm long in a double helix with an MW of 31 million. Brunner & Pootjes (1969) found that release of phage LV1 from the bacterium is suppressed at 35°C but enhanced by a heat shock (30 minutes at 42°C), UV irradiation and treatment with mitomycin C. However, in contrast to the above reports Boyd, Hildebrandt & Allen (1970) failed to detect phages in *A. tumefaciens* and with the exception of the claim by Leff & Beardsley (1970) that the DNA from a phage isolated from tumour tissues can transform normal tissues there is no positive evidence that the phages isolated are in any way involved in tumour induction.

However, as pointed out by Tourneur & Morel (1970) most of the temperate phages investigated have been detected using strain B_6 806 or CV1 which have been cured of phages which are very similar. Several reports suggest that strains of *A. tumefaciens* are polylysogenic (Roussaux *et al.* 1968, Kurkdjian 1968) and it remains possible that tumour transformation is associated with an undiscovered phage.

Hoursangiou-Neubrun & Puiseux-Dao (1969) and Kurkdjian (1968) observed that when *A. tumefaciens* is inoculated into plant tissues phage particles are clearly visible in the bacteria, surrounding cell debris and at the surface of the wound, but never in the intact host cells. Subsequently Kurkdjian (1970) showed that the phage was visible only between the 10th and 28th day after inoculation which was thought to more or less coincide with the period of tumour induction. However, the significance of these observations to tumorigenesis is doubtful since phages were also seen when avirulent bacteria were used. Also the induction period for most crown-gall transformations is completed 4–5 days after inoculation.

Further interesting results have come from the work of Parsons & Beardsley (1968). They reported the presence of phage PS 8 in homogenates of bacteria-free crown-gall tissues of sunflower which had been in culture for 13 years. Electron-microscope studies indicated that PS 8 was morphologically very similar to the temperate phage PB_6 which was isolated from *A. tumefaciens* B_6, the strain which induced the original sunflower tumour. The observation that PS 8 was inactivated by antisera prepared against PB_6 supports the view that the phage detected in the tumour tissue came from the inciting bacterium. The purified phage from the tumour tissues failed to induce tumours on the stems of *Kalanchoë* or on the primary leaves of Pinto beans. However, Leff & Beardsley (1970) have claimed to have induced tumours with DNA preparations from this phage after it had been propagated on a phage-sensitive strain of *Agrobacterium tumefaciens* (see p. 367). Tourneur & Morel (1970) have examined a large number of established bacteria-free crown-gall cultures and have demonstrated the release of phage from all the cultures tested. The phages from the different isolates appeared to be identical and similar to those demonstrated in *A. tumefaciens*. In spite of numerous tests phages were never observed in healthy plant tissues or normal callus cultures.

The possibility that TIP is the bacterium itself which enters the wounded cells and becomes altered in its morphology and physiology and remains undetected in the host cells has not received much attention. However, modified forms of the bacteria do occur and these have been studied for tumour-inducing properties. Rubio-Huertos & Beltra (1962) reported that filterable spheric forms of *A. tumefaciens* induced by high concentrations of glycine retain their ability to induce tumours. On the other hand Beardsley,

Stonier, Lipetz & Parsons (1966) could not induce tumours with permanent spheroplasts of the bacterium. Similarly L-forms of *A. tumefaciens* produced by UV treatment did not induce tumours on *Phaseolus vulgaris*. Furthermore studies with the electron microscope have so far failed to detect any structures in the tumour cells which are likely to be modified forms of the bacterium. Thus there is little positive evidence that modified forms of the bacteria are involved in tumorigenesis.

It may be concluded from the above discussion that the nature of TIP has not been established conclusively. However, the reports that bacterial DNA preparations can induce tumour-like overgrowths together with the evidence from DNA-DNA and DNA-RNA hybridization studies, immuno-diffusion studies and studies of abnormal nitrogen metabolism are highly suggestive. There is also good evidence that bacteriophage is transferred during tumour induction and persists in the transformed cells. However, except for the report of Leff & Beardsley (1970) that phage DNA can induce tumours there is little evidence that phages are directly involved in tumorigenesis.

There is a certain amount of evidence that factors other than the TIP are involved in transformation. Braun & Laskaris (1942) showed that attenuated strains of *Agrobacterium tumefaciens* are capable of inducing typical crown-gall tumours only if auxin is supplied at the time of inoculation. It was suggested that the attenuated strains had the capacity to produce TIP but lacked the ability to produce sufficient auxin or to stimulate its production in the host tissues. Klein & Link (1952) have suggested that auxin acts as a co-carcinogen in crown-gall transformations. Braun (1962) favours the interpretation that the tumour cells themselves acquire as a direct result of the action of TIP the capacity for producing large amounts of growth-promoting substances. El Khalifa & Lippincott (1968) have reported that naphthalene acetic acid, gibberellic acid, (2-chloro-ethyl) trimethylammonium chloride, adenine, tri-iodobenzoic acid and 4-chlorophenoxyisobutyric acid increase the number and size of tumours on primary leaves of Pinto beans when added at the same time as *A. tumefaciens*. The lack of specificity for this effect led them to suggest that these substances cause an imbalance in the host metabolism which renders the cells more vulnerable to transformation.

If growth hormones are required in addition to TIP for crown-gall inductions these could be provided by the bacterium, the host cells or the wound juices. Bertholet & Amoureux (1938) demonstrated that *A. tumefaciens* can produce IAA in culture. More recently substances with gibberellin-like activity (Katznelson & Cole 1965, Galsky & Lippincott 1967) and cytokinin activity (Upper, Helgeson, Kemp & Schmidt 1970) have been detected in bacterial cultures. Wound juices are also known to contain growth-promoting substances but their identities are uncertain. Furthermore it is likely that the burst in cell division associated with the wound-healing cycle would result in a release of growth substances.

Detailed discussions of the changes in cytochemistry that occur during tumour induction have been given by Braun (1962), Braun (1969) and Kupila-Ahvenniemi & Therman (1971). Within 4 days after inoculation potential tumour cells are characterized by increases in nuclear and nucleolar volumes, DNA, RNA and non-histone protein in the nucleolus and cytoplasm. Similar changes are observed in the wounded controls, but they are much less pronounced. Braun (1962) considered that these changes indicate an early activation of cellular mechanisms concerned with protein synthesis and that they are likely to be secondary to the changes concerned with transformation.

Development phase

Following the transormation of normal plant cells to tumour cells the continued abnormal and essentially unregulated proliferation of cells becomes an autonomous process. Histological examination of intact tumours or cultured tumour tissues commonly show highly disorganized masses of small rapidly dividing cells and very large single or multinucleate giant cells interspersed among poorly organized groups of vascular elements. Polyploidy and poly teny are characteristic of many tumour tissues although those of sunflowei often remain diploid (Kupila 1958, Rasch, Swift & Klein 1959). Braun (1962) has drawn attention to the similarities of the abnormal histological and cytological features observed in tumour tissues and those found in normal callus tissues. He considers that these features can be accounted for in terms of abnormal growth hormone physiology and do not result specifically from transformation.

Many of the studies on crown gall have been concerned with comparisons between tumour and callus cells in culture. The ultimate aim of such studies is to discover differences which are significant to the process of transformation. White & Braun (1942a) showed that bacteria-free crown-gall tissue cultures of sunflower grow vigorously on a simple medium containing inorganic salts, three vitamins and sucrose. This was in contrast to comparable normal callus tissues which require auxin and other growth factors. These basic observations have subsequently been confirmed and extended to many other plant species (Braun 1947, De Ropp 1947b, Gautheret 1947), and it is now generally accepted that transformation by *A. tumefaciens* is accompanied by a simplification of the nutrient requirements of the host cells. However, since the nutritional requirements of normal callus tissues differ from species to species the degree of simplification varies (Table 13.2). At one extreme the fully transformed crown-gall tissues of *Vinca rosea* become autonomous as compared with normal tissue for auxin, cytokinin, *meso*-inositol, glutamine, asparagine, cytidylic acid and guanylic acid (Braun 1959) while at the other tissues of sunflower become autonomous only for auxin. Although the number of growth factors involved varies the attainment of auxin independence appears to be a consistent feature of crown-gall transformation.

Several investigators have shown that cultured crown-gall tissues possess relatively high levels of auxin (Kulescha & Gautheret 1948, Link & Eggers 1941, Nitsch 1956). This has led to the suggestion that increased levels of endogenous growth factors particularly auxin are directly responsible for the continuous disorganized growth of tumour tissues. The observed increases in amounts of auxin could arise either from an increased synthesis or a decreased rate of breakdown. The evidence on this point is contradictory. Bitancourt (1949) reported the presence of IAA inactivating enzymes in normal but not crown-gall tumour cells and concluded that the high levels of auxin in the tumour cells result from decreased auxin destruction. In contrast several investigators have concluded that the differences in IAA oxidase and peroxidase activity in normal callus and tumour tissues could not account for the increased amounts of auxin. On the other hand Henderson & Bonner (1952) reported that callus tissues of sunflower but not tumour tissue contained an inhibitor that prevents the conversion of tryptophan to IAA. In addition Robson, Yost & Robison (1961) tested the effects of analogues of IAA on the growth of normal callus and tumour tissues of *Parthenocissus*. They concluded that these analogies acted by inhibiting an IAA oxidase system and that the differences between callus and tumour tissues result from differences in both auxin production and destruction. Recently Stonier (1971) has proposed that labile auxin protectors are involved and act by preventing the enzymic destruction of auxins. He suggests that extremely high levels of auxin protectors in tumour tissues could explain the autonomy of the tissue even when crown-gall tissues contain more auxin destroying enzymes than the callus tissues. However, it will only be possible to test these ideas when the so called auxin protectors have been identified.

There have been few investigations of how far other growth factors may be important in crown-gall transformation. However, Wood (1970), Wood, Lin & Braun (1972) have described two cell division promoting factors, cytokinesins I and II which have been isolated from crown-gall tissues of *Vinca rosea*. Cytokinesin I has been partially identified and is thought to consist of a chromatophore 3,7 dialkyl-2-alkylthio-6-purinone and glucose. Wood *et al.* (1972) suggest that the cytokinesins may play a central role in the autonomous growth of *V. rosea* tumour cells since their continued synthesis could sustain the cells in the dividing state. The requirement for an exogenous source of either a cytokinesin or cytokinin for growth of normal callus was interpreted as indicative of a failure of the normal cells to synthesize cytokinesins in culture. Evidence was, however, obtained that normal callus cells grown in the presence of kinetin synthesize cytokinesin I and it was suggested that kinetin may activate their synthesis of cytokinesins. Wood *et al.* (1972) reported that cytokinesin I is a potent inhibitor of both plant and animal (bovine brain) cyclic AMP phosphodiesterases. A membrane-bound adenylic cyclase was also demonstrated and it was proposed that the cytokinesins exert their biological effects as regulators of c-AMP. It will only be possible to evaluate this work when the active compounds have been properly identified and more is known about the role of c-AMP in controlling plant growth.

Wood & Braun (1961) have compared the growth factor requirements of normal callus, partially transformed tumour cultures and fully transformed tumour tissue cultures of *V. rosea*. They found that the growth rates of the normal callus and partially transformed tissues could equal those of the fully transformed tissues if White's basic medium was supplemented with certain growth factors. The partially transformed cells required an auxin, glutamine and meso-inositol while the normal callus required auxin, cytokinin, meso-inositol, glutamine, asparagine, cytidylic acid and guanylic acid. It was concluded that the growth of normal cells and partially transformed tissues was limited by their inability to synthesize adequate amounts of the required growth factors. Furthermore it was suggested that several quite distinct growth factor synthesizing systems are permanently activated as a result of transformation. They went on to discover that the nutritional requirements of the normal callus and partially transformed tissues were significantly modified by raising the levels of certain inorganic salts in the basic medium. The partially transformed cells grew well without glutamine and meso-inositol when White's basic medium was supplemented with 845 mg/litre of KCl, 1800 mg/litre $NaNO_3$ and 300 mg/litre $NaH_2PO_4H_2O$. With the normal callus these supplements replaced the requirements for glutamine, asparagine, cytidylic acid and guanylic acid. Subsequently it was shown that the normal tissues grew without an exogenous supply of auxin if the medium was further fortified by adding 790 mg/litre of $(NH_4)_2SO_4$ (Braun & Wood 1962). It was also shown that the normal callus tissues cultured on the fortified media contained substances with auxin activity as estimated by the tobacco pith bioassay. Evidence was obtained that the raised salt levels increased the synthesis of meso-inositol in the tissues, but the situation with meso-inositol was complicated by the observation that it influenced the uptake of inorganic ions.

To explain these results Braun & Wood (1962) have postulated that during the transition from a normal cell to a tumour cell an increase in permeability of one or more membrane systems occurs and this permits the penetration of ions and activates a large segment of metabolism which is specifically concerned with cell growth and division on White's medium. Support for this hypothesis was obtained from ion uptake studies (Wood &

N

Braun 1965). The experiments indicated that the tumour cells had an increased capacity for the uptake of $^{42}K^+$ and $^{32}PO_4$. These experiments have recently been criticized on the grounds that only the rate of uptake of the isotopes, but not the rate of release. was measured (Stonier & Yang 1971). Stonier's data indicated that the levels of inorganic salts in the media modify the rate of release of auxin protectors and enzymes into the medium. It was suggested that the observations of Braun and Wood reflect changes in the rate of release of essential growth factors rather than the activation of specific metabolic systems. Although these interpretations are very different from those of Braun they still implicate membrane properties as being very important in transformation.

There are other reasons for being cautious about the general significance of the results obtained by Wood and Braun. First *V. rosea* normal callus is exceptional in that it requires seven organic growth supplements in White's basic medium for rapid growth. As can be seen from Table 13.2 the requirements of many calluses are much simpler. Sunflower normal callus for example will grow indefinitely on White's basic medium with auxin as the only extra supplement. However, with the exception of habituated cultures, spontaneous tumour tissues and a few chemically induced tumour tissues, auxins are essential for continuous rapid growth. Cytokinin is an absolute requirement for many tissues and enhances the growth of others. Thus the claim for *V. rosea* that tumour transformation involves the activation of seven specific biosynthetic pathways associated with cell growth and cell division cannot be extended to all crown-gall transformations. At the most it can only be considered universal in respect to auxin and cytokinin biosynthesis. Secondly if ion activation of specific biosynthetic pathways is a general phenomenon it would be expected that the growth-factor requirements of all normal callus tissues would be simplified by raising the salt levels in the media. Although growth is often markedly increased by raising the inorganic salt levels in the medium there is little evidence that the requirements for auxin can be eliminated in this way. It is perhaps significant that Wood & Braun (1965) could not repeat their earlier results with freshly isolated tissue cultures of *V. rosea*. It seems likely that the high salts provide a better growth medium and affect metabolism in a more general fashion than that suggested by Braun and Wood.

As previously discussed studies of the nitrogen metabolism have revealed that crown-gall cultures in general possess abnormal amino acids not found in callus cultures and normal tissues. Lioret (1960) detected and isolated lysopine (N-α-(1-carboxethyl)-L-lysine) from tumour tissues of *Scorzonera hispanica* and *Nicotiana tabacum*. Seitz & Hochester (1964) later reported that small amounts of lysopine occur in normal tissues, but Morel (1971) has not been able to confirm this. Lejeune (1967) and Lejeune & Jubier (1968) working with radioactively labelled compounds showed that lysopine is metabolized by normal callus and tumour cells, but that the dehydrogenase required for its synthesis is present only in homogenates of the tumour tissues. Studies of the arginine metabolism of *Helianthus tuberosus* have shown that arginase activity is extremely low in the tumour tissues and that a guanidine base is accumulated (Morel & Duranton 1958). A systematic survey of crown-gall tissue cultures from more than thirty species showed that all the isolates except that of *Opuntia vulgaris* contained the same guanidine and lysopine. *O. vulgaris* on the other hand contained a different guanidine (Goldman & Morel 1967). Subsequently the compound isolated from *Scorzonera hispanica* was identified as octopine (N-α-(1-carboxyethyl)-L-arginine) and that from *Opuntia vulgaris* as nopaline (N-α-(glutaryl)-L-arginine). *Scorzonera hispanica* also contained the related octopinic acid (N-α-(1-carboxyethyl)-L-ornithine) (Fig. 13.2).

Petit, Delhaye, Tempé & Morel (1970) examined crown-gall tumour isolates from *Datura stramonium* induced by different strains of *Agrobacterium tumefaciens*. Of these isolates twenty produced octopine, nineteen produced nopaline, two produced octopine and nopaline and two contained only arginine. From this it was concluded that the bacterial strain used determines the nature of the abnormal amino acids produced and that this indicates a specific effect of the bacterium on the hosts' metabolism. Goldmann (1971) and Petit *et al.* (1970) showed that octopine-producing tumours possessed a dehydrogenase which in the presence of $NADH_2$ catalyses the condensation of sodium pyruvate and arginine to octopine. The enzyme appeared to be highly specific although homoarginine and canavinine can act as substrates. Similar experiments with *Opunta vulgaris* and *Nicotiana tabacum* crown-gall cultures initiated with strain T_{37} indicated the presence of a dehydrogenase which catalyses the incorporation of arginine into nopaline. It was concluded that the transformation of a normal cell to a tumour cell is accompanied by the appearance of specific dehydrogenases in the tumour cells. Morel (1971) proposed two hypotheses which could explain these observations: (1) The genetic information for the biosynthesis of the abnormal amino acids exists in the plant, but is only manifest after the metabolic disorganization which results from transformation. (2) The genetic information originates in the inciting bacterium and is transmitted during transformation. The first possibility is difficult to eliminate completely, but there is no positive evidence that intact tissues, normal callus, habituated, genetic tumour or wound tumour tissue cultures possess the enzymes necessary for the synthesis of octopine or nopaline. On the other hand the second explanation has been supported by evidence obtained by Petit *et al.* (1970). Although the abnormal amino acids were not detected in the inciting bacteria they obtained evidence that the 'octopine' strains of *Agrobacterium tumefaciens* possessed octopine dehydrogenase while 'nopaline' strains possessed nopaline dehydrogenase. It was concluded that this is consistent with the view that the information for the synthesis of the dehydrogenases originates in the bacteria. The failure to detect octopine and nopaline in the bacteria could be explained by making the assumption that the reactions catalysed by the dehydrogenases are reversible and that the conditions within the tumour cells favour synthesis of the abnormal amino acids while in the bacteria they favour degradation.

Lippincott & Lippincott (1970) and Lippincott, Lippincott & Chang (1972) have suggested that octopine, nopaline and lysopine may be important in determining the growth characteristics of tumours since they found that they increased the growth of tumours which have been induced by *A. tumefaciens* on primary leaves of Pinto beans. These effects seemed to be highly specific. Morel (1971), however, does not consider that the abnormal amino acids are important in the growth of tumour tissues, but suggests that they may be useful markers for investigating crown-gall transformations.

A more extensive treatment of the biochemical differences between crown-gall tissues and normal tissues in culture has been given by Braun (1962, 1969) and Morel (1970).

Transmission of the tumour-inducing principle

The first indication that tumour tissues may pass on their tumorous properties to healthy cells without the participation of *A. tumefaciens* came from the observations that in certain plants, e.g. sunflower, secondary tumours sometimes develop at a distance from the primary tumours and that such tumours are bacteria-free. However, this evidence is not conclusive since it is possible that the secondary tumours are initiated by small numbers of bacteria transported from the primary tumour via the xylem which do not survive. Indeed it has been shown that the bacteria can move through the xylem and that the bacteria do not survive for long periods in living sunflower tissues (De Ropp 1951).

A more direct approach was used by De Ropp (1948). He grafted bacteria-free sunflower tumour tissue onto healthy stem segments and frequently obtained overgrowths near to the graft and on one occasion a new tumour developed at the opposite end of the segment. McEwan (1952) obtained similar results to De Ropp, but suggested that the overgrowths may result from auxin supplied by the growing tumour tissues. Subsequent work by Camus & Gautheret (1948) indicated that bacteria-free crown-gall tissue of *Helianthus tuberosus* will also, when grafted to healthy tissues, cause the production of overgrowths capable of growing on the medium that supports the growth of crown-gall tissues. Kelbitsch (1960) has reported that a transmissible tumorigeneric agent can be passed from a primary crown-gall tumour to a healthy sunflower with use of dodder. However, other attempts to transmit TIP with dodder have failed (Braun 1962).

More convincing evidence that TIP can be transmitted from bacteria-free tumour cells to normal cells has recently been presented by Aaron-Da-Cunha (1969) and Aaron-Da-Cunha & Paupardin (1971). Bacteria-free crown-gall tissues of tobacco irradiated with X-rays (5000–6000 rads) when grafted onto the stems of healthy plants were incapable of growth and eventually necrosed, while the normal tissues in the proximity of the graft proliferated to form tumours. Tissues from these tumours grew on tissue culture media in the absence of auxins and cytokinins and could be grafted onto healthy plants where they continued to develop. They also contained octopine and phages which are characteristic for tumours induced by the original strain of *Agrobacterium tumefaciens*. The only weakness in this work was that it was not possible to establish unequivocally that the tumours arose from newly transformed cells and not from the original graft. In an attempt to overcome this criticism Aaron-Da-Cunha & Paupardin (1971) grafted X-ray irradiated tissues of variety White Burley onto healthy tissues of variety Samsun. Single cell clones were then isolated from the induced tumours. Some clones behaved like the tumour tissues, while others resembled normal callus tissues and required growth hormones for growth. Plants derived from the clones which produced buds had the appearance of Samsun indicating that those clones at least originated in the stock and not the graft. Unfortunately this still does not rule out the possibility that the tumour cell types originate from the graft. Thus it has still not been established conclusively that bacteria-free tissues are capable of transforming normal cell types.

Reversibility of the tumorigenesis

In general fully transformed crown-gall cells give every sign of being permanently changed in character and many tumour isolates have been cultured for many years without showing any tendency to revert. However, with some plant species such as *Nicotiana tabacum* moderately virulent strains of *Agrobacterium tumefaciens*, e.g. strain T_{37} induce tumours which exhibit a capacity for forming abnormal shoot-like structures known as teratomas. Cultures of such tissues consist of a mixture of partly developed buds and leaves and undifferentiated callus tissue. Like other tumour tissues they grow indefinitely on simple culture media. Braun (1959, 1965) showed that teratoma tissues are not a mixture of normal and tumorous cells by obtaining single-cell clones which gave rise to teratoma cultures. Evidence was, however, obtained that tissues within these teratoma cultures could revert to the normal type. Fragments of the tumour tissue were grafted onto the tips of cut stems of tobacco plants where they grew rapidly and developed into a tissue mass containing highly abnormal leaves and buds. The buds were then removed and grafted onto other healthy plants where they developed slowly into abnormal shoots. When these shoots were 4–6 inches in length they were grafted onto a third healthy plant where they grew rapidly and became more normal in appearance. Eventually the shoot flowered normally and set seed. Braun's interpretation of these experiments was that the repeated grafting of the teratoma onto healthy plants forced the teratoma into rapid organized growth which in turn led to a loss of the tumorous properties. Sacristán & Melchers (1969) and Melchers (1971) have questioned these grafting experiments on the grounds that the regenerated normal shoots could have originated from the stock plant. However, recently Braun & Meins (personal communication) have used different tobacco varieties for the graft and stock and shown conclusively that the regenerated shoots originate from the graft teratoma tissues.

Sacristán & Melchers (1969) also reported that 20-year-old crown-gall cultures and habituated cultures of tobacco could be induced to regenerate shoots by adding high concentrations of kinetin to the culture medium. However, these results did not rule out the possibility that the crown-gall cultures contained a mixture of normal callus and tumour cells and that the regenerates came from selected non-autotrophic non-tumour cells. Attempts to eliminate this possibility by showing that the autotrophic single-cell clones can produce shoots have so far been unsuccessful (Melchers 1971).

It appears then that some of the teratoma tumour cells can revert to the normal type, however, the evidence that tumorigenesis in the fully transformed cells is reversible is inconclusive.

VIRUS TUMOURS

The only virus-induced tumour disease which has been studied in detail is the wound tumour disease. The inciting virus is *Aureogenus magnivena* Black which is transmitted by the leaf hoppers *Agallia constricta* and *A. novella*. *A. magnivena* has been isolated from the host plant and the insect vectors and is a polyhedron of 75–80 μm diameter containing double stranded RNA (Wood & Streissle 1970). Comprehensive reviews of the disease have been given by Braun & Stonier (1958) and Black (1965, 1972).

Tumour development

The disease has a wide host range and Black (1945) in a survey of 100 species found that forty-three species from twenty different families exhibited disease symptoms when inoculated with infected vectors. The symptoms of the disease are extremely variable both in morphological effects and severity. On clover the most common symptoms are irregular vein enlargements which explains why the disease was formerly known as clover big vein. However, root tumours, stem tumours, leaf distortions and short thickened stems are also characteristic of infected plants. The most vigorous tumours have been observed on *Rumex acetosa* and *Melilotus alba*. The responses of the host seem to depend very much on the inherent tendency to form tumours and it has been suggested that the virus exacerbates this tendency to form spontaneous tumours.

Tumours usually develop where accidental wounds occur such as at severed veins, lateral root emergence points and abscission scars, which indicates that wound damage plays an important part in tumour initiation. Black (1965) suggested that wounding releases substances which act synergistically with the virus to stimulate growth. This view is supported by the observation that auxin treatment causes a marked increase in number and size of the tumours produced in infected plants (Black & Lee 1957).

Tumours may be experimentally induced by inoculating healthy plants with leaf hoppers which have been previously fed on diseased crimson clover plants. The tumours require from 10 to 40 days to develop depending on the species and age of the host. It is also possible to transmit the disease by grafting a piece of tumour tissue onto a healthy plant or by inoculating multiple punctures in the plant crown with fresh tumour extracts.

Wound tumour tissue cultures

Compared with crown gall there have been few studies of wound tumour tissues in aseptic culture. Black (1944, 1949) isolated tissues from root tumours of *Rumex acetosa* which grew slowly, but indefinitely, on White's basic medium. The virus could be detected in these cultures 14 months after they had been initiated. Subsequently Burkholder & Nickell (1949) modified the medium by adding extra phosphate and vitamin B_1 and achieved higher growth rates. Growth was also increased by supplementing the medium with RNA hydrolysate, uracil, guanine, xanthine and hypoxanthine (Nickell, Greenfield & Burkholder 1950). It is interesting to note that the normal callus from *R. acetosa* did not grow on these synthetic media.

Although the virus multiplies in newly isolated tumour tissue cultures it commonly disappears after prolonged culture. However, the resulting isolates which appear to be virus-free retain their capacity for continuous growth in the absence of growth hormones and may be grafted onto healthy plants. Streissle (1971) has recently re-examined the persistence of the virus in the tumour tissues isolated from root tumours of *Melilotus officinalis*. It was shown that the virus disappeared more rapidly in the fast-growing tissue cultures and no virus particles were detected after 5 months in culture. Also virus particles were not seen in cell suspensions derived from the tumour tissues. Streissle (1971) was able to graft the apparently virus-free tissues onto healthy plants where they formed tumours and it was suggested that the persistence of tumorous properties may indicate the presence of the viral genome in the cells. However, this evidence on its own is insufficient since it is well known that habituated tissue cultures which do not possess an inciting agent can induce tumours when grafted onto healthy plants.

Although it is difficult to generalize from such a limited number of investigations the evidence seems to suggest that the virus triggers off a tumorous state which is inherent in the susceptible host species rather than endow tumour-like properties to the cells. At present there is no substantial evidence that the viral genome is directly responsible for the tumorous properties of the host cells.

GENETIC TUMOURS

The development of spontaneous tumours is a relatively common occurrence on certain plant species. Often no external causal agents such as bacteria, viruses and oncogenic chemicals are involved and the formation of tumours appears to have a genetic basis, e.g. certain hybrids within the genera *Nicotiana*, *Brassica*, *Bryophyllum*, *Lillium* and *Lycopersicum*. The literature on genetic tumours has recently been reviewed in detail by Smith (1972).

Nicotiana *hybrid tumours*

The most thoroughly investigated genetic tumours have been those which occur spontaneously on a number of *Nicotiana* hybrids. When the appropriate species are crossed tumours form on the hybrid progeny. These tumours can arise at any stage during development, but occur most commonly on mature plants when vegetative growth is arrested. They most frequently form on roots, but also develop on stems and leaves. They are often located at points where there is evidence of tissue damage, e.g. lateral root emergence points, sites of wound healing and abscission scars. In addition exposure to ionizing radiations and treatment with certain chemicals accelerate tumour initiation and increase proliferation. For example X-rays, γ-rays (^{60}Co) and endogenous sources of β-radiation (^{32}P, ^{3}H, ^{14}C) cause an early development of tumours (Sparrow, Gunckel, Schairer & Hagen 1956, Hagen, Gunckel & Sparrow 1961, Ahuja & Cameron 1963, Conklin & Smith 1968). Similarly chemicals such as IAA, kinetin, uridine, thymidine, azauracil and mercaptoethanol cause an early development of tumours when applied to the seedling tips of tumour-prone hybrids (Schaeffer 1962, Conklin & Smith 1968, Buiatti 1968, Ames & Smith 1969). Exposure to higher temperatures (24–27°C) have also been shown to induce tumour formation (Schaeffer, Burk & Tso 1966). Smith (1972) suggests that the tumour-prone hybrids maintain a precarious balance between organized and tumorous growth and that various treatments which cause stress can tip the balance towards abnormal growth.

Genetical background

In 1930 Kostoff described nine interspecific hybrids within the genus *Nicotiana* which regularly produced spontaneous tumours. Since then at least another twenty tumour producing combinations have been added to the list (see Smith 1972). Näf (1958) divided the parents of the tumorous hybrids into two groups which he called plus and minus. The plus group consists mainly of the species of the section Alatae whereas the minus group contains species from several sections. Crosses between the species within each group do not produce tumorous offspring while crosses between a plus and a minus species do. The species within each group were arranged in order of the relative strength of their plus and minus characteristics.

Studies involving detailed breeding experiments have indicated that the causes of tumour formation reside in the chromosomes. With *N. glauca* × *N. langsdorfii* hybrids it was possible to obtain various combinations ranging from two *N. langsdorffii* genomes with one of *N. glauca* (GLL) to one of *N. langsdorffii* and three of *N. glauca* (GGGL). All these combinations were tumorous suggesting that the tumour formation did not depend on the ratio of the genomes (Kehr & Smith 1954). Subsequently, however, Ahuja & Hagen (1967) have shown that there are distinct quantitative differences between some of these hybrid combinations. For example the triploid GLL produced smaller tumours than the F_1 hybrid GL or the amphiploid GGLL.

When Kehr & Smith (1954) repeatedly backcrossed the triploid hybrid GLL they succeeded in obtaining plants which retained one or a few *N. glauca* chromosomes on a *N. langsdorffii* background. Such plants failed to form tumours on their shoots and it was concluded that several *N. glauca* chromosomes are necessary for tumour production. More recently, however, Ahuja & Hagen (1967) have shown that plants possessing three *N. glauca* univalents up to forty-eight univalents and a bivalent of *N. glauca* added to a diploid *N. langsdorffii* (LL) are tumorous. A somewhat simpler situation appears to exist with crosses between *N. longiflora* of the plus group and the amphiploid *N. debneyi-tabacum* (Ahuja 1962). By a programme of repeated backcrossing it was possible to obtain a single *N. longiflora* chromosome or even a chromosome fragment on an amphiploid background and it was shown that such combinations produced tumorous plants. On the basis of these results Ahuja (1968) has proposed that a species of the plus group contributes a relatively simple inherited genetic factor (I) essential for tumour initiation which interacts with factors (ee) of the minus species to form tumours. The different degrees of tumour expression were considered to be controlled genetically by factors (ee) which are located in the chromosomes of the minus group.

Kostoff (1930) and Burk & Tso (1960) observed that the development of hybrid tumours was often accompanied by nuclear abnormalities. In 1939 Kostoff also reported that abnormal mitoses occurred more frequently in root-tips in *N. glauca* × *N. langsdorffii* hybrids than in those of the parents. On the other hand Kupila & Therman (1962) failed to observe mitotic abnormalities in normal tissues of similar hybrid plants. Although it is not conclusive the present evidence suggests that the chromosomal abnormalities seen in developing tumours result from rather than cause the tumorous condition.

Physiology and biochemistry of tumorous plants

Vesters & Anders (1960) reported that young plants of the *N. glauca* × *N. langsdorffii* hybrid contained 50% more free amino acids than the parental species. Later Tso, Burk, Sorokin & Engelhaupt (1962) found significant differences in the concentrations of alkaloids, sugars and organic acids as well as the amino acids. They noted that the newly initiated tumours also accumulated high levels of free amino acids.

Kehr & Smith (1954) reported that *N. glauca* × *N. langsdorffii* hybrids possess a higher auxin content and a more effective enzyme system for converting tryptophane to auxin than the parents. Subsequently Bayer (1965) using the *Avena* coleoptile curvature test demonstrated that the auxin content was highest in the hybrid, lower in *Nicotiana glauca* and lowest in *N. langsdorffii*. Later the same author (Bayer 1967) showed that the hybrids which formed large tumours (GL and GGLL) have a higher auxin content than the parental species and the triploid (GLL). The latter form small tumours late in the development of the plant. Similar analyses were made for *N. longiflora*, *N. debneyi* and their tumour-producing hybrid (Bayer & Ahuja 1968). The potentially tumorous combinations again showed higher levels of auxin than the parents and the non-tumorous segregants.

Tso, Burk, Dieterman & Wender (1964) detected larger amounts of scopoletin and scopolin in tumour-prone hybrids (*N. glauca* × *N. langsdorffii*) than in the parent species. Since these compounds are known to reduce the rate of IAA oxidation by peroxidase it is possible that they influence the level of auxin in the tumorous tissues.

Hybrid tumour tissue cultures

White (1939a, 1944) using normal tissue culture methods succeeded in culturing tumour tissues from hybrid plants of *N. glauca* × *N. langsdorffii*.

Such cultures could be maintained indefinitely on White's medium without auxin or other growth factors. After 5 years some of these tissues produced tumours when grafted onto healthy plants indicating that the tumorous properties persist in culture. The hybrid tissue cultures did not produce organs on an agar medium, but when transferred to a liquid medium abnormal leaves developed. One particular isolate retained its ability to form organized structures for more than 144 subcultures (Skoog 1944). It was found that shoot formation was suppressed by 0·2 mg per litre of IAA, but that this effect could be reversed by raising the levels of $KH_2PO_4H_2O$, $Fe_2(SO_4)_3$ and sucrose. Skoog also showed that extracts from hybrid tumour tissues possessed high auxin activity in the *Avena* test.

It is interesting to note that tissue cultures derived from the normal tissues of tumour-prone hybrids grow on media lacking auxins and that such cultures can be grafted on *Nicotiana glauca* plants to produce tumours (Kehr & Smith 1954). In contrast callus tissue isolates from the parent species require auxins for growth and do not produce tumours when grafted onto intact plants. It appears then that the tumorous condition is initiated when explants are taken from normal tissues. Like the above, hybrid tumour tissue cultures derived from the tumour forming *N. sauveolus* × *N. langsdorffii* and *N. debneyi-tabacum* × *N. longiflora* do not require auxin and kinetin for vigorous growth whereas their non-tumorous parent species do (Schaeffer & Smith 1963, Ahuja & Hagen 1966a, b). On the other hand tissues from a non-tumorous mutant of the *N. glauca* × *N. langsdorffii* hybrid requires kinetin for growth. Hagen (1965) reported that the growth rates of hybrid tumour tissue cultures on a basic medium depend both on the ratio of the parental genomes present and the number of *N. glauca* chromosomes on a background of two *N. langsdorffii* genomes. He found that the growth of pith tissues from the triploid hybrid GLL was lower than that of the 2*n* and 4*n* hybrids, while the combinations with *N. glauca* chromosomes on a background of two *N. langsdorffii* genomes had a growth rate intermediate between the GLL hybrid and the *N. langsdorffii* parental tissue.

Buiatti & Bennici (1970) observed that some species of the plus (I) group (*N. alata*, *N. langsdorffii*, *N. longiflora* and *N. plumbaginifolia*) form callus more readily when treated with IAA, 2,4-D and kinetin than do species of the minus (ee) group. These results may indicate significant differences in hormonal mechanisms between the two groups which interact in hybrid combinations.

Most of the reports discussed above implicate auxins and cytokinins as being important in the development of hybrid tumours and it seems well established that both the tumour-prone plants and the tumour tissue cultures possess high levels of auxin. There is strong evidence that this change in hormone metabolism is genetically controlled, but the actual mechanism is still a matter of speculation. Two genetic explanations are possible. First, in the tumour-prone genotypes there may be a tendency for certain genes to regularly produce mutations characterized by a switch to neoplastic growth. Secondly the initiation of tumour growth may be due to the activation of genes normally repressed in different tissues to produce products necessary for renewed cell division and enlargement. Smith (1972) has reviewed the evidence for and against these possibilities and favours the gene regulation hypothesis.

HABITUATED TISSUE CULTURES

Gautheret (1946) observed that callus tissues of *Scorzonera hispanica* which initially had a requirement for auxin occasionally developed sectors which had the capacity to grow indefinitely on an auxin-omitted medium. Similar observations have since been made with callus cultures of other species, e.g. European grape, tobacco, mallow, virginia creeper, sunflower and carrot (Morel 1947b, Kandler 1952, Czosnowski 1952). Gautheret called this phenomenon 'accoutumance à l'auxin' and 'anergie à l'auxin' while White introduced the term habituation. Habituated tissues are often friable in texture, have a low capacity to differentiate and contain relatively large concentrations of auxin (Kulescha & Gautheret 1948, Kulescha 1952). Thus it seems probable that habituation results from either an increased synthesis or a decreased breakdown of endogenous auxin. The nutritional requirements of fully habituated tissues are similar to those of fully transformed crown-gall tissues and genetic tumour tissues and they can be cultured indefinitely on simple synthetic media in the absence of auxins and cytokinins. Their tumour-like properties have been confirmed by Limasset & Gautheret (1950) who were able to graft habituated tobacco tissues onto healthy tobacco plants where they developed into large overgrowths. Successful grafts have also been made for other species (Braun & Morel 1950, Henderson 1954). Habituation does not appear to be an all or nothing process since tissues with varying degrees of auxin independence occur (Morel & Gautheret 1955). There is also some evidence that similar changes can occur in the requirements for other growth factors such as cytokinins.

The mechanisms underlying habituation are not understood. Gautheret did not consider it to be a mutation or due to selection, but rather a type of enzymatic adaptation leading to fundamental changes in metabolism. In support of this view Gautheret describes experiments where he examined the cultures of *Parthenocissus* for habituation at 2-monthly intervals. During a period of 2 years 85% of the cultures had become auxin independent. Similar experiments showed that only 5% of tobacco cultures became habituated.

On the other hand White (1951), De Ropp (1951) and Kandler (1952) have suggested that habituation results from somatic mutations in the gene complex controlling auxin-regulating systems. Attempts have been made to relate habituation to chromosome abnormalities and Fox (1963) found that habituated cultures of tobacco had higher chromosome numbers than callus cultures. However, Sacristán & Melchers (1969) and Melchers (1971) have been unable to establish any direct relationship between chromosome number and habituation in their tobacco cultures. Sacristán & Wendt-Gallitelli (1971) investigated a chromosomal mutation called 'Unisat' which seemed to be associated with the change of *Crepis capillaris* callus cultures to autotrophic teratomas. They followed the mutations in the cells during the development of normal shoots from the teratomas and in the secondary calluses derived from them. They found that the secondary callus possessed the 'Unisat' mutation but had a requirement for auxin and it was concluded that the mutation was not the direct cause of the habituation. However, it is doubtful whether the autotropic teratoma tissues can be regarded as habituated in the normal sense since they possessed organized structures which could have provided a source of auxin for the associated callus cells.

Sacristán & Melchers (1969) and Melchers (1971) have recently considered whether or not habituated cells have the capacity for regeneration. They showed that habituated tobacco tissues which had been in culture for more than 20 years could be induced to form shoots by placing them on a high kinetin medium and exposing them to low light intensities. The regenerated shoots gave rise to plants which were morphologically distinct from the parent variety and sterile. Callus cultures initiated from these plants were auxin dependent. However the authors have so far been unable to rule out the possibility that the original tissues were a mixture of habituated and normal cells. Indeed Lutz (1966) working with the same culture isolated two single-cell clones, one which grew well on an auxin-free medium and another which grew very slowly without auxin. The former isolate, but not the latter, was successfully grafted onto normal plants where it developed into a tumour.

The process of habituation is difficult to study because, unlike crown-gall tumours, it is unpredictable and cannot be induced at will. However, habituated tissues once obtained can be very valuable as controls in studies of tumours caused by specific inciting agents such as *Agrobacterium tumefaciens* and *Aureogenus magnivena*. They are often more suitable than normal callus cultures as controls since they can be cultured on media identical to that used for the tumour tissues.

Most of the chemically induced overgrowths on plants are of the self-limiting type, however, Bednar & Linsmair-Bednar (1971) have recently reported that certain morphactins (amino fluorenes) when added to the culture media can cause tobacco callus to form compact nodules which on subculture will grow indefinitely on media without morphactin or kinetin. Also Ingram & Butcher (1972) have shown that root callus cultures initiated from *Andrographis paniculata* on a medium containing benazolin (a synthetic auxin) and coconut milk will grow indefinitely on media without auxin, cytokinin or coconut milk. These reports indicate that in certain circumstances it is possible chemically to induce callus cultures having the nutritional characteristics of habituated and tumour tissue cultures. The chemical induction of tissue cultures with tumour-like properties could be very valuable in future studies of plant tumours.

CONCLUSIONS

Studies of crown-gall, virus wound tumours and genetic tumours have revealed that these diseases have many common features. In each case a wound appears to be an essential prerequisite for tumour induction, the disorganized tumour tissues are capable of indefinite growth and the tumour tissue cultures grow on simple media without auxins or cytokinins. Intact and cultured tumour tissues contain high concentrations of growth hormones particularly auxins which could account for their disorganized growth behaviour. The fact that the growth patterns of tumour tissues in culture are very similar to those of normal callus supplied with auxins and cytokinins is in line with this interpretation. Further, the tissue cultures of the three types of tumour may be grafted onto healthy plants where they will form tumours. Finally the form and organization of the tumour depend very much on the host, particularly its regenerative capacity, as well as the inciting agent.

The similarity of tumours caused by *Agrobacterium tumefaciens*, wound tumour virus and genetic constitution has led Braun (1969, 1972) to suggest that the tumorous state is caused by epigenetic modifications rather than nuclear changes of the mutational type. For example the inciting agents plus wounding could trigger off a tumorous state in the host cells. The experiments indicating that the tumorous state is reversible would be in accordance with this view. The existence of the phenomenon of habituation where no inciting agent is present also suggests that plant cells possess a potential for tumour-like properties.

On the other hand recent investigations on crown-gall tissues suggest that DNA or perhaps a phage from the bacterium is incorporated into the host genome where it replicates and transcribes bacterial proteins. At present there is no evidence that the bacterial proteins are directly responsible for the tumour-like properties although it has been suggested that they may contribute to the increased production of auxins and cytokinins. A second possibility is that the presence of bacterial (or phage) DNA (or proteins) either in the host genome or in the cytoplasm could trigger off and maintain the tumorous conditions by interfering with the host control mechanisms for auxin and cytokinin biosynthesis. This explanation would be compatible with Braun's interpretations.

The recent introduction of the methods of molecular biology such as nucleic acid hybridization and immuno-diffusion techniques represents a new phase in plant tumour studies. The careful exploitation of these methods is likely to provide answers to some of the principal problems in plant tumorigenesis. Clearly plant tissue culture methods have contributed a lot to our present knowledge of plant tumorigenesis and they will be indispensable in future investigations.

CHAPTER 14

GROWTH OF PLANT PARASITES IN TISSUE CULTURE

D. S. Ingram

Introduction 392
Fungi 393
Viruses 406
Other pathogens 416
Future prospects 420

INTRODUCTION

It has long been realized that there is great potential advantage in the application of tissue culture techniques to the study of plant parasites, for these organisms may then be cultured together with their hosts in a controlled chemical and physical environment, free from contaminant micro-organisms. This offers the possibility of a simplified experimental system for investigating the structure and physiology of host-parasite interaction and, in the case of specialized parasites, provides a means of maintaining a continuous supply of contaminant-free propagules of pure physiological races and strains.

In experiments where tissue cultures are used large numbers of host cells may be exposed to a parasite without causing tissue injury by artificial removal of protective barriers, and host cell number and parasite inoculum density may be carefully controlled. In addition, metabolic inhibitors and precursors may be added to culture media and diffusible products of interaction may be isolated with comparative ease. Such attributes as these are particularly important in studies of disease resistance, hormonal and enzymatic changes occurring after infection and the maceration and killing of cells and tissues by secreted enzymes and toxins.

Other plant culture methods may also be of value in the study of pathogens. The technique of meristem tip culture has already been used to produce and maintain pathogen-free stocks of commercial crop plants; root organ cultures have been used in studies of plant parasitic nematodes; and isolated protoplasts have been used in infection studies with viruses. Although isolated protoplasts have as yet been little used in investigations of fungi and bacteria, they may prove to be particularly useful in studies of hypersensitive resistance to infection and causes of cell death by secreted toxins and degradative enzymes, where the presence of a cell wall may be a complicating factor.

Despite their many potential advantages, and in contrast to the field of animal pathology where animal tissue cultures have been of great value, plant tissue culture techniques have so far contributed little to research on any plant disease other than crown gall (discussed in Chapter 13). Some of the reasons for this fact are discussed on the following pages. General reviews dealing with the growth of pathogens in plant tissue cultures have been published by Braun & Lipetz (1966), White (1968) and Maheshwari (1969). The fungi and bacteria have not been reviewed in detail, but various aspects of the interaction between viruses and plant tissue cultures have been extensively dealt with by Hollings (1965), Raychaudhuri (1966), Kassanis (1967) and Cocking (1970), and the application of gnotobiotic techniques to plant nematology has recently been reviewed by Zuckerman (1971).

FUNGI

Dual culture of obligate parasites and their hosts

The term physiologically obligate parasite has traditionally been applied to those fungi which can only grow and achieve full development in association with a suitable living host plant and which cannot yet be grown saprophytically on artificial culture media. Many fungi have been considered to be physiologically obligate, including members of the Peronosporaceae (downy mildews), the Albuginaceae (white blisters), the Erysiphales (powdery mildews) and the Uredinales (rusts). Several members of the Uredinales have recently been grown in axenic culture on simple media, and there is every reason to suppose that many other so called obligate parasites will eventually be grown in this way once suitable media have been devised.

It is now almost thirty years since Morel first realized the possible advantages of growing obligate parasites in dual culture with their hosts. In his early experiments (Morel 1944, 1948) he was able to establish good dual cultures of the downy mildew fungus *Plasmopara viticola* and callus of *Vitis vinifera* (the vine) and was able to use sporangia from these cultures in certain fungicide tests (Morel 1946, 1947). However, little further progress has since been made in the development of the dual culture technique. In 1967 Brian was able to list only twelve obligate parasites which had then been grown together with host callus, and although the list can now be extended somewhat (Table 14.1), the total number of species involved is still remarkably small and very few of the dual cultures have actually been used experimentally.

Obtaining contaminant-free inoculum

In his experiments with *Plasmopara viticola* Morel (1944, 1948) obtained

TABLE 14.1. Fungal obligate parasites grown in host tissue cultures

Fungus	Host	References
CLASS PHYCOMYCETES Order Plasmodiophorales *Plasmodiophora brassicae* Woron. (Clubroot of Brassicas)	*Brassica* spp.	Williams & Yukawa 1967, Keen, Reddy & Williams 1969, Williams, Reddy & Strandberg 1969, Ingram 1969a, b, Reddy & Williams 1970, Dekhuijzen & Overeem 1971, Tommerup & Ingram 1971, Ingram & Tommerup 1972
Order Chytriadiales *Synchytrium endobioticum* (Schilb.) Perc. (Potato wart disease)	*Solanum tuberosum* L.	Mirzabekyan 1969, Ingram 1971
Order Peronosporales (Downy mildews and white blisters) *Peronospora farinosa* (Fr.) Fr. f.sp. *betae*	*Beta vulgaris* L.	Ingram & Joachim 1971, Ingram & Tommerup 1973
Peronospora parasitica (Fr.) Tul.	*Brassica* spp.	Nakamura 1965, Ingram 1969c
Peronospora tabacina Adams.	*Nicotiana tabacum* L.	Izard, Lacharpagne & Schiltz 1964
Plasmopara viticola (Berk. & Curt.) Bebl. de Toni.	*Vitis vinifera* L.	Morel 1944, 1948
Pseudoperonospora humuli (Miy. & Tak.) Wilson.	*Humulus lupulus* L.	Griffin & Coley-Smith 1968
Sclerospora graminicola (Sacc.) Schroet.	*Pennisetum typhoides* Stapf. & Hubb.	Tiwari & Arya 1967, 1969, Arya & Tiwari 1969
Albugo ipomoeae-panduratae	*Ipomoeae pentaphylla*	Singh 1963, 1966
CLASS ASCOMYCETES Order Erysiphales (Powdery mildews) *Erysiphe cichoracearum* D.C. *Unicula necator* (Schw.) Burr.	*Helianthus annuus* L. *Vitis vinifera* L.	Heim & Gries 1953 Morel 1948
CLASS BASIDIOMYCETES Order Uredinales (the Rusts) *Puccinia antirrhini* Diet. & Holw.	*Antirrhinum* cultivars	Rossetti & Morel 1958
Puccinia helianthi Schw.	*Helianthus annuus* L.	Nozzolillo & Craigie 1960

Fungus	Host	References
Puccinia minutissima Arth.	*Decodon verticillatus* (L.) Ell.	Rossetti & Morel 1958
Puccinia tatarica Tranzsch.	*Lactuca tatarica* (L.) C.A.Meyer	Bauch & Simon 1957, Witkowski & Grümmer 1960
Gymnosporangium juniperae-virginianae Schw.	*Juniperus virginiana* L. *Pyrus* spp.	Hotson & Cutter 1951, Hotson 1953, Rossetti & Morel 1958, Cutter 1959
Gymnosporangium spp.	*Juniperus* spp.	Constabel 1957
Melampsora lini (Pers.) Lev.	*Linum usitatissimum* L.	Turel & Ledingham 1957, Turel 1969
Cronartium ribicola J. C. Fisch. ex. Rabenh.	*Pinus monticola* Dougl.	Koenigs 1968, Harvey & Grasham 1969a, b and c, 1970a, b, c and d, 1971a, b, Harvey, Grasham & Waldron 1971, Harvey & Woo 1971
Uromyces ari-triphylli (Schw.) Sieler	*Arisaema triphyllum* (L.) Schott	Cutter 1951, 1960a

contaminant-free inoculum by first surface sterilizing infected vine leaves and then allowing the fungus to grow through the leaf surface and sporulate. Sporangia were removed from the leaves with sterile distilled water and allowed to form zoospores, which were used to inoculate established callus. Similar techniques have been used with other downy mildew fungi, by Nakamura (1965) and Griffin & Coley-Smith (1968). A modification of the surface sterilization procedure involving detached infected cotyledons has proved useful for obtaining contaminant-free inoculum of *Peronospora parasitica*, *P. farinosa* and *Bremia lactucae*, and could be successfully used with many other downy mildew fungi (Ingram 1969c, Ingram & Joachim 1971).

Surface sterilization-sporulation procedures have also been useful in securing contaminant-free uredial inocula of rust fungi (Williams, Scott & Kuhl 1966, Maheshwari, Hildebrandt & Allen 1967, Ingram & Tommerup 1973). Infected host leaves were surface sterilized immediately prior to the bursting of uredospore pustules and screened for contaminants on a salts + sucrose medium. Contaminant-free spores were then removed from the pustules either with the tip of a sterile scalpel or with a camel-hair brush. The problem of obtaining contaminant-free inoculum of powdery mildew fungi is more difficult because of the superficial habit of their mycelium. Morel (1948) removed conidia of *Uncinula necator* from non-sterile infected vine leaves with a pipette, and used these to inoculate large numbers of vine calluses. This method was obviously unsatisfactory because of the high contamination rate, although a small number of successful cultures was eventually obtained. Heim & Gries (1953) used a more refined technique with *Erysiphe cichoracearum* and *Helianthus annuus* (sunflower). They inoculated aseptic seedlings of *H. annuus* with non-sterile conidia and allowed infection and sporulation to occur. Conidia produced on these plants were then successively transferred to other aseptic plants until, after several transfers, mildew suitable for the inoculation of callus was obtained. Surface sterilization-sporulation procedures have not yet been used with powdery mildew fungi, although it is likely that they would be successful.

Infection of host cultures

Successful infection of callus has been achieved with several downy mildew fungi simply by placing contaminant-free spores or mycelium directly onto the tissue surface (e.g. Morel 1944, 1948, Izard, Lacharpagne & Schiltz 1964, Nakamura 1965, Griffin & Coley-Smith 1968, Tiwari & Arya 1967, Ingram 1969c, Ingram & Joachim 1971). Direct infection of host callus by spores of powdery mildew fungi has also been possible (Morel 1948, Heim & Gries 1953) although Mence & Hildebrandt (1966) could not infect rose callus with spores of *Sphaerotheca pannosa*.

In the case of the rust fungi direct infection of calluses has proved more difficult to achieve. Morel (1948) attempted to produce dual cultures by inoculating appropriate host calluses with uredospores and basidiospores; these attempts were unsuccessful, partly because germ tubes were apparently unable to penetrate the cells. Similarly, Rossetti & Morel (1958) were unable to infect either *Antirrhinum* callus, or detached *Antirrhinum* stem segments with the epidermis removed, with uredospores of *Puccinia antirrhini*, although they were able to infect stem segments which possessed an epidermis. They suggested that the epidermis and cuticle play an essential role in the infection of host tissues by uredospores, perhaps by inducing the differentiation of germ tubes into infection structures (appressoria, vesicles and infection hyphae). This possibility was strongly supported by the observation of Dickinson (1949) that although uredospores of *Puccinia recondita* f.sp. *tritici* would germinate on wheat leaves from which the epidermis had been removed, they would not infect. Infection did occur, however, in intact leaves or in stripped leaves provided with an artificial collodian membrane in place of an epidermis. In contrast, some other workers have been able to infect exposed cereal leaf mesophyll cells with uredospores of rust fungi (e.g. Chakravarti 1966). In 1967, Maheshwari, Hildebrandt and Allen showed that callus of the sunflower (*Helianthus annuus*) could not be infected with aseptic uredospores of *Puccinia helianthi*, although 20 to 30% of the spores produced germ tubes on the callus surface. Moreover, they were unable to infect calluses with uredo-spores by overlaying the tissue with pieces of host cuticle or with pieces of paraffin-oil-nitrocellulose membrane on which uredospores had been allowed to form infection structures. Other workers, including Turel & Ledingham (1957) and Nozzolillo & Craigie (1960), have experienced difficulty in infecting callus tissues with rust fungi, although Hotson & Cutter (1951) and Cutter (1959, 1960a) infected appropriate host calluses with homogenized mycelium of *Gymnosporangium juniperi-virginianae* and *Uromyces ari-triphylli*, and Harvey & Grasham (1969b, 1970c) infected callus of *Pinus monticola* with aerial mycelium and basidiospores of *Cronartium ribicola*. It is quite clear that the importance of differentiation of the host into cuticle, epidermis and other organized tissues, and of the fungal germ tube into infection structures, needs to be investigated further. Experiments where culture media are varied to alter the nature of the callus surface, or where natural or artificial membranes are used in place of a differentiated epidermis, might be useful in this context, as might experiments where infection of tissue cultures in various states of organ differentiation is attempted.

Other factors apart from the absence of an epidermis and cuticle may be important in preventing the infection of callus cultures by obligate parasites. Maheshwari, Hildebrandt & Allen (1967) partly attributed their failure to infect callus of *Antirrhinum majus* with uredospores of *Puccinia antirrhini* to inhibition of germination by an abnormal extracellular metabolite of the callus tissues which could not be detected in the mesophyll tissues of the intact plant. Similarly Mason (personal communication) showed that although callus derived from a susceptible lettuce variety became infected with the downy mildew fungus *Bremia lactucae*, the lesion failed to develop because the cells produced an abnormal fungitoxic metabolite in response to the infection. Saad and Boone (1971) found that apple callus produced an abnormal metabolite which was fungistatic to *Venturia inequalis*, and there are in the literature many other reports of callus cultures producing antimicrobial and other metabolites which are not detectable in the intact plant (e.g. Campbell, Chan & Barker 1965, Mathes 1963, 1967, Khanna & Staba 1968, Furuya 1968, Butcher & Connolly, 1971).

Morel's failure to infect callus tissues with rust spores (Morel 1948) prompted him to suggest that better results might be achieved by initiating dual cultures directly from explants of systemically infected plants. This method was used by Hotson & Cutter (1951) who, despite a very high contamination rate, were able to establish a few dual cultures of the heteroecious rust *Gymnosporangium juniperi-virginianae* and *Juniperus virginiana*, its telial host, by incubating explants of surface sterilized infected gall tissue on callus culture media. The callus tissue arising from the explants contained systemic mycelium and could in some instances be subcultured to give stable lines, although very often the fungus died out with repeated transfer. Cutter (1960a) used similar techniques to obtain stable dual cultures of *Uromyces ari-triphylli* and *Arisaema triphyllum* callus, and the technique has since been used successfully with other fungi, notably *Plasmodiophora brassicae* and *Brassica* spp. (Williams, Reddy & Strandberg 1969, Ingram 1969a) and *Cronartium ribicola* and *Pinus monticola* (Koenigs 1968, Harvey & Grasham 1969a).

The maintenance of stable lines

Although a number of dual cultures of obligate parasites and their hosts have now been initiated (Table 14.1), it has been possible to establish stable lines capable of unlimited growth with only a few species because of the difficulty of maintaining a balance between the growth rates of the parasite and of the host tissues. Under natural conditions obligate parasites usually cause slowly developing damage to their hosts (Brian 1967) although in most cases diseased organs are eventually killed; in culture this balance may be tipped in favour of the host or the parasite. For example, although Turel & Ledingham (1957) obtained abundant aerial mycelium of *Melampsora lini* on flax leaves and on callus cultures, the mycelium became sparse and was eventually left behind when callus growth was vigorous. In contrast, Ingram (1969c) found that the downy mildew fungus *Peronospora parasitica* when growing in Brassica tissue cultures often behaved like an aggressive parasite, forming both inter- and intra-cellular hyphae, resulting in the death of calluses within a few days of infection. The kind of balance existing in natural conditions was more closely approximated by Morel's cultures of *Plasmopara viticola* and *Vitis vinifera* callus, where the mycelium was mainly intercellular and apparently normal haustoria were produced. Dual growth, with production of viable sporangia, lasted for 4 to 6 weeks, but after this time the callus tissues were killed (Morel 1944, 1948). In such cases infected cultures may be perpetuated by inoculating fresh parasite-free calluses with pieces of moribund infected tissue at periodic intervals (e.g. Griffin & Coley-Smith 1968, Ingram 1969c). Even more balanced equilibrium between host and parasite may be obtained in culture with some species, so that the host callus is never killed completely. Hotson & Cutter (1951), Hotson (1953) and Cutter (1951, 1959) obtained such growth with their dual cultures of the rust *Gymnosporangium juniperi-virginianae* and gall callus of *Juniperus virginiana*; abundant aerial mycelium was produced, although teleutospores were formed only occasionally. The equilibrium broke down only in those cases where saprophytic growth of the mycelium was about to occur (see below). Other workers have been less successful in maintaining dual cultures of *Gymnosporangium* and its hosts (Constabel 1957, Rossetti & Morel 1958, Maheshwari *et al.* 1967). Cutter (1960a) obtained good dual growth, but only limited sporulation, with another rust fungus, *Uromyces ari-triphylli*, and callus of the monocotyledon *Arisaema triphyllum*. Recently Koenigs (1968) and Harvey & Grasham (1969a) have grown balanced dual cultures of the heteroecious rust fungus *Cronartium ribicola* and callus of the aecial host *Pinus monticola*. Intercellular hyphae with apparently normal haustoria were formed in the callus, and in some cultures pycnia and aecia-like sori were produced, although aeciospores were not found. In one series of experiments Harvey, Grasham & Waldron (1971) showed that increasing concentrations of auxins in the culture medium enhanced growth of both the *P. monticola* callus and the fungus. Kinetin did not affect growth at low levels, but it did increase the longevity of both host and parasite, and in combination with high auxin led to formation of aecia-like structures. Gibberellic acid up to 1·0 mg/litre medium did not affect growth of the host tissues, but did inhibit growth of the fungus.

Two downy mildew fungi which cause morphological disturbance of host tissues have been a source of balanced dual cultures, namely *Sclerospora graminicola* which causes flower proliferation in the green ear of *Pennisetum typhoides* (Millet) (Tiwari & Arya 1967, Arya & Tiwari 1969), and *Peronospora farinosa* which causes thickening and distortion in the leaves of *Beta vulgaris* (sugar beet) (Ingram & Joachim 1971). Both fungi formed considerable aerial mycelium with viable and pathogenic conidia on host callus tissues (Fig. 14.1C), and occasionally produced oospores. Host callus growth was apparently not reduced, intercellular fungal growth was sparse and haustoria, when formed, caused little cell damage.

Perhaps the most studied balanced dual system so far established is that of *Plasmodiophora brassicae* and host *Brassica* species. *Plasmodiophora brassicae* is the causal organism of clubroot disease of Crucifers, and during the major part of its life history takes the form of multinucleate plasmodia inhabiting the cytoplasm of the living cells of the stele and cortex of the host root (Ingram & Tommerup 1972). These cells are stimulated to divide and enlarge abnormally in response to the parasite and a root gall (the club) is the result. Callus cultures have been readily established from explants of surface sterilized clubs on a variety of media, both complex and defined (Williams, Reddy & Strandberg 1969, Ingram 1969a). On media containing NAA and kinetin the parasite plasmodia, which are apparently structurally and physiologically normal (Williams & Yukawa 1967, Williams *et al.* 1969), produced viable resting spores which germinated to produce other stages of the life history in culture (Tommerup & Ingram 1971), but on media containing 2,4-D and coconut milk the parasite life cycle was arrested at the vegetative plasmodial stage and fungus and host cells lived in complete harmony (Ingram 1969a). Up to 35% of the callus cells contained one or more plasmodia, which divided synchronously with them (Fig. 14.1A, B).

Recent studies by Ingram (1969b), Williams, Reddy & Strandberg (1969), Reddy & Williams (1970) and Dekhuijzen & Overeem (1971) have shown that *Brassica* callus tissues infected with plasmodia of *Plasmodiophora brassicae* grow more rapidly than non-infected tissues on media containing kinetin and NAA; doubling times of about 5 days for non-infected tissues and about 3 days for infected tissues were quoted. Comparable growth of infected and

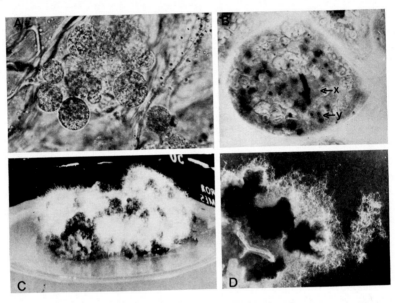

FIG. 14.1   *Fungal obligate parasites growing in tissue culture.*
A. Living plasmodia of *Plasmodiophora brassicae* released from the cells of infected turnip callus tissue (after Ingram 1969a).
B. Nuclear division in a turnip callus cell infected with a multinucleate plasmodium of *P. brassicae*. The host nucleus has been labelled *x* and a group of plasmodial nuclei *y*. (Tommerup & Ingram unpublished).
C. A sugar-beet callus heavily infected with mycelium of *Peronospora farinosa* (after Ingram & Joachim 1971).
D. Growth of *P. farinosa* away from parent sugar-beet callus on the surface of a chemically defined culture medium (petri dish photographed from the underside). After Ingram & Joachim (1971).

non-infected tissues was only obtained when the medium contained high levels of 2,4-D and coconut milk (Ingram 1969b). It was also shown that infected tissues, unlike non-infected tissues, were independent of an exogenous supply of cytokinin, and substances having cytokinin-like activity were extracted from infected calluses grown on kinetin-free media (Reddy & Williams 1970, Dekhuijzen & Overeem 1971). Moreover, there was some evidence that infected calluses, unlike non-infected calluses, could grow very slowly even in the absence of an exogenous supply of both cytokinin and auxin (Ingram 1969b, Dekhuijzen & Overeem 1971). It is not yet known whether growth substances are synthesized by the parasite, or whether their formation is stimulated by its presence. It is quite clear, even from the small amount of evidence available, that the auxin and cytokinin components of culture media, together with other chemical constituents and physical factors such as light and temperature, may have very marked effects upon the growth and sporulation of obligate parasites grown on host tissue cultures. With the possible exceptions of work with *P. brassicae* and its host, and *Melampsora lini* and flax callus (Turel & Ledingham 1957) the effects of such factors on dual growth have, however, not been widely investigated.

Experimental studies with dual cultures

Apart from investigations concerned with the effect of obligate parasites on the hormonal control of host tissue growth, dual cultures have as yet been little used in studies of host-parasite interaction (Ingram & Tommerup 1973) although they have been used in other ways. For example, Tommerup & Ingram (1971) and Ingram & Tommerup (1972a) used infected *Brassica napus* tissue cultures in their cytological studies of *Plasmodiophora brassicae* and as a result were able to elucidate certain stages of the life history of the organism as it occurs in intact plants. Some physiological experiments with contaminant-free isolated mycelium or spores obtained from dual cultures have also been done: Morel (1946, 1947a) used sporangia of *Plasmopara viticola* from vine callus in fungicide tests; Ingram & Joachim (unpublished data) compared conidia of *Peronospora farinosa* from beet callus with conidia from beet plants in spore germination and inhibitor tests; Turel & Ledingham (1957), Mitchell & Shaw (1968) and Williams & Shaw (1968) used excised aseptic mycelium of *Melampsora lini* from flax leaf cultures in radioactive feeding experiments, and Keen, Reddy & Williams (1969) used plasmodia of *Plasmodiophora brassicae* isolated from callus cultures in a similar way. However, the greatest potential value of dual cultures probably lies in axenic culture studies. There have been many attempts to grow fungal obligate parasites away from host tissues either by seeding culture media with contaminant-free spores, or by growing dual cultures of host tissues and parasite in the hope that saprophytic mycelium will arise from them and become established in the tissue culture medium. Axenic culture attempts have now been successful with a number of rust fungi (Uredinales), and in many such instances dual cultures have been of direct value.

The first saprophytic cultures of a rust fungus were obtained from dual cultures of *Gymnosporangium juniperi-virginianae* and *Juniperus virginiana* (Hotson & Cutter 1951, Cutter 1951, Cutter 1959). In one out of many hundreds of isolates of infected telial gall callus, and in six of the many subcultures of this dual culture, mycelium grew out on to the surrounding medium and became established independently of the host tissues. Emergence of the saprophytic hyphae was preceded by production of a pink mycelium on the callus surface and a necrotic reaction by the tissues themselves reminiscent of the effect of a destructive parasite. Cutter suggested that some change in the host callus tissues selected for variant fungal nuclei with the potential for saprophytic growth. Some of the axenic isolates could be inoculated into callus or plants of *Pyrus* spp., the alternate host of *Gymnosporangium juniperi-virginianae*, and typical aecia were produced by some strains and pycnia by others. However, callus and plants of *Crataegus* spp., the alternate host of *Gymnosporangium globosum*, could also be infected, and aecia typical of *Crataegus globosum* were produced (Hotson & Cutter 1951). This observation led to some uncertainty concerning the correct identification of the saprophytic isolates as *Gymnosporangium juniperi-virginianae*, although Cutter (1959) advanced strong evidence in support of his original identification. On reinoculation into host tissues all axenic isolates of *G. juniperi-virginianae* lost their power of saprophytic growth and Cutter suggested that this may have been due to a reverse mutation.

Cutter grew other rust fungi saprophytically; in 1960 he reported saprophytic hyphal growth arising from five dual culture isolates of *Uromyces ari-triphylli* and *Arisaema triphyllum* callus (Cutter 1960a), and in the same year he obtained a saprophytic colony of *Puccinia malvacearum* from a basidiospore arising from infected callus of *Althea* (Cutter 1960b). Unfortunately, other workers were not able to obtain axenic cultures of *Gymnosporangium* and other rust fungi from dual cultures (e.g. Rossetti & Morel 1958, Constabel 1957, Maheshwari *et al.* 1967), and the work of Hotson and Cutter remained the subject of controversy for a number of years (Brian 1967) until Williams, Scott & Kuhl (1966) and Williams, Scott, Kuhl & Maclean (1967) obtained vegetative colonies of *Puccinia graminis tritici* from uredospores on a simple medium supplemented with yeast extract and Evans' peptone. This and subsequent axenic cultures of rust fungi have mostly been obtained by seeding culture media with contaminant-free uredospores obtained from surface sterilized infected host leaves (see Scott & Maclean 1969, Ingram & Tommerup 1972b). However, Turel (1969) grew axenic cultures of *Melampsora lini* using uredospores from flax leaf tissue cultures, and Harvey and Grasham (1970a, d) had limited success with attempts to obtain axenic cultures of *Cronartium ribicola* from dual cultures of this fungus and *Pinus monticola*. In one series of experiments (Harvey & Grasham 1970a) they transferred aerial mycelium of *Cronartium ribicola* from a dual culture to the surface of a membrane placed over a non-infected callus of *Pinus monticola*. After 90 days of incubation extensive mycelial growth had developed from the separated inoculum and it was concluded that *Cronartium ribicola* was able to grow saprophytically on substances diffusing from host tissues. In later experiments Harvey & Grasham (1971a, b) inoculated tissue cultures of the non-host *Pseudotsuga menziesii* with mycelium of *Cronartium ribicola*, and also grew the fungus on a membrane separating it from callus of this species. In another experiment (Harvey & Grasham 1970d) limited saprophytic growth of *C. ribicola* mycelium on a defined medium which had formerly supported tissue cultures of *Pinus monticola* was obtained, but it was not possible to establish saprophytic cultures capable of unlimited transfer on artificial media.

Axenic cultures of powdery mildew fungi (Erysiphales) have not yet been obtained, but tissue culture studies have yielded promising results with certain of the downy mildew fungi (Peronosporales). It has been observed that some members of this group are sufficiently independent of host metabolism to grow out for short distances from host tissue cultures on the surface of their culture media. Such a phenomenon was observed by Griffin & Coley-Smith (1968) in their experiments with *Pseudoperonospora humuli*. Similarly, mycelium of *Peronospora farinosa* grew for distances of 1·5 to 2·0 cm out from infected sugar-beet calluses on the surface of a modified Murashige & Skoog (1962) medium (Ingram & Joachim 1971). In some instances peninsulas of mycelium with conidiophores and conidia were formed (Fig. 14.1D) and microscopic examination showed that these peninsulas were often connected to the parent dual culture by only a single hypha; unfortunately the isolated mycelium died when this hypha was cut. In other experiments Tiwari & Arya (1969) obtained sporulation and axenic growth of *Sclerospora graminicola* mycelium arising from infected callus of *Pennisetum typhoides*, and this survived two transfers in the absence of living host tissues. It is probable that by supplementing culture media with additional nutrients, or even mutagens, or by using host callus tissues or axenic cultures of rust fungi or of less specialized oomycetes as nurse cultures. it will eventually be possible to grow some of the downy mildew fungi saprophytically.

Disease resistance

Although tissue cultures would appear to have many advantages for the study of resistance to pathogenic fungi in plants, the limited number of experiments so far carried out have yielded confusing and sometimes contradictory results. One interaction that has received particular attention involves the potato blight fungus *Phytophthora infestans* and its host *Solanum tuberosum* (potato). It is known from a number of studies that hypersensitive resistance of *Solanum* spp. to *Phytophthora infestans* is controlled by a series of major genes called R-genes, designated R1, R2, etc.; each gene confers resistance to some, but not all, races of *P. infestans*. It is clear that the operation of these R-genes is associated with biochemical changes occurring in the host following infection. Ingram & Robertson (1965) and Ingram (1967) investigated the expression of R-genes in potato tissue cultures using the variety Majestic (rr), which contains no R-genes, and the variety Orion which as an intact plant contains the R1 gene conferring resistance in both tubers and leaves to race 4, but not race 1, of *P. infestans*. Tissue culture aggregates of var. Majestic were readily invaded by *P. infestans* race 4 on agar plates, but the tissue aggregates of var. Orion and of other R-gene resistant *Solanum* lines were not invaded, thus behaving in a similar way in response to *Phytophthora infestans* as the intact plants from which they were derived. Moreover, tissue aggregates of var. Orion supported growth of *P. infestans* race 1, to which the intact Orion plant is susceptible. It was shown that toxicity to *P. infestans* zoospores was developed in var. Orion, but not in var. Majestic, tissue suspension cultures in response to infection with race 4. The development of toxicity was apparently rather slow, but it was suggested that this may have been due to an inability to detect the toxic principle in bioassays until it had been produced in considerable quantities, because the ratio between the volume of the tissue culture suspension medium and the infected tissues was very high. Ingram (1967) concluded that R-genes may be expressed in potato tissue cultures and that the failure of the var. Orion tissue aggregates to support growth of *P. infestans* race 4 may have been due to an expression of the R1 gene operating through the development of post infection toxicity.

Later Robertson, Friend, Aveyard, Brown, Huffee & Homans (1968) extracted into ether the toxic materials produced in var. Orion tissue cultures following infection, and identified salicylic, vanillic and *p*-hydroxybenzoic acids. However, Friend (1973) has reported that when similar experiments were done with potato tuber slices it was not possible to detect any hydroxybenzoic acids, although traces were found in intact leaf tissues. After hydrolysis of bound phenolic compounds in tubers, only cinnamic acids were found, although some benzoic acids were found in leaf hydrolysates. Friend concluded that the results for tissue cultures did not necessarily apply to tubers and leaves. There is evidence that sesquiterpenoid compounds (e.g. rishitin) may be important in the expression of some R-genes in potato (Katsui, Murai, Tagasuki, Imaizumi & Masamune 1968, Sato, Tomiyama, Katsui & Masamune 1968, Tomiyama, Sakuma, Ishizaka, Sato, Katsui, Tagasuki & Masamune 1968, Varns, Kuc & Williams 1971) but Friend suggested that this may not be so in var. Orion, where other compounds may be involved. Thus the mechanism of R-gene resistance in the potato is still not fully understood, and it is not yet possible to assess the validity of resistance reactions expressed in tissue cultures.

Certain other studies of resistance have also been carried out. Warren &

Routley (1970) showed that a major gene for resistance to *P. infestans* was partially expressed in undifferentiated tomato callus, and Ah-Sun (1970) obtained good correlation between intact potato tubers and tissue aggregate suspension cultures in their resistance and susceptibility to four plant pathogenic fungi. In contrast, Ingram (1969c) found little correlation between Brassica plants and tissue cultures derived from them in their interactions with *Peronospora parasitica*, and Ingram & Tommerup (1973) showed that only some tissue culture isolates from a sugar beet variety having a major gene for resistance to *P. farinosa* actually excluded the fungus, while many others supported prolific mycelial growth; many callus isolates from a susceptible line of sugar beet were immune to *P. farinosa* and only some were susceptible. Bailey (1970) showed that although callus cultures of *Pisum sativum* (pea) produced the phytoalexin pisatin, this occurred in the absence of non-pathogenic fungi; the agent responsible for stimulating such abnormal production was the coconut milk contained in the culture medium. After continued growth and establishment of fast-growing callus, the capacity of the pea tissues to synthesize pisatin decreased and, since pisatin was shown to be inhibitory to pea callus growth, it was concluded that the loss of ability to produce the compound may have been due to selection of fast-growing tissue at the time of transfer of cultures. Keen & Horsch (1972) have recently shown that another phytoalexin, the pterocarpan hydroxyphaseollin, was not produced normally in soybean callus challenged with races of *Phytophthora megasperma* var. *sojae*.

The reasons for the confused and contradictory results obtained from studies of resistance to fungal infection in plant tissue culture may lie in the lack of genetical uniformity in the plants used, and in the nature of the cultured tissues themselves. In contrast to animal tissue cultures, where the cells retain a good deal of their individual identity, plant tissue cultures usually form undifferentiated suspensions or masses of callus which lack chlorophyll and are often in a state of constant meristematic activity: their secondary metabolism may be very different from that of the plants from which they are derived (e.g. Butcher & Connolly 1971) and may vary from medium to medium. Although some of these differences may be temporary, it is likely that many are due to permanent genetic changes such as the loss or addition of chromosomes and the development of mutations (e.g. Ingram & Tommerup 1973, Heinz & Mee 1971). The populations of such altered cells might be expected to increase rapidly in culture as a result of artificial selection for fast growing or other types of tissue at the time of transfer to fresh media, and this would inevitably be reflected in the reaction of tissues to infection (for further discussion of nuclear changes occurring in tissue cultures see Chapter 7).

Many more comparative studies are necessary, but the findings from resistance studies so far suggest that in their interactions with fungi tissue cultures may be very different from the plants from which they are derived, and results obtained using tissue culture techniques may be very misleading, and should be treated with considerable caution. Clearly the nature and extent of the differences between tissue cultures and intact plants in their interactions with parasites must be investigated further; the recent work of Helgeson, Kemp, Haberlach & Maxwell (1972) may be particularly important in this context. They are studying the expression in tobacco tissue cultures of major genes for resistance to *Phytophthora parasitica* var. *nicotianae*, cause of black shank disease. Tobacco tissue cultures are ideal experimental tools for they have for many years been widely used in growth substance assays and in studies of differentiation and metabolism, while many of the metabolic differences between cultured and normal tissues have been investigated and are at least partially understood. A completely defined culture medium has been developed and tissue growth and differentiation can be closely controlled by changes in the concentrations of carbohydrates and growth substances. The genetical make up of both tobacco and *P. parasitica* is well understood and can be controlled. Preliminary studies involving two races of *P. parasitica* and tissue cultures from near isogenic lines of resistant and susceptible tobacco plants have shown that resistance is at least partially retained in culture, and that the degree of gene expression can be controlled by changes in the growth substance balance of the culture medium or in the incubation temperature. Moreover, normally resistant and susceptible shoots of tobacco can be regenerated from the tissue cultures. These results appear to be very promising, and it will be interesting to see whether they will lead to an elucidation of the mechanism of major gene resistance in the intact tobacco plant. The ability to control gene expression in tissue culture by changes in the cultural environment could be very useful in future investigations.

VIRUSES

Since viruses cannot be grown in the absence of living cells and tissues it is not surprising that considerable effort should have been directed towards perfecting methods for growing them in host tissue cultures. Animal tissue culture techniques have been of great value in studies of animal viruses and are regularly used in the commercial production of certain vaccines such as those against poliomyelitis, measles and adenovirus (Rapp & Melnick 1966). In sharp contrast, the contribution of plant tissue culture techniques to the study of plant-virus physiology has been very limited, although the technique of apical meristem culture is of great practical value in the production of virus-free stocks of economically important plants (Hollings 1965). There are many possible reasons why tissue culture techniques should have contributed so little to plant virology, but two are particularly important. First, it has until recently been very difficult to obtain synchronous and consistently good infection of large numbers of cultured plant cells with virus, and secondly, even when dual cultures have been established, the virus titre has frequently been very low and has often decreased still further with increasing time in culture. Recently Black (1969) has turned to insect vector tissue cultures as alternatives to plant tissue cultures for studies of plant viruses.

Infection of cell and tissue cultures

The procedures used for obtaining contaminant-free virus inoculum suitable for infection of plant cell and tissue cultures are relatively straightforward. After extraction from infected leaves or tissues the virus suspension may be purified by centrifugation and then freed of contaminant micro-organisms by filtration. For example, Murakishi, Hartmann, Pelcher & Beachy (1970) prepared tobacco mosaic virus (TMV) by centrifuging for three cycles of low speed (12,000 × g) for 15 minutes and high speed (62,000 × g) for 1 hour; the virus was then suspended in 0·1 M phosphate buffer at pH 7·3 and filtered through an ultrafine sintered glass filter (pore size approx. 1·0 μm). Millipore or other membrane filters of suitable pore size may also be used.

Much of the experimental work on the infection of cultured plant cells and tissues has been concerned with tobacco mosaic virus (TMV) and various *Nicotiana* species (tobacco). Most workers have found it difficult to obtain more than very low levels of infection of cultured cells and tissues, with correspondingly low virus titres, often less than 5% of those in intact leaves. Morel (1948) infected tobacco callus cultures with TMV by rubbing them with an aseptic virus inoculum containing an abrasive powder, a technique similar to that used with intact leaves, while Kassanis, Tinsley & Quak (1958) infected tobacco callus cultures by pricking them with a fine needle dipped in TMV inoculum. Kassanis (1967) suggested that this method was better because internal cells became infected, thus increasing the likelihood of obtaining infected progenies when the inoculated tissues were subcultured. Micro-injection methods of inoculation have been used by Hirth & Lebeurier (1965) and by Nims, Halliwell & Rosberg (1967). Segretain (1943) and Kassanis *et al.* (1958) claimed to have infected unwounded tobacco callus cultures with TMV by pouring inoculum over established cultures growing on agar medium, and other workers (Bergmann & Melchers 1959, Kassanis *et al.* 1958, Wu, Hildebrandt & Riker 1960, Quak 1965) obtained various degrees of infection with TMV by gently shaking unwounded cell and tissue suspensions in liquid medium containing the virus inoculum. Kassanis *et al.* (1958) pointed out, however, that the possibility of making small wounds in the cells of tissue clumps which were being broken up even by gentle agitation could not be ruled out. Wu, Hildebrandt & Riker (1960) claimed that young tobacco tissue cultures (mainly meristematic) were more resistant to infection by and subsequent multiplication of TMV than were older cultures composed mainly of enlarging and senescent cells.

Recently Murakishi, Hartmann, Pelcher & Beachy (1970) have described a method of inoculating tobacco tissue cultures with TMV which results in a high virus titre and crystal formation, and which makes possible synchronous inoculation of large numbers of cells. Cell aggregates and single cells of tobacco (*N. tabacum*) were transferred to large tubes containing liquid medium and a contaminant-free TMV suspension. The tubes were then vibrated vigorously (800 rpm) for 20 seconds on a Vortex mixer which gave a rapid whirling motion. The cells were finally washed, and transferred to solid medium in petri dishes. Inclusions of TMV were first seen in the infected cells after 44 hours, although maximum numbers of cells with inclusions were not seen until 6 to 8 days. In one experiment as many as 94% of cells were recorded as having inclusions at 7 days, while in other experiments infection levels of 76, 74 and 45% were obtained. The inclusions from the cultured cells were comparable with those from tobacco leaf hair cells, and were verified as TMV by inoculation back into intact leaves. In a further series of experiments (Murakishi, Hartmann, Beachy & Pelcher 1971) these authors repeated their previous findings, and suggested that the mode of entry of the virus particles into the mechanically dispersed cells may have been through the exposed ends of ruptured plasmodesmata (Spencer & Kimmins 1969, have recently demonstrated the existence of plasmodesmata in tobacco tissue cultures), or through other small non-lethal injuries caused during agitation (Murakishi 1968), and discounted the possibility of uptake by pinocytosis, such as may occur with isolated protoplasts (Cocking 1970). Cells grown on agar before inoculation were less satisfactory than cells grown in suspension, although incubation of cells on agar after infection gave a higher virus titre and yield than cells incubated in suspension culture; this may have been associated with drying of the virus suspension on the cell surfaces. The growth curve of TMV in cell suspensions was similar to the increment curve from tobacco leaves; in cell suspensions an initial drop in virus titre was noted 10 hours after inoculation, and this may have been due to uncoating of virus particles resulting in free viral nucleic acid. Virus yields from inoculated cells incubated on agar for 7 days were equal to those obtained from the leaves of intact tobacco plants.

In 1967 Kassanis suggested that the difficulty of inoculating cultured plant cells with virus might be overcome by use of isolated protoplasts. This would enable infection to occur in the absence of a wall barrier and would allow synchronous inoculation of large numbers of individuals. Techniques for the isolation and manipulation of plant cell protoplasts are now well advanced; protoplasts can be isolated from fruit and tissue culture cells and from leaf mesophyll cells, and can be induced to regenerate walls, (see Chapter 5, and Cocking 1970). Evidence that tomato fruit protoplasts could be infected by TMV was first reported by Cocking as long ago as 1966, while Cocking & Pojnar (1969) showed electron microscopically that the processes probably occurred pinocytotically into vesicles in the protoplast cytoplasm. Following disappearance of the virus from the vesicles and the regeneration of the isolated protoplasts into cells the virus appeared in the cytoplasm as aggregates characteristic of TMV infection. The process of virus uptake by isolated fruit protoplasts has been reviewed by Cocking (1970). Protoplasts isolated from mesophyll cells can also be infected with virus. Aoki & Takebe (1969) infected isolated tobacco mesophyll protoplasts with infectious RNA from TMV, and showed that a synchronous multiplication of TMV occurred. Subsequently Takebe & Otsuki (1969) were able to obtain infection of up to 30% of mesophyll protoplasts by intact TMV, and in later experiments (Takebe, Otsuki & Aoki 1971) even greater success was claimed. Cocking's earlier findings on the mechanism of virus uptake were supported by the observation that infection of mesophyll protoplasts was enhanced by poly-L-ornithine which is known to enhance strongly the pinocytotic uptake of proteins by cultured animal cells. Following infection the TMV multiplied in the infected protoplasts and reached a level of 10^6 particles per protoplast in 24 hours. Prior to this built-up a drop in virus levels was observed which may have reflected uncoating of the virus particles (see Cocking 1970). Now that efficient methods for the infection of both cultured plant cells and isolated protoplasts are available it should be possible to use tissue culture systems to greater advantage in studies of the penetration and subsequent multiplication of viruses in plants, and to determine more precisely their effects on and interaction with cells at the molecular level.

Infection by vectors

In nature many viruses are known to be transmitted to host plants by vectors such as nematodes, aphids, leaf hoppers and fungi, but these agents have been little used in infection studies with tissue cultures, possibly because of the problems of maintaining sterility. In 1964 Mitsuhashi and Maramorosch transmitted Aster yellows virus (AYV) from aseptic China aster plants (*Callistephus chinensis* Nees.) to carrot tissue cultures by means of aseptically reared leafhoppers (*Macrosteles fascifrons* Stal.), and Kassanis & Macfarlane (1964) were able to transmit tobacco necrosis virus (TNV) to callus tissues of tobacco and *Parthenocissus tricuspidata* via the zoospores of the fungal obligate parasite *Olpidium brassicae*. The tissue cultures were first transferred

O

to glass vials and then covered with a nutrient solution containing zoospores of the fungus, together with a suspension of purified virus particles. It was found that much lower levels of virus inoculum were needed to infect the tobacco callus in the presence of *O. brassicae* than were needed to obtain infection by mechanical means. These experiments could, however, not be continued because of the great difficulty in freeing *O. brassicae* of unwanted contaminants. It is clear from these two studies that much interesting work on the infection of plant tissue cultures by vectors is possible, if problems of sterility can be overcome.

The growth of dual cultures

Because of the difficulties of obtaining infection, much of the research on the growth of virus infested tissue cultures has been carried out on isolates obtained from systemically infected plants. As with infection studies, the most widely used combination has been *Nicotiana* species (tobacco) and TMV. In contrast to the situation in animal tissue cultures, where lysis and plaque formation may follow infection, the presence of virus particles in plant cells and tissues usually has little effect on their growth, although some changes in growth rate have been recorded, and a local lesion reaction to TMV in tobacco tissue cultures has been demonstrated by Beachy & Murakishi (1971).

The concentration of virus in artificially infected tissue and cell cultures is usually very low, probably due in part to the inefficiency of infection techniques. However, in cultures derived from systemically infected plants virus titres may also be very low indeed. For example, Morel (1948) showed that the titres of potato viruses X and Y were lower in cultured potato cells than in intact plants, and Augier de Montgremier and Morel (1948) showed that callus tissues from crown gall infected with TMV had a virus titre thirty to forty times lower than infected leaves of intact plants. Similarly Kassanis (1957a) showed that in infected tobacco crown-gall cultures, obtained from Morel, the TMV level was about one-thirtieth of that usual in sap from infected tobacco leaves. Even when virus levels are relatively high at the time of dual culture initiation, they may fall markedly with increasing time in culture, and may even disappear altogether (Kassanis *et al.* 1958, Reinert 1966, Chandra & Hildebrandt 1967, Hirth & Durr 1971). The reasons for low virus concentrations in tissue culture are not yet clear. In the intact plant virus distribution is known to be irregular. Some viruses are often absent from meristems (e.g. Solberg & Bald 1963), although Walkey & Webb (1968, 1970), Walkey & Cooper (1972), Appiano & Pennazio (1972) and others have demonstrated the presence of some viruses in stem apices. By analogy with the intact plant situation it has been suggested that the low virus levels of tissue cultures may be due to their meristematic nature (Morel 1948, Braun & Lipetz 1966, Hirth & Segretain 1956, Wu *et al.* 1960, Streissle 1971) although Augier de Montgremier & Morel (1948) did find higher titres of TMV in rapidly growing crown-gall tissues than in slower growing habituated cultures. Kassanis (1957, 1967) showed that non-infected tobacco tissue cultures contained only one-twentieth as much protein nitrogen as leaves and suggested that since viruses use some of the same cellular systems that produce normal cell proteins the smaller virus concentration in tissue culture might reflect their less active protein metabolism, while Wu *et al.* (1960) found from inhibitor experiments that the nuclear protein synthesis required for cell division in tobacco tissue cultures was competitive with TMV synthesis.

Normal constituents of tissue culture media probably have significant effects on virus multiplication, either directly, or indirectly by affecting the metabolism of the cultured cells. For example, Kutsky & Rawlins (1950) showed that 1-naphthalene acetic acid at 10^{-6} g/ml markedly increased the levels of TMV in tobacco tissue cultures without apparently affecting tissue growth. Segretain & Hirth (1953) reported that coconut milk favoured cell division in tobacco callus and lowered its TMV titre, while aspartic acid lowered virus titre and slightly inhibited tissue growth, and glutamic acid favoured virus increase at concentrations not affecting tissue growth. In contrast, Kassanis (1957a) reported that coconut milk increased crown-gall tissue growth and titre of TMV. Recently Milo & Srivastava (1969a, b) have shown that cytokinins may be inhibitory to TMV multiplication both in intact tobacco plants and in infected tissue cultures of pith origin. Other normal media constituents as well as physical conditions (light, pH and temperature) which affect the metabolism of host cells may affect virus multiplication and survival in culture, as may added factors such as anti-metabolites. A review of some such studies, which are often conflicting and inconclusive but which have important applications in chemotherapy and production of virus-free plants (see section below) has been published by Raychaudhuri (1966).

Distribution of viruses in tissue cultures

It is known that there are large differences in virus levels between the various tissues of systemically infected plants. Studies by Hirth & Lebeurier (1965) provided some evidence that inoculated tobacco tissue cultures consist of a network of infected cells among healthy cells. Similarly, Hansen & Hildebrandt (1966) showed that about 40% of callus cells carried virus when single cells from infected tissue cultures were mechanically separated and individually assayed for virus by application of a micro-homogenate to indicator hosts. In another study (Chandra & Hildebrandt 1967) tobacco callus cells from TMV-infected *N. tabacum* plants were studied individually in microchambers by phase contrast microscopy. Cells infected with TMV frequently had inclusions, and the number of inclusion-bearing cells declined during successive passages on culture media, although two callus isolates still showed inclusions even after sixteen passages. Out of a total of 100 inclusion-bearing cells and 150 inclusion-free cells only ten and seventy cells respectively divided in the microchambers, and only five and sixty-five formed colonies. The amount of infective virus in these different clones varied considerably, and three of the inclusion-free clones were also virus-free. Both the virus-containing and the virus-free single cell clones were induced to differentiate, and both healthy and infected plants were obtained from infected single-cell clones.

Recently Hirth & Durr (1971) have shown that 30 to 40% of cells from TMV-infected tobacco tissue cultures contained virus, the remaining cells being healthy. Most of the healthy cells present in the infected cultures were, however, susceptible to infection by TMV, although they did not become infected during growth of the tissues. Some of the virus-free cells were apparently resistant to infection, a phenomenon which could be valuable in obtaining virus-resistant plant clones by regeneration from tissue culture. It is as yet difficult to see why susceptible cells in tissue cultures should remain free of virus, for Kassanis, Tinsley & Quak (1958) showed that TMV moved through tobacco callus tissue at the same rate as through tobacco leaves, although they failed to demonstrate the presence of plasmodesmata in the tissues. Plasmodesmata have, however, recently been shown to be present in some tobacco tissue cultures (Spencer & Kimmins 1969).

Production of virus-free plants

Meristem tip culture

Most commercial crop plants, particularly those which are propagated vegetatively, contain systemic viruses which affect performance or depress yield. Clearly, therefore, it is desirable to produce stocks of virus-free or near virus-free plants which can be clonally multiplied before being released commercially. In many species this can best be achieved by heat treatment of various organs or of actively growing plants (Hollings 1965). However, certain viruses have so far resisted all attempts at eradication by this means, and other methods of therapy are necessary. The most successful alternative so far devised is that of meristem-tip culture, often linked with chemotherapy or heat treatment. When these methods are employed plant stocks are not only freed of viruses, but of fungal and other pathogens too.

It is now 20 years since Morel & Martin (1952, 1955) were able to show that certain viruses could be eliminated from potato and dahlia by aseptic culture of meristem tips; unfortunately the shoots produced did not form roots and had to be grafted on to healthy seedlings. Since that time, however, methods have been devised to produce virus-free plants with roots from a wide range of species, and clones so derived are now widely used commercially (Hollings 1965, Hollings & Stone 1968 and Table 14.2). The basis of the meristem-tip culture technique is aseptic excision of a small segment of a stem or axillary bud apex which is then transferred to a suitable culture medium and grown on to form a plant. The term stem apex is open to a number of interpretations, but in most cases the tissue unit used is the meristem dome plus

TABLE 14.2. Some examples of of plants of economic importance which have been freed of viruses by meristem-tip culture techniques, sometimes linked with thermotherapy or chemotherapy

Host Plant	Viruses	Authors
Carnation (*Dianthus* spp.)	Carnation mottle, ringspot, vein mottle, latent, etched ring and streak viruses	Quak 1957, 1961, Hollings 1961, 1962, 1964, Baker & Philips 1962, Hollings & Stone 1964, 1965, Vermeulen & Haen 1964, Philips 1968, Stone 1968
Chrysanthemum (*Chrysanthemum motifolium* Ramat.)	Chrysanthemum virus B, vein mottle and greenflower viruses	Hollings 1963, Hakkaart & Quak 1964
Gooseberry (*Ribes uva-crispa* L.)	Vein banding virus	Jones & Vine 1968
Hop (*Humulus lupulus* L.)	Necrotic ring spot and hop latent viruses	Vine & Jones 1969
Potato (*Solanum tuberosum* L.)	Potato viruses A, S, X and Y, paracrinkle, and spindle tuber	Morel & Martin 1955, Kassanis 1957b, Thompson 1957, Quak 1961, Yora & Tsuchizaki 1962, Kassanis & Varma 1967, Morel, Martin & Muller 1968, Stace-Smith & Mellor 1968, 1970, Pennazio 1971
Strawberry (*Fragaria xananassa* Duchesne)	Strawberry yellow edge, crinkle, vein banding, latent A and vein chlorosis viruses	Belkengren & Miller 1962, Miller & Belkengren 1963, Vine 1968, Smith, Hilton & Frazier 1970
Rhubarb (*Rheum rhaponticum* L.)	Arabis mosaic, turnip mosaic, cherry leaf roll and strawberry latent ringspot viruses	Walkey 1968, Walkey & Woolfitt 1968a

the first pair of leaf primordia (Fig. 14.2, Hollings 1965). Depending on plant species the meristem tip may be 0·1 to 0·5 mm long; some workers claim to have cultured the meristem dome only, and others have obtained virus-free plants from apical segments up to 5·0 mm long (Vine & Jones 1969). The apical buds or the excised meristem tips may be surface sterilized, but according to Hollings (1965) this is not usually necessary. The actual techniques of meristem-tip culture vary from species to species; some practical procedures have been outlined by Hollings (1965) and by Hollings & Stone (1968), and a handbook dealing exclusively with techniques for carnation meristem-tip culture has been produced by The Colorado State University Experiment Station (Philips 1968).

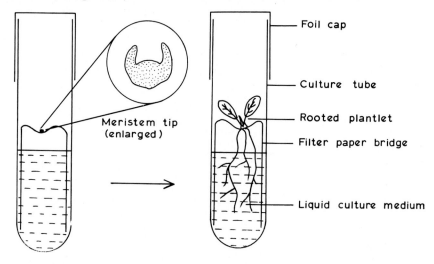

FIG. 14.2. *A diagram illustrating the technique of meristem-tip culture.*

Agar media have been found to be unsuitable for growing plants from meristem tips, and filter paper bridges, with their bases immersed in liquid nutrient medium contained in culture tubes, have been found to be more satisfactory (Fig. 14.2). On such bridges good root systems with root hairs are produced, and plants can readily be transferred to soil. A wide range of culture media of varying complexity have been used for meristem-tip culture, but no general purpose medium is yet available. Hollings (1965) and Hollings & Stone (1968) found that a medium consisting of the major and minor elements of Knop and Berthelot's solutions (beryllium and titanium omitted), glucose (40·0 g/litre), thiamine (10^{-3} g/litre) and meso-inositol (10^{-3} g/litre) at pH 5·5 could be used for a wide range of species. 1-Naphthalene acetic acid (NAA) at 10^{-3} g/litre is often necessary to induce the formation of root initials, but cultures are then transferred to medium without NAA. There is little information regarding ideal temperature and lighting conditions, although Hollings & Stone (1968) recommended incubation at 20°C below a bank of fluorescent lamps giving 22 hours' illumination per day. The time taken for meristem tips to form roots and shoots may vary from a few weeks to several months, depending on species.

Meristem-tip culture alone has enabled healthy clones to be obtained with *Narcissus* and *Cymbidium* sp., although improved results have been obtained with other species when meristems from heat-treated plants were used (Quak 1957, Hollings 1965). It was found by Hollings & Stone (1968) that only 5% of surviving meristem tips of chrysanthemum were free of leaf mottling viruses, although about 98% success was obtained when inoculum was taken immediately after heat treatment. Elimination of virus has also been enhanced by incorporation of inhibitors of virus multiplication into the culture medium; such compounds as malachite green, 2,4-D and thiouracil have been used in this way (Hollings 1965). Once virus-free plants have been produced a long probationary period with rigorous testing is necessary before plants can be released into commercial production. Eventually it may be possible to produce large numbers of virus-free plants from only a few apices using methods of clonal multiplication such as those devised by Walkey & Woolfitt (1968b, 1970).

It is not yet clear how viruses are eliminated during meristem-tip culture. The irregular distribution of some viruses in systemically infected plants is well known; Limasset & Cornuet (1949) showed that TMV concentrations increased away from the apex of tobacco plants, and Solberg & Bald (1963) showed a downward gradient of TMV in *Nicotiana glauca* plants. In contrast, Hollings & Stone (1964) detected carnation mottle virus in meristem tips and meristem domes of carnation using an inoculation procedure, and Walkey & Webb (1968, 1970) showed virus particles in the apical meristems of *N. rustica* by electron microscopy. Walkey, Fitzpatrick & Woolfitt (1969) and Walkey & Cooper (1972) were later able to produce virus-free plants of *N. rustica* from meristem tips known to contain some viruses, without thermotherapy. Recently Appiano & Pennazio (1972) have found potato virus particles in potato meristems, although potato virus X-free stocks have sometimes been obtained from meristems without thermotherapy (Morel, Martin & Muller 1968, Pennazio 1971). It is not known whether excised apices produce their own virus inactivation factors, or whether inactivation is brought about by the direct or indirect effects of some constituent of the culture medium or the culture environment.

Regeneration procedures

The regeneration of plants from both tissue and cell cultures is now an established procedure with many species (see Chapter 12), and there is every reason to suppose that it will eventually become so with many more. Whilst regeneration of plants from cell and tissue cultures has obvious practical implications in plant breeding, it may also be of value in the future for the production of virus-free clones. The levels of virus in tissue cultures from systemically infected plants may be very low, and viruses may sometimes be lost altogether. Moreover, not all cells in a culture are infected, and although some of the non-infected cells may be virus susceptible, others may be resistant. It is possible, therefore, that virus-free plants or even virus-resistant plants may eventually be produced from tissue culture. Work in this field has so far been limited: Chandra & Hildebrandt (1967) were able to produce virus-free tobacco plants from infected tissue cultures; Pillai & Hildebrandt (1968) and Abo-El Nil & Hildebrandt (1971b) obtained virus-symptomless plants from *Pelargonium hortorum* callus derived from stem-tips and anthers; and Simonsen & Hildebrandt (1971) differentiated virus-free gladiolus plants with corms from gladiolus tissue cultures derived from cormel stem-tips.

OTHER PATHOGENS

Bacteria

Since most bacteria have relatively simple nutritional requirements almost all attempts to grow combinations of these pathogens and their host tissue cultures have led to extensive colonization of the medium. Although incorporation of antibiotics or other bacteriocidal agents into culture media is theoretically possible, it would probably be of little practical value. With the exception of studies of the crown-gall disease, therefore, plant tissue culture has contributed little to our understanding of the interaction between bacteria and their hosts. Some limited studies have been carried out, however.

In 1953 Volcani, Hildebrandt and Riker showed that soft rot bacteria, including *Erwinia caratovorum*, were able to cause typical soft-rot injuries to callus cultures of carrot, *Vinca rosea*, potato and *Calendula officinalis*, and that bacteria-free culture filtrates contained tissue-rotting activity. Garber & Goldman (1956) found that twenty-seven mutants of *Erwinia caratovorum* were capable of attacking *Vitis* spp. tissue cultures. More recently Bajaj & Saettler (1970) have shown that growth of *Phaseolus vulgaris* callus tissue is inhibited by culture filtrates of *Pseudomonas phaseolicola*, *P. syringae* and *P. morsprunorum*. There were certain similarities in the physiological effects of toxin-containing filtrates of *P. phaseolicola* on bean callus and on green leaf tissue, and it was suggested that callus tissues might, therefore, be used to study the mode of action of the bacterial toxin. Some studies of callus induction and symptom expression in stem segments of *Populus cardicus* infected with *Aplanobacter populi* have been carried out by De Lange (1968).

Perhaps the most exciting work with bacteria and tissue cultures reported recently is that of Holsten, Burns, Hardy & Hebert (1971). These authors were able to establish a symbiotic nitrogen-fixing relationship between *Rhizobium* species and soybean root-cell calluses, whereas previously it had only been possible to grow *Rhizobium* in root-organ cultures. The morphology of the symbiosis *in vitro* was analogous to that of intact soybean root nodules, and a sensitive acetylene-ethylene assay was used to demonstrate nitrogenase activity. It was suggested that the system could provide a new tool for elucidation of the mechanism controlling nodulation and subsequent nitrogen fixation in plant symbiotic associations. A previous attempt to infect soybean calluses by *R. japonicum* (Veliky & La Rue 1967) met with only limited success. It would be interesting to see whether callus derived from non-leguminous plants could also be infected with *Rhizobium*, and whether systemically infected leguminous and non-leguminous plants capable of nitrogen fixation could be regenerated from infected callus.

Insects

In studies of the interaction between insects and plants, tissue culture techniques are likely to be of most value for the investigation of gall induction and formation. These galls, unlike those caused by bacteria and fungi, are often very complex in structure, being of constant size and morphology and showing considerable internal differentiation. Insect galls are usually self limiting, and cease to develop if the inciting insect or its larvae die. Many attempts to culture complex insect galls, such as those incited by Cynipids, have failed (e.g. Pelet 1959), but Pelet, Hildebrandt, Riker & Skoog (1960) were able to compare growth in culture of healthy grape-stem tissue with tissue of the simple *Phylloxera* gall, and were also able to culture *Phylloxera* itself on grape callus. Other studies of the *Phylloxera* gall in culture have been made by Hildebrandt (1963), Arya (1965) and Warwick & Hildebrandt (1966). Although tissue culture has as yet contributed little to knowledge of insect galls, it is an area of study which warrants further attention.

Nematodes

Most plant parasitic nematodes complete their life cycle only in the presence of living cells, and very few species (e.g. see Myers 1967) can be grown in axenic culture. Dual cultures are, therefore, of great potential value for studies of the physiology and biochemistry of host-nematode interaction and for the maintenance of pure stocks of aseptic nematodes for experimental and plant breeding studies. Nematodes were first grown aseptically in excised root cultures in 1955, when Mountain established dual cultures of *Pratylenchus minyus* and tobacco and red clover roots. Since then a number of dual cultures of parasitic nematodes and detached host roots have been grown and successfully used in the maintenance of contaminant-free stocks and in studies of host parasite interaction (e.g. Braun & Lipetz 1966, McClure & Viglierchio 1966, Johnson & Viglierchio 1969a, b, Dasgupta, Nand & Seshadri 1970). Excised roots or aseptic seedlings grown on agar have generally been found to be better substrates for nematode infection and multiplication than roots grown in liquid medium (e.g. Sayre 1958).

Infection of callus cultures

Callus cultures have already been of value in the study of nematodes, and their use is increasing, although it has been reported that some nematodes will not infect or achieve full development in undifferentiated tissue. For example, Sayre (1958) found that attached tomato roots were more attractive to *Meloidogyne incognita* than excised roots, and that these in turn were more attractive than callus. Adult females of *M. incognita* were observed in the differentiated tissue only, and it was concluded that the presence of vascular tissue was necessary for complete development of the parasite. Darling, Faulkner & Wallendel (1957) reported that *Ditylenchus destructor*, the potato-rot nematode, could be grown in undifferentiated callus of potato, carrot, clover and tobacco, and Dolliver, Hildebrandt & Riker (1962) found that the chrysanthemum nematode *Aphelenchoides ritzemabosi* grew and multiplied on callus of tobacco, carrot, periwinkle and marigold. Similarly, Webster (1966) and Webster & Lowe (1966) found that *A. ritzemabosi* grew on callus of a number of species, including oat, rose, potato and red clover, all non-host of the pathogen, and also on callus of lucerne, a normal host (Webster 1967a). Techniques for producing aseptic nematodes by surface sterilization procedures, and for establishing and maintaining dual cultures, have been described in detail by Krusberg (1961), Bingefors & Eriksson (1968) and Eriksson (1972). Corbett (1970) has described techniques for maintaining dual cultures under oil and Webb (1971) has designed an apparatus for the efficient extraction of large numbers of nematodes from aseptic cultures.

Maintenance of aseptic stock cultures

Callus tissue cultures of both hosts and non-hosts may be of value for maintaining monoxenic stocks of many plant parasitic nematode species ready for experimental and other studies. Mai & Thistlethwayte (1967) discussed the feasibility of using nematodes produced in monoxenic culture for nematocide sceening and Lownsbery, Huang & Johnson (1967) described the long-term maintenance of the root lesion nematode *Pratylenchus vulvus* in host callus cultures. An example of a species which has been studied in detail from the point of view of long-term maintenance is the stem nematode *Ditylenchus dipsaci*. Its growth in callus of alfalfa was first described by Krusberg in 1961, and Bingefors & Eriksson (1963, 1968) discussed the use of such dual cultures as a means of producing pure strains of aseptic nematodes for experimental studies. In fact, *D. dipsaci* produced in callus tissue was used in crossing experiments by Eriksson (1965) and in race studies by Eriksson & Granberg (1969). The propagation of *D. dipsaci* in callus tissue is now a routine procedure, and a nematode bank supplying plant breeders and researchers in Scandinavia has been developed, although this is apparently not always completely free of problems (Bingefors 1970, Eriksson 1972).

Host specificity and resistance expression in culture were also studied by Eriksson (1972). In experiments involving several strains of *D. dipsaci* and other nematode species, and callus cultures of many host and non-host legume species it was found that nematode reproduction in callus was quite different from that observed in intact plants. Factors causing a plant to be a host or a non-host for *D. dipsaci* races were apparently partly retained in culture, and significantly influenced nematode reproduction on the tissues. However, resistance expression in culture was quite abnormal, for callus tissues isolated from resistant and susceptible cultivars of the same plant species supported equal numbers of both pathogenic and non-pathogenic stem nematode races. Moreover, nematode reproduction on callus was the same irrespective of the plant organ from which the tissue had been isolated. Loss of resistance in callus culture apparently arose immediately the tissues were isolated. These findings resemble very closely the findings of Ingram (1969c) and Ingram & Tommerup (1973) concerning the expression of resistance to fungi in tissue culture, and cast further doubt on the value of tissue cultures for studies of host-parasite interaction.

Hormonal studies

Tissue cultures have been used to some extent in studies of the role of plant hormones in the interaction between nematodes and their hosts. Sandstedt & Schuster (1963) showed that eggs of *Meloidogyne incognita* (root-knot nematode) hatched on the surface of carrot-root discs on a basal White's medium, and that the larvae then entered the tissues and induced formation of lumps of callus. Giant cells formed in both the callus and the original tissue, and all stages of the life cycle of the nematode were represented in culture. Later Sandstedt & Schuster (1965) demonstrated that root-knot nematodes induced callus more readily on the radical ends on the foliar ends of carrot pieces grown in the absence of exogenous growth substances, and suggested that polar transport of auxin endogenous to the carrot tissue may have been involved in callus formation. It was also shown (Sandstedt & Schuster 1966a) that the nematodes increased growth of excised tobacco-pith tissue and induced typical syncytia only when an auxin and a cytokinin were added to the medium, suggesting that the nematode-induced growth was dependent on these two kinds of growth substance, and that the nematode did not supply them. Further experiments (Sandstedt & Schuster 1966b) were carried out with peeled tobacco stem segments cultured *in vitro* on a basal medium, to which were applied aseptic nematodes or indol-3yl-acetic acid (IAA) or 2,3,5-triiodobenzoic acid (TIBA) in agar blocks. The results of these experiments suggested that *M. incognita* may induce tissue proliferation in a manner analogous to that of TIBA, by causing infected tissues to retain and use endogenous auxin that would otherwise be transported away.

It was shown by Dolliver, Hildebrandt & Riker (1962) that certain plant parasitic nematodes reproduced fastest on seedlings callused by 2,4-D. Similarly Schroeder & Jenkins (1963), Krusberg & Blickenstaff (1964) and Webster & Lowe (1966) showed that calluses formed on media containing 2,4-D were good substrates for nematode growth. It was suggested by Webster & Lowe (1966) that 2,4-D acted indirectly on nematode reproduction by increasing the susceptibility of the plant tissue to infection. These authors also showed that the susceptibility of plant tissues to infection by *Aphelenchoides ritzemabosi* was decreased by CCC, and Webster (1966) reported that gibberellic acid increased and CCC decreased the number of nematodes in culture. In further studies Webster (1967a) examined the effect of a number of exogenously applied plant growth substances and their inhibitors on the host-parasite relationships of *A. ritzemabosi* in culture, and showed that the nematodes multiplied fastest when plant growth or cell division or enlargement was greatest. *In vitro* studies were also related to field studies on the effect of herbicides on nematode infected cereals (Webster 1972b). Plant tissue cultures may prove to be useful tools in other studies with plant parasitic nematodes, particularly those which cause morphological disturbance to host tissues or are influenced by plant hormones.

FUTURE PROSPECTS

Plant tissues culture techniques clearly have a possible future role in the maintenance of stocks of specialized parasites. Contaminant-free fungi and nematodes grown in tissue culture have already been used experimentally and in axenic culture studies, and it seems likely that this trend will continue, once the problems of dual growth have been overcome with more host parasite combinations.

In studies of the interaction between hosts and parasites the future of tissue cultures as providing a simplified experimental system is much less certain. Except in studies of tumour physiology (see Chapter 13) the limited results so far obtained, particularly from resistance studies with fungi and nematodes, have been both disappointing and confusing, and have only served to emphasize the marked genetical and metabolic differences between normal and cultured plant tissues. The nature and extent of such differences as they apply to plant pathology must be investigated further, so as to determine which host-parasite situations may best be investigated using tissue culture techniques, and to what extent abnormal interactions may be exploited in attempts to define normal ones. A possible role for tissue cultures in the study of gall physiology, particularly where insects and fungi are concerned, has yet to be fully explored.

The technique of isolated protoplast culture, still in its infancy, may have an important part to play in studies of plant pathogens, and warrants further attention. It is already being applied with some success to studies of the uptake and multiplication of viruses and, together with the new methods for synchronous infection of cells in suspension, may be of still further use in defining the interaction of viruses and their hosts at a physiological and molecular level. Protoplasts may also have a part to play in studies of other pathogens; for example, as wall-free assay systems for investigation of the killing of host cells by fungal and bacterial toxins and enzymes.

The production of pathogen-free plant stocks will continue to be very important commercially, and the technique of meristem-tip culture still has a valuable contribution to make in this field. If the regeneration of greater numbers of species from tissue culture can be achieved, then this technique too could have a future role in the production of pathogen-free and pathogen-resistant crop plants, and might even be used for growing plants systemically infected with beneficial micro-organisms such as nitrogen-fixing bacteria.

CHAPTER 15

OLD PROBLEMS AND NEW PERSPECTIVES

H. E. Street

An important element in scientific progress is the development of new techniques which either enable more critical experiments to be undertaken or which render certain problems accessible to experimental study for the first time. The tissue and cell culture techniques described in this volume must, therefore, be evaluated in terms of the scientific progress they have made or will make possible in the future.

Although the authors of the separate chapters of this book have seen it as part of their task to attempt such evaluations, they have concentrated attention upon those particular facets of plant science to which the techniques they consider are most directly applicable. This concluding chapter, however, provides an opportunity to contemplate the prospects for advances in plant science and in cell biology which might follow when these various techniques are brought together. The importance of this multi-disciplinary approach was formulated in 1965 (Street, Henshaw & Buiatti) as follows 'Whilst stressing the existence of formidable technical obstacles, a broad survey, such as this, of current work on the culture of isolated plant cells and suspensions of such cells, indicates exciting prospects in the fields of cell biochemistry, biochemical genetics and plant breeding from both fundamental and practical standpoints. The immediate need is for research into the culture of higher plant cells by a team embracing workers in microbiology, plant physiology, biochemistry, cytology and genetics.' The changed situation, which is the central theme of this volume, is that many of the technical obstacles mentioned in 1965 have now been overcome; recognition and exploitation of the possibilities awaits only the necessary application and imagination of appropriately trained scientists. That these possibilities are rapidly being recognized is eloquently demonstrated by the rapid rise (particularly during the last 5 years) in the number and spread amongst journals of published papers where plant tissue and cell culture techniques have been central to the 'Experimental Materials and Methods'.

Major advances in cellular physiology based upon studies with micro-organisms (bacteria, fungi and certain unicellular algae) followed from the perfection of continuous culture systems, the availability of a wide range of biochemical mutants and the use of genetical techniques. Higher plant cells can now be grown in similar continuous cultures and their growth kinetics analysed by procedures developed in microbiology (Chapters 4 and 11). With the availability of haploid cells from anther culture (Chapter 9) and the techniques of single-cell cloning, mutation and mutant selection (Chapter 8) we may expect to see, in the immediate future, advances in the cellular physiology of higher plants comparable with those which developed two decades ago in the field of micro-organisms.

The demonstration that division in plant cell cultures (Chapter 11) can be highly synchronized and retain this synchrony should, using modern techniques of cytology and biochemistry, enable the cell cycle in plant cells to be examined at a comparable depth to that hitherto only achieved with yeasts and certain other micro-organisms.

There are many central problems in plant cell physiology in which, despite a great deal of effort, only very limited progress has been made, primarily because no really appropriate experimental system has been available. This is perhaps particularly so in the field of the plant growth hormones (phytohormones). Although it is clear that these phytohormones play a key role in the control of cell division and cell expansion in plant cells, we are still very far from understanding how they act at the molecular level.

Hitherto most experiments designed to study the mechanism of action of phytohormones have been carried out using organized multicellular plant structures—experimental objects of considerable complexity and not readily replicable. A 4-litre continuous liquid culture of plant cells can present the experimenter with 4×10^9 cells in a steady state of growth and metabolism. The metabolic and structural events which follow the addition to such a culture of a hormone perturbing the steady state can be followed with precision over a short or a longer period. Not only can changes in cell growth, morphology and physiology be continuously monitored but sufficient cells can be harvested from the population for full biochemical analysis, histology and electron microscopy without significantly disturbing the cultural environment. Such a system is capable of contributing to many other classical problems in plant physiology whose study has hitherto been handicapped by our inability to work at the level of the individual cell. Studies of physiology and biochemistry at this level provide the essential foundation, at present extremely limited, upon which to build an understanding of organ and whole plant physiology.

There is at present a surge of research activity in the whole field of developmental biology. Molecular biologists, having built up a sound framework of knowledge concerning the molecular events characteristic of prokaryotes and having shown that this framework enshrines the common denominators of all living cells are now turning their attention to the special features of eukaryote cells, to the differences which can occur between cells of common inheritance and to the origin of these differences in multicellular organisms. Animal and plant tissue and cell cultures are clearly going to play a key role in this new wave of research activity and the unique features of plant tissue and cell cultures are only now beginning to be fully appreciated by developmental biologists. These features include the ability of a wide range of plant cells to be cultured indefinitely in fully defined media and their capacity to retain, at least for long periods, their competency for morphogenesis, as well as the availability of haploid cells and their capacity to regenerate the whole organism either via organogenesis or embryogenesis (Chapter 12).

The heading morphogenesis in any recent textbook of plant physiology now leads the reader to an account of the control of organogenesis based upon work with organ and tissue culture (classical work on shoot bud and root initiation in callus cultures, culture of stem and root apices, culture of immature leaves, flowers and fruits—Street 1966b, 1969). Classical work with cell cultures of carrot (Steward 1958, 1961) is cited to introduce the concept of totipotency in plant cells. Similarly any account of cytodifferentiation refers to work on vascular tissue differentiation in callus cultures (Wetmore & Rier 1963, Gautheret 1966, Torrey 1971) and in cultured roots (Loomis & Torrey 1964; Torrey, 1967b; Street 1968b). It is this capacity of cell, callus and meristem cultures to regenerate whole plants that has been exploited to obtain virus-free stocks of a number of horticultural plants (Hollings 1965, Vine 1968, Jones & Vine 1968, Vine & Jones 1969) (see Chapter 14) and to effect the micropropagation of orchids and some ornamentals and vegetables (Morel 1964, Hackett & Anderson 1967, Pow 1969). The possibility that embryogensis might also be used as a practical method of propagation has also been explored (Wilmar & Hellendoorn 1968, Smith 1972) but there is now evidence that the initial assumption that plants derived from tissue and cell cultures by way of embryogenesis would be 'genetically identical' is by no means the case (see later reference to this chapter to the implications of work on the nuclear cytology of cultured cells discussed in Chapter 7).

It must be emphasized that although tissue cultures from many species, when first established, have been reported to give rise to roots, the number of species in which it has been possible to induce shoot bud initiation or embryogenesis in culture remains very limited indeed. Furthermore, it is an almost universal experience that the morphogenetic potential of callus and suspension cultures declines and ultimately reaches zero as they are serially propagated in culture. Any anticipation, following the work with tobacco tissue (Skoog & Miller 1957), that the induction of organogenesis would merely require the application, via the culture medium, of an appropriate balance of known phytohormones has proved an illusion. Apparently only in a very few cultured tissues is the endogenous regulator complex adjustable to the required state by an exogenous supply of auxin, cytokinin or gibberellin either separately or in combination. Progress in our knowledge of the control of organogenesis now demands more intensive studies of the endogenous phytohormone levels in plant cultures, the identification of the chemical nature of the natural auxin(s), cytokinin(s) and gibberellin(s) in each particular tissue and a recognition that there may well be new phytohormones to be discovered. Such research is at a much more sophisticated level than that represented by just adding a range of concentrations of available (and to the tissue probably foreign) phytohormones via the culture medium.

Even in those cases where organogenesis can be readily induced there is a need for much more intensive anatomical and histochemical examination by both light and electron microscopy to trace the emergence of organization within the initially relative uniform cell mass of the callus or suspension culture aggregate.

One aspect of possible importance here is that of the degree of cellular contact and metabolite interchange within the culture. In individual cases it is often possible to predict morphogenetic potential from observation of the friability or compactness of the tissue but we neither understand the nature and significance of such intercellular contact in terms of middle lamella and cell-wall composition, formation and functioning of plasmo-desmata nor of development of tensions and compressions.

Almost identical problems will have to be solved to advance our under-standing of embryogenesis in tissue and cell cultures. Of particular interest here are two physiological problems: (1) the 'dedifferentiation' process brought about during callus initiation which restores, to previously differen-tiated cells, an ability to again express their totipotency and (2) the need to identify the factors which induce, in individual superficial cells of the callus or cell aggregate, the required polarity to initiate the orderly sequence of segmentations characteristic of embryology. The observation that cultures derived from zygotic or tissue culture embryos (Steward, Mapes, Kent & Holsten 1964), Vasil, Hildebrandt & Riker 1964, Konar & Nataraja 1965b) are of particularly high embryogenic potential suggests that the further cytodifferentiation progresses, as plant development procedes, the more difficult it may become to achieve the required 'dedifferentiation.' Further, we know that changes in nuclear cytology are often concomitant with differentia-tion, suggesting that totipotency may be lost during certain pathways of differentiation and may, in other cases, be retained only imperfectly (as suggested by the frequent occurrence of clearly imperfect embryos in culture). There is certainly no evidence to justify a blanket statement that all plant cells retain their totipotency during cytodifferentiation.

Cells in the plant body are what they are because of the cell lineage which connects them to a plant meristem and because of where they are within a plant organ. The characteristic morphology and physiology of higher plant cells *in vivo* is determined by the changing internal environment to which they are exposed both during division and subsequent expansion and maturation. Only when organized structures are formed in culture does normal cytodifferentiation occur resulting in the emergence of characteristic patterns of tissues. Various degrees of cellular differentiation can be observed in callus (Chapter 10), occasionally even in the aggregates of coarse suspensions (Chapters 11 and 12) indicating that cell association is probably an essential component towards creating the required environment for cytodifferentiation. Cells in fine suspension culture do not correspond morphologically to any of the various types of tissue cells of the plant body and such cultures cannot yet be shunted into normal pathways of cytodifferentiation. Where such suspension cultures have been examined for the secondary plant products characteristic of the species, either no such products have been detected, or related but different products have been found, or the normal products have been detectable but at levels far below those occurring either in the whole organism or its structural components. Future work may resolve whether it will be possible to switch the cells of a free-cell suspension into normal pathways of cytodifferentiation or activate the particular biosynthetic sequences leading to the production of desired metabolic products.

Various workers have suggested that it may be possible to induce cytodifferentiation in cell suspensions by a planned sequence of cultural treatments spanning both the growth stage and a subsequent maintenance (nongrowing stage) of the culture (Kent & Steward 1965, Steward, Kent and Mapes 1967, Fosket and Torrey 1969). This approach, however, is at present so problematical that it presents a rather daunting prospect to the investigator. An alternative concept is that the opening up of particular pathways of morphological and/or biochemical differentiation would follow inactivation of particular repressor molecules (by, for instance, accumulation of particular primary metabolites) or the inactivation of the regulator genes responsible for the synthesis of the repressors. Hence, appropriate mutant cells might be expected to express these pathways without being subjected to the very special micro-environments needed for their activation in normal cells (Street 1973). Work with micro-organisms further suggests that once a pathway is opened up it can often be intensified both by modification of the cultural conditions and by further selection of mutants. (A mutation causing a selective change in permeability might lead to continuous release of a desired metabolite into the culture medium—a very effective 'sink'). This approach, via mutant selection, may prove more immediately rewarding— both for the study of normal pathways of cytodifferentiation and for exploiting the biosynthetic capacities of cultured cells—than trying to reproduce, in culture apparatus and in an appropriately timed sequence, the complex of interacting factors which operate to control differentiation in developing plant organs. The approach certainly has the merit of being unexplored and full of promise.

The cytological instability of plant tissue and cell cultures (Chapter 7) presents both problems and opportunities. At least some success has almost always been achieved whenever attempts have been made to obtain some desired culture variant (e.g. one more friable, more autotrophic for a growth hormone, more resistant to an antimetabolite, better able to utilize a wider spectrum of carbon sources, etc.) by applying the appropriate selection pressure either with or without associated single cell cloning (Chapter 8). Whenever old-established cultures have been subjected to single cell cloning, they have yielded a great diversity of colonies even when the criteria of difference have been limited to growth rate, colour, colony texture and resistance to necrosis and death when held in stationary phase. There is, therefore, in tissue and cell cultures a continuously renewed range of variation; a not unexpected finding where very large cell populations (often up to 10^{10} cells) are maintained under conditions where all cells which retain a capacity to grow and divide will survive and where any variant with even a marginal advantage in culture will quickly become a significant component of the population. This situation can be contrasted with that in the whole organism in that cell division and cell growth are there restricted to the meristems and that any variant, if it is to survive and give rise to a chimera, must be able to continue division and growth in a way which does not disrupt the essential organization of the meristem.

As our knowledge of the cytology of cultured plant cells grows it becomes increasingly clear that a major cause of this variation is at the level of changes in chromosome number and chromosomal mutations. Whilst, therefore, cytological instability may provide us with a wide range of viable genotypes, at the same time, it poses the problem of whether it will be possible to preserve a particular genotype in continuous culture. Such preservation is essential not only for tissue and cell cultures to find application in the clonal propagation of economic plants but also for all the physiological and metabolic researches envisaged earlier in this chapter. Cytological stability will be critical for the isolation and subsequent study of point mutations and for the exploitation of the potential of haploid cultures.

There is now a considerable amount of data indicating that the cytological stability of plant cultures is profoundly affected by cultural conditions and particularly by the composition of culture media (Demoise & Partanen 1969, Van't Hof and McMillan 1969, Bennici, Biuatti, D'Amato & Pagliai 1970, Singh, Harvey, Kao & Miller 1972). A particular instance of this is the reported preservation of the haploid state in tobacco cell cultures by incorporation of parafluorophenylalanine into the culture medium (Gupta &

Carlson 1972). Further critical work is now required to determine how far different culture media enhance or diminish the incidence of chromosomal mutations and how far the apparent stabilization or breakdown of nuclear cytology in different media is due to such media being selective or non-selective for particular genotypes. It seems, however, that an alternative approach to the conservation of genotypes which avoids the need for continuous serial subculture is freeze preservation (Quatrano 1968, Latta 1971, Bannier & Steponkus 1972).

Very early in the history of plant tissue culture, there were reports of the culture of cells derived from crown-gall tumours; these tumour tissues had simpler growth requirements than the normal tissues of the host species, grew rapidly and usually produced a uniformly parenchymatous callus (Chapter 13). This enhanced synthetic ability of the tumour cells led to the hypothesis that during tumour induction there occurred an inheritable activation in the cells of growth-substance synthesizing systems by virtue of the transfer from the specific bacterium of a tumorigenic principle (TIP) (Braun 1961). This immediately suggests the transfer of genetic information from the bacterium to the plant cell, but does not exclude the alternative possibility that TIP is a derepressor of certain important metabolic pathways related to cell growth. Crown-gall tissues are reported to release certain enzymes not normally released by untransformed cells, but so far there is no convincing evidence of the synthesis by tumour cells of additional enzymes or isozymes. Nevertheless, recent studies involving RNA–DNA hybridization seem to favour the hypothesis that tumour induction is the result of the transfer of genetic information. The possibility that bacterial DNA can be transferred to higher plant cells and become incorporated into their genome is also raised by experiments with *Arabidopsis thaliana* (Ledoux, Huart & Jacob 1971). The possibility of DNA transfer to cultured cells should now be tested using bacteriophages carrying particular new operons.

A new possibility of transferring DNA, or even new organelles to plant cells is opened up by recent progress in preparing viable protoplasts from plant cells and by the demonstration that such protoplasts can be induced to reform their cell walls and embark upon division to give single cell callus clones (Chapter 5). One of many difficulties encountered in trying to incorporate foreign DNA into any cell is that the act of isolating the DNA may inactivate it. However, it is now possible to cause protoplasts, during the actual process of their release, to take up, by an infolding of their plasmalemma, nuclei, bacteria (e.g. N-fixing bacteria) or organelles (e.g. chloroplasts); these structures are initially enclosed in a vesicle of plasmalemma. The later breakdown of the ingested 'foreign' structure might conceivably release DNA in an active form capable of incorporation into the cell's genome.

Isolated protoplasts may undergo spontaneous and multiple fusion.

and powerful approaches to many major problems in plant physiology, biochemistry and genetics.

The approach of this chapter seems to be well summarized by the following short quotations:

'*The basic texture of research consists of dreams into which the threads of reasoning, measurement and calculation are woven*'

A. Szent-Györgyi

'*I am a firm believer that without speculation there is no good or original observation*'

Charles Darwin in a letter
to Alfred Russell Wallace

LITERATURE CITED

Immediately below each reference and on a separate line are given the page number(s) on which the reference is cited in the text.

AARON-DA-CUNHA M.I. (1969) *Recherches sur la permanence et la transmission des propriétés tumorales des cellules de crown-gall.* Thèse Sci. Université de Paris.
381

AARON-DA-CUNHA M.I. & PAUPARDIN C. (1971) Remarques sur le phénomène d'induction tumorale provoquée par les tissus de crown-gall de tabac, pp. 473–477, in *Les Cultures de Tissus de Plantes.* Colloques Internationaux du C.N.R.S., no. 193, Paris.
381

ABO EL-NIL M.M. & HILDEBRANDT A.C. (1971a) Geranium plant differentiation from anther callus. *Am. J. Bot.* **58**, 475.
217

ABO EL-NIL M.M. & HILDEBRANDT A.C. (1971b) Differentiation of virus-symptomless geranium plants from anther callus. *Pl. Dis. Reptr.* **55**, 1017–1020.
416

ADAMSON D. (1962) Expansion and division in auxin-treated plant cells. *Can. J. Bot.* **40**, 719–744.
50, 251

AGHION-PRAT D. (1965) Neoformation de fleurs *in vitro* chez *Nicotiana tabacum* L. *Physiol. Végét,* **3**, 229–303.
342, 346

AH-SUN R.A.A. (1970) *Active resistance of potato tubers and potato tissue cultures to fungal infection.* Ph.D. Thesis, University of East Anglia, England.
405

AHUJA M.R. (1962) A cytological study of heritable tumors in *Nicotiana* species hybrids. *Genetics* **47**, 865–880.
385

AHUJA M.R. (1968) An hypothesis and evidence concerning the genetic components controlling tumor formation in *Nicotiana. Mol. and Gen. Genetics* **103**, 176–184.
385

AHUJA M.R. & CAMERON D.R. (1963) The effects of X-irradiation on seedling tumor production in *Nicotiana* species and hybrids. *Radiat. Bot.* **3**, 55–57.
384

AHUJA M.R. & HAGEN G.L. (1966a) Chromosomes and nutritional requirements of a tumor forming *Nicotiana* hybrid and its derivatives. (Abstract) *Am. J. Bot.* **53**, 609.
387

AHUJA M.R. & HAGEN G.L. (1966b) Morphogenesis in *Nicotiana debneyi-tabacum, N. longiflora* and their tumor forming hybrid derivatives *in vitro. Devl. Biol.* **13**, 408–423.
387

AHUJA M.R. & HAGEN G.L. (1967) Cytogenetics of tumor-bearing interspecific triploid *Nicotiana glauca—langsdorffii. J. Heredity* **58**, 103–108.
385

AITCHISON P.A. (1972) Unpublished results.
56, 247

ALBERGONI F. (1964) Processi de neoformasione di mitochondri nella cellula vegetale. *Nouvo G. bot. ital.* **71**, 243–251.
142

AMES I. & MITRA J. (1966). The mitotic cycle time of *Haplopappus gracilis* root tip cells as measured with tritiated thymidine. *The Nucleus* **9**, 61–66.
300

AMES I.H. & SMITH H.H. (1969) Effects of mercaptoethanol on tumor induction in a *Nicotiana* amphiploid. *Can. J. Bot.* **47**, 921–924.
384

ANDERSON R.A. & SOWERS J.A. (1968) Optimum conditions for bonding of plant phenols to insoluble polyvinyl-pyrrolidone. *Phytochem.* **7**, 293–301.
106

AOKI S. & TAKEBE I. (1969) Infection of tobacco mesophyll protoplasts by tobacco

433

mosaic virus ribonucleic acid. *Virology* **39**, 439–448.
409

APPIANO A. & PENNAZIO S. (1972) Electron microscopy of potato meristem tips infected with potato virus X. *J. gen. Virol.* **14**, 273–276.
410, 415

ARYA H.C. (1965) Cultural behaviour of insect gall and normal plant stem single cell clones, pp. 293–309, in *Tissue Culture* (Proc. Seminar Baroda). Ed. C.V.Ramakrishnan. W.Jank, The Hague.
417

ARYA H.C., HILDEBRANDT A.C. & RIKER A.J. (1962) Growth in tissue culture of single-cell clones from grape-stem and *Phylloxera*-gall. *Pl. Physiol., Lancaster* **37**, 387–397.
201

ARYA H.C. & TIWARI M.M. (1969) Growth of *Sclerospora graminicola* on callus tissues of *Pennisetum typhoides* and in culture. *Indian Phytopathology* **22**, 446–452.
394, 399

AUGIER DE MONTGREMIER H. & MOREL G. (1948) Sur la diminution de la teneur en virus (*Marmor tabaci* Holmes) de tissus de tabac cultivés *in vitro*. *C.r. hebd Séanc. Acad. Sci. Paris* **227**, 688–689.
411

BACKS-HUSEMANN D. & REINERT J. (1970) Embryobilding durch isolierte Einzelzellen aus Gewebekulturen von *Daucus carota. Protoplasma* **70**, 49–60.
352

BAGSHAW V. (1969) *Changes in ultrastructure during the development of callus cells*. Ph.D. Thesis, University of Edinburgh.
123, 125, 127, 129, 130, 131, 134, 135, 137

BAGSHAW V., BROWN R. & YEOMAN M.M. (1969). Changes in the mitochondrial complex accompanying callus growth. *Ann. Bot.* **33**, 35–44.
123, 137, 141, 142

BAILEY I.W. (1920) The cambium and its derivative tissues. III. A reconnaissance of cytological phenomena in the cambium. *Am. J. Bot.* **7**, 417–434.
123

BAILEY J.A. (1970) Pisatin production by tissue cultures of *Pisum sativum* L. *J. gen. Microbiol.* **61**, 409–415.
405

BAJAJ Y.P.S. & SAETTLER A.W. (1970) Effect of halo toxin-containing filtrates of *Pseudomonas phaseolicola* on the growth of bean callus tissue. *Phytopathology* **60**, 1065–1067.
416

BAKER N., FEINBERG H. & HILL R. (1954) Analytical procedures using a combined combustion–diffusion vessel. Simple wet-combustion method suitable for routine carbon-14 analyses. *Analyt. Chem.* **26**, 1504–1506.
94

BAKER R. & PHILLIPS D.J. (1962) Obtaining pathogen-free stock by shoot tip culture. *Phytopathology* **52**, 1242–1244.
413

BALL E. (1950) Differentiation in a callus culture *Sequoia sempervirens*. *Growth* **14**, 295–325.
341

BANNIER L.J. & STEPONKUS P.L. (1972) Freeze preservation of callus cultures of *Chrysanthemum morifolium* Ramat. *Hort. Sci.* **7**, 194.
428

BARNES R.L. & NAYLOR A.W. (1958) Culture of pine root callus and the use of *Pinus clausa* callus in preliminary metabolic studies. *Bot. Gaz.* **120**, 63–66.
57

BAUCH R. & SIMON U. (1957) Kulturversuche mit Rostpilzen. *Ber. dt. bot. Ges.* **70**, 145–156.
395

BAYER M.H. (1965) Paper chromatography of auxins and their inhibitors in two *Nicotiana* species and their hybrid. *Am. J. Bot.* **52**, 883–890.
386

BAYER M.H. (1967) Thin layer chromatography of auxin and inhibitors in *Nicotiana glauca*, *N. langsdorffii* and three of their tumor-forming hybrids. *Planta* **72**, 329–337.
386

BAYER M.H. & AHUJA M.R. (1968) Tumour formation in *Nicotiana*: Auxin levels and auxin inhibitors in normal and tumour-prone genotypes. *Planta* **72**, 292–298.
386

BAYLEY J.M., KING J. & GAMBORG O.L. (1972) The ability of amino compounds and conditioned medium to alleviate the reduced nitrogen requirement of soybean cells grown in suspension culture. *Planta* **105**, 25–32.
291

BEACHY R.N. & MURAKISHI H.H. (1971) Local lesion formation in tobacco tissue cultures. *Phytopathology* **61**, 877–878. 410

BEARDSLEY R.E. (1955) Phage production by crown gall bacteria and the formation of plant tumors. *Am. Nat.* **89**, 175–176. 358, 372

BEARDSLEY R.E. (1960) Lysogenicity in *Agrobacterium tumefaciens*. *J. Bact.* **80**, 180. 372

BEARDSLEY R.E. (1972) The inception phase in the crown gall disease. *Prog. exp. Tumor Res*, **15**, 1–75. 363, 364, 369

BEARDSLEY R.E., STONIER T., LIPETZ J. & PARSONS C.L. (1966) Mechanisms of tumor induction in crown gall. I. Production and pathogenicity of spheroplasts of *Agrobacterium tumefaciens*. *Cancer Res.* **26**, 1606–1610. 373

BEDNAR T.W. & LINSMAIR-BEDNAR E.M. (1971) Induction of cytokinin-independent tobacco tissues by substituted fluorenes. *Proc. natn. Acad. Sci. U.S.A.* **68**, 1178–1179. 389

BEIDERBECK R. (1970) Untersuchungen an crown-gall IV. Rifampicin und ein resistenter Klon von *Agrobacterium tumefaciens* bei der Tumorinduktion. *Z. Naturf.* **25b**, 1458. 365, 366

BEIDERBECK R. (1971) Untersuchungen an crown-gall V. Der Einfluss von polyornithin auf die Tumorinduktion durch *Agrobacterium tumefaciens*. *Z. Pflanzenphysiol.* **64**, 199–205. 365

BELKENGREN R.O. & MILLER P.W. (1962) Culture of apical meristems of *Fragaria vesca* strawberry plants as a method of excluding latent A virus. *Pl. Dis. Reptr.* **46**, 119–121. 413

BELL P.R. & MUHLETHALER K. (1964) The degeneration and reappearance of mitochondria in the egg cells of a plant. *J. Cell. Biol.* **20**, 235–248. 142

BELLAMY A. & BIELESKI R. (1966) Some salt-uptake and tissue-ageing phenomena studied with cultured tobacco cells. *Aust. J. biol. Sci.* **19**, 23–26. 318

BELTRÁ R. & RODRIQUES D.L. (1971) Aseptic induction of crown gall tumours by the nucleic acid fraction from *Agrobacterium tumefaciens*. *Phytopath. Z.* **70**, 351–358. 367

BENBADIS A. (1965) Croissance de cellules isolées de Ronce sur les milieux mitritifs 'conditionnés' par quantités variables de tissu. *C.r. hebd. Séanc. Acad. Sci., Paris* **261**, 4829–4832. 89

BENBADIS A. (1968) Culture des cellules isolées: Le problème du conditionnement des milieux de culture, pp. 121–129 in *Les Cultures de Tissus de Plantes*. Strasbourg 1967. C.N.R.S., Paris. 193

BENNETT M.D. (1969) Induced and developmental variation in chromosomes of meristematic cells. *Chromosoma* **27**, 226–244. 93

BENNICI A. BUIATTI M. & D'AMATO F. (1968) Nuclear conditions in haploid *Pelargonium in vivo* and *in vitro*. *Chromosoma* **24**, 194–201. 183

BENNICI A., BUIATTI M., D'AMATO F. & PAGLIALI M. (1971) Nuclear behaviour in *Haplopappus gracilis* callus grown *in vitro* on different culture media, pp. 245–250, in *Les Culture de Tissus de Plantes*. Colloques Internationaux du C.N.R.S., no. 193, Paris. 343, 427

BERGMANN L. (1960) Growth and division of single cells of higher plants *in vitro*. *J. gen. Physiol.* **43**, 841–851. 9, 95, 153, 192, 195

BERGMANN L. & MELCHERS G. (1959) Infektionsversuche an submers kultivierten Geweben mit Tabakmosaikvirus. *Z. Naturforsh. B* **14**, 76–78. 407

BERTHOLET A. & AMOUREUX G. (1938) Sur la formation d'acide indol-3-acétique dans l'action de *Bacterium tumefaciens* sur le tryptophane. *C.r. hebd. Séanc. Acad. Sci., Paris* **206**, 537–540. 374

BHANDARY R., COLLIN H.A., THOMAS E. & STREET H.E. (1969) Root callus and cell suspension cultures from *Atropa belladonna*, L. and *Atropa belladonna*, cultivar *lutea* Doll. *Ann. Bot.* **33**, 647–656. 261

BIEBER J. & SARFET E. (1968) Zur Frage der Tumorbildung durch Desoxyribonuklein säure ans *Agrobacterium tumefaciens* (Smith and Townsend) Conn. *Phytopath. Z.* **62**, 323–326.
369

BINGEFORS S. (1970) Resistance against stem nematodes, *Ditylenchus dipsaci* (Kühn) Filipjev. *EPPO Publications Ser. A* no. **54**, 63–75.
419

BINGEFORS S. & ERIKSSON K.B. (1963) Rearing stem nematode inoculum on tissue culture. Preliminary report. *Lantbrukshögsk. Annaler* **27**, 385–398.
419

BINGEFORS S. & ERIKSSON K.B. (1968) Some problems connected with resistance breeding against stem nematodes in Sweden. *Z. PflZücht.* **59**, 359–375.
418, 419

BIRNSTIEL M. (1967) The nucleolus in cell metabolism. *Ann. Rev. Pl. Physiol.* **18**, 25–58.
123

BITANCOURT A.A. (1949) Mecanismo genetico da tumorisacao nos vegetais. *Segunda semana de Genetica Piracicaba, Sao Paulo, Feb. 8–12, 1949,* p. 7.
376

BLACK L.M. (1944) Some viruses transmitted by agallian leafhoppers. *Proc. Am. Phil. Soc.* **88**, 132–144.
383

BLACK L.M. (1945) A virus tumor disease of plants. *Am. J. Bot.* **32**, 408–415.
382

BLACK L.M. (1949) Virus tumours. *Survey of Biological Progress* **1**, 155–231.
383

BLACK L.M. (1965) Physiology of virus induced tumors in plants, pp. 236–266, in *Handbuch der Pflanzenphysiologie* **XV(2).** Ed. W.Ruhland. Springer Verlag, Berlin.
382

BLACK L.M. (1969) Insect tissue cultures as tools in plant virus research. *Ann. Rev. Phytopathology,* **7**, 73–100.
407

BLACK L.M. (1972) Plant tumors of viral origin. *Progr. exp. Tumor Res.* **15**, 110–137.
382

BLACK L.M. & LEE C.L. (1957) Interaction of growth-regulating chemicals and tumefacient virus on plant cells. *Virology* **3**, 146–159.
383

BLACKMAN V.H. (1919) The compound interest law and plant growth. *Ann. Bot.* **33**, 353–360.
259

BLACKWELL S.J., LAETSCH W.M. & HYDE B.B. (1969) Development of chloroplast fine structure in aspen tissue culture. *Am. J. Bot.* **56**, 457–463.
139

BLAKELY L.M. (1964) Growth and organized development of cultured cells. VI. The behaviour of individual cells in nutrient agar. *Am. J. Bot.* **51**, 792–807.
193

BLAKELY L.M. & STEWARD F.C. (1962) The growth of free cells, II. observations on individual cells and their subsequent patterns of growth, III. The observation and isolation of variant strains. *Am. J. Bot.* **49**, 653.
95.

BLAKELY L.M. & STEWARD F.C. (1964a) Growth and organized development of cultured cells. V. The growth of colonies from free cells in nutrient agar. *Am. J. Bot.* **51**, 780–791.
87

BLAKELY L.M. & STEWARD F.C. (1964b) Growth and organized development of cultured cells. VI. The behaviour of individual cells on nutrient agar. *Am. J. Bot.* **51**, 792–807.
87

BLAKELY L.M. & STEWARD F.C. (1964c) Growth and organized development of cultured cells. VII. Cellular variation. *Am. J. Bot.* **51**, 809–820.
162, 189, 193, 202

BLAYDES D.F. (1966) Interaction of kinetin and various inhibitors in the growth of soybean tissues. *Physiologia Pl.* **19**, 748–753.
216, 218

BLAKELY L.M. & STEWARD F.C. (1961) Growth induction in cultures of *Happlopappus gracilis.* I. The behaviour of cultured cells. *Am. J. Bot.* **48**, 351–358.
152, 260

BODEN G., GORMAN M., JOHNSON I. & SIMPSON P. (1964) Tissue culture studies of *Catharanthus roseus* crown gall. *Lloydia* **27**, 328–333.
295

BONNETT H.T. & TORREY J.G. (1966) Comparative anatomy of endogenous bud and lateral root formation in *Convolvulus arvensis* roots cultured *in vitro*. *Am. J. Bot.* **53**, 496–507.
341

BOONE D.M. (1971) Genetics of *Venturia inequalis*. *Ann. Rev. Phytopathology* **9**, 297–318.
397

BOUCK G.B. & CRONSHAW J. (1965) The fine structure of differentiating sieve tube elements. *J. Cell Biol.* **25**, 79–96.
129

BOURGIN J.P. & NITSCH J.P. (1967) Obtention de *Nicotiana* haploides à partir d'étamines cultivées *in vitro*. *Ann. Physiol. vég.* **9**, 377–382.
9, 215, 216, 218

BOWES B.G. (1969) The fine structure of wall modifications and associated structures in callus cultures of *Andrographis paniculata* Ness. *New Phytol.* **68**, 619–626.
159

BOWES B.G. (1971) The occurrence of shoot teratoma in tissue cultures of *Taraxacum officinale*. *Planta* **100**, 272–276.
341

BOWES B.G. & BUTCHER D.N. (1967) Electron microscopic observations on cell wall inclusions in *Andrographis paniculata* cells. *Z. Pflanzenphysiol.* **58**, 86–89.
49

BOYD R.J., HILDEBRANDT A.C. & ALLEN O.N. (1970) Specificity patterns of *Agrobacterium tumefaciens* phages. *Arch. Mikrobiol.* **73**, 324–330.
372

BRAUN A.C. (1943) Studies on tumour inception in the crown-gall disease. *Am. J. Bot.* **30**, 674–677.
360

BRAUN A.C. (1947) Thermal studies on the factors responsible for tumor initiation in crown gall. *Am. J. Bot.* **34**, 234–240.
361, 363, 364, 375

BRAUN A.C. (1950) Thermal inactivation studies on the tumor-inducing principle in crown gall. *Phytopathology* **40**, 3.
364

BRAUN A.C. (1951) Cellular anatomy in crown gall. *Phytopathology* **41**, 963–966.
363

BRAUN A.C. (1952) Conditioning of the host cell as a factor in the transformation process in crown gall. *Growth* **16**, 65–74.
361

BRAUN A.C. (1958) A physiological basis for autonomous growth of the crown-gall tumour cell. *Proc. natn. Acad. Sci. U.S.A.* **45**, 344–349.
363

BRAUN A.C. (1959) A demonstration of the recovery of the crown-gall tumour cell with the use of complex tumours of single-cell origin. *Proc. natn. Acad. Sci. U.S.A.* **45**, 932–938.
265, 341, 381

BRAUN A.C. (1961) Plant tumours as an experimental model. *Harvey Lectures Ser.* **56**, 191–220.
428

BRAUN A.C. (1962) Tumour inception and development in the crown gall disease. *Ann. Rev. Pl. Physiol.* **13**, 533–558.
265, 363, 364, 367, 374, 375, 380

BRAUN A.C. (1965) The reversal of tumor growth. *Scient. Am.* **213**, 75–83.
381

BRAUN A.C. (1969) Abnormal growth in plants, pp. 379–420, in *Plant Physiology, a Treatise*, vol. **VB**. Ed. F.C.Steward. Academic Press, New York.
375, 380, 390

BRAUN A.C. (1972) The relevance of plant tumor systems to an understanding of the basic cellular mechanisms underlying tumorigenesis. *Progr. exp. Tumor Res.* **15**, 165–187.
390

BRAUN A.C. & LASKARIS T. (1942) Tumour formation by attenuated crown-gall bacteria in the presence of growth-promoting substances. *Proc. natn. Acad. Sci. U.S.A.* **28**, 468–477.
374

BRAUN A.C. & LIPETZ J. (1966) The use of tissue culture in phytopathology, pp. 691–722, in *Cells and Tissues in Culture*, vol. **3**, Methods, Biology and Physiology. Ed. E.W.Willmer. Academic Press, London and New York.
265, 393, 411, 418

BRAUN A.C. & MANDLE R.J. (1948) Studies on the inactivation of the tumor-inducing principle in crown-gall. *Growth* **12**, 255–269.
363, 364

BRAUN A.C. & MOREL G. (1950) A comparison of normal, habituated and crown-

gall tumor tissue implants in the European grape. *Am. J. Bot.* **37**, 499–501.
388

BRAUN A.C. & STONIER T. (1958) Morphlogy and physiology of plant tumours. *Protoplasmatologia* **10**, (5a) 1–93.
357, 364, 382

BRAUN A.C. & WHITE P.R. (1943) Bacteriological sterility of tissues derived from secondary crown-gall tumors. *Phytopathology* **33**, 85–100.
7

BRAUN A.C. & WOOD H.N. (1961) The plant tumour problem. *Adv. Cancer Res.* **6**, 81–109.
265

BRAUN A.C. & WOOD H.N. (1962) On the activation of certain essential biosynthetic systems in cells of *Vinca rosea* L. *Proc. natn. Acad. Sci. U.S.A.* **48**, 1776–1782.
194, 362, 377

BRAUN A.C. & WOOD H.N. (1966) On the inhibition of tumour inception in the crown-gall disease with the use of ribonuclease A. *Proc. natn. Acad. Sci. U.S.A.* **56**, 1417–1782.
365

BREWBAKER J.L. (1957) Pollen cytology and self-incompatibility systems in plants. *J. Hered.* **48**, 271–277.
221

BREWBAKER J.L. & EMERY G.C. (1962) Pollen radiobotany. *Radiat. Bot.* **1**, 101–154.
208

BRIAN P.W. (1967) Obligate parasitism in fungi. *Proc. Roy. Soc. B* **168**, 101–118.
393, 398, 402

BRINKLEY B.R. (1965) The fine structure of the nucleolus in mitotic divisions of Chinese hamster cells *in vitro*. *J. Cell Biol.* **27**, 411–422.
123

BROWN D.J., CANVIN D.T. & ZILKEY B.F. (1970) Growth and metabolism of *Ricinus communis* endosperm in tissue culture. *Can. J. Bot.* **48**, 2323–2331.
261

BROWN E.G. & SHORT K.C. (1969) The changing nucleotide pattern of sycamore cells during culture in suspension. *Phytochem.* **8**, 1365–1372.
281, 284

BROWN R. (1951) The effects of temperature on the duration of the different stages of cell division in the root tip. *J. exp. Bot.* **2**, 96–110.
254

BROWN R. & DYER A. (1972) Cell division in higher plants, pp. 49–50, in *Plant Physiology, a Treatise*, vol. **VIC**. Ed. F.C.Steward. Academic Press, New York.
241

BROWN R. & RICKLESS P.A. (1949) A new method for the study of cell division and cell extension with preliminary observations on the effect of temperature and nutrients. *Proc. Roy. Soc. B* **136**, 110–125.
51, 90

BROWN S.A. (1966) Lignins. *Ann. Rev. Pl. Physiol.* **17**, 223–244.
267

BRUNNER M. & POOTJES C.F. (1969) Bacteriophage release in a lysogenic strain of *Agrobacterium tumefaciens*. *J. Virol.* **3**, 181–186.
372

BRYSON V. (1952) Microbial selection. II. The turbidostatic selector—a device for automatic isolation of bacterial variants. *Science N.Y.* **116**, 48–51.
318

BRYSON V. (1959) Application of continuous culture to microbial selection, pp. 371–380, in *Recent Progress in Microbiology* (VIIth Int. Cong. for Microbiology). Ed. G.Tunerall. Stockholm.
320

BÜCHNER S.A. & STABA E.J. (1964) Preliminary chemical examination of *Digitalis* tissue culture for cardenolides. *J. Pharm. Pharmac.* **16**, 733–737.
267, 295

BUIATTI M. (1968) The induction of tumors in the hybrid *Nicotiana glauca × N. langsdorffii* plants by 8-azauracil and its reversal by uracil and actinomycin D. *Cancer Res.* **28**, 166–169.
384

BUIATTI M. & BENNICI A. (1970) Callus formation and habituation in Nicotiana species in relation to the specific ability for dedifferentiation. *Lincei-Rend. Sci. fis. mat. nat.* **48**, 261–269.
387

BURK L.G. & TSO T.C. (1960) Genetic tumors in *Nicotiana* associated with chromosome loss. *J. Hered.* **51**, 184–187.
386

BURKHOLDER P.R. & NICKELL L.G. (1949) Atypical growth of plants. I. Cultivation

of virus tumours of *Rumex* on nutrient agar. *Bot. Gaz.* **110,** 426–437.
383

BURTON K. (1956) A study of the conditions and mechanisms of the diphenylamine reaction for the colorimetric estimation of deoxyribosenucleic acid. *Biochem. J.* **62,** 315–323.
54

BUTCHER D.N. & CONNOLLY J.D. (1971) An investigation of factors which influence the production of abnormal terpenoids by callus cultures of *Andrographis paniculata* Nees. *J. exp. Bot.* **22,** 314–322.
397, 405

BUTCHER D.N. & STREET H.E. (1960) The effects of gibberellins on the growth of excised tomato roots. *J. exp. Bot.* **11,** 206–216.
90

BUTCHER D.N. & STREET H.E. (1964) Excised root culture. *Biol. Rev.* **30,** 513–586.
5

BUTENKO R.G. (1964) *Plant Tissue Culture and Plant Morphogenesis.* Translated from Russian. Israel Program for Scientific Translation, Jerusalem, 1968.
31, 45, 264, 268

BUTENKO R.G., STROGONOV B.P. & BABAEVA J.A. (1967) Somatic embryogenesis in carrot tissue culture under conditions of high salt concentration in the medium. *Dokl. Akad. Nauk SSSR* **175,** 1179–1181.
351

BUTENKO R.G. & VOLODARSKY A.D. (1968) Analyse immunochimique de la differentiation cellulaire dans les cultures de tissus de tabac. *Physiol. Vég.* **6,** 299–309.
261

BUVAT R. (1945) Recherches sur la dédifferentiation des cellules végétales. *Annls Sci. nat. (Bot. Bio. Vég.)* **6,** 1–119.
339

CAMERON I.L. & PADILLA G.M. (1966) Ed. *Cell Synchrony—Studies in Biosynthetic Regulation.* Academic Press, New York.
298

CAMERON I.L., PADILLA G.M. & ZIMMERMAN A. (1971) Ed. *Developmental Aspects of the Cell Cycle.* Academic Press, New York.
298

CAMPBELL A. (1957) Synchronization of cell division. *Bact. Rev.* **21,** 263–272.
336

CAMPBELL G., CHAN E.C.S. & BARKER W.G. (1965) Growth of lettuce and cauliflower tissues *in vitro* and their production of antimicrobial metabolites. *Can. J. Microbiol.* **11,** 785–789.
397

CAMUS G. (1949) Recherches sur le rôle des bourgeons dans les phénomènes de morphogénèse. *Rev. Cytol. Biol. Vég.* **11,** 1–195.
8, 341

CAMUS G. & GANTHERET R.J. (1948) Nouvelles recherches sur le greffage des tissus normaux et tumoraux sur des fragments des fragments des racines de Scorsonère cultivés *in vitro*. *C.r. Séanc. Soc. Biol.* **142,** 769–771.
380

CAPLIN S.M. (1947) Growth and morphology of tobacco tissue cultures *in vitro*. *Bot. Gaz.* **108,** 379–393.
257, 258, 259

CAPLIN S.M. (1959) Mineral oil overlay for conservation of plant tissue cultures. *Am. J. Bot.* **46,** 324–329.
48, 204

CAPLIN S.M. (1963) Effect of initial size on growth of plant tissue cultures. *Am. J. Bot.* **50,** 91–94.
36, 37

CARCELLER M., DAVEY M.R., FOWLER M.W. & STREET H.E. (1971) The influence of sucrose, 2,4-D and kinetin on the growth, fine structure and lignin content of cultured sycamore cells. *Protoplasma* **73,** 367–385.
147, 154, 159, 160, 291, 294

Carlson P.S. (1969) Production of auxotropic mutants in ferns. *Genet. Res. (Camb.)* **14,** 337–9.
203

CARLSON P.S. (1970) Induction and isolation of auxotrophic mutants in somatic cell cultures of *Nicotiana tabacum*. *Science, N.Y.* **168,** 487–9.
203, 221, 237, 238

CARLSON P.S., SMITH H.H. & DEARING R.D. (1972) Parasexual interspecific plant hybridization. *Proc. natn. Acad. Sci. U.S.A.* **69,** 2292–2294.
429

CARTER O., YAMADA Y. & TAKAHASHI E. (1967) Tissue culture of oats. *Nature, Lond.* **214,** 1029–1030.
31

CASPERSSON T., ZECH L., MODEST E.J.,

FOLEY G.E., WAGH U. & SIMONSSON E. (1969) DNA-binding fluorochromes for the study of the organization of the metaphase nucleus. *Exp. Cell Res.* **58**, 141–152.
177

CHADHA K.C. & SRIVASTAVA B.I. (1971) Evidence for the presence of bacteria-specific proteins in sterile crown gall tumor tissue. *Pl. Physiol., Lancaster* **48**, 125–129.
371

CHAKRAVARTI B. (1966) Attempts to alter infection processes and aggressiveness of *Puccinia graminis* var. *tritici*. *Phytopathology* **56**, 223–229.
396

CHAN W.N. & STABA E.J. (1965) Alkaloid production by *Datura* callus and suspension tissue cultures. *Lloydia* **28**, 55–62.
267, 362

CHANDRA N. & HILDEBRANDT A.C. (1967) Differentiation of plants from TMV inclusion-bearing and inoculum-free tobacco cells. *Virology* **31**, 414–421.
410, 411, 416

CHERRY J.H. (1963) Nucleic acid, mitochondria and enzyme changes in cotyledons on peanut seeds during germination. *Pl. Physiol. Lancaster* **38**, 440–446.
129

CHUPEAU Y. & MOREL G. (1970) Obtention de protoplastes de plantes supérieur à partir de tissus cultivés *in vitro*. *C.r. hebd. Séanc. Acad. Sci., Paris* **270**, 2659–2662.
104

CLAPHAM D. (1971) *In vitro* development of callus from the pollen of *Lolium* and *Hordeum*. *Z. PflZücht.* **65**, 285–292.
214, 215, 216, 221, 225

CLEAVER J.S. (1967) Thymidine metabolism and cell kinetics, in *Frontiers of Biology*, vol. 6. Ed. A.Neuberger & E.Tatum. North-Holland Publ. Co., Amsterdam.
899, 301, 302

COCKING E.C. (1960) A method for the isolation of plant protoplasts and vacuoles. *Nature, Lond.* **187**, 927–929.
9

COCKING E.C. (1966) An electron microscopic study of the initial stages of infection of isolated tomato fruit protoplasts by tobacco mosaic virus. *Planta* **68**, 206–214.
409

COCKING E.C. (1970) Virus uptake, cell wall regeneration and virus multiplication in isolated plant protoplasts. *Inter. Rev. Cytol.* **28**, 89–124.
113, 393, 408, 409

COCKING E.C. (1972) Plant cell protoplasts—isolation and development. *Ann. Rev. Pl. Physiol.* **23**, 29–50.
101

COCKING E.C. (1973a) Fine structure of cultured plant cells, in *Dynamic Aspects of Plant Ultrastructure*. Ed. A.W.Robards. McGraw-Hill, New York. (In press).
101

COCKING E.C. (1973b) Plant cell modification; problems and perspectives, in *Protoplastes et fusion de cellules somatiques végétales*, pp. 327–341, in Colloques Internationaux du C.N.R.S., no. 212, Paris.
120

COCKING E.C. & POJNAR E. (1969) An electron microscope study of the infection of isolated tomato fruit protoplasts by tobacco mosaic virus. *J. gen. Virol.* **4**, 305–312.
409

CONKLIN M.E. & SMITH H.H. (1968) Endogenous beta irradiation and tumor production in an amphiploid *Nicotiana* hybrid. *Am. J. Bot.* **55**, 473–476.
384

CONSTABEL F. (1957) Ernährungsphysiologische und manometrische Untersuchungen zur gewebekultur der *Gymnosporangium*—Gallen von *Juniperus*—Arten. *Biol. Zbl.* **76**, 385–413.
395, 398, 402

CONSTABEL F. (1968) Gerbstoffproduktion der calluskulturen von *Juniperus communis* L. *Planta* **79**, 58–64.
267

CONSTABEL F., MILLER R.A. & GAMBORG O.L. (1971) Histological studies on embryos produced from cell cultures of *Bromus inermis*. *Can. J. Bot.* **49**, 1415–1417.
347

CONSTABEL F., SHYLUK J. & GAMBORG O. (1971) The effect of hormones on anthocyanin accumulation in cell cultures of *Haplopappus gracilis*. *Planta* **96**, 306–316.
270, 294, 324

COOPER L.S., COOPER C.C., HILDEBRANDT A.C. & RIKER A.J. (1964) Chromosome numbers in single cell clones of tobacco tissue. *Am. J. Bot.* **51**, 284–90.
165, 169, 171, 173

COPPING L.G. & STREET H.E. (1972) Properties of the invertases of cultured sycamore cells and changes in their activity during culture growth. *Physiologia Pl.* **26**, 346–354.
291

CORBETT D.C.M. (1970) Maintaining nematode cultures under mineral oil. *Nematologica* **16**, 156.
418

CORDUAN G. (1970) Autotrophe gewebekulturen von *Ruta graveolens* und deren $^{14}CO_2$ markerungsprodukte. *Planta* **91**, 291–301.
262

COUTTS R.H.A., COCKING E.C. & KASSANIS B. (1972) Infection of tobacco mesophyll protoplasts with tobacco mosaic virus. *J. gen Virol.* **17**, 289–294.
101, 111, 113, 114

COX B.J. (1972) *RNA metabolism in cultured plant cells.* Ph.D. Thesis, Univ. Leicester, England.
307

COX B.J., TURNOCK G. & STREET H.E. (1973) Studies on the growth in culture of plant cells. XV. Uptake and utilization of uridine during the growth of *Acer pseudoplatanus* L. cells in suspension culture. *J. exp. Bot.* **24**, 159–174.
297

CRONSHAW J. (1964) Crystal-containing bodies of plant cells. *Protoplasma* **59**, 319–325.
154

CRONSHAW J. & BOUCK G.B. (1965) The fine structure of differentiating xylem elements. *J. Cell Biol.* **24**, 415–431.
125

CRONSHAW J. & ESAU K. (1968) Cell division in leaves of *Nicotiana*. *Protoplasma* **65**, 1–24.
135

CUTTER V.M. (1951) The isolation of plant rusts upon artificial media and some speculations on the metabolism of obligate plant parasites. *Trans. N.Y. Acad. Sci.* **14**, 103–108.
395, 398, 401

CUTTER V.M. (1959) Studies on the isolation and growth of plant rusts in host tissue cultures and upon synthetic media. I. *Gymnosporangium. Mycologia* **51**, 248–295.
395, 397, 398, 401, 402

CUTTER V.M. (1960a) Studies on the isola-

tion and growth of plant rusts in host tissue cultures and upon synthetic media. II. *Uromyces ari-triphylli. Mycologia* **52**, 726–742.
395, 397, 398, 402

CUTTER V.M. (1960b) An axenic culture of *Puccinia malvacearum. Bull. Assoc. Southern Biologists*, **7**, 26 (abstr.).
402

CZOSNOWSKI J. (1952) Physiological features of three types of tissue of *Vitis vinifera*: healthy, crown-gall and chemical tumor, grown *in vitro. Poznán Towarz. Przyjaciol. Nauk. Wydzial Mat.-Pryzyrod. Prace Komisji Biol.* **13**, 189–208.
388

D'AMATO F. (1952) (i) Endopolyploidy in differentiated plant tissues, *Caryologia* **4**, 115–117; (ii) New evidence on endopolyploidy in differentiated plant tissues. *Caryologia* **4**, 121–144.
343

D'AMATO F. (1965) Endopolyploidy as a factor in plant tissue development, pp. 449–462, in *Proc. Intern. Conf. Plant Tissue Cult. Penn. State Univ.* 1963. Ed. P.R.White & A.R.Grove. McCutchan Publ. Corp., Berkeley, California.
182, 184, 204

D'AMATO F., DEVREUX M. & SCARASCIA MUGNOZZA G.T. (1965) The DNA content of the nuclei of the pollen grains in tobacco and barley. *Caryologia* **18**, 377–382.
209

DANNENBERG W.N. & LIVERMAN J.L. (1957) Conversion of tryptophan-2^{14}C to IAA by watermelon tissue slices. *Pl. Physiol. Lancaster* **32**, 263–269.
264, 265

DARLING H.M., FAULKNER L.R. & WALLENDEL P. (1957) Culturing the potato rot nematode. *Phytopathology* **47**, 7 (abstr.).
418

DARLINGTON C.D. & LA COUR L.F. (1970) *The Handling of Chromosomes* (5th edn). Allen & Unwin, London.
165

DARLINGTON C.D. & WYLIE A.P. (1952) A dicentric cycle in *Narcissus*. Symposium on chromosome breakage. *Heredity* (Suppl.) **6**, 197–213.
178

DASGUPTA D.R., NAND S. & SESHADRI A.R. (1970) Culturing, embryology and

life history studies on the lance nematode, *Haplolaimus indicus*. *Nematologica* **16**, 235–248.
418

DAVEY M.R. (1970) *Growth and fine structure of cultured plant cells*. Ph.D. Thesis, Univ. Leicester, England.
154, 157

DAVEY M.R. & COCKING E.C. (1972) Uptake of bacteria by isolated higher plant protoplasts. *Nature, Lond.* **239**, 455–456.
120

DAVEY M.R., FOWLER M.W. & STREET H.E. (1971) Cell clones contrasted in growth, morphology and pigmentation isolated from a callus culture of *Atropa belladonna* var. *lutea*. *Phytochem.* **10**, 2559–2575.
202, 266, 271

DAVEY M.R. & STREET H.E. (1971) Studies on the growth in culture of plant cells. IX. Additional features of the fine structure of *Acer pseudoplantanus* L. cells cultured in suspension. *J. exp. Bot.* **22**, 90–95.
95, 154, 159

DAVIDSON A.W. (1971) *Effect of light on developing Helianthus tuberosus L. callus cultures*. Ph.D. Thesis, University of Edinburgh, Scotland.
57, 243, 251

DAVIES M. (1971) Multisample enzyme extraction from cultured plant cell suspensions. *Pl. Physiol., Lancaster* **47**, 38–42.
105, 289

DAVIES M. (1972) Polyphenol synthesis in cell suspension cultures of Paul's Scarlet Rose. *Planta* **104**, 50–65.
294

DAVIES M.E. (1971) Regulation of histidine biosynthesis in cultured plant cells: evidence from studies on amitrole toxicity. *Phytochem.* **10**, 783–788.
105

DAVIS B.J. (1964) Disc electrophoresis—2. Method and application to human serum proteins. *Ann. N.Y. Acad. Sci.* **121**, 404–427.
55

DE JONG D., JANSEN E. & OLSEN A. (1967) Oxidoreductive and hydrolytic enzyme patterns in plant suspension culture cells. *Exp. Cell Res.* **47**, 138–156.
270, 279

DEKHUIJZEN H.M. & OVEREEM J.C. (1971) The role of cytokinins in clubroot formation. *Physiol. Pl. Path.* **1**, 151–162.
394, 400, 401

DE LANGE A. (1968) *Pathogenesis of Aplanobacter populi in cuttings and explants of Populus candicus*. Dissertation Amsterdam 1968, Phytopathologisch Laboratorium 'Willie Commelin Scholten', Baarn, Mededeling no. 70, pp. 10–71.
416

DELMER D.P. & MILLS S.E. (1968) Tryptophan biosynthesis in cell cultures of *Nicotiana tabacum*. *Pl. Physiol., Lancaster* **43**, 81–87.
267

DÉMÉTRIADES S.D. (1954) Sur le phénomène de differentiation des tissus végétaux tumoraux cultivés *in vitro*. *Ann. Phytopathol. Benaki* **8**, 88–95.
342

DEMOISE C.F. & PARTANEN C.R. (1969) Effects of subculturing and physical condition of medium on the nuclear behaviour of a plant tissue culture. *Am. J. Bot.* **56**, 147–152.
166, 427

DE ROPP R.S. (1947a) The isolation and behavior of bacteria-free crown-gall tissue from primary galls of *Helianthus annuus*. *Phytopathology* **37**, 201–206.
359

DE ROPP R.S. (1947b) The growth-promoting and tumefacient factors of bacteria-free crown-gall tumor tissue. *Am. J. Bot.* **34**, 248–260.
360, 375

DE ROPP R.S. (1948) The interaction of normal and crown-gall tumor tissue in *in vitro* grafts. *Am. J. Bot.* **35**, 372–377.
380

DE ROPP R.S. (1951) The crown-gall problem. *Bot. Rev.* **17**, 629–670.
380, 389

DE ROPP R.S. (1955) The growth and behaviour *in vitro* of isolated plant cells. *Proc. Roy. Soc. B* **144**, 86, 93.
97

DEVREUX M. (1970) New possibilities for the *in vitro* cultivation of plant cells. *Euro-spectra.* **9**, 105–110.
233, 354

DEVREUX M. & SACCARDO F. (1971) Mutazioni sperimentali osservate su piante aploidi di tabacco ottenute per

colture *in vitro* di antere irradiate. *Atti. Ass. Genet. Ital.* **16,** 69–71.
236

DEVREUX M., SACCARDO F. & BRUNORI A. (1971) Plantes haploides et lignes isogeniques de *Nicotiana tabacum* obtenues par cultures d'anthères et de tiges *in vitro. Caryologia* **24,** 141–148.
215

DICKINSON S. (1949) Studies on the physiology of obligate parasitism. II. The behaviour of the germ tubes of certain rusts in contact with various membranes. *Ann. Bot.* **13,** 219–236.
396

DOLLIVER J.S., HILDEBRANDT A.C. & RIKER A.J. (1962) Studies of reproduction of *Aphelenchoides ritzemabosi* Schwartz on plant tissues in culture. *Nematologica* **7,** 294–300.
418, 420

DONACHIE W. & MASTERS M. (1969) Temporal control of gene expression in bacteria, pp. 37–76, in *The Cell Cycle— gene-enzyme interactions.* Ed. G.Padilla, G.Whitson & I.Cameron. Academic Press, New York.
298

DORÉE M., LEGUAY J-J., TERRINE C., SADORGE P., TRAPY F. & GUERN J. (1971) Adaptation à l'adénine des cellules d'Acer pseudoplatanus: modalités d'utilisation de l'adénine exogène, pp. 345–365, in *Les Cultures de Tissus de Plantes.* Colloques Internationaux du C.N.R.S., no. 193, Paris.
297, 312

DOUGALL D.K. (1965) The biosynthesis of protein amino acids in plant tissue culture. I. Isotope competition experiments using glucose-U-C^{14} and the protein amino acids. *Pl. Physiol., Lancaster* **40,** 891–897.
55, 318

DOUGALL D.K. (1970) Threonine deaminase from Paul's Scarlet Rose tissue cultures. *Phytochem.* **9,** 959–964.
56

DOUGALL D.K. (1971) Isotope competition experiments using mixtures of protein amino acids, pp. 367–371, in *Les Cultures de Tissus de Plantes.* Colloques Internationaux du C.N.R.S., no. 193, Paris.
297

DRAVNIEKS P.E., SKOOG F. & BURRIS R.H. (1969) Cytokinin activation of de novo thiamine biosynthesis in tobacco tissue culture. *Pl. Physiol., Lancaster* **44,** 866–870.
267

DUHAMET L. (1957) Sur la possibilité de stimuler la prolifération de tissus végétaux en les associant en culture *in vitro* à des tissus appartenant à une autre espece. *C.r. hebd. Séanc. Acad. Sci., Paris* **245,** 2359–2361.
193

EARLE E. (1965) Cell colony formation from isolated plant cells, pp. 401–409, in *Proc. Int. Conf. on Plant Tissue Culture,* 1963. Ed. R.R.White & A.R.Grove. McCutchan Publ. Co., Berkeley, California.
274

EARLE E.D. & TORREY J.G. (1965) Colony formation by isolated *Convolvulus* cells plated on defined media. *Pl. Physiol., Lancaster* **40,** 520–528.
87, 193, 194, 195, 197, 201

EARLE W.R., SANFORD K.K., EVANS H.K.W. & SHANNON J.E. (1951) The preparation and handling of replicate tissue cultures for quantitative studies. *J. natn. Cancer Inst.* **11,** 907–927.
87

EDELMAN J. & HALL M.A. (1965) Enzyme formation in higher-plant tissues. Development of invertase and ascorbate-oxidase activities in mature storage tissue of *Helianthus tuberosus. Biochem. J.* **95,** 403–410.
129

EL KHALIFA M.D. & EL NUR E.E. (1970) Crown-gall on castor bean leaves. 1. Crown-gall bioassay on primary castor bean leaves. *Angew. Bot.* **44,** 29–37.
366

EL KHALIFA M.D. & LIPPINCOTT J.A. (1968) The influence of plant-growth factors on the initiation and growth of crown-gall tumours on primary pinto bean leaves. *J. exp. Bot.* **19,** 749–759.
374

ELLIOT C. (1951) *Manual of Bacterial Plant Pathogens.* 2nd edn. Chronica Botanica, Waltham, Mass.
357

ENDERLE W. (1951) Tagesperiodische wachstums- und turgoreschwankungen an gewebekulturen. *Planta* **39,** 570–588.
257

ENGELBERG J. (1961) A method of measuring the degree of synchronization of cell populations. *Exp. Cell Res.* **23**, 218–227. 303

ENGELBERG J. (1964) Measurement of degrees of synchrony in cell populations, pp. 497–508, in *Synchrony in Cell Division and Growth*. Ed. E.Zeuthen. Interscience, New York. 303

ENGVILD K.C. (1972) Callus and cell suspension cultures of carnation. *Physiologia Pl.* **26**, 62–66. 280

ERICKSON R.O. (1948) Cytological and growth correlations in the flower bud and anther of *Lilium longiflorum*. *Am. J. Bot.* **35**, 729–739. 239

ERICKSON R.O. (1964) Synchronous cell and nuclear division in tissues of the higher plants, pp. 11–37, in *Synchony in Cell Division and Growth*. Ed. E.Zeuthen. Interscience, New York. 298, 299

ERIKSSON K.B. (1965) Crossing experiments with races of *Ditylenchus dipsaci* on callus tissue cultures. *Nematologica* **11**, 244–248. 419

ERIKSSON K.B. (1972) *Studies on Ditylenchus dipsaci (Kühn) with reference to plant resistance*. Ph.D. Thesis, University of Uppsala, Sweden. 418, 419

ERIKSSON K.B. & GRANBERG J. (1969) Studies of *Ditylenchus dipsaci* races using electrophoresis in acrylamide gel. *Nematologica* **15**, 520–534. 419

ERIKSSON T. (1965) Studies on the growth requirements and growth measurements of cell cultures of *Haplopappus gracilis*. *Physiologia Pl.* **18**, 976–993. 312, 318, 319

ERIKSSON T. (1966) Partial synchronization of cell division in suspension cultures of *Haplopappus gracilis*. *Physiologia Pl.* **19**, 900–910. 303, 304, 305, 312

ERIKSSON T. (1967a) Duration of the mitotic cycle in cell cultures of *Haplopappus gracilis*. *Physiologia Pl.* **20**, 348–354. 278, 300, 301, 303, 312

ERIKSSON T. (1967b) Effects of ultraviolet and X-ray radiation on *in vitro* cultivated cells of *Haplopappus gracilis*. *Physiologia Pl.* **20**, 507–518. 189, 202, 312

ERIKSSON T. & JONASSEN K. (1969) Nuclear division in isolated protoplasts from cells of higher plants grown *in vitro*. *Planta* **89**, 85–89. 9, 103

EVANS G.M. & REES H. (1971) Mitotic cycles in dicotyledons and monocotyledons. *Nature, Lond.* **233**, 350–351. 278

EVANS H. & SCOTT D. (1964) Influence of DNA synthesis on the production of chromatid aberrations by X-rays and maleic hydrazide in *Vicia faba*. *Genetics, Princeton*, **49**, 17–38. 300

EVANS P.K. (1967) *Studies on cell division during early callus development in tissue isolated from Jerusalem artichoke tubers*. Ph.D. Thesis, University of Edinburgh, Edinburgh, Scotland. 54, 129, 246

EVANS P.K., KEATES A.G. & COCKING E.C. (1972) Isolation of protoplasts from cereal leaves. *Planta* **104**, 178–181. 109

FELLENBERG G. (1963) Über die Organbildung an *in vitro* kultiviertem Knollengewebe von *Solanum tuberosum*. *Z. Bot.* **51**, 113–141. 341

FENCL Z. (1966) Theoretical analysis of continuous culture systems, pp. 67–156, in *Theoretical and Methodological Basis of Continuous Culture of Microorganisms*. Ed. I.Málek & Z.Fencl. Academic Press, New York. 334

FERGUSON J.D., STREET H.E. & DAVID S.B. (1958) The carbohydrate nutrition of tomato roots. V. The promotion and inhibition of excised root growth by various sugars and sugar alcohols. *Ann. Bot.* **22**, 513–524. 19, 38, 83

FILNER P. (1965) Semi-conservative replication of DNA in a higher plant cell. *Exp. Cell Res.* **39**, 33–39. 274, 275, 278, 297

FILNER P. (1966) Regulation of nitrate reductase in cultured tobacco cells. *Biochim. biophys. Acta* **118**, 299–310. 290, 291

FLETCHER J.S. & BEEVERS H. (1970) Acetate metabolism in cell suspension cultures. *Pl. Physiol., Lancaster* **45**, 765–772. 279, 287, 288

FOSKET D.E. (1970) The time course of

xylem differentiation and its relation to DNA synthesis in cultured *Coleus* stems egments. *Pl. Physiol., Lancaster* **46**, 64–68.
298

FOSKET D.E. & ROBERTS L.W. (1965) A histochemical study of callus initiation from carrot taproot phloem cultivated *in vitro. Am. J. Bot.* **52**, 929–937.
243, 245

FOSKET D.E. & TORREY J.G. (1969) Hormonal control of cell proliferation and xylem differentiation in cultured tissues of *Glycine max* var. Biloxi. *Pl. Physiol., Lancaster* **44**, 871–880.
298, 351, 426

FOWKE L.C. & SETTERFIELD G. (1968) Cytological responses in Jerusalem artichoke tuber slices during ageing and subsequent auxin treatment, pp. 581–602, in *Physiology and Biochemistry of Plant Growth Substances.* Ed. F.Wightman & G.Setterfield. Runge Press, Ottawa.
127

FOWLER C.W., HUGHES H.G. & JANICK J. (1971) Callus formation from strawberry anthers. *Hort. Res.* **11**, 116–117.
217

FOWLER M.W. (1971) Studies on the growth in culture of plant cells. XIV. Carbohydrate oxidation during the growth of *Acer pseudoplatanus* L. cells in suspension culture. *J. exp. Bot.* **22**, 715–724.
56, 279, 287

FOX D.P. (1969) Some characteristics of the cold hydrolysis technique for staining plant tissues by the Feulgen reaction. *J. Histochem. Cytochem.* **17**, 266–72.
165

FOX J.E. (1963) Growth factor requirements and chromosome number in tobacco tissue cultures. *Physiologia Pl.* **16**, 793–803.
172, 177, 189, 201, 343, 389

FRASER R.S.S. (1968) *The synthesis and properties of ribonucleic acid in dividing plant cells.* Ph.D. Thesis, University of Edinburgh, Edinburgh, Scotland.
55, 56, 250

FREARSON E.M., POWER J.B. & COCKING E.C. (1973) The isolation and culture of *Petunia* leaf protoplasts. *Devel. Biology* **33** (in press).
101, 110, 118

FREDERICK S.E. & NEWCOMB E.H. (1969) Cytochemical localization of catalase in leaf microbodies (peroxisomes). *J. Cell Biol.* **43**, 343–353.
154

FREDERICK S.E., NEWCOMB E.H., VIGIL E.L. & WERGIN W.P. (1968) Fine structural characterisation of plant microbodies. *Planta* **81**, 229–252.
123, 129

FRENCH C.S. & MILNER H.W. (1955) in *Methods in Enzymology*, vol. I, p. 64. Ed. S.P.Colowick & N.O.Kaplan. Academic Press, Inc., New York.
30

FREY-WYSSLING A. & MUHLETHALER K. (1965) *Ultrastructural Plant Cytology.* Elsevier, Amsterdam.
135

FRIEND J. (1973) Resistance of potato to *Phytophthora*, pp. 383–396, in *Fungal Pathogenicity and the Plant's Response*, Ed. R.J.V.Byrde & C.V.Cutting. Academic Press, London.
404

FRITIG B., HIRTH L. & OURISSON G. (1970) Biosynthesis of the coumarins: Scopoletin formation in tobacco tissue cultures. *Phytochem.* **9**, 1963–1975.
267, 268

FUJII T. (1970) Callus formation in wheat anthers. *Japanese National Institute of Genetics Annual Report* no. 20, 91–92.
217

FUKUMI T. & HILDEBRANDT A.C. (1967) Growth and chlorophyll formation in edible green plant callus tissues *in vitro* on media with limited sugar supplements. *Bot. Mag., Tokyo* **80**, 199–212.
262

FURUHASHI K. & YATAZAWA M. (1970) Methionine-lysine-threonine-isoleucine interrelationships in the amino-acid nutrition of rice callus tissue. *Pl. Cell Physiol.* **11**, 569–578.
267

FURUYA T. (1968) Metabolic products and their chemical regulation in plant tissue cultures. *Kitasato Archives Exp. Medicine* **41**, 47–63.
397

FURUYA T., HIROTANI M. & KAWAGUCHI K. (1971) Biotransformation of progesterone and pregnenolone by plant suspension cultures. *Phytochem.* **10**, 1013–1017.
296

FURUYA T., HIROTANI M. & SHINOHANA T. (1970) Biotransformation of digitoxin by suspension callus cultures of *Digitalis purpurea. Chem. Pharm. Bull.* **18**, 1080–1081.
296

FURUYA T., KOJIMA H., & SYONO K. (1971) Regulation of nicotine biosynthesis by auxins in tobacco callus tissues. *Phytochem.* **10**, 1529–1532.
262, 267

GADGIL V.N. & ROY S.K. (1961a) Studies on crown gall tumour. I. Host susceptibility of the causal organism, *Agrobacterium tumefaciens* Strain B-23. *Trans. Bose Res. Inst.* **24**, 141–146.
357

GADGIL V.N. & ROY S.K. (1961b) Studies on crown gall tumour. II. Isolation of bacteria-free crown gall tumour tissue. *Trans. Bose Res. Inst.* **24**, 147–153.
360

GALSKY A.G. & LIPPINCOTT J.A. (1967) Production of a gibberellin-like substance by strains of *Agrobacterium tumefaciens*. *Pl. Physiol. Lancaster* **42**, 5–29.
374

GALSTON A.W. (1959) Studies on indole-acetic acid oxidase inhibitor and its relation to photomorphogenesis, pp. 137–157, in *Photoperiodism and related Phenomena in Plants and animals*. Ed. R.B.Withrow. Amer. Assoc. Adv. Sci., Washington.
264

GALSTON A.W. & HILLMAN W.S. (1961) The degradation of auxin. *Handbuch der Pflanzenphysiologie Bd.* **14**, 647–670.
264

GAMBORG O.L. (1966) Aromatic metabolism in plants. II. Enzymes of the shikimate pathway in suspension cultures of plant cells. *Can. J. Biochem.* **44**, 791–799.
297

GARBER E.D. & GOLDMAN M. (1956) The response of grape tissue cultures to inoculation with biochemical mutants of *Erwinia aroideae*. *Bot. Gaz.* **118**, 128–130.
416

GAUTHERET R.J. (1934) Culture du tissu cambial. *C.r. hebd. Séanc. Acad. Sci., Paris* **198**, 2195–2196.
6, 246

GAUTHERET R.J. (1937) Nouvelles recherches sur la culture du tissu cambial *C.r. hebd. Séanc. Acad. Sci., Paris* **205**, 572–574.
6

GAUTHERET R.J. (1938) Sur le repiquage des cultures de tissu cambial de *Salix*

capraea. *C.r. hebd. Séanc. Acad. Sci., Paris* **206**, 125-127.
6

GAUTHERET R.J. (1939) Sur la possibilité de réaliser la culture indéfinie des tissu de tubercules de carotte. *C.r. hebd. Séanc. Acad. Sci., Paris* **208**, 118–121.
6, 255

GAUTHERET R.J. (1941) Sur le repiquage des cultures de tissues d'Endive, de Salsifis et de Topinambour. *C.r. hebd. Séanc. Acad. Sci., Paris* **213**, 317–318.
362

GAUTHERET R.J. (1942) *Manuel technique de culture des tissus végétaux*. Masson, Paris.
39, 43, 362

GAUTHERET R.J. (1945) *Une voie nouvelle en Biologie végétale, La culture des tissus végétaux*. Gallimard, Paris.
344

GAUTHERET R.J. (1946) Comparison entre l'actions de l'acide indole-acetique et celle du *Phytomonas tumefaciens* sur la croissance des tissus végétaux. *C.r. Séanc. Soc. Biol.* **140**, 169–171.
388

GAUTHERET R.J. (1947) Action de l'acide indole-acetique sur le développement des tissus normaux et des tissus de crown-gall de Topinambour cultivé *in vitro*. *C.r. hebd. Séanc. Acad. Sci., Paris* **224**, 1728–1730.
375

GAUTHERET R.J. (1948) Sur la culture de trois types de tissus de Scorsonère: tissus normaux, tissus de crown-gall et tissus accoutumés à l'hetero-auxin. *C.r. hebd. Séanc. Acad. Sci., Paris* **226**, 270–271.
388

GAUTHERET R.J. (1953) Recherches anatomiques sur la culture des tissus de rhizomes de topinambour et d'hybrides de soleil et de Topinambour. *Rev. Gen. Bot.* **60**, 129–193.
243

GAUTHERET R.J. (1957) Histogenesis in plant tissue cultures. *J. natn. Cancer Inst.* **19**, 555–573.
7

GAUTHERET R.J. (1959) *La Culture des Tissus Végétaux, Techniques et Realisations*. Masson, Paris.
31, 43, 256, 360

GAUTHERET R.J. (1966) Factors affecting differentiation of plant tissues grown *in vitro*, pp. 55–71, in *Cell Differentiation and Morphogenesis*. Ed. W.Beermann. North Holland Publ. Co.
7, 137, 256, 339, 346, 349, 424

GAUTHERET R.J. (1971) Action de variations de température sur la rhizogenèse des tissus de Topinambur cultivés *in vitro*, pp. 187–199, in *Les Cultures de Tissus de Plantes*, Colloques Internationaux du C.N.R.S., no. 193, Paris.
344

GEITLER L. (1939) Die Entstehung der polyploiden Somakerne der Heteropteren durch Chromosomenteilung ohne Kernteilung. *Chromosoma* 1, 1–22.
167

GEROLA F.M. & BASSI M. (1964) Sui cristalleidi proteice delle cellule vegetali. *Caryologia* 17, 399–407.
127

GEROLA F.M. & DASSU G., (1960) L'evoluzione dei chloroplasti durante l'inverdimento sperimentale di frammenti di tuberi di Topinambour. (*Helianthus tuberosus*). *Nuovo Giorn. Bot. Ital.*, n.s. 67, 63–78.
125, 140

GIBBS J.L. & DOUGALL D.K. (1963) Growth of single plant cells. *Science, N.Y.* 141, 1059.
195

GILES K.L. (1972) An interspecific aggregate cell capable of cell wall regeneration. *Pl. Cell Physiol.* 13, 207–210.
104

GIVAN C.V. & COLLIN H.A. (1967) Studies on the growth in culture of plant cells. II. Changes in respiration rate and nitrogen content associated with the growth of *Acer pseudoplatanus* cells in suspension culture. *J. exp. Bot.* 18, 321–331.
281

GOH C.J. (1971) Respiratory enzyme systems in cultured callus tissues. *New Phytol.* 70, 389–395.
56

GOLDACRE P.L., UNT H. & KEFFORD N.P. (1962) Cultivation of isolated tissue derived from the pericycle of roots. *Nature, Lond.* 193, 1305–1306.
32

GOLDACRE P.L., GALSTON A.W. & WEINTRAUB R.L. (1953) The effect of substituted phenols on the activity of the indoleacetic acid oxidase of peas. *Archs. Biochem. Biophys.* 43, 358–373.
264

GOLDMAN A. & MOREL G. (1967) Cultures et particularités biochimiques de differents tissus d'*Opuntia vulgaris*, pp. 251–257, in *Les cultures de tissus de plantes*. Colloque C.N.R.S., Strasbourg.
379

GOLDMANN-MÉNAGÉ A. (1971) *Recherches sur le métabolisme azoté des tissus de crown-gall cultivé in vitro*. Thèse Sci. Univ. Paris.
362, 379

GONZALEZ-FERNANDEZ A., GIMENEZ-MARTIN G., DIEZ J.L., TORRE C. DE LA & LOPEZ-SAEZ J.F. (1971) Interphase development and beginning of mitosis in the different nuclei of polynucleate homokaryotic cells. *Chromosoma* 36, 100–111.
171

GRAEBE J.E. & NOVELLI G.D. (1966) A practical method for large-scale plant tissue culture. *Exp. Cell Res.* 41, 509–520.
67, 314, 315

GRAMBOW H.J., KAO K.W., MILLER R.A. & GAMBORG O.L. (1972) Cell division and plant development from protoplasts of carrot cell suspension cultures. *Planta* 103, 348–355.
104

GRANT M.E. & FULLER K.W. (1968) Tissue culture of root cells of *Vicia faba*. *J. exp. Bot.* 19, 667–680.
260, 263

GRAVES J. & SMITH W.K. (1967) Transformation of pregnenolone and progesterone by cultured plant cells. *Nature, Lond.* 214, 1248–1249.
296

GRIBNAU A.G.M. & VELDSTRA H. (1969) The influence of mitomycin C on the induction of crown-gall tumors. *FEBS Letters* 3, 115–117.
369

GRIFFIN M.J. & COLEY-SMITH J.R. (1968) The establishment of hop tissue cultures and their infection by downy mildew *Pseudoperonospora humuli* (Miy. & Tak.) Wilson under aseptic conditions. *J. gen. Microbiol.* 53, 213–236.
394, 395, 396, 398, 403

GROUT B.W.W., WILLISON J.H.M. & COCKING E.C. (1972) Interactions at the surface of plant cell protoplasts: an

electrophoretic and freeze-etch study. *J. Bioenergetics* **4**, 585–602.
101

GUHA S., IYER R.D., GUPTA N. & SWAMINATHAN M.S. (1970) Totipotency of gametic cells and the production of haploids in rice. *Current Sci.* **39**, 174–176.
215, 219

GUHA S. & MAHESHWARI S.C. (1964) *In vitro* production of embryos from anthers of *Datura*. *Nature, Lond.* **204**, 497.
205, 214, 354

GUHA S. & MAHESHWARI S.C. (1966) Cell division and differentiation of embryos in the pollen grains of *Datura in vitro*. *Nature, Lond.* **212**, 97–98.
214

GUHA S. & MAHESHWARI S.C. (1967) Development of embryoids from pollen grains of *Datura in vitro*. *Phytomorphology* **17**, 454–461.
206, 214, 217, 354

GUILLÉ & GRISVARD J. (1971) Modification of genetic information in crown gall tissue cultures. *Biochem. biophys. Res. Commun.* **44**, 1402.
370

GUNNING B.E.S. & PATE J.S. (1969) 'Transfer cells': Plant cells with wall invaginations specialized in relation to short distance transport of solutes. Their occurrence, structure and development. *Protoplasma* **68**, 107–133.
159

GUPTA N. & CARLSON P.S. (1972) Preferential growth of haploid plant cells *in vitro*. *Nature New. Biol.* **239**, 86.
427

HABERLANDT G. (1902) Kultinversuche mit isolierten Pflanzellen. *Sber. Akad. Wiss. Wien* **111**, 69–92.
5, 243

HACCIUS B. (1971) Zur derzeitigen Situation der Angiospermen-Embryologie. *Bot. Jb.* **91**, 309.
347

HACKETT W.P. & ANDERSON J.M. (1967) Aseptic multiplication and maintenance of differentiated carnation shoot tissue derived from shoot apices. *Proc. Am. Soc. hort. Sci.* **90**, 365–369.
424

HAGEN G.L. (1965) Genetic and metabolic problems in tobacco tumor growth and morphogenesis, pp. 61–68, in *International Conference on Plant Tissue Culture*. Ed. P.R.White & A.R.Grove. McCutchan Publ. Co., Berkeley, California.
387

HAGEN G.L., GUNCKEL J.E. & SPARROW A.H. (1961) Morphology and histology of tumor types induced by X, gamma and beta irradiation of a tobacco hybrid. *Am. J. Bot.* **48**, 691–699.
384

HAHLBROCK K., KUHLEN E. & LINDL T. (1971) Änderungen von Enzymaktivitäten Während des Wachstums von Zellsuspensionskulturen von *Glycine max*: Phenylalanin Ammonia Lyase and p-Cumurat: CoAligase. *Planta* **99**, 311–318.
289

HAKKAART F.A. & QUAK F. (1964) Effect of heat treatment of young plants on freeing chrysanthemums from virus B by means of meristem culture. *Neth. J. Plant Pathol.* **70**, 154–157.
413

HALL M.D. & COCKING E.C. (1971) The bursting response of isolated *Avena* coleoptile protoplasts to indol-3yl-acetic acid. *Biochem. J.* **124**, 33P.
102

HALPERIN W. (1966a) Alternative morphogenetic events in cell suspensions. *Am. J. Bot.* **53**, 443–453.
347

HALPERIN W. (1966b) Single cells, coconut milk and embryogenesis *in vitro*. *Science, N.Y.* **153**, 1287–1288.
351

HALPERIN W. & JENSEN W.A. (1967) Ultrastructural changes during growth and embryogenesis in carrot cell cultures. *J. Ultrastruct. Res.* **18**, 426–443.
127, 137

HALPERIN W. & WETHERELL D.F. (1964) Adventive embryony in tissue cultures of the wild carrot, *Daucus carota*. *Am. J. Bot.* **51**, 274–283.
349

HAMERS-CASTERMAN C. & JEENER R. (1957) An initial ribonuclease-sensitive phase in the multiplication of tobacco mosaic virus. *Virology* **3**, 197.
112

HANSEN A.J. & HILDEBRANDT A.C. (1966) The distribution of tobacco mosaic virus in plant callus cultures. *Virology* **28**, 15–21.
411

HANSON A.D. & EDELMAN J. (1972) Photosynthesis by carrot tissue cultures. *Planta* **102**, 11–25.
262

HARLAND J. (1971) *Changes in the pattern of enzyme activities during the cell division cycle.* Ph.D. Thesis, University of Edinburgh, Edinburgh, Scotland.
55, 56, 242

HARLAND J., JACKSON J. & YEOMAN M.M. (1973) Changes in some enzymes involved in DNA biosynthesis following induction of division in cultured plant cells. *J. Cell Sci.* (in press).
247

HARN C. (1969) Studies on the anther culture of rice. *Korean J. Breeding* **1**, 1–6.
216

HARN C. (1970) Studies on the anther culture of rice. *Korean J. Breeding* **13**, 17–23.
216

HARN C. (1971) Studies on anther culture in *Solanum nigrum. SABRAO, Newsletter* **3**, 39–42.
216, 234

HARN C. (1972) II. Cytological and histological observations. *SABRAO, Newsletter* **4**, 27–32.
216

HARRIS H. (1970) *Cell Fusion.* Oxford University Press.
171

HARVEY A.E. & GRASHAM J.L. (1969a) Growth of the rust fungus *Cronartium ribicola* in tissue cultures of *Pinus monticola. Can. J. Bot.* **47**, 663–666.
395, 398, 399

HARVEY A.E. & GRASHAM J.L. (1969b) The relative susceptibility of needle-and-stem-derived white pine tissue cultures to artificial inoculation with mycelium of *Cronartium ribicola. Can. J. Bot.* **47**, 1789–1790.
395, 397

HARVEY A.E. & GRASHAM J.L. (1969c) Procedure and media for obtaining tissue cultures of twelve conifer species. *Can. J. Bot.* **47**, 547–549.
395

HARVEY A.E. & GRASHAM J.L. (1970a) Growth of *Cronartium ribicola* in the absence of physical contact with its host. *Can. J. Bot.* **48**, 71–73.
395, 402

HARVEY A.E. & GRASHAM J.L. (1970b) *In vivo* verification of *Cronartium ribicola* propagated on tissue cultures of *Pinus monticola. Can. J. Bot.* **48**, 1429–1430.
395

HARVEY A.E. & GRASHAM J.L. (1970c) Inoculation of western white pine tissue cultures with basidiospores of *Cronartium ribicola. Can. J. Bot.* **48**, 1309–1312.
395, 397

HARVEY A.E. & GRASHAM J.L. (1970d) Survival of *Cronartium ribicola* on a medium containing host tissue culture diffusates. *Mycopath. Mycol. appl.* **40**, 245–248.
395, 402, 403

HARVEY A.E. & GRASHAM J.L. (1971a) Inoculation of a nonhost tissue culture with *Cronartium ribicola. Can. J. Bot.* **49**, 881–882.
395, 403

HARVEY A.E. & GRASHAM J.L. (1971b) Production of the nutritional requirements for growth of *Cronartium ribicola* by a nonhost species. *Can. J. Bot.* **49**, 1517–1519.
395, 403

HARVEY A.E., GRASHAM J.L. & WALDRON C.C. (1971) The effects of growth regulating compounds on healthy and blister rust-infected tissue cultures of *Pinus monticola. Phytopathology* **61**, 507–509.
395, 399

HARVEY A.E. & WOO J.Y. (1971) Histopathology and cytology of *Cronartium ribicola* in tissue cultures of *Pinus monticola. Phytopathology* **61**, 773–779.
395

HEBLE M.R., NARAYANASWAMI S. & CHADHA M.S. (1968) Diosgenin and β-sitosterol isolated from *Solanum xanthocarpum* tissue cultures. *Science N.Y.* **161**, 1145.
267, 268, 295

HEIM J.M. & GRIES G.A. (1953) The culture of *Erysiphe cichoracearum* on sunflower tumour tissue. *Phytopathology* **43**, 343–344.
394, 396

HEINZ D.J. & MEE G.W.P. (1971) Morphological, cytogenetic and enzymatic variation in *Saccharum* species hybrid clones derived from callus tissue. *Ann. J. Bot.* **58**, 257–262.
405

HEINZ D.J., MEE G.W.P. & NICKELL L.G.

(1969) Chromosome numbers of some *Saccharum* species hybrids and their cell suspension cultures. *Am. J. Bot.* **56,** 450–456.
173, 174, 175, 177

HELGESON J.P., KEMP J.D., HABERLACH G.T. & MAXWELL D.P. (1972) A tissue culture system for studying disease resistance: the black shank disease in tobacco callus cultures. *Phytopathology* **62,** 1439–1443.
406

HELLER R. (1953) Recherches sur la nutrition minérale des tissus végétaux cultivés *in vitro*. Thèse, Paris, et *Annls Sci. Nat.* (*Bot. Biol. Veg.*), Paris **14,** 1–223.
39, 43, 104, 217, 362

HELLER R. & GAUTHERET R.J. (1949) Sur l'emploi de papeir filtre sans centres comme support pour les cultures de tissus vegetaux. *C.r. Séanc. Soc. Biol.* **143,** 335–337.
45

HELLMANN S. & REINERT J. (1971) Protoplasten aus Zellkulturen von *Daucus carota*. *Protoplasma* **72,** 479–484.
104

HENDERSON J.H.M. (1954) The changing nutritional pattern from normal to habituated sunflower callus tissue *in vitro*. *L'Année Biol. 3ᵉ Sér.* **30,** 329–348.
388

HENDERSON J.H.M. & BONNER J. (1952) Auxin metabolism in normal and crown gall tissue of sunflower. *Am. J. Bot.* **39,** 444–451.
264, 376

HENDERSON J.H.M., DURRELL M.E. & BONNER J. (1952) The culture of normal sunflower stem callus. *Am. J. Bot.* **39,** 467–473.
257

HENDRICKSON A.A., BALDWIN I.L. & RIKER A.J. (1934) Studies on certain physiological characters of *Phytomonas tumefaciens Phytomonas rhizogenes* and *Bacillus radiobacter*, Part II. *J. Bact.* **28,** 597–618.
358

HENSHAW G.G., JHA K.K., MEHTA A.R. SHAKESHAFT D.J. & STREET H.E. (1966) Studies on the growth in culture of plant cells. I. Growth patterns in batch propagated suspension cultures. *J. exp. Bot.* **17,** 362–377.
65, 149, 152, 198, 272, 273, 277, 297, 312

HEPLER P.K. & NEWCOMB E.H. (1964) Microtubules and fibrils in the cytoplasm of *Coleus* cells undergoing secondary wall deposition. *J. Cell Biol.* **20,** 529–533.
125

HERBERT D. (1959) Some principles of continuous culture, pp. 381–396, in *Recent Progress in Microbiology* (VIIth Int. Cong. of Microbiology). Ed. G. Tunevall. Stockholm.
323

HERBERT D. (1961) The chemical composition of micro-organisms as a function of their environment. *Symp. Soc. gen. Microbiol.* **11,** 391–416.
336

HERBERT D., ELSWORTH R. & TELLING R. (1956) The continuous culture of bacteria; a theoretical and experimental study. *J. gen. Microbiol.* **14,** 601–622.
323, 334

HILDEBRANDT A.C. (1962) Tissue and single cell cultures of higher plants as a basic experimental method, pp. 383–421, in *Modern Methods of Plant Analyses*, Vol. **5.** Ed. H.F.Binskens & M.V.Tracey. Springer Verlag, Berlin.
215

HILDEBRANDT A.C. (1963) Growth of single cell clones of diseased and normal tissue origins, pp. 1–22, in *Plant Tissue Culture and Morphogenesis*. Scholar's Library, New York.
417

HILDEBRANDT A.C. (1965) Growth *in vitro* of plant single cell clones of normal and diseased origins, pp. 411–427, in *Proc. Int. Conf. on Plant Tissue Culture*. Ed. P.R. White and A.R.Grove. McCutchan Publ. Co., Berkeley, California.
274

HILDEBRANDT A.C., RIKER A.J. & DUGGAR, B.M. (1946) The influence of the composition of the medium on growth *in vitro* of excised tobacco and sunflower tissue cultures. *Am. J. Bot.* **33,** 591–597.
39, 43

HILLARY B.B. (1939) Use of Feulgen reaction in cytology. I. Effect of fixatives on the reaction. *Bot. Gaz.* **101,** 276–300.
93

HILLARY B.B. (1940) Use of the Feulgen reaction in cytology. II. New techniques and special applications. *Bot. Gaz.* **102,** 225–235.
93

HILLMAN W.S. & GALSTON A.W. (1957) Inductive control of indoleacetic acid oxidase activity by red and near infra red light. *Pl. Physiol., Lancaster* **32,** 129–135.
204

HIRTH L. & DURR A. (1971) Données récentes sur la Multiplication du Virus de la Mosaique du Tabac dans des Tissus de Tabac cultivés *in vitro,* pp. 481–498, in *Les Cultures de Tissus de Plantes,* Colloques Internationaux du C.N.R.S., no. 193, Paris.
32, 410, 412

HIRTH L. & LEBEURIER G. (1965) Remarques sur la sensibilité des cellules des cultures de tissus de tabac à l'infection par le virus de la mosaique du tabac ou son acide ribonucleique. *Rev. Gen. Bot.* **72,** 5–20.
407, 411

HIRTH L. & SEGRETAIN G. (1956) Quelques aspects de la multiplication du virus de la mosaique du tabac en culture de tissus. *Ann. Inst. Pasteur* **91,** 523–536.
411

HOLCOMB G.E., HILDEBRANDT A.C. & EVERT R.F. (1967) Staining and acid phosphatase reactions of spherosomes in plant tissue culture cells. *Am. J. Bot.* **53,** 1204–1209.
142, 147, 148

HOLLINGS M. (1961) Virology. *Glasshouse Crops Res. Inst. Rept.* 1960, 72–78.
413

HOLLINGS M. (1962) Heat treatment in the production of virus-free ornamental plants. *Natl. Agric. Advis. Ser. Quart. Rev.* **57,** 31–34.
413

HOLLINGS M. (1963) Virology. *Glasshouse Crops Res. Inst. Ann. Rept.* 1962, 86–93.
413

HOLLINGS M. (1964) Virology. *Glasshouse Crops Res. Inst. Rept.* 1963, 87–92.
413

HOLLINGS M. (1965) Disease control through virus-free stock. *Ann. Rev. Phytopathology* **3,** 367–396.
393, 406, 412, 413, 414, 415, 424

HOLLINGS M. & STONE O.M. (1964) Investigation of carnation viruses. I. Carnation mottle. *Ann. appl. Biol.* **53,** 103–118.
413, 415

HOLLINGS M. & STONE O.M. (1965) Investigation of carnation viruses. II. Carnation ringspot. *Ann. appl. Biol.* **56,** 73–86.
413

HOLLINGS M. & STONE O.M. (1968) Techniques and problems in the production of virus-tested planting material. *Sci. Hort.* **20,** 57–72.
413, 414, 415

HOLSTEN R.D., BURNS R.C., HARDY R.W.F. & HEBERT R.R. (1971) Establishment of symbiosis between *Rhizobium* and plant cells *in vitro. Nature, Lond.* **232,** 173–175.
417

HOTSON H.H. (1953) The growth of rust in tissue culture. *Phytopathology* **43,** 360–363.
395, 398

HOTSON H.H. & CUTTER V.M. (1951) The isolation and culture of *Gymnosporangium juniperi-virginianae* Schw. upon artificial medium. *Proc. nat. Acad. Sci., U.S.A.* **37,** 400–403.
395, 397, 398, 401, 402

HOURSANGIOU-NEUBRUN & PUISIEUX-DAO S. (1969) Premières observations mettant en evidence la possibilité pour des phages de l'*Agrobacterium tumefaciens* d'être responsables du crown-gall. *C.r. hebd. Séanc. Acad. Sci., Paris* **268,** 1493–1494.
373

HOWARD A. & PELC S.R. (1951) Nuclear incorporation of P^{32} as demonstrated by autoradiographs. *Exp. Cell Res.* **2,** 178–187.
299

HOWARD A. & PELC S.R. (1953) Synthesis of desoxyribonucleic acid in normal and irradiated cells and its relation to chromosome breakage. *Heredity, Lond.* **6** (Suppl.), 261–273.
301

HYDE B.B., HODGE A.J., KAHN A. & BIRNSTIEL M.L. (1963) Studies on phytoferritin. I. Identification and localization. *J. Ultrastruct. Res.* **9,** 248–258.
160

INGRAM D.S. (1967) The expression of R-gene resistance to *Phytophthora infestans* in tissue cultures of *Solanum tuberosum. J. gen. Microbiol.* **49,** 99–108.
404

INGRAM D.S. (1969a) Growth of *Plasmodiophora brassicae* in host callus. *J. gen. Microbiol.* **55,** 9–18.
394, 398, 399, 400

INGRAM D.S. (1969b) Abnormal growth of tissues infected with *Plasmodiophora brassicae. J. gen. Microbiol.* **56,** 55–67.
394, 400, 401

INGRAM D.S. (1969c) The susceptibility of *Brassica* callus to infection by *Peronospora parasitica*. *J. gen. Microbiol.* **58**, 391–401.
394, 395, 396, 398, 405, 419

INGRAM D.S. (1971) An attempt to establish dual cultures of *Synchytrium endobioticum* and *Solanum tuberosum* callus. *Phytopath. Z.* **71**, 21–24.
394, 401

INGRAM D.S. & BUTCHER D.N. (1972) Benazolin (4-chloro-2-oxobenzothiazolin-3yl-acetic acid) as a growth factor for initiating and maintaining plant tissue cultures. *Z. Pflanzenphysiol.* **66**, 206–214.
390

INGRAM D.S. & JOACHIM I. (1971) The growth of *Peronospora farinosa* f.sp. *betae* and sugar beet callus tisues in dual culture. *J. gen. Microbiol.* **69**, 211–220.
394, 395, 396, 399, 400, 403

INGRAM D.S. & ROBERTSON N.F. (1965) Interaction between *Phytophthora infestans* and tissue cultures of *Solanum tuberosum*. *J. gen. Microbiol.* **40**, 431–437.
404

INGRAM D.S. & TOMMERUP I.C. (1972) The life history of *Plasmodiophora brassicae* Woron. *Proc. Roy. Soc. B*, **180**, 103–112.
394, 399, 401

INGRAM D.S. & TOMMERUP I.C. (1973b) The study of obligate parasites *in vitro*, pp. 121–140, in *Fungal Pathogenicity and the Plant's Response*, Ed. R.J.W.Byrde & C.V.Cutting. Academic Press, London.
394, 395, 401, 402, 405, 419

ISRAEL H.W. & STEWARD F.C. (1966) The fine structure of quiescent and growing carrot cells: its relation to growth induction. *Ann. Bot.* **30**, 63–79.
49, 123, 127, 129

ISRAEL H.W. & STEWARD F.C. (1967) The fine structure and development of plastids in cultured cells of *Daucus carota*. *Ann. Bot.* **31**, 1–18.
49, 125, 127, 137, 139, 140

IZARD C., LACHARPAGNE J. & SCHILTZ P. (1964) Behaviour of *Peronospora tabacina* in tissue cultures, and role of the leaf epidermis, pp. 95–99, in *Annales de la Direction des Études et l'Équipment*. Service d'Exploitation Industrielle des Tabacs et des Allumettes (Section 2).
394, 396

JACQUIOT C. (1951) Action du meso-inositol et de l'adenine sur la formation de bourgeons par le tissu cambial. *C.r. hebd. Séanc. Acad. Sci., Paris* **233**, 815–817.
339

JAMES T.W. (1966) Cell synchrony, a prologue to discovery, pp. 1–13, in *Cell Synchrony—Studies in Biosynthetic Regulation*. Ed. I.Cameron & G.Padilla. Academic Press, New York.
303

JASPARS E.M.J. & VELDSTRA H. (1965a) An α-amylase from tobacco crown-gall tissue cultures. I. Purification and some properties of the enzyme pattern of α-amylase isoenzymes in different tobacco tissue. *Physiologia Pl.* **18**, 604–625.
56

JASPARS E.M.J. & VELDSTRA H. (1965b) An α-amylase from tobacco crown-gall tissue cultures. II. Measurements of the activity in media and tissues. *Physiologia Pl.* **18**, 626–634.
56

JEFFS R.A. & NORTHCOTE D.H. (1967) The influence of indol-3yl-acetic acid and sugar in the pattern of induced differentiation in plant tissue cultures. *J. Cell Sci.* **2**, 77–88.
8

JENSEN R.G., FRANCKI R.I.B. & ZAITLIN M. (1971) Metabolism of separated leaf cells. I. Preparation of photosynthetically active cells from tobacco. *Pl. Physiol. Lancaster* **48**, 9–13.
109

JENSEN W.A. (1962) *Botanical Histochemistry*. Freeman, London.
48, 50

JENSEN W.A. (1965) Composition and ultrastructure of the nucellus in cotton. *J. Ultrastruct. Res.* **13**, 112–128.
137, 158

JOHANNEAU J.P. & PÉAUD-LENOËL C. (1967) Croissance et synthèse des proteines de suspensions cellulaires de Tabac sensible à la kinétine. *Physiologia Pl.* **20**, 834–850.
93

JOHANSEN D.A.J. (1940) *Plant Microtechnique*. McGraw-Hill, New York and London.
48

JOHNSON R.N. & VIGLIERCHIO D.R. (1969a) Sugar beet nematode (*Heterodera schachtii*) reared on axenic *Beta vulgaris* root

explants. I. Selected environmental factors affecting penetration. *Nematologica* **15**, 129–143.
418

JOHNSON R.N. & VIGLIERCHIO D.R. (1969b) Sugar beet nematode (*Heterodera schachtii*) reared on axenic *Beta vulgaris* root explants. II. Selected environmental and nutritional factors affecting development and sex ratio. *Nematologica* **15**, 144–152.
418

JONES L.E., HILDEBRANDT A.C., RIKER A.J. & WU J.H. (1960). Growth of somatic tobacco cells in microculture. *Am. J. Bot.* **47**, 468–475.
97, 99, 123, 145, 147, 152, 153, 193

JONES O.P. & VINE S.J. (1968) The culture of gooseberry shoot tips for eliminating viruses. *J. Hort. Sci.* **43**, 289–292.
413, 424

JORDAN E.G. & CHAPMAN J.M. (1971) Ultrastructural changes in the nucleoli of Jerusalem artichoke (*Helianthus tuberosus*) tuber discs. *J. exp. Bot.* **22**, 627–634.
131

JOSHI P.C. & BALL E. (1968a). Growth values and regenerative potentiality in mesophyll cultures of *Arachis hypogaea*. *Z. Pflanzenphysiol.* **59**, 109–123.
32

JOSHI P.C. & BALL E. (1968b) Growth of isolated palisade cells of *Arachis hypogaea in vitro*. *Developmental Biology* **17**, 308–325.
108

JOUANNEAU J-P. (1971) Contrôle par les cytokinines de la synchronisation des mitoses dans les cellules de Tabac. *Exp. Cell Res.* **67**, 329–337.
304

KADO C.I., HESKETT M.G. & LANGLEY R.A. (1972) Studies on *Agrobacterium tumefaciens*: characterization of strains 1D135 and B6, and analysis of the bacterial chromosome, transfer RNA and ribosomes for tumor-inducing ability. *Physiol. Pl. Path.* **2**, 47–57.
367, 369

KAMEYA T. & HINATA K. (1970) Induction of haploid plants from pollen grains of *Brassica*. *Jap. J. Breeding* **20**, 82–87.
210, 214, 215, 221, 354

KANDLER O. (1952) Über eine physiologische Umstimmung von Sonnenblumenstengelgewebe durch Dauereinwirkung von β-Indolylessigsäure. *Planta* **40**, 346–349.
388, 389

KANEKO T. (1967) Comparative studies on various calluses and crown gall of Jerusalem artichoke. II. Some biochemical properties. *Pl. Cell Physiol. Tokyo*, **8**, 375–384.
261

KAO K.N., GAMBORG O.L., MILLER R.A. & KELLER W.A. (1971) Cell divisions in cells regenerated from protoplasts of soybean and *Haplopappus gracilis*. *Nature, New Biol.* **232**, 124.
103

KAO K.N., MILLER R.A., GAMBORG O.L. & HARVEY B.L. (1970) Variations in chromosome number and structure in plant cells grown in suspension cultures. *Can. J. Genet. Cytol.* **12**, 297–301.
162, 166, 177, 178, 181, 189

KASPERBAUER M.J. & COLLINS G.B. (1972). Reconstitution of diploids from leaf tissue of anther-derived haploids in tobacco. *Crop Sci.* **12**, 98–101.
235

KASPERBAUER M.J. & REINERT R.A. (1967) Photometrically assayable phytochrome *in vivo* in callus tissue cultured from *Nicotiana tabacum*. *Physiologia Pl.* **20**, 977–981.
266

KASSANIS B. (1957a) The multiplication, of tobacco mosaic virus in cultures of tumerous tobacco tissues. *Virology* **4**, 5–13.
410, 411

KASSANIS B. (1957b) The use of tissue culture to produce virus-free clones from infected potato varieties. *Ann. appl. Biol.* **45**, 422–427.
413

KASSANIS B. (1967) Plant tissue culture, pp. 537–566, in *Methods in Virology*, vol. 1. Ed. K.Maramorosch & H.Koprowski. Academic Press, London.
393, 407, 408, 411

KASSANIS B. & MACFARLANE I. (1964) Transmission of tobacco necrosis virus to tobacco callus tissues by zoospores of *Olpidium brassicae*. *Nature, Lond.* **201**, 218–219.
409

KASSANIS B., TINSLEY T.W. & QUAK F. (1958) The inoculation of tobacco callus tissue with tobacco mosaic virus. *Ann. appl. Biol.* **46**, 11–19.
407, 410, 412

KASSANIS B. & VARMA A. (1967) The pro-

duction of virus-free clones of some British potato varieties. *Ann. appl. Biol.* **59**, 447–450.
413

KATO H. & TAKEUCHI M. (1963) Morphogenesis *in vitro* starting from single cells of carrot roots. *Pl. Cell Physiol.* **4**, 243–245.
347

KATO Y. (1963) Physiological and morphogenetic studies of fern gametophytes in aseptic culture. I. Callus tissues from dark-cultured *Pteris vittata. Bot. Gaz.* **124**, 413–416.
31

KATO Y. (1964) Physiological and morphogenetic studies of fern gametophyte by aseptic culture. III. Growth and differentiation of single cells isolated from callus tissues of *Pteris vittata. Cytologia* **29**, 79–85.
31

KATSUI N., MURAI A., TAGASUKI M., IMAIZUMI K. & MASAMUNE T. (1968) The structure of rishitin, a new antifungal compound from diseased potato tubers. *Chem. Comm.* 43–44.
404

KATZ J., ABRAHAM S. & BAKER N. (1954) Analytical procedures using a combined combustion-diffusion vessel. Improved method for combustion of organic compounds in aqueous solution. *Analyt. Chem.* **26**, 1503–1504.
94

KATZNELSON H. & COLE S. (1965) Production of gibberellin-like substances by bacteria and actinomycetes. *Can. J. Microbiol.* **11**, 733–741.
374

KAUL B. & STABA E.J. (1967) *Ammi visnaga* L. Lam. tissue cultures: multiliter suspension growth and examination for furanochromones. *Planta Medica* **15**, 145–156.
293, 296

KAUL B. & STABA E.J. (1968) *Dioscorea* tissue cultures. I. Biosynthesis and isolation of diosgenin from *Dioscorea deltoidea* callus and suspension cells. *Lloydia* **31**, 171–179.
295

KAUL B., STOHS S.J. & STABA E.J. (1969) *Dioscorea* tissue cultures. III. Influence of various factors of diosgenin production by *Dioscorea deltoidea* callus and

suspension cultures. *Lloydia* **32**, 347–359.
267

KEEGSTRA K. & ALBERSHEIM P. (1970) The involvement of glycosidases in cell wall metabolism of suspension cultures of *Acer pseudoplatanus* cells. *Pl. Physiol., Lancaster* **45**, 675–678.
291

KEEN N.T. & HORSCH R. (1972) Hydroxyphaseollin production by various soybean tissues: a warning against use of 'unnatural' host-parasite systems. *Phytopathology* **62**, 439–442.
405

KEEN N.T., REDDY M.N. & WILLIAMS P.H. (1969) Isolation and properties of *Plasmodiophora brassicae* plasmodia from infected Crucifer tissues and from tissue culture callus. *Phytopathology* **59**, 637–644.
394, 401

KEHR A.E. & SMITH H.H. (1954) Genetic tumors in *Nicotiana* hybrids, pp. 57–78, in *Abnormal and Pathological Plant Growth, Brookhaven Symp. Biol.* **6**.
385, 386, 387

KELBITSCH H. (1960) Cuscuta als Uberträger des crown gall-erzengeden Prinzips. *Protoplasma* **52**, 437–445.
380

KELLER W.A., HARVEY B., GAMBORG O.L., MILLER R.A. & EVELEIGH D.E. (1970) Plant protoplasts for use in somatic cell hybridization. *Nature, Lond.* **226**, 280–282.
103

KENT A.E. & STEWARD F.C. (1965) Morphogenesis in free cell cultures of carrot as affected by sequential treatments with naphthalene acetic acid and with coconut milk. *Am. J. Bot.* **52**, 619.
426

KENT G.C. (1937) Some physical, chemical and biological properties of a specific bacteriophage of *Pseudomonas tumefaciens. Phytopathology* **27**, 871–902.
358

KHANNA P. & STABA E.J. (1968) Antimicrobials from plant tissue cultures. *Lloydia* **31**, 180–189.
295, 397

KIHLMAN B.A. (1966) *Action of Chemicals on Dividing Cells.* Prentice-Hall, New Jersey.
171

KING P.J. (1973) *The continuous culture of plant cells.* Ph.D. Thesis, Univ. Leicester, England.
282

KING P.J., MANSFIELD K.J. & STREET H.E. (1973) Control of growth and metabolism of cultured plant cells. *Can. J. Bot.* (in press).
305

KIRBY K.S. (1965) Isolation and characterization of ribosomal ribonucleic acid. *Biochem. J.* **96**, 266–269.
55

KLEIN A.O. & HAGEN C.W. JR. (1961) Anthocyanin production in detached petals of *Impatiens balsamina* L. *Pl. Physiol., Lancaster* **36**, 1–9.
262

KLEIN R.M. (1955) Resistance and susceptibility of carrot roots to crown-gall tumor formation. *Proc. natn. Acad. Sci. U.S.A.* **41**, 271–274.
361

KLEIN R.M. (1965) The physiology of bacterial tumors in plants and of habituation, pp. 209–235, in *Handbuch der Pflanzenphysiologie*, vol. **XV(2)**. Ed. W.Ruhland. Springer Verlag, Berlin.
364

KLEIN R.M. & BEARDSLEY R.E. (1957) On the role of omega bacteriophage in formation of crown gall tumor cells. *Am. Nat.* **91**, 330–331.
372

KLEIN R.M. & LINK G.K.K. (1952) Studies on the metabolism of plant neoplasms. V. Auxin as a promoting agent in the transformation of a normal to crown-gall tumour cells. *Proc. natn. Acad. Sci. U.S.A.* **38**, 1066–1072.
374

KLEIN R.M. & MANOS G.E. (1960) Use of metal chelates for plant tissue cultures. *Ann. N.Y. Acad. Sci.* **88**, 416–425.
38, 83

KLEIN R.M. & TENENBAUM I.L. (1955) A quantitative bioassay from crown gall tumor formation. *Am. J. Bot.* **42**, 709–712.
365

KOCHBAR T. SABHARWAL P. & ENGELBERG J. (1971) Production of homozygous diploid plants by tissue culture technique. *J. Hered.* **62**, 59–61.
235

KOENIGS J.W. (1968) Culturing the white pine blister rust fungus in callus of western white pine. *Phytopathology* **58**, 46–48.
395, 398, 399

KOHLENBACH H.W. (1959) Streckungsund Teilungswachstum isolierter Mesophyllzellen von *Macleaya cordata. Naturwissenschaften* **46**, 116–117.
108

KOHLENBACH H.W. (1966) Die entwicklungspotenzen explantierter und isolierter Dauer-Zellen. I. Das streckungs-und Teilungswachstum isoliertier Mesophyllzellen von *Macleaya cordata. Z. Pflanzenphysiol.* **55**, 142–157.
108

KONAR R.N. (1963) A haploid tissue from the pollen of *Ephedra foliata* Boiss. *Phytomorphology* **13**, 170–174.
214, 225

KONAR R.N. & NATARAJA, K. (1965a) Production of embryoids from the anthers of *Ranunculus sceleratus* L. *Phytomorphology* **15**, 245–248.
217, 234

KONAR R.N. & NATARAJA K. (1965b) Experimental studies in *Ranunculus sceleratus* L. plantlets from freely suspended cells and cell groups. *Phytomorphology* **15**, 206–211.
425

KONAR R.N., THOMAS E. & STREET H.E. (1972a) The diversity of morphogenesis in suspension cultures of *Atropa belladonna* L. *Ann. Bot.* **36**, 123–145.
347

KONAR R.N., THOMAS E. & STREET H.E. (1972b) Origin and structure of embryoids arising from epidermal cells of the stem of *Ranunculus sceleratus* L. *J. Cell Sci.* **11**, 77–93.
349

KORANT B.D. & POOTJES C.F. (1970) Physicochemical properties of *Agrobacterium tumefaciens* phage LVI and its DNA. *Virology* **40**, 48–54.
372

KORDAN H.A. (1959) Proliferation of excised juice vesicles of lemon *in vitro. Science, N.Y.* **129**, 779–780.
41, 246

KOSTOFF D. (1930) Tumors and other malformations on certain *Nicotiana* hybrids. *Zbl. Bakter, usw Abt. 11* **81**, 244–260.
385, 386

KOSTOFF D. (1939) Abnormal mitosis in tobacco plants forming hereditary tumours. *Nature, Lond.* **144**, 599.
386

KOTTÉ W. (1922a) Wurzelmeristem in

Gewebekultur. *Ber. dt. bot. Ges.* **40**, 269–272.
6

KOTTÉ W. (1922b) Kulturversuch isolierten Wurzelspitzen. *Beitr. allg. Bot.* **2**, 413–434.
6

KOVOOR A. (1967) Sur la transformation de tissus normaux de Scorzonère provoquée *in vitro* par l'acide désoxyribonucleique d'*Agrobacterium tumefaciens*. *C.r. hebd. Séanc. Acad. Sci.*, *Paris* **265**, 1623–1626.
367

KRIKORIAN A.D. & BERQUAM D.L. (1969) Plant cell and tissue cultures—the role of Haberlandt. *Bot. Rev.* **35**, 59–88.
5

KRIKORIAN A.D. & STEWARD F.C. (1969) Biochemical differentiation: the biosynthetic potentialities of growing and quiescent tissue, pp. 227–326, in *Plant Physiology*, vol. **VB**. Ed. F.C.Steward. Academic Press, New York.
293

KRUSBERG L.R. (1961) Studies on the culturing and parasitism of plant-parasitic nematodes, in particular *Ditylenchus dipsaci* and *Aphelenchoides ritzemabosi* on alfalfa tissues. *Nematologica* **6**, 181–200.
418, 419

KRUSBERG L.R. & BLICKENSTAFF M.L. (1964) Influence of plant growth regulating substances on reproduction of *Ditylenchus dipsaci* and *Pratylenchus penetrans* and *Pratylenchus zeae* on alfalfa tissue cultures. *Nematologica* **10**, 145–150.
420

KUBITSCHEK H.E. (1970) *Introduction to Research with Continuous Cultures*. Prentice-Hall, Hemel Hempstead.
323

KULESCHA Z. (1952) *Recherches sur l'elaboration de substances de croissance par les tissus végétaux*. Thèse Sci. Université de Paris. *Revue gén. Bot.* **59**, 1–264.
388

KULESCHA Z. & GAUTHERET R.J. (1948) Sur l'elaboration de substances de croissance par trois types de cultures de tissus de Scorsonère: cultures normales, cultures de crown-gall et cultures accontumées à l'hetero-auxine. *C.r. hebd. Séanc. Acad. Sci.*, *Paris* **227**, 292–294.
376, 388

KUPILA S. (1958) Anatomical and cytological comparison of the development of crown gall in three host species. *Ann. Bot. Soc. Zool. Bot. Fennicae* '*Vanamo*' **30**, 1–89.
375

KUPILA S., BRYAN A.M. & STERN H. (1961) Extractability of DNA and its determination in tissues of higher plants. *Pl. Physiol.*, *Lancaster* **36**, 212–215.
54

KUPILA S. & THERMAN E. (1962) Anatomical observations on genetic tumours and crown gall in amphiploid *Nicotiana glauca* × *langsdorffii*. *Suomal. elain-ja kasvit. Seur. van. Julk* **32**, 1–21.
386

KUPILA-AHVENNIEMI S. & THERMAN E. (1971) First DNA synthesis around sterile and crown-gall-inoculated wounds in *Vicia faba*. *Physiologia Pl.* **24**, 23–26.
375

KURKDJIAN A. (1968) Apparition de phages au cours de l'induction des tumeurs du crown-gall. *J. Microscopie* **7**, 1039–1044.
373

KURKDJIAN A. (1970) Observations sur la présence des phages dans les plaies infectées par différentes souches d'*Agrobacterium tumefaciens* (Smith and Townsend) conn. *Ann. Inst. Pasteur*, *Paris* **118**, 690–696.
373

KURZ W.G.W. (1971) A chemostat for growing higher plant cells in single cell suspension cultures. *Exp. Cell Res.* **64**, 476–479.
67, 68, 324

KÜSTER E. (1928) Das Verhalten pflanzlicher Zellen *in vitro* und *in vivo*. *Arch. exp. Zellforsch.* **6**, 28–41.
6

KUTSKY R.J. & RAWLINS T.E. (1950) Inhibition of virus multiplication by naphthaleneacetic acid in tobacco tissue cultures as revealed by a spectrophotometric method. *J. Bact.* **60**, 763–765.
411

LA COUR L.F. (1949) Nuclear differentiation in the pollen grain. *Heredity*, *Lond.* **3**, 319–337.
207, 209

LAFONTAINE J.C. & CHOUINARD L.A. (1963) A correlated light and electron microscope study of the nucleolar material

during mitosis in *Vicia faba. J. Cell Biol.*
17, 167–201.
123

LAMPORT D.T.A. (1964) Cell suspension cultures of higher plants: isolation and growth energetics. *Exp. Cell Res.* **33**, 195–206.
27, 65

LANCE C. (1957) Remarques sur l'emploi de differents criteres de croissance dans le cas dés tissus vegetaux cultivés *in vitro. Rev. gén. Bot.* **64**, 123–130.
257

LANDBOUWHOGESCHOOL WAGENINGEN (1971) *Effects of sterilisation on components in nutrient media.* Miscellaneous Papers of, Vol 9 (pp. 145), Veenman, Wageningen, Netherlands.
19

LANGE H. & ROSENSTOCK G. (1970) Zur Function hypothetischer Wundsaft-faktoren bei der Mitoseanslosung und pflanzlichen Tumorbildung durch *Agrobacterium tumefaciens. Phytopath. Z.* **67**, 367–369.
361

LA RUE C.D. (1933) Regeneration in mutilated seedlings. *Proc. natn. Acad. Sci. U.S.A.* **19**, 53–63.
6

LA RUE D.D. (1947) Growth and regeneration of the endosperm of maize in culture. *Am. J. Bot.* **34**, 585–586.
32

LA RUE D.D. (1949) Cultures of the endosperm of maize. *Am. J. Bot.* **36**, 798.
41

LA RUE T. & GAMBORG O. (1971) Ethylene production by plant cell cultures. *Pl. Physiol., Lancaster* **48**, 394–398.
287, 288

LATIES G.G. (1962) Controlling influence of thickness on development and type of respiratory activity in potato slices. *Pl. Physiol., Lancaster* **37**, 679–690.
243

LATTA R. (1971) Preservation of suspension cultures of plant cells by freezing. *Can. J. Bot.* **49**, 1253–1254.
189, 204, 428

LAWN A.M. (1960) The use of potassium permanganate as an electron-dense stain for sections of tissue embedded in epoxy resin. *J. biophys. biochem. Cytol.* **7**, 197–198.
49

LAYNE E. (1957) Spectrophotometric and turbidometric methods for measuring proteins, protein estimation with Folin-Crocalteu reagent, pp. 448–50, in *Methods in Enzymology*, vol. 3. Ed. C.Kaplan. Academic Press, New York.
91, 92

LEDBETTER M.C. & PORTER K.B. (1963) A 'microtubule' in plant fine structure. *J. Cell Biol.* **19**, 329–250.
125

LEDBETTER M.C. & PORTER K.B. (1964) Morphology of microtubules of plant cells. *Science, N.Y.* **144**, 873–874.
125

LEDOUX L., HUART R. & JACOBS M. (1971) Fate of exogenous DNA in *Arabidopsis thaliana.* Translocation and integration. *Eur. J. Biochem.* **23**, 96–108.
428

LEE T.T. (1971) Cytokinin controlled indoleacetic acid oxidase isoenzymes in tobacco callus cultures. *Pl. Physiol., Lancaster*, **47**, 181–185.
264

LEFF J. & BEARDSLEY R.E. (1970) Action tumorigène de l'acide nucléique d'un bacteriophage présent dans les cultures de tissu tumoral de Tournesol (*Helianthus annuus*). *C.r. hebd. Séanc. Acad. Sci., Paris* **270**, 2505–2507.
366, 367, 372, 373, 374

LEIS E. & RALPH B.J. (1960) A grinding device for the disintegration of fungal hyphae. *Aust. J. Sci.* **22**, 348–349.
30

LEJEUNE B. (1967) Etude de la synthèse de lysopine *in vitro* par les extraits de crown-gall. *C.r. Séanc. Acad. Sci. Sér. O.* **265**, 1753–1755.
378

LEJEUNE B. & JUBIER M.F. (1968) Etude de la synthèse de lysopine par les tissus de crown gall. *C.r. Séanc. Acad. Sci. Sér. O.* **266**, 1189–1191.
378

LEPPARD G.G., COLVIN R.J., ROSE D. & MARTIN S.M. (1971) Lignofibrils on the external cell wall surface of cultured plant cells. *J. Cell Biol.* **50**, 63–80.
104

LESCURE A.M. (1966) Etude quantitative de la croissance d'une culture *d'Acer pseudoplantanus* L. *Physiol. Vég.* **4**, 365–378.
93

LESCURE A.M. (1970) Mutagénèse de cellules végètales cultivés *in vitro*: Methodes et résultats. *Soc. bot. Fr. memoires*, 353–365.
201

LETHAM D.S. (1964) Isolation of a kinin from plum fruitlets and other tissues, pp. 109–117, in *Regulateurs Naturels de la Croissance Végétale*. Colloques Internationaux du C.N.R.S., Paris.
42

LEVINE M. (1951) The effect of growth substances and chemical carcinogens on fibrous roots of carrot tissue grown *in vitro*. *Am. J. Bot.* **38**, 132–138.
341

LEWIS K.R. & JOHN B. (1963) *Chromosome Marker*. Churchill, London.
171

LIAU D.F. & BOLL W. (1971) Growth and patterns of growth and division in cell suspension cultures of bush bean (*Phaseolus vulgaris* cv. Contender). *Can. J. Bot.* **49**, 1131–1139.
270, 273, 278, 312

LIMASSET P. & CORNUET P. (1949) Recherche du virus de la mosaique du tabac dans des méristèmes des plantes infectées. *C.r. hebd. Séanc. Acad. Sci. Paris*, **228**, 1971.
415

LIMASSET P. & GAUTHERET R. (1950) Sur le caractère tumoral des tissus de tabac ayant subi le phenomène d'accontumance aux hétéro-auxines. *C.r. hebd. Séanc. Acad. Sci., Paris* **230**, 2043–2045.
388

LINK G.K.K. & EGGERS V. (1941) Hyperauxiny in crown gall of tomato. *Bot. Gaz.* **103**, 87–106.
376

LINSMAIER E.M. & SKOOG F. (1965) Organic growth factor requirements of tobacco tissue cultures. *Physiologia Pl.* **18**, 100–127.
216, 218, 237

LIORET C. (1960) *Recherches sur le metabolisme de deux tissus vegetaux cultivés in vitro*. Thèse Sci. Université de Paris, 1960.
378

LIPETZ J. (1966) Crown gall tumorigenesis. II. Relations between wound healing and the tumorigenic response. *Cancer Res.* **26**, 1597–1605.
364

LIPPINCOTT J.A. & HEBERLEIN G.T. (1965a) The induction of leaf tumors by *Agrobacterium tumefaciens*. *Am. J. Bot.* **52**, 396–403.
366

LIPPINCOTT J.A. & HEBERLEIN G.T. (1965b) The quantitative determination of the infectivity of *Agrobacterium tumefaciens*. *Am. J. Bot.* **52**, 856–863.
357, 366

LIPPINCOTT J.A. & LIPPINCOTT B.B. (1970) Lysopine and octopine promote crown-gall tumor growth. *Science N.Y.* **170**, 176–177.
380

LIPPINCOTT J.A., LIPPINCOTT B.B. & CHI CHENG CHANG (1972) Promotion of crown gall tumor growth by lysopine, octopine, nopaline and carnosine. *Pl. Physiol., Lancaster* **49**, 131–137.
380

LOENING U.E. (1965) The synthesis of messenger and ribosomal RNA in pea seedlings as detected by electrophoresis. *Proc. Roy. Soc. Lond. B.* **162**, 121–136.
55

LOENING U.E. (1967) The fractionation of high-molecular weight ribonucleic acid by polyacrylamide gel electrophoresis. *Biochem. J.* **102**, 251–257.
55

LOOMIS R.S. & TORREY J.G. (1964) Chemical control of vascular cambium initiation in isolated radish roots. *Proc. natn. Acad. Sci. U.S.A.* **52**, 3–11.
424

LOWNSBERY B.F., HUANG G.S. & JOHNSON R.N. (1967) Tissue culture and maintenance of the root lesion nematode *Pratylenchus vulvus*. *Nematologica* **13**, 390–394.
418

LOWRY O.H., ROSEBROUGH N.J., FARR A.L. & RANDALL R.J. (1951) Protein measurement with folin phenol reagent. *J. biol. Chem.* **193**, 265–275.
55, 91

LUTZ A. (1966) Obtention de plantes de Tabac à partir de cultures unicellulaires provenans d'une souche anergiée. *C.r. hebd. Séanc. Acad. Sci., Paris* **262**, 1856–1858.
389

MCCLINTOCK B. (1951) Chromosome organization and genic expression. *Cold Spr. Harb. Symp. Quant. Biol.* **21**, 13–47.
178, 179

MCCOWN B.H., MCCOWN D.D., BECK G.E. & HALL T.C. (1970) Isoenzyme comple-

ments of *Dianthus* callus cultures: Influence of light and temperature. *Am. J. Bot.* **57**, 148–152.
262

McEwen D.M. (1952) Cancerous response in plants. *Nature, Lond.* **169**, 839.
380

Mackenzie I.A. & Street H.E. (1970) Studies on the growth in culture of plant cells. VIII. The production of ethylene by suspension cultures of *Acer pseudoplatanus* L. *J. exp. Bot.* **21**, 824–834.
287

McLaren I. & Thomas D.R. (1967) CO_2 fixation, organic acids and some enzymes in green and colourless tissue cultures of *Kalanchoe crenata*. *New Phytol.* **66**, 683–695.
262

McLeish J. (1963) Quantitative relationships between deoxyribonucleic acid and ribonucleic acid in isolated plant nuclei. *Proc. Roy. Soc. Lond. B* **158**, 261–278.
93

McLure M.A. & Viglierchio D.R. (1966) Penetration of *Meloidogyne incognita* in relation to growth and nutrition of sterile, excised cucumber roots. *Nematologica* **12**, 237–247.
418

Maheshwari R. (1969) Applications of plant tissue and cell culture in the study of physiology of parasitism. *Proc. Indian Acad. Sci. B* **59**, 152–172.
393

Maheshwari R., Hildebrandt A.C. & Allen P.J. (1967) Factors affecting the growth of rust fungi on host tissue cultures. *Bot. Gaz.* **128**, 153–159.
395, 396, 297, 398, 402

Mai W.F. & Thistlethwayte B. (1967) The feasibility of using plant pathogenic nematodes produced in monoxenic culture for nematocide screening. *Phytopathology* **57**, 820 (abstr.).
418

Málek I. & Fencl Z. (1966) Ed. *Theoretical and Methodological Basis of Continuous Culture of Microorganisms*. Academic Press, New York.
317, 323, 334

Manasse R.J. & Lipetz J. (1971) A simplified method for isolating bacteria-free crown gall tissue from *Vinca rosea*. *Can. J. Bot.* **49**, 1255–1257.
360

Manton I. (1961) Some problems of mitochondrial growth. *J. exp. Bot.* **12**, 421–429.
142

Marchant R. & Roberts A.W. (1968) Membrane systems associated with the plasmalemma of plant cells. *Ann. Bot.* **32**, 457–471.
160

Maretzki A., dela Cruz, A. & Nickell L.G. (1971) Extracellular hydrolysis of starch in sugarcane cell suspensions. *Pl. Physiol., Lancaster* **48**, 521–525.
291

Maretzki A. & Thom M. (1972) The existence of two membrane transport systems for glucose in suspensions of sugarcane cells. *Biochem. biophys Res. Commun.* **47**, 44–50.
297

Maretzki A., Thom M. & Nickell L.G. (1969) Products of arginine catabolism in growing cells of sugarcane. *Phytochem.* **8**, 811–818.
297

Marks G.E. (1957) The cytology of *Oxalis dispar* (Brown). *Chromosoma* **8**, 650–670.
175

Marks G.E. & Sunderland N. (1966) John Innes Annual Report, no. 57. pp. 22–23.
163, 165, 177, 178

Mathes M.C. (1963) Antimicrobial substances from Aspen tissue grown *in vitro*. *Science N.Y.* **140**, 1101–1102.
397

Mathes M.C. (1967) The secretion of antimicrobial materials by various isolated plant tissues. *Lloydia* **30**, 177–181.
397

Matthysse A.G. & Torrey J.G. (1967a) Nutritional requirements for polyploid mitoses in cultured pea root segments. *Physiologia Pl.* **20**, 661–672.
35

Matthysse A.G. & Torrey J.G. (1967b) DNA synthesis in relation to polyploid mitoses in excised pea root segments cultured *in vitro*. *Exp. Cell Res.* **48**, 484–498.
35

Mehta A.R. (1963) *Nutritional and morphogenetic studies with callus and suspension cultures derived from roots*. Ph.D. Thesis, Univ. Wales.
193

Mehta A.R., Henshaw G.G. & Street H.E. (1967) Aspects of growth in sus-

pension culture of *Phaseolus vulgaris* L. and *Linum usitatissimum* L. *Ind. J. Pl. Physiol.* **10**, 44–53.
272

MELCHERS G. (1971) Transformation or habituation to autotrophy and tumor growth and recovery, pp. 229–234, in *Les Cultures de Tissus de Plantes.* Colloques Internationaux du C.N.R.S., no. 193, Paris.
382, 389

MELCHERS G. (1972) Haploid higher plants for plant breeding. *Z. PflZücht.* **67**, 19–32.
236

MELCHERS G. & BERGMANN L. (1959) Untersuchungen an kulturen von haploid Geweben von *Antirrhinum majus. Ber. dt. bot. Ges.* **71**, 459–73.
9, 28, 67, 269, 270

MELCHERS G. & ENGELMANN U. (1955) Die Kultur von Pflanzegewebe en flüssigen medium mit Dauerbelüftung *Naturwissenschaften* **42**, 564–565.
67

MELCHERS G. & LABIB G. (1970) Die Bedeutung haploider höheren Pflanzen für Pflanzenphysiologie und Pflanzenzüchtung. *Ber. dt. bot. Ges.* **83**, 129–50.
215, 236

MENCE M.J. & HILDEBRANDT A.C. (1966) Resistance to powdery mildew in rose. *Ann. appl. Biol.* **58**, 309–320.
396

MILLER C.O. (1969) Control of deoxyisoflavone synthesis in soybean tissue. *Planta* **87**, 26–35.
267

MILLER C.O & SKOOG F. (1953) Chemical control of bud formation in tobacco stem segments. *Am. J. Bot.* **40**, 768–773.
8

MILLER C.O., SKOOG F., OKUMURA F.S., VON SALTZA M.H. & STRONG F.M. (1956) Isolation, structure and synthesis of kinetin, a substance promoting cell division. *J. Am. chem. Soc.* **78**, 1375–1380.
8

MILLER C.O., SKOOG F., VON SALTZ M.H. & STRONG F.M. (1955) Kinetin, a cell division factor from desoxyribonucleic acid. *J. Am. chem. Soc.* **77**, 1392.
42

MILLER P.W. & BELKENGREN R.O. (1963) Elimination of yellow edge, crinkle, and vein banding viruses and certain other virus complexes from strawberries by excision and culturing apical meristems. *Pl. Dis. Reptr.* **47**, 298–300.
413

MILLER R.A., GAMBORG O.L., KELLER W.A. & KAO K.N. (1971) Fusion and division of nuclei in multinucleated soybean protoplasts. *Can. J. Genet. Cytol.* **13**, 347–353.
117

MILLER R.A., SHYLUK J.P., GAMBORG O.L. & KIRKPATRICK J.W. (1968) Phytostat for continuous culture and automatic sampling of plant-cell suspensions. *Science N.Y.* **159**, 540–542.
28, 67, 69, 73, 74, 324, 335

MILO G.E. & SRIVASTAVA S.B.I. (1969a) Effects of cytokinins on tobacco mosaic virus production in local lesion and systemic hosts. *Virology* **38**, 26–31.
411

MILO G.E. & SRIVASTAVA S.B.I. (1969b) Effect of cytokinins on tobacco mosaic virus production in tobacco pith tissue cultures. *Virology* **39**, 621–623.
411

MILO G.E. & SRIVASTAVA B.I. (1969c) RNA-DNA hybridization studies with the crown gall bacteria and the tobacco tissue. *Biochem. biophys. Res. Comm.* **34**, 196–199.
370

MIRZABEKYAN R.O. (1969) Kul'tura tkani klubnevoi formy raka Kartofetya (Tissue culture of tuber wart of potato). *Mikol. i Fitopatol.* **3**, 337–342.
394

MITCHELL D. & SHAW M. (1968) Metabolism of glucose-C-14, pyruvate-C-14 and mannitol-C-14 by *Melampsora lini.* II. Conversion to soluble products. *Can. J. Bot.* **46**, 435–460.
401

MITCHELL J.P. (1967) DNA synthesis during the early division cycles of Jerusalem artichoke callus cultures. *Ann. Bot.* **31**, 427–435.
50, 242

MITCHELL J.P. (1968) The pattern of protein accumulation in relation to DNA replication in Jerusalem artichoke callus cultures. *Ann. Bot.* **32**, 315–326.
50, 242, 243

MITCHELL J.P. (1969) RNA accumulation in relation to DNA and protein accumulation in Jerusalem artichoke callus cultures. *Ann. Bot.* **33**, 25–34.
50, 242, 243, 247

MITCHISON J.M. (1971) *The Biology of the Cell Cycle.* Cambridge University Press. 298

MITRA J., MAPES M.O. & STEWARD F.C. (1960) Growth and organized development of cultured cells. IV. The behaviour of the nucleus. *Am. J. Bot.* **47**, 357–368. 162, 177, 343

MITRA J. & STEWARD F.C. (1961) Growth induction in cultures of *Haplopappus gracilis*. III. The behaviour of the nucleus. *Am. J. Bot.* **48**, 358–368. 162, 172, 177

MITSUHASHI J. & MARAMOROSCH K. (1964) Inoculation of plant tissue cultures with aster yellows virus. *Virology* **23**, 277–279. 409

MOHR W.P. & COCKING E.C. (1968) A method of preparing highly vacuolated, senescent or damaged plant tissue for ultrastructural study. *J. Ultrastruct. Res.* **21**, 171–181. 95

MOLLENHAUER H.H. & MORRÉ D.J. (1966) Golgi apparatus and plant secretion. *Ann. Rev. Pl. Physiol.* **17**, 27–46. 123, 135

MONOD J. (1950) La technique de culture continuée. Théorie et application. *Ann. Inst. Pasteur, Paris* **79**, 390–410. 321, 323

MONTANT C. (1957) Essais de culture *in vitro* de fragments de tige d'*Euphorbia charcias* L. *C.r. Séanc. Soc. Biol.* **151**, 391–392. 32

MOREL G. (1944) Le développement du mildiou sur des tissus de Vigne cultures *in vitro*. *C.r. hebd. Séanc. Acad. Sci., Paris* **218**, 50–52. 393, 394, 396, 398

MOREL G. (1946) Essais de laboratoire sur le mildiou de la vigne. *Rev. Vitic. (Paris)* **93**, 210–213. 393, 401

MOREL G. (1947a) Methode d'essai en serre des produits de lutte contre le mildiou de la vigne. *Ann. Épiphyt. (sér Pathologie végétale)* **13**, 57–66. 401

MOREL G. (1947b) Transformations des cultures de vigne produites par l'hetero-auxine. *C.r. Séanc. Soc. Biol.* **141**, 280–282. 388

MOREL G. (1948) Recherches sur la culture associée de parasites obligatoires et de tissus végétaux. *Ann. Épiphyt. (sér Pathologie Végétale)* **14**, 1–112. 393, 394, 395, 396, 397, 398, 407, 410, 411

MOREL G. (1964) La culture *in vitro* du méristème apical. *Rev. cytol. cytophysiol. végét.* **27**, 307–314. 424

MOREL G. (1970) Le problème de la transformation tumorale chez les végétaux. *Ann. Physiol. Vég.* **8**, 189–204. 380

MOREL G. (1971) Deviations du métabolisme azoté des tissus de crown-gall, pp. 463–471, in *Les Cultures de Tissus de Plantes*. Colloques Internationaux du C.N.R.S., no. 193, Paris. 367, 371, 378, 379, 380

MOREL G. & DURANTON H. (1958) Le métabolism de l'arginine par les tissus végétaux. *Bull. Soc. Chim. biol.* **40**, 2155–2167. 379

MOREL G. & GAUTHERET R.J. (1955) Recherches sur les phénomènes d'histogenèse provoqués par l'acide naphtalène-acetique et le lait de Coco dans les cultures de tissus d'*Amophophallus riviera*. *Revue gén. Bot.* **62**, 437–456. 388

MOREL G. & MARTIN G. (1952) Guérison de dahlias atteints d'une maladie à virus. *C.r. hebd. Séanc. Acad. Sci., Paris* **235**, 1324–1325. 412

MOREL G. & MARTIN G. (1955) Guérison de pommes de terre atteints de maladies à virus. *C.r. hebd. Séanc. Acad. Agric. Fr.* **41**, 472–475. 412, 413

MOREL G., MARTIN G. & MULLER J.F. (1968) La guérison de pommes de terre atteints de maladies à virus. *Ann. Physiol. Vég.* **10**, 113–139. 413, 415

MOREL G. & WETMORE R.H. (1951) Tissue culture of monocotyledons. *Am. J. Bot.* **38**, 138–140. 342

MOSES M.J. & TAYLOR J.H. (1955) Desoxypentose nucleic acid synthesis during microsporogenesis in *Tradescantia*. *Exp. Cell Res.* **9**, 474–488. 209

MOTA M., HILDEBRANDT A.C. & RIKER A.J. (1964) Movements of cytoplasm

between and during nuclear divisions in living tobacco cells of tissue cultures. A motion picture film. *Agronomia Lusitana* **26**, 205–212.
131, 147, 153

MOUNTAIN W.B. (1955) A method of culturing plant parasitic nematodes under sterile conditions. *Proc. Helminth. Soc. (Washington)* **22**, 49–52.
417

MUIR W.H. (1953) *Cultural conditions favouring the isolation and growth of single cells from higher plants in vitro.* Ph.D. Thesis. Univ. Wisconsin, U.S.A.
8, 62, 192, 269

MUIR W.H., HILDEBRANDT A.C. & RIKER A.J. (1954) Plant tissue cultures produced from single isolated plant cells. *Science, N.Y.* **119**, 877–878.
8

MUIR W.H., HILDEBRANDT A.C. & RIKER A.J. (1958) The preparation, isolation and growth in culture of single cells from higher plants. *Am. J. Bot.* **45**, 589–597.
144, 145, 199, 200, 269

MURAKISHI H.H. (1968) Infection of tomato callus cells in suspension with TMV-RNA. *Phytopathology* **58**, 993–996.
408

MURAKISHI H.H., HARTMAN J.X., BEACHY R.N. & PELCHER L.E. (1971) Growth curve and yield of tobacco mosaic virus in tobacco callus cells. *Virology* **43**, 62–68.
408

MURAKISHI H.H., HARTMAN J.X., PELCHER L.E. & BEACHY R.N. (1970) Improved inoculation of cultured plant cells resulting in high virus titre and crystal formation. *Virology* **41**, 365–367.
407

MURASHIGE T. & NAKANO R. (1965) Morphogenetic behaviour of tobacco tissue cultures and implication of plant senescence. *Am. J. Bot.* **52**, 819–827.
201

MURASHIGE T. & NAKANO R. (1967) Chromosome complement as a determinant of the morphogenic potential of tobacco cells. *Am. J. Bot.* **54**, 963–970.
185, 344

MURASHIGE T. & SKOOG F. (1962) A revised medium for rapid growth and bioassays with tobacco tissue cultures. *Physiologia Pl.* **15**, 473–497.
39, 42, 43, 145, 218, 362, 366, 403

MYERS R.F. (1967) Axenic cultivation of

plant parasitic nematodes. *Nematologica* **13**, 323.
417

NÄF V. (1958) Studies on tumor formation in *Nicotiana* hybrids. I. The classification of the parents into two etiological significant groups. *Growth* **21**, 167–180.
385

NAGATA T. & TAKEBE I. (1970) Cell wall regeneration and cell division in isolated tobacco mesophyll protoplasts. *Planta* **92**, 301, 308.
117, 118

NAGATA T. & TAKEBE I. (1971) Plating of isolated tobacco mesophyll protoplasts on agar medium. *Planta* **99**, 12–20.
101, 104, 110, 116, 117, 118

NAIK G.G. (1965) *Studies on the effect of temperature on the growth of plant tissue cultures.* M.Sc. Thesis, University of Edinburgh, Edinburgh, Scotland.
252

NAKAMURA H. (1965) The use of tissue cultures in the study of obligate parasites, pp. 535–540, in *Proc. Int. Conf. Plant Tissue Culture.* Ed. P.R.White and A.R. Grove. McCutchan Publ. Co., Berkeley, California.
394, 395, 396

NAKATA K. & TANAKA M. (1968) Differentiation of embryoids from developing germ cells in anther culture of tobacco. *Jap. J. Genet.* **43**, 65–71.
9, 215, 221

NAKATA K. & TANAKA M. (1970) Methods of producing high-grade haploid plants from anther culture of tobacco (in Japanese). *Jap. J. Breeding* **20**, Suppl. 1, 7–8.
235

NARAYANASWAMY S. & CHANDY L.P. (1971) *In vitro* production of haploid, diploid, and triploid androgenic embryoids and plantlets in *Datura metel* L. *Ann. Bot.* **35**, 535–542.
214, 225, 233, 234

NASH D.T. (1968) *A preliminary study of the metabolism of Paul's scarlet rose cells in suspension culture.* Ph.D. Thesis, Birmingham University, Birmingham.
106

NASH D. & DAVIES M. (1972) Some aspects of growth and metabolism of Paul's Scarlet Rose cell suspensions. *J. exp. Bot.* **23**, 75–91.
270, 275, 276, 278, 279, 280, 282, 284, 293

NAYLOR J., SANDER G. & SKOOG F. (1954)

Mitosis and cell enlargement without cell division in excised tobacco pith tissue. *Physiologia Pl.* **7**, 25–9.
171, 184, 185

NEWCOMB E.H. (1951) Effect of auxin on ascorbic oxidase activity in tobacco pith cells. *Proc. Soc. exp. Biol. Med.* **76**, 504–509.
56

NEWCOMB E.H. (1960) Dissociation of the effects of auxin on metabolism and growth of cultured tobacco pith. *Physiologia Pl.* **13**, 459–467.
56

NEWCOMB E.H. (1967) Fine structure of protein storing plastids in bean root tips. *J. Cell Biol.* **33**, 143–163.
160

NICHOLSON M.O. & FLAMM W.G. (1965) Properties and significance of free and bound ribosomes from cultured tobacco cells. *Biochim. biophys. Acta* **108**, 266–274.
127

NICKELL L.G. (1955) Nutrition of pathological tissues caused by plant viruses. *Année Biol.* **31**, 107–121.
339, 342

NICKELL L.G. (1956) The continuous submerged cultivation of plant tissues as single cells. *Proc. natn. Acad. Sci. U.S.A.* **42**, 848–850.
8, 144, 269

NICKELL L.G., GREENFIELD P. & BURKHOLDER P.R. (1950) Atypical growth of plants. III. Growth responses of virus tumors of *Rumex* to certain nucleic acid components and related compounds. *Bot. Gaz.* **112**, 42–52.
383

NICKELL L.G. & TULECKE W. (1959) Responses of plant tissue cultures to gibberellin. *Bot. Gaz.* **120**, 245–250.
201

NICKELL L.G. & TULECKE W. (1960) Submerged growth of cells of higher plants. *J. biochem. microbiol. Techn. & Engng.* **2**, 287–297.
269

NICKELL L.G. & TULECKE W. (1961) Growth substances and plant tissue cultures, pp. 675–685, in *Plant Growth Regulation*. 4th International Conference, Iowa State University Press.
267

NIIZEKI M. & GRANT W.F. (1971) Callus, plantlet formation, and polyploidy from cultured anthers of *Lotus* and *Nicotiana*. *Can. J. Bot.* **49**, 2041–2051.
215, 217

NIIZEKI H. & OONO K. (1968) Induction of haploid rice plant from anther culture. *Proc. Jap. Acad.* **44**, 554–557.
216

NIIZEKI H. & OONO K. (1971) Rice plants obtained by anther culture, pp. 251–257, in *Les Cultures de Tissus de Plantes*. Colloques Internationaux du C.N.R.S., no. 193, Paris.
221, 233, 234, 354

NILLSON-TILLGREN T. & VON WETTSTEIN-KNOWLES P. (1970) When is the male plastome eliminated? *Nature, Lond.* **227**, 1265–1266.
236

NIMS R.C., HALLIWELL R.S. & ROSBERG D.W. (1967) Disease development in cultured cells of *Nicotiana tabacum* L. var. Samsun NN injected with tobacco mosaic virus. *Cytologia* **32**, 224–235.
407

NISHI T. & MITSUOKA S. (1969) Occurrence of various ploidy plants from anther and ovary culture of rice plant. *Jap. J. Genet.* **44**, 341–346.
234

NITSCH C. (1968) Induction *in vitro* de la floraison chez une plante de jours courts: *Plumbago indica* L. *Annls Sci. nat. (Bot.)* **9**, 1–92.
342, 344, 346

NITSCH J.P. (1956) Methods for the investigation of natural auxins and growth inhibitors, pp. 3–31, in *The Chemistry and Mode of Action of Plant Growth Substances*. Ed. R.L.Wain & F.Wightman. Butterworth, London and Washington.
376

NITSCH J.P. (1951) Growth and development *in vitro* of excised ovaries. *Am. J. Bot.* **38**, 566–576.
216, 217

NITSCH J.P. (1970) Experimental androgenesis in *Nicotiana*. *Phytomorphology* **19**, 389–404.
213, 214, 218, 353

NITSCH J.P. (1972) Haploid plants from pollen. *Z. PflZücht.* **67**, 3–18.
214, 218, 236, 238

NITSCH J.P. & NITSCH C. (1964) Néoformation de boutons floraux sur cul-

tures *in vitro* de feuilles et de racines de *Cichorium intybus* L. Existence de un état vernalisé en l'absence de bourgeons. *Bull. Soc. bot. Fr.* **111**, 299–304.
342

NITSCH J.P. & NITSCH C. (1956) Auxin-dependent growth of excised *Helianthus tuberosus* tissues. *Am. J. Bot.* **43**, 839–851.
32, 39, 43, 243

NITSCH J.P. & NITSCH C. (1969) Haploid plants from pollen grains. *Science N.Y.* **163**, 85–87.
213, 214, 215, 221

NITSCH J.P., NITSCH C. & HAMON S. (1968) Réalisation expérimentale de l'androgenèse chez divers *Nicotiana*. *C.r. Séanc. Soc. Biol.* **162**, 369–372.
221

NITSCH J.P., NITSCH C. & HAMON S. (1969) Production de *Nicotiana* diploides a partir de cals haploides cultivés *in vitro*. *C.r. hebd. Séanc. Acad. Sci., Paris* **269**, 1275–1278.
235

NITSCH J.P., NITSCH C. & PÉREAU-LEROY P. (1969) Obtention de mutants à partir de *Nicotiana* haploides issus de grains de pollen. *C.r. hebd. Séanc. Acad. Sci., Paris* **269**, 1650–1652.
236

NITZSCHE W. (1970) Herstellung haploider Pflanzen aus *Festuca-Lolium* Bastarden. *Náturwissenschaften* **57**, 199–200.
216

NOBÉCOURT P. (1937) Culture en série de tissus vegetaux sur milieu artificiel. *C.r. hebd. Séanc. Acad. Sci., Paris* **205**, 521–523.
6

NOBÉCOURT P. (1938a) Sur les proliférations spontanées de fragments de tubercules de carotte et leur culture sur milieu synthétique. *Bull. Soc. bot. Fr.* **85**, 1–7.
6

NOBÉCOURT P. (1938b) Sur la prolifération *in vitro* du parenchyme amylifère du tubercule de *Solanum tuberosum* L. *Bull. Soc. bot. Fr.* **85**, 480–493.
6

NOBÉCOURT P. (1939a) Sur la perennite et l'augmentation de volume des cultures de tissus vegetaux. *C.r. Séanc. Soc. Biol.* **130**, 1270–1271.
255

NOBÉCOURT P. (1939b) Sur les radicelles naissant des cultures de tissus végétaux. *C.r. Séanc. Soc. Biol.* **130**, 1271.
339

NORRIS J. & RIBBONS D. (1971) Eds. *Methods in Microbiology*, vol. **2**. Academic Press, New York.
317, 323

NORSTOG K. & RHAMSTINE E. (1967) Isolation and culture of haploid and diploid Cycad tissues. *Phytomorphology* **17**, 374–381.
347

NORSTOG K., WALL W.E. & HOWLAND G.P. (1969) Cytological characteristics of ten-year-old rye-grass endosperm tissue cultures. *Bot. Gaz.* **130**, 83–86.
162, 177, 178, 181

NORTHROP J.H. (1954) Apparatus for maintaining bacterial cultures in the steady state. *J. gen. Physiol.* **38**, 105–115.
318

NOTH M.H. & ABEL W.O. (1971) Zur Entwicklung haploider Pflanzen aus unreifen Mikrosporen verschiedener *Nicotiana*—Arten. *Z. PflZücht.* **65**, 277–284.
215

NOVICK A. & SZILARD L. (1950) Description of the chemostat. *Science, N.Y.* **112**, 715–716.
321

NOZZOLILLO C. & CRAIGIE J.H. (1960) Growth of the rust fungus *Puccinia helianthi* on tissue cultures of its host. *Can. J. Bot.* **38**, 227–233.
394, 397

O'BRIAN T.P. & THIMANN K.V. (1967) Observations on the fine structure of the oat coleoptile. II. The parenchyma cells of the apex. *Protoplasma* **63**, 417–442.
129

OGUTUGA D.B.A. & NORTHCOTE D.H. (1970a) Biosynthesis of caffeine in tea callus tissue. *Biochem. J.* **117**, 715–720.
267, 268

OGUTUGA D. & NORTHCOTE D. (1970b) Caffeine formation in tea callus tissue. *J. exp. Bot.* **21**, 258–273.
296

OLSEN A.C. (1971) Secreted polysaccharides and proteins from *Nitotiana tabacum* suspension cultures, pp. 411–420, in *Les Cultures de Tissus de Plantes*. Colloques Internationaux du C.N.R.S., no. 193, Paris.
291, 292

OLSON A.C., EVANS J.J., FREDERICK D.P.

& JANSEN E.E. (1969) Plant suspension culture media macromolecules—pectic substances, protein and peroxidase. *Pl. Physiol. Lancaster* **44**, 1594–1600.
292

ORNSTEIN L. (1964) Disc electrophoresis—1 Background and theory. *Ann. N.Y. Acad. Sci.* **121**, 321–349.
55

OSBURN O.L. & WERKMAN L.H. (1932) Determination of carbon in fermented liquors. *Ind. Eng. Chem. analyt Edn.* **4**, 421–423.
94

OTSUKI Y. & TAKEBE I. (1969) Isolation of intact mesophyll cells and their protoplasts from higher plants. *Pl. Cell Physiol.* **10**, 917–201.
108

PADILLA G.M., WHITSON G.L. & CAMERON I.L. (1969) *The Cell Cycle: Gene Enzyme Interactions.* Academic Press, New York and London.
242, 298

PARSONS L.C. & BEARDSLEY R.E. (1968) Bacteriophage activity in homogenate of crown gall tissue. *J. Virol.* **2**, 651.
373

PARTANEN C.R. (1963a) The validity of auxin-induced divisions in plants as evidence of endopolyploidy. *Exp. Cell Res.* **31**, 597–599.
185

PARTANEN C.R. (1963b) Plant tissue culture in relation to developmental cytology. *Int. Rev. Cytol.* **15**, 215–243.
169, 171, 172, 344

PARTANEN C.R. (1965a) On the chromosomal basis for cellular differentiation. *Am. J. Bot.* **52**, 204–209.
185

PARTANEN C.R. (1965b) Cytological behaviour of plant tissues *in vitro* as a reflection of potentialities *in vivo*, pp. 463–71, in *Proc. Int. Conf. Plant Tissue Culture.* Ed. P.R.White and A.R.Grove. McCutchan Publ. Corp., Berkeley, California.
169

PARTANEN C.R., SUSSEX I.M. & STEEVES T.A. (1955) Nuclear behaviour in relation to abnormal growth in fern prothalli. *Am. J. Bot.* **42**, 245–256.
201

PATAU K. & DAS N.K. (1961) The relation of DNA synthesis and mitosis in tobacco pith tissue cultured *in vitro. Chromosoma* **11**, 553–572.
165, 167, 184

PATAU K., DAS N.K. & SKOOG F. (1957) Induction of DNA synthesis by kinetin and indoleacetic acid in excised tobacco pith tissue. *Physiologia Pl.* **10**, 949–966.
33, 183

PATTERSON B.D. & CAREW D.P. (1969) Growth and alkaloid formation in *Catharanthus* tissue cultures. *Lloydia* **32**, 131–140.
262, 267

PEARCE R. (1972) *Culture of isolated higher plant protoplasts.* Ph.D. Thesis, University of Nottingham, England.
105

PEARSE A.G.E. (1961) *Histochemistry: Theoretical and Applied.* 2nd edn. J. & A. Churchill Ltd., London.
50

PEARSON G.G. & INGLE J. (1972) The origin of stressed-induced satellite DNA in plant tissues. *Cell Differentiation* **1**, 43–51.
371

PÉAUD-LENOEL C. & JOUANNEAU J.-P. (1971) Contrôle du cycle mitotique dans les suspensions de cellules de Tabac cultivées en milieu liquide, pp. 95–102, in *Les Cultures de Tissus de Plantes.* Colloques Internationaux du C.N.R.S., no. 193, Paris.
304, 306

PELET F. (1959) *Growth in vitro of grape, elm, poplar, willow and oak tissues isolated from normal stems and insect galls.* Ph.D. Thesis, University of Wisconsin, Madison, U.S.A.
417

PELET F., HILDEBRANDT A.C., RIKER A.J. & SKOOG F. (1960) Growth *in vitro* of tissues isolated from normal stems and insect galls. *Am. J. Bot.* **47**, 186.
417

PELLETIER G., RAQUIN C. & SIMON G. (1972) La culture *in vitro* d'anthères d'Asperge (*Asparagus officinalis*). *C.r. hebd. Séanc. Acad. Sci., Paris* **274**, 848–851.
215

PENNAZIO S. (1971) Terapia di viroxi della patata (*Solanum tuberosum* L.): cultura di apici meristematici abbinata a termoterapia. *Revista dell'Ortoflorofrutticoltura Italiana* **5**, 446–452.
413, 415

PETERSEN D., TOBEY R. & ANDERSON E. (1969) Essential biosynthetic activity in synchronized mammalian cells, pp. 311–359, in *The Cell Cycle: gene-enzyme interactions.* Ed. G.Padilla, G.Whitson & I.Cameron. Academic Press, New York. 298, 303

PETIT A., DELHAYE S., TEMPÉ J. & MOREL G. (1970) Recherches sur les guanidines des tissus de crown-gall. Mise en évidence d'une relation biochimique spécific entre les sourches d'*Agrobacterium tumefaciens* et les tumeurs qu'elles induisant. *Ann. Physiol. Vég.* **8**, 205–213. 371, 379

PFEIFFER H. (1931) Beobachtungen an Kulturen nackter Zellen aus pflanzenlichen Bieienperikarpen. *Arch. exp. Zellforsch.* **11**, 424–434. 6

PFEIFFER H. (1933) Über das migrationsvermögen pflanzlicher Zellen *in situ* und *in vitro. Arch. exp. Zellforsch.* **14**, 152–170. 6

PHILIPS D.J. (1968) *Carnation Shoot Tip Culture.* Technical Bulletin No. 102 of Colorado State University Experiment Station, Fort Collins. 22 pages. 413, 414

PHILIPSON L. & KAUFMAN M. (1964) The efficiency of ribonuclease inhibitors tested with viral ribonucleic acid as substrate. *Biochim. biophys. Acta* **80**, 151–154. 112

PHILLIPS H.L. (1970) *Metabolic changes associated with the division and differentiation in cells of Acer pseudoplatanus L. in suspension culture.* Ph.D. Thesis, Univ. Birmingham, England. 282

PHILLIPS H.L. & TORREY J.G. (1972) Duration of cell cycles in cultured roots of *Convolvulus. Am. J. Bot.* **59**, 183–188. 278, 298

PICKETT-HEAPS J.D. (1967) Further observations of the Golgi apparatus and its functions in cells of the wheat seedlings. *J. Ultrastruct. Res.* **18**, 287–303. 135

PICKETT-HEAPS J.D. & FOWKE L.C. (1969) Cell division in *Oedogonium.* 1. Mitosis, cytokinesis, and cell elongation. *Aust. J. biol. Sci.* **22**, 857–894. 133

PICKETT-HEAPS J.D. & NORTHCOTE D.H. (1966a) Relationship of cellular organelles to the formation and development of the plant cell wall. *J. exp. Bot.* **17**, 20–26. 130, 135

PICKETT-HEAPS J.D. & NORTHCOTE D.H. (1966b) Organization of microtubules and endoplasmic reticulum during mitosis and cytokinesis in wheat meristems. *J. Cell Sci.* **1**, 109–120. 130, 135

PICKETT-HEAPS J.D. & NORTHCOTE D.H. (1966c) Cell division in the formation of the stomatal complex of the young leaves of wheat. *J. Cell Sci.* **1**, 121–128. 135

PIERIK R.L.M. (1967) Regeneration, vernalization and flowering in *Lunaria annua in vivo* and *in vitro. Meded. Landb Hoogesch., Wageningen* **67**, 60–71. 342, 344, 346

PILET P.E. (1961) Culture *in vitro* de tissus de Carotte et organogenèse. *Ber. Schweiz bot. Ges.* **71**, 189–208. 8

PILET P.E. (1971) Effets de quelques auxines sur les protoplastes racinaires. *C.r. hebd. Séanc. Acad. Sci., Paris* **273**, 2253–2256. 100

PILET P.E., PRAT R. & ROLAND J.C. (1972) Morphology, RNAase and transaminase of root protoplasts. *Pl. Cell Physiol.* **13**, 297–309. 100

PILLAI S.K. & HILDEBRANDT A.C. (1968) Geranium plants differentiated *in vitro* from stem tip and callus cultures. *Pl. Dis. Reptr.* **52**, 600–601. 416

POW J.J. (1969) Clonal propagation *in vitro* from cauliflower curd. *Hort. Res.* **9**, 2, 151–152. 424

POWER J.B. (1971) *The isolation, behaviour and fusion of higher plant protoplasts.* Ph.D. Thesis, University of Nottingham, England. 112

POWER J.B. & COCKING E.C. (1970) Isolation of leaf protoplasts: macromolecule uptake and growth substance response. *J. Exp. Bot.* **21**, 64–70. 108

POWER J.B. & COCKING E.C. (1971) Fusion of plant protoplasts. *Sci. Prog., Oxf.* **59**, 181–198. 101, 104

POWER J.B., CUMMINS S.E. & COCKING E.C.

(1970) Fusion of isolated plant proto-plasts. *Nature, Lond.* **225**, 1016–1018.
108, 114, 120

POWER J.B., FREARSON E.M. & COCKING E.C. (1971) The preparation and culture of spontaneously fused tobacco leaf spongy-mesophyll protoplasts. Proceedings of the Biochem. Soc., *Biochem. J.* **123**, 29–30P.
117

PUCK T.T. (1964) Studies of the life cycle of mammalian cells. *Cold Spring Harb. Symp. Quant. Biol.* **29**, 167–176.
303

PUCK T.T. (1971) Biochemical genetics studies on mammalian cells *in vitro*, in *Control Mechanisms in the Expression of Cellular Phenotypes. Symp. Soc. Cell Biol.* **9**, 135–146. Academic Press, New York.
202

PUCK T.T. & KAO F.T. (1967) Genetics of somatic mammalian cells. V. Treatment with 5-bromodeoxyuridine and visible light for isolation of nutrionally deficient mutants. *Proc. natn. Acad. Sci. U.S.A.* **58**, 1227–1234.
202

PUCK T.T., MARCUS P.I. & CIECURA S.J. (1956) Clonal growth of mammalian cells *in vitro*. II. Growth characteristics of colonies from single *Hela* cells with and without a 'feeder' layer. *J. exp. Med.* **103**, 273–284.
87

PUHAN Z. & MARTIN S.M. (1971) Industrial potential of plant cell culture. *Prog. Ind. Microbiol.* **9**, 13–19.
293

QUAK F. (1957) Meristeem cultuur, gecombineerd met warmtebehandeling, vor het verkrijgen van virus—vrijeanjerplanten. *Tijdschr. Plantenziekten* **63**, 13–14.
413, 415

QUAK F. (1961) Heat treatment and substances inhibiting virus multiplication, in meristem culture to obtain virus-free plants. *Adv. Hort. Sci. Appl.* **1**, 144–148.
413

QUAK F. (1965) Infection of tobacco callus tissue with tobacco mosaic virus and multiplication of virus in such tissue, pp. 513–519, in *Proc. Int. Conf. Plant Tissue Culture*. Ed. P.R.White and A.R.Grove. McCutchan Publ. Corp., Berkeley, California.
407

QUATRANO R.S. (1968) Freeze-preservation of cultured flax cells utilizing dimethyl sulfoxide. *Pl. Physiol., Lancaster* **43**, 2057–2061.
204, 428

QUETIER F., HUGUET T. & GUILLÉ E. (1969) Induction of crown gall: partial homology between tumor cell DNA, bacterial DNA and the G+C rich DNA of stressed normal cells. *Biochem. biophys. Res. Comm.* **34**, 128–133.
370

RAJASEKHAR E.W., EDWARDS M., WILSON S.B. & STREET H.E. (1971) Studies on the growth in culture of plant cells. XI. The influence of shaking rate on the growth of suspension cultures. *J. exp. Bot.* **22**, 107–117.
4, 27, 63, 64, 94

RAPP F. & MELNICK J.L. (1966) Cell, tissue and organ cultures in virus research, pp. 263–306, in *Cells and Tissues in Culture*, vol. **3**, Ed. E.N.Willmer. Academic Press, London.
406

RAQUIN C. & PILET V. (1972) Production de plantules à partir d'anthères de Pétunias cultivées *in vitro*. *C.r. hebd. Séanc. Acad. Sci., Paris* **274**, 1019–1022.
215, 234

RASCH E., SWIFT H. & KLEIN R.M. (1959) Nucleoprotein changes in plant tumor growth. *J. biophys. biochem. Cytol.* **6**, 11–34.
375

RAYCHAUDHURI S.P. (1966) Plant viruses in tissue culture. *Adv. Virus Res.* **12**, 175–206.
393, 411

REDDI K.K. (1966) Ribonuclease induction in cells transformed by *Agrobacterium tumefaciens. Proc. natn. Acad. Sci. U.S.A.* **56**, 1207–1214.
265

REDDY M.N. & WILLIAMS P.H. (1970) Cytokinin activity in *Plasmodiophora brassicae*—infected cabbage tissue cultures. *Phytopathology* **60**, 1463–1465.
394, 400

REINERT J. (1956) Dissociation of cultures of *Picea glauca* into small tissue fragments and single cells. *Science, N.Y.* **123**, 457–458.
269

REINERT J. (1958) Untersuchungen über die Morphogenese an Gewebekulturen. *Ber. dtsch. Bot. Ges.* **71**, 15.
8, 347

REINERT J. (1959) Über die Kontrolle der Morphogenese und die Induktion von Adventivembryonen an Gewebekulturen aus Karotten. *Planta* **53**, 318–333.
8, 345, 347, 351

REINERT J. (1962) Morphogenesis in plant tissue cultures. *Endeavour* **21**, 85–90.
341

REINERT J. (1963a) Growth of single cells from higher plants on synthetic media. *Nature, Lond.* **200**, 90–91.
195

REINERT J. (1963b) Experimental modification of organogenesis in plant tissue cultures, p. 168, in *Plant Tissue and Organ Cultures—a Symposium.* Int. Soc. Plant Morphologists, Delhi.
349

REINERT J. (1964) Differenzierung und Bildung von Adventivembryonen aus einzelnen Zellen von Karottengewebe. Abstr. 10th Internat. Bot. Congr. Edinburgh, 209–210.
349

REINERT J. (1965) Growth of single cells from *Haplopappus gracilis* and *Vitis vinifera* on synthetic media, pp. 393–400, in *Proc. Int. Conf. on Plant Tissue Culture.* Ed. P.R.White & A.R.Grove. McCutchan Publ. Corp., Berkeley, California.
274

REINERT J. (1968) Morphogenese in Gewebe- und Zellkulturen. *Naturwissenschaften* **55**, 170–175.
339

REINERT J., BACKS-HÜSEMANN D. & ZERBAN H. (1971) Determination of embryo and root formation in tissue cultures from *Daucus carota*, pp. 261–268, in *Les Cultures de Tissus de Plantes.* Colloques Internationaux du C.N.R.S., no. 193, Paris.
340, 342, 343, 349, 352

REINERT J. & KÜSTER H.J. (1966) Diploide, chlorophyllhaltige Gewebekulturen aus Blätten von *Crepis capillaris* (L) Wallr. *Z. Pflanzenphysiol.* **54**, 213–222.
189

REINERT J. & VON ARDENNER R. (1964) Abhangigkeit des Teilungswachsums isolierter Einzelzellen aus *Vitis vinifera* kulturen von benachbarten wachsender Geweben. *Z. Naturf.* **19(b)**, 1150–1156.
195

REINERT J. & WHITE P.R. (1956) The cultivation *in vitro* of tumor tissues and normal tissues of *Picea glauca. Physiologia Pl.* **9**, 177–189.
31, 43, 217

REINERT R.A. (1966) Virus activity and callus growth of infected and healthy callus tissues of *Nicotiana tabacum* grown *in vitro. Phytopathology* **56**, 731–733.
410

REYNOLDS B.S. (1963) The use of lead citrate at high pH as an electron-opaque stain in electron microscopy. *J. Cell Biol.* **17**, 208–212.
49

RICHARDS E.G. & LECANIDOU R. (1971) Quantitative aspects of the electrophoresis of RNA in polyacrylamide gels. *Analyt. Biochem.* **40**, 43–71.
55

RICHARDS E.G. & TEMPLE C.J. (1971) Some properties of polyacrylamide gels. *Nature Phys. Sci.* **230**, 92–96.
55

ŘIČICA J. (1966) Continuous systems, pp. 32–66, in *Theoretical and Methodological Basis of Continuous Culture of Microorganisms.* Ed. I.Málek & Z.Fencl. Academic Press, New York.
315, 316, 334

RIKER A.J. (1923) Some relations of the crown-gall organism to its host tissue. *J. agric. Res.* **25**, 119–132.
358

RIKER A.J. (1926) Studies on the influence of some environmental factors on the development of crown gall. *J. agric. Res.* **32**, 83–96.
358

RIKER A.J., BANFIELD W.M., WRIGHT W.H., KEITT G.W. & SAGEN H. (1930) Studies on infectious hairy root of nursery apple trees. *J. agric. Res.* **41**, 507–540.
358

ROBARDS A.W. (1968) A new interpretation of plasmodesmatal ultrastructure. *Planta* **82**, 200–210.
135

ROBBINS E. & SCHARFF M. (1966) Some macromolecular characteristics of synchronized Hela cells, pp. 353–374, in *Cell Synchrony: Studies in Biosynthetic Regulation.* Eds. I.Cameron & G.Padilla. Academic Press, New York.
298

ROBBINS W.J. (1922a) Cultivation of ex-

cised root tips and stem tips under sterile conditions. *Bot. Gaz.* **73**, 376–390.
6

ROBBINS W.J. (1922b) Effect of autolysed-yeast and peptone on growth of excised corn root tips in the dark. *Bot. Gaz.* **74**, 59–79.
6

ROBERTS K. & NORTHCOTE D.H. (1970) The structure of sycamore callus cells during division in a partially synchronised suspension culture. *J. Cell Sci.* **6**, 299–231.
147, 153, 154, 157, 303, 304, 307

ROBERTSON A.I. (1966) *Metabolic changes during callus development in tissue isolated from Jerusalem artichoke tubers.* Ph.D. Thesis, University of Edinburgh, Edinburgh, Scotland.
250, 253, 254, 255, 257

ROBERTSON N.F. (1959) Experimental control of hyphal branching and branch form in hyphomycetous fungi. *J. Linn. Soc. Bot.* **56**, 207–211.
239

ROBERTSON N.F., FRIEND J., AVEYARD M., BROWN J., HUFFEE M. & HOMANS A.L. (1968) The accumulation of phenolic acids in tissue culture pathogen combinations of *Solanum tuberosum* and *Phytophthora infestans. J. gen. Microbiol.* **54**, 261–268.
404

ROBSON H.H., YOST H.T. & ROBISON M. M. (1961) Growth of *Parthencissus tricuspidata* tissue cultured on media containing aza analogues of indole-3-acetic acid and indole-3-propionic acid. *Pl. Physiol., Lancaster* **36**, 621–625.
376

ROGERS A.W. (1967) *Techniques of Autoradiography.* Elsevier Pub. Co., Amsterdam.
50

ROSE D., MARTIN S.M. & CLAY P.P.F. (1972) Metabolic rates for major nutrients in suspension cultures of plant cells. *Can. J. Bot.* **50**, 1301–1308.
286, 297

ROSE R.J., SETTERFIELD G. & FOWKE L.C. (1972) Activation of nucleoli in tuber slices and the function of nucleolar vacuoles. *Exp. Cell Res.* **71**, 1–16.
123, 131

ROSSETTI V. & MOREL G. (1958) Le développement du *Puccinia antirrhini* sur tissu de muflier cultivés *in vitro. C.r. hebd. Séanc. Acad. Sci., Paris* **247**, 1893–1895.
394, 395, 396, 398, 402

ROUSSAUX J., KURKDJIAN A. & BEARDSLEY R.E. (1968) Bacteriophages d'*Agrobacterium tumefaciens* (Smith et Town.) Conn. 1. Isolement et caractères. *Ann. Inst. Pasteur, Paris* **114**, 237–247.
358, 372, 373

RUBIO-HUERTOS M. & BELTRA R. (1962) Fixed pathogenic L forms of *Agrobacterium tumefaciens. Nature, Lond.* **195**, 101.
373

RUESINK A.W. (1971) Protoplasts of plant cells, pp. 197–209, in *Methods in Enzymology*, vol. **23**. Academic Press, New York.
102

RYSER H.J.P. (1967) Uptake of protein by mammalian cells: an under-developed area. *Science N.Y.* **159**, 390–396.
113

SACRISTÁN M.D. (1967) Auxin-Autotrophie und Chromosomenzahl. *Molec. Gen. Genet.* **99**, 311–321.
162

SACRISTÁN M.D. (1971) Karyotypic changes in callus cultures from haploid and diploid plants of *Crepis capillaris* (L) Wallr. *Chromosoma* **33**, 273–283.
162, 170, 175, 177, 185, 189

SACRISTÁN M.D. & MELCHERS G. (1969) The caryological analysis of plants regenerated from tumorous and other callus cultures of tobacco. *Molec. Gen. Genet.* **105**, 317–333.
117, 118, 162, 187, 188, 342, 344, 382, 389

SACRISTÁN M.D. & WENDT-GALLITELLI M.F. (1971) Transformation to auxin-autotrophy and its reversibility in a mutant line of *Crepis capillaris* callus culture. *Molec. Gen. Genet.* **110**, 355–360.
162, 389

SANDSTEDT R. & SCHUSTER M.L. (1963) Nematode-induced callus on carrot discs grown *in vitro. Phytopathology* **53**, 1309–1312.
419

SANDSTEDT R. & SCHUSTER M.L. (1965) Host-parasite interaction in root-knot nematode-infected carrot tissue. *Phytopathology* **55**, 393–395.
419

SANDSTEDT R. & SCHUSTER M.L. (1966a) Excised tobacco pith bioassays for root-knot nematode-produced plant growth

substances. *Physiologia Pl.* **19**, 99–104.
420

SANDSTEDT R. & SCHUSTER M.L. (1966b)
The role of auxins in root-knot nematode-
induced growth on excised tobacco stem
segments. *Physiologia Pl.* **19**, 960–
967.
420

SANGER J.M. & JACKSON W.T. (1971) Fine
structure study of pollen development in
Haemanthus Katherinae Baker. *J. Cell
Sci.* **8**, 289–301.
207

SARGENT J.A. & SKOOG F. (1960) Effects of
indoleacetic acid and kinetin on scopoletin and scopolin levels in relation to
growth of tobacco tissues *in vitro*. *Pl.
Physiol., Lancaster* **35**, 934–941.
262

SASTRI R.L.N. (1963) Morphogenesis in
plant tissue cultures, pp. 105–107, in
*Plant Tissue and Organ Culture—a
Symposium.* Int. Soc. Plant Morphologists, Delhi.
345

SATO N., TOMIYAMA K., KATSUI N. &
MASAMUNE T. (1968) Isolation of rishitin
from tubers of interspecific potato
varieties containing late-blight resistance
genes. *Ann. Phytopath. Soc. Japan* **34**,
140–142.
404

SAX K. & EDMONDS H.W. (1933) Development of the male gemetophyte in *Tradescantia. Bot. Gaz.* **95**, 156–163.
207, 208

SAYRE R.M. (1958) *Plant tissue culture as
a tool in the study of the physiology of
root-knot nematode Meloidogyne incognita
Chit.* Ph.D. Thesis, University of Nebras-
ka, Lincoln, U.S.A.
418

SCALA J. & SEMENSKY F. (1971) An induced
fructose-1,6-diphosphatase from cultured
cells of *Acer pseudoplatanus. Phytochem.*
10, 567–570.
291

SCHAEFFER G. W. (1962) Tumour induction
by an indolyl-3-acetic acid-kinetin interaction in a *Nicotiana* hybrid. *Nature,
Lond.* **196**, 1326–1327.
384

SCHAEFFER G.W., BURK L.G. & TSO T.C.
(1966) Tumors of interspecific *Nicotiana*
hybrids. I. Effect of temperature and
photoperiod upon flowering and tumor

formation. *Am. J. Bot.* **53**, 928–932.
385

SCHAEFFER G.W. & SMITH H.H. (1963)
Auxin-kinetin interaction in tissue culture
of *Nicotiana* species and tumour conditioned hybrids. *Pl. Physiol., Lancaster* **38**,
291–297.
387

SCHEITTERER H. (1931) Versuche zur kultur
von Pflanzengeweben. *Arch. exp. Zell.-
forsch.* **12**, 141–176.
6

SCHENK R.U. & HILDEBRANDT A.C. (1969)
Production of protoplasts from plant
cells in liquid culture using purified
commercial cellulases. *Crop Sci.* **9**,
629–631.
103, 107

SCHILDE-RENTSCHLER L. (1972) A simple
method for the preparation of plant
protoplasts. *Z. Naturforsch.* **27b**, 208–209.
108

SCHILPEROORT R.A. (1969) *Investigation on
plant tumors crown gall. On the biochemistry of tumor-induction by Agrobacterium tumefaciens.* Thesis, University
of Leyden, Belgium.
364

SCHILPEROORT R.A., MEIJS W.H., PIPPEL
G.M.W. & VELDSTRA H. (1969) *Agrobacterium tumefaciens* cross-reacting antigens in sterile crown-gall tumors. *FEBS
Letters* **3**, 173–176.
371

SCHILPEROORT R.A., VELDSTRA H., WAR-
NAAR S.O., MULDER G. & COHEN J.A.
(1967) Formation of complexes between
DNA isolated from tobacco crown gall
tumours and RNS complementary to
Agrobacterium tumefaciens DNA. *Biochim. biophys. Acta* **145**, 523–525.
370

SCHMIDT R. (1969) Control of enzyme
synthesis during the cell cycle of *Chlorella*,
pp. 159–177, in *The Cell Cycle: gene-
enzyme interactions.* Ed. G.Padilla, G.
Whitson & I.Cameron, Academic Press
New York.
298

SCHMUCKER T. (1929) Isolierte Gewebe und
Zellen von Blutenflanzen. *Planta* **9**,
339–340.
6

SCHNEIDER H.A.W. (1970) Aktivatat und
eigenschaften von aminolavulinatdehy-
dratase in ergunenden zellkulturen von

Nicotiana tabacum var. Samsum. *Z. Pflphysiol.* **62**, 133–145.
266

SCHNEIDER W.C. (1945) Phosphorus compounds in animal tissues. I. Extraction and estimation of desoxypentose nucleic acid and of pentose nucleic acid. *J. biol. Chem.* **161**, 293–303
54

SCHNEPF E. (1964) Zur Cytologie und Physiologie der Pflanzlichen Drusen. IV. Licht- und elektronenmikroskopische Untersuchungen an Septalnekratien. *Protoplasma* **58**, 137–171.
129

SCHROEDER P.H. & JENKINS W.R. (1963) Reproduction of *Pratylenchus penetrans* on root tissues grown on three media. *Nematologica* **9**, 327–331.
420

SCOTT K.J. & MACLEAN D.J. (1969) Culturing of rust fungi. *Ann. rev. Phytopathology* **7**, 123–146.
402

SCOTT K.J., SMILLIE R.M. & KROTKOV G. (1962) Respiration and phosphorus-containing compounds in normal and tumour tissues of red beet roots. *Can. J. Bot.* **40**, 1251–1256.
265

SEALEY P. (1973) Unpublished observations.
127

SEGRETAIN G. (1943) Culture d'un virus et son inoculation sur fragments de tige de tabac cultivés *in vitro*. *Ann. Inst. Pasteur* **69**, 61–63.
407

SEGRETAIN G. & HIRTH L. (1953) Action de substances azotées sur la multiplication du virus de la mosaique du tabac en culture de tissus. *C.r. Séanc. Soc. Biol.* **147**, 1042–1043.
411

SEITZ E.W. & HOCHESTER R.M. (1964) Lysopine in normal and crown-gall tumor tissue of tomato and tobacco. *Can. J. Bot.* **42**, 999–1004.
378

SETTERFIELD G. (1963) Growth regulation in excised slices of Jerusalem artichoke tuber tissue. *Symp. Soc. Exp. Biol.* **17**, 98–126.
251

SEVENSTER P. & KARSTENS W.K.H. (1955) Observations on the proliferation of stem pith parenchyma *in vitro*. II. The internal structure of stem pith cylinders of *Helianthus tuberosus* L. cultivated *in vitro*. *Acta Botan. Neerl.* **4**, 188–192.
256

SHAMINA Z.B. (1966) Cytogenetic study of tissue culture of *Haplopappus gracilis*, pp. 337–380, in *Proc. Symp. The mutational process: mechanism of mutation and inducing factors*. Ed. Z.Landa. Academia, Prague.
170, 172

SHARP W.R., DOUGALL D.K. & PADDOCK, E.F. (1971). Haploid plantlets and callus from immature pollen grains of *Nicotiana* and *Lycopersicon*. *Bull. Torrey Bot. Club* **98**, 219–222
216, 218

SHARP W.R. & DOUGAL D.K. (1970) Growth and morphogenetic potencies of pith and anther callus cultures derived from several varieties of *Lycopersicon*. *In Vitro* **6**, 237.
216

SHARP W.R., RASKIN R.S. & SOMMER H.E. (1972) The use of nurse culture in the development of haploid clones in tomato. *Planta* **104**, 357–361.
210, 214

SHEAT D.E.G., FLETCHER B.H. & STREET H.E. (1959) Studies on the growth of excised roots. VIII. The growth of excised tomato roots supplied with various inorganic sources of nitrogen. *New Phytol.* **58**, 128–141.
38, 83

SHIMADA T. (1971) Chromosome constitution of tobacco and wheat callus cells. *Jap. J. Genet.* **46**, 235–241.
162

SHIMADA T., SASAKUMA T. & TSUNEWAKI K. (1969) *In vitro* culture of wheat tissues. Callus formation, organ redifferentiation and single cell culture. *Can. J. Genet. Cytol.* **11**, 294–304.
162

SHIMADA T. & TABATA M. (1967) Chromosome numbers in cultured pith tissues of tobacco. *Japan J. Genet.* **42**, 195–201.
175, 186

SHORT K.C., BROWN E.G. & STREET H.E. (1969a) Studies on the growth in culture of plant cells. V. Large-scale culture of *Acer pseudoplatanus* cell suspensions. *J. exp. Bot.* **20**, 579–590.
27, 65, 66

SHORT K.C., BROWN E.G. & STREET H.E.

(1969b) Studies on the growth in culture of plant cells. VI. Nucleic acid metabolism of *Acer pseudoplatanus* cell suspensions. *J. exp. Bot.* **20**, 579–590.
92, 280, 282, 283, 284

SIDORENKO P.G. & KUNAKH V.A. (1970) The character of karyotype variability in the cell population of a tissue culture of *Haplopappus gracilis* on prolonged subculturing (in Russian). *Tsitol. Genet.* **4**, 235–241.
170, 172

SIEGAL A. & ZAITLIN M. (1964) Infection process in plant virus diseases. *Ann. Rev. Phytopathol.* **2**, 179–202.
111

SIEVERT R.C. & HILDEBRANDT A.C. (1965) Variation within single cell clones of tobacco tissue cultures. *Am. J. Bot.* **52**, 742–750.
200

SIEVERT R.C., HILDEBRANDT A.C., BURRIS R.H. & RIKER A.J. (1961) Growth *in vitro* of single cell clones isolated from a single cell clone. *Pl. Physiol., Lancaster* **36**, Suppl. XXIX.
200

SIMON E.W. & CHAPMAN J.A. (1961) The development of mitochondria in *Arum* spadix. *J. exp. Bot.* **12**, 414–420.
129

SIMONSEN J. & HILDEBRANDT A.C. (1971) *In vitro* growth and differentiation of *Gladiolus* plants from callus cultures. *Can. J. Bot.* **49**, 1817–1819.
416

SIMPKINS I., COLLIN H.A. & STREET H.E. (1970) The growth of *Acer pseudoplatanus* cells in a synthetic liquid medium: response to the carbohydrate, nitrogenous and growth hormone constituents. *Physiologia Pl.* **23**, 385–396.
145, 150, 271

SIMPKINS I. & STREET H.E. (1970) Studies on the growth in culture of plant cells. VII. Effects of kinetin on the carbohydrate and nitrogen metabolism of *Acer pseudoplatanus* L. cells grown in suspension culture. *J. exp. Bot.* **21**, 170–185.
281, 284

SINGH B.D., HARCEY B.L., KAO K.N. & MILLER R.A. (1972) Selection pressure in cell populations of *Vicia hajastana* cultured *in vitro. Can. J. Genet. Cytol.* **14**, 65–70.
427

SINGH H. (1963) The growth of *Albugo* in the callus culture of *Ipomoea. Curr. Sci.* **32**, 472–473.
394

SINGH H. (1966) On the variability of the callus of *Ipomoea* infected with *Albugo* grown under *in vitro* conditions. *Phytomorphology* **16**, 189–192.
394

SINNOTT E.W. & BLOCH R. (1940) Cytoplasmic behaviour during division of vacuolate plant cells. *Proc. natn. Acad. Sci. U.S.A.* **26**, 223–227.
123, 131, 133

SINNOTT E.W. & BLOCH R. (1941) Division in vacuolated plant cells. *Am. J. Bot.* **28**, 225–232.
123, 131, 153

SJOLUND R.D. & WEIER T.E. (1971) An ultrastructural study of chloroplast structure and dedifferentiation in tissue cultures of *Streptanthus tortuosus* (Cruciferae). *Amer. J. Bot.* **58**, 172–181.
140

SKOOG F. (1944) Growth and organ formation in tobacco tissue cultures. *Am. J. Bot.* **31**, 19–24.
344, 387

SKOOG F. (1971) Aspects of growth factor interactions in morphogenesis in tobacco tissue cultures, pp. 115–136, in *Les Cultures de Tissus de Plantes*. Colloques Internationaux du C.N.R.S., no. 193, Paris.
346

SKOOG F. & MILLER C.O. (1957) Chemical regulation of growth and organ formation in plant tissues cultured *in vitro. Symp. Soc. exp. Biol.* **11**, 118–130.
138, 263, 339, 345, 424

SKOOG F. & MONTALDI E. (1961) Auxin-kinetin interaction regulating the scopoletin and scopolin levels in tobacco tissue cultures. *Proc. natn. Acad. Sci. U.S.A.* **74**, 36–49.
263, 264

SKOOG F. & TSUI C. (1948) Chemical control of growth and bud formation in tobacco stem segments and callus cultured *in vitro. Am. J. Bot.* **35**, 782–787.
32, 255

SLEPYAN L.I., VOLLOSOVICH A.G. & BUTENKO R.G. (1968) Tissue cultures of medicinal herbs and prospects for their use. *Rast. Resur.* **4**, 457–467.
261

SMITH C.O. (1943) Crown gall on species of Taxaceae, Taxodiaceae and Pinaceae as determined by artificial inoculations. *Phytopathology* **32**, 1005.
357

SMITH H.H. (1972) Plant genetic tumors. *Progr. exp. Tumor Res.* **15**, 138–164.
384, 385, 388

SMITH S.M. (1973) *Embryogenesis in tissue cultures of the domestic carrot, Daucus carota* L. Ph.D. Thesis. Univ. Leicester, England.
424

SMITH S.H., HILTON R.E. & FRAZIER N.W. (1970) Meristem culture for elimination of strawberry viruses. *Californian Agric.* **24**, 9–10.
413

SMITHERS A.G. & SUTCLIFFE J.F. (1967) A method for the removal of mineral salts from coconut milk and comparison of some of the growth-promoting properties of whole and desalted coconut milk in carrot root tissue cultures. *Ann. Bot.* **31**, 333–350.
243

SOLBERG R.A. & BALD J.G. (1963) Distribution of a natural and an alien form of tobacco mosaic virus in the shoot apex of *Nicotiana glauca* Grah. *Virology* **21**, 300–308.
410, 415

SOROKIN H.P. (1955) Mitochondria and spherosomes in the living epidermal cell. *Am. J. Bot.* **42**, 225–231.
142

SPARROW A.H., GUNCKEL J.E., SCHAIRER L.A. & HAGEN G.L. (1956) Tumor formation and other morphogenetic responses in an amphiploid tobacco exposed to chronic gamma irradiation. *Am. J. Bot.* **43**, 377–388.
384

SPEAKE T., MCCLOSKEY P., SMITH W., SCOTT T. & HUSSEY H. (1964) Isolation of nicotine from cell cultures of *Nicotiana tabacum*. *Nature, Lond.* **201**, 614–615.
295

SPENCER D.F. & KIMMINS W.C. (1969) Presence of plasmodesmata in callus cultures of tobacco and carrot. *Can. J. Bot.* **47**, 2049–2050.
408, 412

SRIVASTAVA B.I.S. (1968) Patterns of nucleic acids synthesis in normal and crown gall tumour tissues culture of tobacco. *Archs. Biochem. Biophys.* **125**, 817–823.
265

SRIVASTAVA B.I. (1970) DNA-DNA hybridization studies between bacteria DNA, crown gall DNA and the normal cell DNA. *Life Sciences* **9**, 889–892.
370

SRIVASTAVA B.I. & CHADHA, K.C. (1970) Liberation of *Agrobacterium tunefaciens* DNA from the crown gall tumor cell DNA by shearing. *Biochem. biophys. Res. Comm.* **40**, 968–972.
371

STABA E.J. & LAURSEN P. (1966) Morning glory tissue culture: growth and examination of indole alkaloids. *J. Pharm. Sci.* **55**, 1099–1101.
267

STACE-SMITH R. & MELLOR F.C. (1968) Eradication of potato virus X and Y by thermotherapy and axillary bud culture. *Phytopathology* **58**, 199–203.
413

STACE-SMITH R. & MELLOR F.C. (1970) Eradication of potato spindle tuber virus by thermotherapy and axillary bud culture. *Phytopathology* **60**, 1857–1858.
413

STAPP C. (1927) Der bakterielle Pflanzenkrebs und seine Beziehungen zum tierischen und menschlichen Krebs. *Ber. dt. bot. Ges.* **45**, 480–504.
358

STECK W., BAILEY B., SHYLUK J. & GAMBORG O. (1971) Coumarins and alkaloids from cell cultures of *Ruta graveolens*. *Phytochem.* **10**, 191–194.
295

STERLING C. (1951) Origin of buds in tobacco stem segments cultured *in vitro*. *Am. J. Bot.* **38**, 761–767.
341

STEWARD F.C. (1958) Growth and development of cultivated cells. III. Interpretations of the growth from free cell to carrot plant. *Am. J. Bot.* **45**, 709–713.
8, 424

STEWARD F.C. (1961) Vistas in plant physiology: problems or organisation, growth and morphogenesis. *Can. J. Bot.* **39**, 441–460.
122, 424

STEWARD F.C., BIDWELL R.G.S. & YEMM E.W. (1958) Nitrogen metabolism,

respiration, and growth of cultured plant tissue. *J. exp. Bot.* **9,** 11–51.
129

STEWARD F.C. & CAPLIN S.M. (1951) A tissue culture from potato tubers. The synergetic action of 2,4-D and coconut milk. *Science, N.Y.* **113,** 518–520.
244

STEWARD F.C. & CAPLIN S.M. (1952) Investigations on growth and metabolism of plant cells. IV. Evidence on the role of the coconut-milk factor in development. *Ann. Bot.* **16,** 491–504.
42, 43

STEWARD F.C. & CAPLIN S.M. (1954) The growth of carrot tissue explants and its relation to the growth factors present in coconut milk. I. The development of the quantitative method and the factors affecting the growth of carrot tissue explants. *Année. Biol.* **30,** 385–395.
243, 252, 253

STEWARD F.C., CAPLIN S.M. & MILLAR F.K. (1952) Investigations of growth and metabolism of plant cells. I. New techniques for the investigation of metabolism, nutrition and growth in undifferentiated cells. *Ann. Bot.* **16,** 58–77.
8, 27, 32, 35, 46, 60, 61, 129, 253

STEWARD F.C., ISRAEL H. & MAPES M. (1968) Growth regulating substances: their roles observed at different levels of cellular organization, pp. 875–892, in *The Biochemistry and Physiology of Plant Growth Substances.* Ed. F.Wightman & G.Setterfield. Runge Press, Ottawa.
270

STEWARD F.C., KENT A.E. & MAPES M.O. (1966) The culture of free plant cells and its signification for embryology and morphogenesis, pp. 113–154, in *Current Topics in Developmental Biology,* Academic Press, New York.
349

STEWARD F.C., KENT A.E. & MAPES M.O. (1967) Growth and organization in cultured cells: sequential and synergistic effects of growth-regulating substances. *Ann. N.Y. Acad. Sci.* **144,** 326–334.
351, 426

STEWARD F.C., MAPES M.O., KENT A.E. & HOLSTEN R.D. (1964) Growth and development of cultured plant cells. *Science, N.Y.* **143,** 20–27.
47, 241, 250, 347, 425

STEWARD F.C., MAPES M.O. & MEARS K.

(1958) Growth and organized development of cultured cells. II. Organization in cultures grown from freely suspended cells. *Am. J. Bot.* **45,** 705–708.
8, 341, 347

STEWARD F.C., MAPES M.O. & SMITH J. (1958) Growth and organized development of cultured cells. I. Growth and division of freely suspended cells. *Am. J. Bot.* **45,** 693–703.
123, 144, 150, 152

STEWARD F.C. & MOHAN RAM H.Y. (1961) Determining factors in cell growth: some implications for morphogenesis in plants. *Advan. Morphogenesis* **1,** 189–265.
297, 298

STEWARD F.C. & SHANTZ E.M. (1956) The chemical induction of growth in plant tissue cultures, pp. 165–186, in *The Chemistry and Mode of Action of Plant Growth Substances.* Ed. R.L.Wain & F.Wightman. Butterworths Ltd., London.
8, 60, 61, 247, 257, 269

STICHEL E. (1959) Gleichzeitige Induktion von Sprossen und Wurzeln an *in vitro* kultivierten Gewebestückchen von *Cyclamen persicum. Planta* **53,** 293–317.
345

STICKLAND R.G. & SUNDERLAND N. (1972a) Production of anthocyanins, flavonols, and chlorogenic acids by cultured callus tissues of *Haplopappus gracilis. Ann. Bot.* **36,** 443–457.
187

STICKLAND R.G. & SUNDERLAND N. (1972b) Photocontrol of growth, and of anthocyanin and chlorogenic acid production in cultured callus tissues of *Haplopappus gracilis. Ann. Bot.* **36,** 671–685.
187

STOBART A.K., McLAREN I. & THOMAS D.R. (1967) Chlorophylls and carotenoids of colourless callus, green callus and leaves of *Kalanchoe crenata. Phytochem.* **6,** 1467–1474.
266

STOBART A.K. & THOMAS D.R. (1968) δ-aminolaevulinic acid dehydratase in tissue cultures of *Kalanchoe crenata. Phytochem.* **7,** 1313–1316.
266

STOBART A.K., WEIR N.R. & THOMAS D.R. (1968) Phytol in tissue culture of *Kalanchoe crenata. Phytochem.* **8,** 1089–1100.
266

STOHS S.J. & STABA E.J. (1965) Production

of cardiac glycosides by plant tissue cultures. IV. Biotransformation of digitoxigenines and related substances. *J. Pharm. Sci.* **54**, 56–58.
296

STONE O. (1968) The elimination of four viruses from carnation and sweet-william by meristem tip culture. *Ann. appl. Biol.* **62**, 119–122.
413

STONIER T. (1971) The role of auxin protectors in autonomous growth, pp. 423–435, in *Les Cultures de Tissus de Plantes*. Colloques Internationaux du C.N.R.S., no. 193, Paris.
376

STONIER T., MCSHARRY J. & SPEITAL T. (1967) *Agrobacterium tumefaciens* Conn. IV. Bacteriophage $PB2_1$ and its inhibitory effect on tumor induction. *J. Virol.* **1**, 268–273.
358, 372

STONIER T. & YANG H. (1971) Studies on auxin protectors. X. Protector levels and lignification in sunflower crown gall tissue. *Physiologia Pl.* **25**, 474–481.
378

STRAUS J. (1954) Maize endosperm tissue grown *in vitro*. II. Morphology and cytology. *Am. J. Bot.* **41**, 833–839.
31, 32

STRAUS J. (1962) Invertase in cell walls of plant tissue cultures. *Pl. Physiol., Lancaster* **37**, 342–348.
264

STRAUS J. & LA RUE C.D. (1954) Maize endosperm tissue grown *in vitro*. I. Culture requirements. *Am. J. Bot.* **41**, 687–694.
41, 257

STREET H.E. (1957) Excised root culture. *Biol. Rev.* **32**, 117–155.
7

STREET H.E. (1959) Special problems raised by organ and tissue culture. Correlation between organs of higher plants as a consequence of specific metabolic requirements, pp. 153–178, in *Handbuch der Pflanzenphysiologie*, vol. **11**. Ed. W.Ruhland. Springer Verlag, Berlin.
7

STREET H.E. (1966a) The nutrition and metabolism of plant tissue and organ cultures, pp. 533–629, in *The Biology of Cells and Tissues in Culture*, vol. 3. Ed.

E.N.Willmer. Academic Press, New York.
7

STREET H.E. (1966b) Growth, differentiation and organogenesis in plant tissue and organ cultures, pp. 631–689, in *The Biology of Cells and Tissues in Culture*, vol. **3**. Ed. E.N.Willmer. Academic Press, New York.
137, 270, 271, 343, 424

STREET H.E. (1968a) The induction of cell division in plant cell suspension cultures, pp. 177–193, in *Les Cultures de Tissus de Plantes*. Colloques Internationaux du C.N.R.S., Strasbourg.
54, 87, 194

STREET H.E. (1968b) Factors influencing the initiation and activity of meristems in roots, pp. 20–41, in *Root Growth*, 15th Easter School, School of Agriculture, University of Nottingham. Ed. W.J. Whittington. Butterworths Ltd., London.
424

STREET H.E. (1969) Growth in organized and unorganized systems—knowledge gained by culture of organs and tissue explants, pp. 3–224, in *Plant Physiology*, vol. **5B**. Ed. F.C.Steward. Academic Press, New York.
5, 424

STREET H.E. (1973) Plant cell cultures: their potential for metabolic studies, pp. 93–125, in *Biosynthesis and its Control in Plants*. Ed. B.V.Milborrow. Academic Press, London.
274, 298, 310, 333, 334, 426

STREET H.E. & HENSHAW G.G. (1966) Introduction and methods employed in plant tissue culture, pp. 459–532, in *The Biology of Cells and Tissues in Culture*, vol 3. Ed. E.N.Willmer. Academic Press, New York.
5

STREET H.E., HENSHAW G.G. & BUIATTI M.C. (1965) The culture of isolated plant cells. *Chem. Ind.* 27–33.
152, 273, 422

STREET H.E., KING P.J. & MANSFIELD K. (1971) Growth control in plant cell suspension cultures, pp. 17–40, in *Les Cultures de Tissus de Plantes*. Colloques Internationaux du C.N.R.S., no. 193, Paris.
78, 272, 274, 297

STREISSLE G. (1971) The persistence of virus in wound tumor tissue cultures, pp.

499–501, in *Les Cultures de Tissus de Plantes*, Colloques Internationaux du C.N.R.S., no. 193, Paris.
32, 383, 384, 411

STROUN M., ANKER P., CHARLES P. & LEDOUX L. (1967) Translocation of DNA of bacterial origin in *Lycopersicum esculentum* by ultracentrifugation in caesium chloride gradient. *Nature, Lond.* **215**, 975–976.
366

STROUN M., ANKER P., GAHAN P., ROSSIER A. & GREPPIN H. (1971) *Agrobacterium tumefaciens* ribonucleic acid synthesis in tomato cells and crown gall induction. *J. Bact.* **106**, 634–639.
369

STUART R. (1969) *The induction of cell division in plant suspension cultures.* Ph.D. Thesis, Univ. Leicester, England.
98

STUART R. & STREET H.E. (1969) Studies on the growth in culture of plant cells. IV. The initiation of division in suspensions of stationary phase cells of *Acer pseudoplantanus* L. *J. exp. Bot.* **20**, 556–571.
86, 88, 193, 194

STUART R. & STREET H.E. (1971) Studies on the growth in culture of plant cells. X. Further studies on the conditioning of culture media by suspensions of *Acer pseudoplantanus* L. *J. exp. Bot.* **22**, 96–106.
65, 193, 195, 196

SUNDERLAND N. (1966) Pigmented plant tissues in culture. I. Auxins and pigmentation in chlorophyllous tissues. *Ann. Bot.* **30**, 253–268.
266

SUNDERLAND N. (1970) Pollen plants and their significance. *New Scient.* **47**, 142–144.
234, 235, 236

SUNDERLAND N. (1971) Anther culture: a progress report. *Sci. Prog., Oxf.* **59**, 527–549.
218, 221, 222, 236, 238, 354

SUNDERLAND N. & DUNWELL J.M. (1971) *John Innes Annual Report*, no. 62, pp. 60–61.
211

SUNDERLAND N., DUNWELL J.M. & LAWES A.C. (1971) *John Innes Annual Report*, no. 62, pp. 58–60.
217, 218

SUNDERLAND N. & WELLS B. (1968) Plastid structure and development in green callus tissues of *Oxalis dispar*. *Ann. Bot.* **32**, 327–346.
49, 139, 262, 266

SUNDERLAND N. & WICKS F.M. (1969) Cultivation of haploid plants from tobacco pollen. *Nature, Lond.* **224**, 1227–1229.
215, 226, 228

SUNDERLAND N. & WICKS F.M. (1971) Embryoid formation in pollen grains of *Nicotiana tabacum*. *J. exp. Bot.* **22**, 213–226.
215, 217, 219, 225, 226, 231, 232

SUNDERLAND N., WICKS F.M. & STOREY B. (1970) *John Innes Annual Report*, no. 61, pp. 26–27.
217

SUSSEX I.M. (1965) The origin and morphogenesis of eucalyptus cell populations, pp. 383–391, in *Proc. Int. Conf. on Plant Tissue Culture*. Ed. P.R.White & A.R. Grove. McCutchan, Berkeley, California.
272

SUTTON-JONES B. & STREET H.E. (1968) Studies on the growth in culture of plant cells. III. Changes in fine structure during the growth of *Acer pseudoplantanus* L. cells in suspension culture. *J. exp. Bot.* **19**, 114–118.
94, 129, 154, 157

SYONO K. (1965) Changes in organ forming capacity of carrot root callus during subcultures. *Pl. Cell Physiol.* **6**, 403–419.
352

TABATA M., YAMAMOTO H. & HIRAOKA N. (1971) Alkaloid production in the tissue cultures of some solonaceous plants, pp. 390–402, in *Les Cultures de Tissus de Plantes*. Colloques Internationaux du C.N.R.S., no. 193, Paris.
295, 296

TABATA M., YAMAMOTO H., HIRAOKA N., MARUMOTO Y. & KONOSHIMA M. (1971) Regulation of nicotine production in tobacco tissue culture by plant growth regulators. *Phytochem.* **10**, 723–729.
261, 267

TAIZ L. & JONES R.L. (1971) The isolation of barley-aleurone protoplasts. *Planta* **101**, 95–100.
102

TAKEBE I. & OTSUKI Y. (1969) Infection of tobacco mesophyll protoplasts by tobacco

mosaic virus. *Proc. natn. Acad. Sci. U.S.A.* **64,** 843–848.
111, 113, 409

TAKEBE I., OTSUKI Y. & AOKI S. (1968) Isolation of tobacco mesophyll cells in intact and active state. *Pl. Cell Physiol.* **9,** 115–124.
108, 112

TAKEBE I., OTSUKI Y. & AOKI S. (1971) Infection of isolated tobacco mesophyll protoplasts by tobacco mosaic virus, pp. 503–511, in *Les Cultures de Tissus de Plantes.* Colloques Internationaux du C.N.R.S., no. 193, Paris.
409

TANAKA M. & NAKATA K. (1969) Tobacco plants obtained by anther culture and experiment to get diploid seeds from haploids. *Jap. J. Genet.* **44,** 47–54.
235

TAZAWA M. & REINERT J. (1969) Extracellular and intracellular chemical environments in relation to embryogenesis *in vitro. Protoplasma* **68,** 157–173.
351

THEIS T.N., RIKER A.J. & ALLEN O.N. (1950) The destruction of crown-gall bacteria in periwinkle by high temperature with high humidity. *Am. J. Bot.* **37,** 792–801.
360

THERMAN E. & KUPILA-AHVENNIEMI S. (1971) Crown gall development stimulated by wound washing. *Physiologia Pl.* **25,** 178–180.
361, 363

THIELMANN M. (1924, 1925) Über Kulturversuche mit spaltoffmingzellen. *Ber. dt. bot. Ges.* **42,** 429–434; *Arch. exp. Zellforsch.* **50,** 66–108.
6

THOMAS E., KONAR R.N. & STREET H.E. (1972) The fine structure of the embryogenic callus of *Ranunculus sceleratus* L. *J. Cell Sci.* **11,** 95–109.
139, 140, 141, 147, 349

THOMAS D.R. (1970) Mevalonate activating enzymes in greening tissue culture. *Phytochem.* **9,** 1443–1451.
266

THOMSON A.D. (1957) Elimination of potato virus Y from a potato variety. *New Zealand J. Sci. Technol.* **38,** 482–490.
413

THORPE T.A. & MURASHIGE T. (1968) Starch accumulation in shoot forming

tobacco callus cultures. *Science, N.Y.* **160,** 421–422.
263

THORPE T.A. & MURASHIGE T. (1970) Some histochemical changes underlying shoot initiation in tobacco callus cultures. *Can. J. Bot.* **48,** 277–285.
263

TIWARI M.M. & ARYA H.C. (1967) Growth of normal and diseased *Pennisetum typhoides* tissues infected with *Sclerospora graminicola* in tissue culture. *Indian Phytopathology* **20,** 356–368.
394, 396, 399

TOMIYAMA K., SAKUMA T., ISHIZAKA N., SATO N., KATSUI N., TAGASUKI M. & MASAMUNE T. (1968) A new antifungal substance isolated from resistant potato tuber tissue infected by pathogens. *Phytopathology* **58,** 115–116.
404

TIWARI M.M. & ARYA H.C. (1969) *Sclerospora graminicola* axenic culture. *Science, N.Y.* **163,** 291–293.
394, 403

TOMMERUP I.C. & INGRAM D.S. (1971) The life-cycle of *Plasmodiophora brassicae* Woron. in Brassica tissue cultures and in intact roots. *New Phytol.* **70,** 327–332.
394, 399

TORREY J.G. (1957) Cell division in isolated single cells *in vitro. Proc. natn. Acad. Sci. U.S.A.* **43,** 887–891.
97, 99

TORREY J.G. (1959) Experimental modification of development in the root, pp. 189–222, in *Cell, Organism and Milieu.* Ed. D.Rudnick. Ronald Press, New York.
162, 171, 175, 177, 204, 339, 343

TORREY J.G. (1961) Kinetin as trigger for mitosis in mature endomitotic plant cells. *Exp. Cell Res.* **23,** 281–299.
35, 162, 171, 175, 204

TORREY J.G. (1965) Cytological evidence of cell selection by plant tissue culture media, pp. 473–484, in *Proc. Int. Conf. Plant Tissue Culture.* Ed. P.R.White & A.R.Grove. McCutchan Publ. Corp., Berkeley, California.
162, 175, 185

TORREY J.G. (1966) The initiation of organized development in plants. *Adv. Morphogenesis* **5,** 39–91.
351

TORREY J.G. (1967a) Morphogenesis in relation to chromosomal constitution in long-term plant tissue cultures. *Physiologia Pl.* **20**, 265–275.
162, 171, 175, 343

TORREY J.G. (1967b) *Development in Flowering Plants.* Macmillan Ltd., New York.
424

TORREY J.G. (1971) Cytodifferentiation in plant cell and tissue culture, pp. 177–186, in *Les Cultures de Tissus de Plantes.* Colloques Internationaux du C.N.R.S., no. 193, Paris.
298, 424

TORREY J.G. & REINERT J. (1961) Suspension culture of higher plant cells in synthetic medium. *Pl. Physiol., Lancaster,* **36**, 483–491.
81, 271, 273

TORREY J., REINERT J. & MERKEL N. (1962) Mitosis in suspension cultures of higher plant cells in synthetic medium. *Am. J. Bot.* **49**, 420–425.
271, 272, 277

TORREY J.G. & SHIGOMURA J. (1957) Growth and controlled morphogenesis in pea root callus tissue grown in liquid media. *Am. J. Bot.* **44**, 334–344.
41, 260

TOURNEUR J. & MOREL G. (1970) Sur la présence de phages dans les tissus de crown-gall cultivés *in vitro. C.r. hebd. Séanc. Acad. Sci., Paris* **270**, 2810–2812.
367, 373

TOURNEUR J. & MOREL G. (1971) Bacteriophages et crown-gall. Résultats et hypothèses. *Physiol. Vég.* **9**, 527–539.
372

TRIBE H.T. (1955) Studies on the physiology of parasitism. XIX. On the killing of plant cells by enzymes from *Botrytis cincerea* and *Bacterium aroideae. Ann. Bot.* **19**, 351–368.
101

TRIONE E.J., JONES L.E. & METZGER R.J. (1968) *In vitro* culture of somatic wheat callus tissue. *Am. J. Bot.* **55**, 529–531.
31

TRYON K. (1956) Scopoletin in differentiating and nondifferentiating cultured tobacco tissue. *Science, N.Y.* **123**, 590.
263

TSO T.C., BURK L.G., DIETERMAN L.J. & WENDER S.H. (1964) Scopoletin, scopolin and chlorogenic acid in tumours of interspecific *Nicotiana* hybrids. *Nature, Lond.* **204**, 779–780.
386

TSO T.C., BURK L.G., SOROKIN T.P. & ENGELHAUPT M.E. (1962) Genetic tumours of *Nicotiana. Pl. Physiol., Lancaster* **37**, 357–360.
386

TULECKE W. (1953) A tissue derived from the pollen of *Ginkgo biloba. Science, N.Y.* **117**, 599–560.
214

TULECKE W. (1954) Preservation and germination of the pollen of *Ginkgo* under sterile conditions. *Bull. Torrey, Bot. Club.* **81**, 509–512.
210

TULECKE W. (1957) The pollen of *Ginkgo biloba: in vitro* culture and tissue formation. *Am. J. Bot.* **44**, 602–608.
214, 222, 223, 231, 343

TULECKE W. (1959) The pollen cultures of C. D. La Rue: a tissue from the pollen of *Taxus. Bull. Torrey Bot. Club* **86**, 283–289.
210, 214

TULECKE W. (1960) Arginine-requiring strains of tissue obtained from *Ginkgo* pollen. *Pl. Physiol. Lancaster* **35**, 19–24.
209

TULECKE W. (1966) Continuous cultures of higher plant cells in liquid media; the advantages and potential use of a phytostat. *Ann. N.Y. Acad. Sci.* **139**, 162–175.
67, 270, 335

TULECKE W. & NICKELL L.G. (1959) Production of large amounts of plant tissue by submerged culture. *Science, N.Y.* **130**, 863–864.
269

TULECKE W. & NICKELL L.G. (1960) Methods, problems and results of growing plant cells under submerged conditions. *Trans. N.Y. Acad. Sci.* **22**, 196–206.
67

TULECKE W. & SEHGAL N. (1963) Cell proliferation from the pollen of *Torreya nucifera. Contr. Boyce Thompson Inst. Pl. Res.* **22**, 153–163.
209, 214, 231

TULECKE W., TAGGART R. & COLAVITO L. (1965) Continuous cultures of higher plant cells in liquid media. *Contr. Boyce Thompson Inst. Pl. Res.* **23**, 33–46.
271, 323, 324

TULETT A.J., BAGSHAW V. & YEOMAN

M.M. (1969) Arrangement and structure of plastids in dormant and cultured tissue from artichoke tubers. *Ann. Bot.* **33**, 217–226.
48, 125, 127, 129, 137, 140

TUREL F.L.M. (1969) Saprophytic development of the flax rust *Melampsora lini*, race no. 3. *Can. J. Bot.* **47**, 821–823.
395, 402

TUREL F.L.M. & LEDINGHAM G.A. (1957) Production of aerial mycelium and uredospores by *Melampsora lini* (Pers.) Lév on flax leaves in tissue culture. *Can. J. Microbiol.* **3**, 813–819.
395, 397, 398, 401

TURNER T.D. (1971) Pharmaceutical applications of plant tissue culture. *Pharm. J.* **206**, 341–344.
293

UMBREIT W.W., BURRIS R.H. & STAUFFER J.F. (1957) In *Manometric Techniques*. 3rd edn. Burgess Publishing Co., Minneapolis, U.S.A.
57

UPPER C.D., HELGESON J.P., KEMP J.D. & SCHMIDT C.J. (1970) Gas–liquid chromatographic isolation of cytokinins from natural sources. 6-(3-methyl-2-butenylamino) purine from *Agrobacterium tumefaciens*. *Pl. Physiol., Lancaster* **45**, 543–547.
374

VAN'T HOF J. & MCMILLAN B. (1969) Cell population kinetics in callus tissues of cultured pea root segments. *Am. J. Bot.* **56**, 42–51.
427

VARDJAN M. & NITSCH J.P. (1961) Le Régéneration chez *Cichorium endiva* L.: étude auxines et des 'kinines' endogènes. *Bull. Soc. bot. Fr.* **108**, 363–374.
345

VARNS J.L., KUC J. & WILLIAMS E.G. (1971) Terpenoid accumulation as a biochemical response of the potato tuber to *Phytophthora infestans*. *Phytopathology* **61**, 174–177.
404

VASART B. (1971a) Premiere division haploide et formation de la cellule generatrice dans le pollen de tabac. *Ann. Univ. et ARERS.* **9**, 179–187.
207

VASART B. (1971b) Infrastructure de microspores de *Nicotiana tabacum* L. susceptibles de se développer en embryoides

après excision et mise en culture des anthères. *C.r. hebd. Séanc. Acad. Sci., Paris* **272**, 549–552.
228

VASIL I.K. & ALDRICH H.C. (1970) A histochemical and ultrastructural study of the ontogeny and differentiation of pollen in *Podocarpus macrophyllus* D. *Protoplasma* **71**, 1–37.
209

VASIL V. & HILDEBRANDT A.C. (1965) Differentation of tobacco plants from single isolated cells in microculture. *Science, N.Y.* **150**, 889–892.
201

VASIL V., HILDEBRANDT A.C. & RIKER A.J. (1964) Plantlets from free suspended cells and cell groups *in vitro*. *Science N.Y.* **146**, 76–77.
425

VELIKY I. & LA RUE T.A. (1967) Changes in soybean root cultures invaded by *Rhizobium japonicum*. *Naturwissenschaften* **54**, 96.
417

VELIKY I.A. & MARTIN S.M. (1970) A fermenter for plant cell suspension cultures. *Can. J. Microbiol.* **16**, 223–226.
67, 69, 71, 271, 314

VELIKY I., SANDKVIST A. & MARTIN S. (1969) Physiology of, and enzyme production by, plant cell cultures. *Biotechnol. Bioeng.* **11**, 1247–1254.
279, 291

VENABLE J.H. & COGGESHALL R. (1965) A simplified lead citrate stain for use in electron microscopy. *J. Cell Biol.* **25**, 407–408.
49

VENKETESWARAN S. (1962) Tissue culture studies on *Vicia faba* L. I. Establishment of culture. *Phytomorphology* **12**, 300–306.
32

VENKETESWARAN S. (1965) Studies on the isolation of green pigmented callus tissue of tobacco and its continued maintenance in suspension cultures. *Physiologia Pl.* **18**, 776–789.
262

VENKETESWARAN S. & SPIESS E.B. (1963) Tissue culture studies on *Vicia faba*. III. Effect of growth factors on chromosome morphology. *Cytologia*, **28**, 201–212.
162

VENKETESWARAN S. & SPIESS E.B. (1964) Tissue culture studies on *Vicia faba*. IV.

Effect of growth factors on mitotic activity. *Cytologia* **29**, 298–310.
162

VERMA D. & VAN HUYSTEE R. (1970a) Cellular differentiation and peroxidase isozymes in cell cultures of peanut cotyledons. *Can. J. Bot.* **48**, 429–431.
270

VERMA D. & VAN HUYSTEE R. (1970b) Relationship between peroxidase, catalase, and protein synthesis during cellular development in cell cultures of peanut. *Can. J. Biochem.* **48**, 444–449.
270

VERMA D. & VAN HUYSTEE R. (1971) Derivation, characteristics and large-scale culture of a cell line from *Arachis hypogea* L. cotyledons. *Expl. Cell Res.* **69**, 402–408.
315

VERMEULEN H. & HAEN W.M. (1964) Additional remarks on the production of virus-free carnations by means of meristem culture. *Neth. J. Plant Pathol.* **70**, 185–186.
413

VESTER F. & ANDERS F. (1960) Der Gehalt an Freien Aminosäure des spontan tumorbildenden Artbastands von *Nicotiana glauca* and *N. langsdorffii. Biochem. Z.* **332**, 396–402.
386

VINE S.J. (1968) Improved culture of apical tissues for production of virus-free strawberries. *J. Hort. Sci.* **43**, 292–297.
413, 424

VINE S. & JONES O.P. (1969) The culture of shoot tips of hop (*Humulus lupulus* L.) to eliminate virus. *J. Hort. Sci.* **44**, 281–284.
413, 414, 424

VOLCANI Z., RIKER A.J. & HILDEBRANDT A.C. (1953) Destruction of various tissues in culture by certain bacteria. *Phytopathology* **43**, 92–94.
416

VOSA C.G. (1970) Heterochromatin recognition with fluorochromes. *Chromosoma* **30**, 366–372.
177

WALKEY D.G.A. (1968) The production of virus-free rhubarb by apical tip-culture. *J. Hort. Sci.* **43**, 283–287.
413

WALKEY D.G.A. & COOPER V.C. (1972) Some factors affecting the behaviour of

plant viruses in tissue culture. *Physiol. Pl. Path.* **2**, 259–264.
410, 415

WALKEY D.G.A., FITZPATRICK J. & WOOLFITT J.M.G. (1969) The inactivation of virus in cultured shoot tips of *Nicotiana rustica. J. gen. Virol.* **5**, 237–241.
415

WALKEY D.G.A. & WEBB M.J.W. (1968) Virus in plant apical meristems. *J. gen. Virol.* **3**, 311–313.
410, 415

WALKEY D.G.A. & WEBB M.J.W. (1970) Tubular inclusion bodies in plants infected with viruses of the NEPO type. *J. gen. Virol.* **7**, 159–166.
410, 415

WALKEY D.G.A. & WOOLFITT J.M.G. (1968a) Plant Pathology. *Rept. Nat. Veg. Res. Stn.* **20**, 102–113.
413

WALKEY D.G.A. & WOOLFITT J.M.G. (1968b) Clonal multiplication of *Nicotiana rustica* L. from shoot meristems in culture. *Nature, Lond.* **220**, 1346–1347.
415

WALKEY D.G.A. & WOOLFITT J.M.G. (1970) Rapid clonal multiplication of cauliflower by shake culture. *J. Hort. Sci.* **45**, 205–206.
415

WARD M. (1960) Callus tissues from the mosses *Polytrichum* and *Atrichum. Science N.Y.* **132**, 1401–1402.
31

WARREN K.S. & ROUTLEY D.G. (1970) The use of tissue culture in the study of single gene resistance of tomato to *Phytophthora infestans. J. Amer. Soc. Hort. Sci.* **95**, 266–269.
404

WARWICK R.P. & HILDEBRANDT A.C. (1966) Free amino acid contents of stem and *Phylloxera* gall tissue cultures of grape. *Pl. Physiol. Lancaster* **41**, 573–578.
417

WEBB R.M. (1971) Extraction of nematodes from sterile culture. *Nematologica* **17**, 172–174.
418

WEBSTER J.M. (1966) Production of oat callus and its susceptibility to a plant parasitic nematode. *Nature, Lond.* **212**, 1472.
31, 418, 420

WEBSTER J.M. (1967a) The influence of

plant growth substances and their inhibitors on the host-parasite relationships of *Aphelenchoides ritzemabosi* in culture. *Nematologica* **13**, 256–262.
418, 420

WEBSTER J.M. (1967b) Some effects of 2,4-dichlorophenoxyacetic acid herbicides on nematode-infested cereals. *Plant Pathology* **16**, 23–26.
420

WEBSTER J.M. & LOWE D. (1966) The effect of the synthetic plant growth substance, 2,4-dichlorophenoxyacetic acid, on the host-parasite relationships of some plant-parasitic nematodes in monoxenic callus culture. *Parasitology* **56**, 313–322.
418, 420

WENT F.W. & THIMANN K.V. (1937) *Phytohormones*. Macmillan Co., New York.
6

WERNER D. & GOGOLIN D. (1970) Keninzeichnung der bildung und altering von wurzeln in callus-und organ-kulturen von *Daucus carota* durch die aktivitat der glutamatdehydrogenase. *Planta* **91**, 155–164.
263

WEST F.R. & MIKA S. (1957) Synthesis of atropine by isolated roots and root callus culture of *Belladonna*. *Bot. Gaz.* **119**, 50–54.
267

WETMORE R.H. & RIER J.P. (1963) Experimental induction of vascular tissues in callus of angiosperms. *Am. J. Bot.* **50**, 418–430.
8, 256, 424

WETMORE R.H. & SOROKIN S. (1955) On the differentiation of xylein. *J. Arnold Arboretum* (Harvard Univ.) **36**, 305–317.
8

WHALEY W.G. & MOLLENHAUER H.H. (1963) The golgi apparatus and cell plate formation—a postulate. *J. Cell Biol.* **17**, 216–219.
135

WHITE P.R. (1934) Potentially unlimited growth of excised tomato root tips in a liquid medium. *Pl. Physiol., Lancaster* **9**, 585–600.
6

WHITE P.R. (1937) Vitamin B_1 in the nutrition of excised tomato roots. *Pl. Physiol., Lancaster* **12**, 803–811.
6

WHITE P.R. (1939a) Potentially unlimited growth of excised plant callus in an artificial medium. *Am. J. Bot.* **26**, 59–64.
7, 255, 386

WHITE P.R. (1939b) Controlled differentiation in a plant tissue culture. *Bull. Torrey bot. Club* **66**, 507–513.
7, 339, 344

WHITE P.R. (1941) Plant tissue culture. *Biol. Rev.* **16**, 34–48.
7

WHITE P.R. (1943) *A Handbook of Plant Tissue Culture*. J. Cattell, Lancaster, Pa.
110, 359, 362

WHITE P.R. (1944) Transplantation of plant tumors of genetic origin. *Cancer Res.* **4**, 791–794.
386

WHITE P.R. (1945) Metastatic (graft) tumors of bacteria free crown-galls of *Vinca rosea*. *Am. J. Bot.* **32**, 237–241.
360

WHITE P.R. (1951) Neoplastic growth in plants. *Quart. Rev. Biol.* **26**, 1–16.
389

WHITE P.R. (1953) A comparison of certain procedures for the maintenance of plant tissue cultures. *Am. J. Bot.* **40**, 517–524.
45

WHITE P.R. (1954) *The Cultivation of Animal and Plant Cells*. Ronald Press, New York.
351, 353

WHITE P.R. (1963) *The Cultivation of Animal and Plant Cells*, 2nd edn. Ronald Press, New York.
31, 39, 41, 43, 217, 246

WHITE P.R. (1967) Promises and challenges of tissue culture for biology and for mankind, pp. 12–19, in *Plant Cell, Tissue and Organ Cultures*. U.G.C. Centre of Advanced Study in Botany, Delhi, India.
256, 261

WHITE P.R. (1968) Tissue culture in the service of phytopathology. *Indian Phytopath.* **21**, 14–22.
393

WHITE P.R. & BRAUN A.C. (1941) Crown gall production by bacteria-free tumor tissues. *Science, N.Y.* **94**, 239–241.
359

WHITE P.R. & BRAUN A.C. (1942) A cancerous neoplasm of plants. Autonomous bacteria-free crown gall tissue. *Cancer Res.* **2**, 597–617.
265, 375

WILLIAMS P.G., REDDY M.N. & STRAND-

BERG J.O. (1969) Growth of non-infected and *Plasmodiophora brassicae* infected cabbage callus in culture. *Can. J. Bot.* **47**, 1217–1221.
394, 398, 399, 400

WILLIAMS P.G., SCOTT K.J. & KUHL J.L. (1966) Vegetative growth of *Puccinia graminis* f.sp. *tritici in vitro. Phytopathology* **56**, 1418–1419.
395, 402

WILLIAMS P.G., SCOTT K.J., KUHL J.L. & MACLEAN D.J. (1967) Sporulation and pathogenicity of *Puccinia graminis* f.sp. *tritici* grown on an artificial medium. *Phytopathology* **57**, 326–327.
402

WILLIAMS P.G. & SHAW M. (1968) Metabolism of glucose-C-14, pyruvate-C-14 and mannitol-C-14 by *Melampsora lini*. I. Uptake. *Can. J. Bot.* **46**, 435–440.
401

WILLIAMS P.H. & YUKAWA Y.B. (1967) Ultrastructural studies on the host-parasite relations of *Plasmodiophora brassicae*. *Phytopathology* **57**, 682–687.
394, 399

WILLISON J.H.M. & COCKING E.C. (1972) The production of microfibrils at the surface of isolated tomato fruit protoplasts. *Protoplasma* **75**, 397–403.
102

WILLISON J.H.M., GROUT B.W.W. & COCKING E.C. (1971) A mechanism for the pinocytosis of latex spheres by tomato fruit protoplasts. *J. Bioenergetics* **2**, 371–382.
101

WILMAR C. & HELLENDOORN M. (1968) Growth and morphogenesis of Asparagus cells cultured *in vitro*. *Nature, Lond.* **217**, 369–370.
424

WILSON G. (1971) *The nutrition and differentiation of cells of Acer pseudoplatanus L. in suspension culture*. Ph.D. Thesis, Univ. Birmingham, England.
270, 297, 312, 313, 314, 326, 329, 330, 332, 335

WILSON S.B., KING P.J. & STREET H.E. (1971) Studies on the growth in culture of plant cells. XII. A versatile system for the large scale batch or continuous culture of plant cell suspensions. *J. exp. Bot.* **21**, 177–207.
28, 29, 67, 69, 71, 72, 73, 74, 77, 79, 80, 81, 83, 85, 268, 316, 377, 318, 324

WIMBER D.E. (1960) Duration of the nuclear cycle in *Tradescantia paludosa* root tips as measured with H^3-thymidine. *Am. J. Bot.* **47**, 828–834.
300, 301

WIMBER D.E. & QUASTLER H. (1963) A ^{14}C- and ^{3}H-thymidine double labelling technique in the study of cell proliferation in *Tradescantia* root tips. *Expl Cell Res.* **30**, 8–22.
300

WINKLER H. (1902) Besprechung der Arbert G. Haberlandt's Culture versuche mit isolienten Pflanzenzellen 1902. *Bot. Ztg.* **60**, 262–264.
6

WITHAM F.H. (1968) Effect of 2,4-dichlorophenoxyacetic acid on the cytokinin requirement of soybean cotyledon and tobacco stem pith callus tissues. *Pl. Physiol., Lancaster* **43**, 1455–1457.
32

WITHERS L.A. & COCKING E.C. (1972) Fine structural studies on spontaneous and induced fusion of higher plant protoplasts. *J. Cell Sci.* **11**, 59–75.
101, 108, 114

WITKOWSKI R. & GRÜMMER G. (1960) Beobachtungen an *Puccinia tatarica* Tranzsch. auf gewebekulturen von *Lactuca tatarica* (L.). C.A.Meyer. *Z. allg. Mikrobiol.* **1**, 79–82.
395

WOOD H.A. & STREISSLE G. (1970) Wound tumour virus: purification and fractionation of the double-stranded ribonucleic acid. *Virology* **40**, 329–334.
382

WOOD H.N. (1970) Revised identification of the chromophore of a cell division factor from crown gall tumor cells of *Vinca rosea* L. *Proc. natn. Acad. Sci. U.S.A.* **67**, 1283–1287.
376

WOOD H.N. & BRAUN A.C. (1961) Studies on the regulation of certain essential biosynthetic systems in normal and crown-gall tumor cells. *Proc. natn. Acad. Sci. U.S.A.* **47**, 1907–1913.
194, 377

WOOD H.N. & BRAUN A.C. (1965) Studies on the net uptake of solutes by normal and crown-gall tumor cells. *Proc. natn. Acad. Sci. U.S.A.* **54**, 1532–1538.
377, 378

WOOD H.N., LIN M.C. & BRAUN A.C. (1972)

The inhibition of plant and animal 3':5'-cyclic monophosphate phosphodiesterases by a cell division promoting substance from tissues of higher plant species. *Proc. natn. Acad. Sci. U.S.A.* **69**, 403–406.
376, 377

WOODARD J.W. (1958) Intracellular amounts of nucleic acids and protein during pollen grain growth in *Tradescantia. J. bioph. biochem. Cyt.* **4**, 383–390.
209

WOODING F.B.P. (1968) Ribosome helices in mature cells. *J. Ultrastruct. Res.* **24**, 157–164.
123

WOODING F.B.P. & NORTHCOTE D.H. (1965) Association of the endoplasmic reticulum and the plastids in *Acer* and *Pinus. Am. J. Bot.* **52**, 526–531.
125

WRIGHT W.H., HENDRICKSON A.A. & RIKER A.J. (1930) Studies on the progeny of single cell isolations from the hairy-root and crown-gall organisms. *J. agric. Res.* **41**, 541–547.
358

WU J.H., HILDEBRANDT A.C. & RIKER A.J. (1960) Virus-host relationships in plant tissue culture. *Phytopathology* **50**, 587–594.
407, 408, 411

WU L. & LI M.W. (1970) Esterase isoenzyme patterns in rice somatic organs and the 2,4-D induced callus tissues. *Bot. Bull. Acad. sin., Shanghai* **11**, 113–117.
261

YAMADA T., NAKAGAWA & SINOBO Y. (1967) Studies on the differentiation in cultured cells. I. Embryogenesis in three strains of *Solanum* callus. *Bot. Mag. Tokyo* **80**, 68–74.
347

YEOMAN M.M. (1970) Early development in callus cultures. *Int. Rev. Cyt.* **29**, 383–409.
246, 247, 253

YEOMAN M.M. & BROWN R. (1971) Effects of mechanical stress on the plane of cell division in developing callus cultures. *Ann. Bot.* **35**, 1101–1112.
256

YEOMAN M.M. & DAVIDSON A.W. (1971) Effect of light on cell division in developing callus cultures. *Ann. Bot.* **35**, 1085–1100.
243

YEOMAN M.M., DYER A.F. & ROBERTSON A.I. (1965) Growth and differentiation of plant tissue cultures. I. Changes accompanying the growth of explants from *Helianthus tuberosus* tubers. *Ann. Bot.* **29**, 265–276.
35, 57, 123, 243, 247, 253, 256, 257

YEOMAN M.M. & EVANS P.K. (1967) Growth and differentiation of plant tissue cultures. II. Synchronous cell divisions in developing callus cultures. *Ann. Bot.* **31**, 323–332.
54, 246

YEOMAN M.M., EVANS P.K. NAIK G.G. (1966) Changes in mitotic activity during early callus development. *Nature, Lond.* **209**, 1115–1116.
54, 246

YEOMAN M.M. & MITCHELL J.P. (1970) Changes accompanying the addition of 2,4-D to excised Jerusalem artichoke tuber tissue. *Ann. Bot.* **34**, 799–810.
36, 54, 55, 241, 243, 245

YEOMAN M.M., NAIK G.G. & ROBERTSON A.I. (1968) Growth and differentiation of plant tissue cultures. III. The initiation and pattern of cell division in developing callus cultures. *Ann. Bot.* **32**, 301–313.
36, 243, 245, 253, 257

YEOMAN M.M., TULETT A.J. & BAGSHAW V. (1970) Nuclear extensions in dividing vacuolated plant cells. *Nature, Lond.* **226**, 557–558.
127, 133

YORA K. & TSUCHIZAKI T. (1962) The elimination of latent viruses from potatoes of Danshaku variety by use of tissue culture. *Ann. Phytopathol. Soc. Japan*, **27**, 219–221.
413

ZEIBUR N.K. & SHRIFT A. (1971) Response to selenium by cell culture derived from *Astragalus* species. *Pl. Physiol., Lancaster* **47**, 545–550.
261

ZENKTELER M.A. (1971) *In vitro* production of haploid plants from pollen grains of *Atropa belladonna* L. *Experientia* **27**, 1087.
214, 353

ZEUTHEN E. (1964) *Synchrony in Cell Division and Growth.* Interscience, New York.
242, 298

ZIMMERER R.P., HAMILTON R.H. & POOTJES C. (1966) Isolation and morphology of temperate *Agrobacterium tumefaciens* bacteriophage. *J. Bact.* **92,** 746–750. 358, 372

ZUCKERMAN B.M. (1971) Gnotobiology, pp. 159–184, in *Plant Parasitic Nematodes,* vol. **2.** Ed. B.M.Zuckerman, W.F.Mai & R.A.Rhode. Academic Press, New York. 393

ZWAR J.A., & BROWN R. (1968) Distribution of labelled plant growth regulators within cells. *Nature, Lond.* **220,** 500–501. 241

SUBJECT INDEX

Latin species names shown in italics. Pages which mark the commencement of a detailed discussion of the subject are shown in bold type.

A chromosomes, 163, 172, 178
α-amylase, 56
Aberrant variants, 187
α-bromonaphthalene, 166
'Accoutumance à l'auxin', 388
Acentric fragments, 175
Acer pseudoplatanus, see Sycamore
Acetate: CoA ligase, 289
Acetate metabolism, 287, 288
Acetocarmine, 219, 223, 224, 226
Aceto-orcein, 307
Acetylene-ethylene assay, 417
Acid phosphatase, 142, 262
Acid-resistant surfaces, 13
Acriflavin, 202
Actinomycin D, 264
Adenine, 297, 312, 345, 346, 374
Adventive embryos, *see* Embryoids
Aecia, 399
Aecial host, 399
Aeciospore, 399
Aeration of cultures, 67, 71, 210
Agallia spp., *see* Leaf hoppers
Agar, 18, 44, 95, 197
Age distribution, in cell cultures, 299
Aggregates, 95, 193, 198
Agrobacterium radiobacter, 357, 358, 370, 371
Agrobacterium tumefaciens, 188, **356,** 357, 363, 365, 366, 367, 371, 375, 379, 380, 381, 389, 390
  L-forms of, 374
  strains of, **358,** 366, 369, 370, 371, 372, 379, 381
Air filtration, 22, 71
Air-line filters, 71
Akuammicine, 262
Alanine, 327
Albinos, 233, 236
Albuginaceae, 393
Albugo ipomoeae-panduratae, 394
Alfalfa, 419
Algae, 298, 422
Alkaloids, 261, 295, 355, 386
Allium spp., 142, 185, 300

Allopolyploid, 204
Althea spp., 402
Aluminium foil, 20, 63
Amino acids, 43, 195, 203, 217, 237, 270, 287, 288, 297, 345, 351, 386
  metabolism of, **267**
  pools of, 267, 287
Aminolaevulinate dehydrase, 266
5-aminouracil, 303
Ammi visnaga, 294, 295, 296
Ammonium, 291, 351, 377
Amorphophallus, 342
Amphiploid, 385
Amylases, 291
Amyloplasts, 156, 159, 266
Anaphase, 134, 138, 153, 163, 164, 182, 184
Anatabine, 262
Androgenesis, 353
Andrographis paniculata, 49, 159, 390
Anemone, 219
'Anergie à l'auxin', 388
Aneuploidy, 145, 172, 173, 185, 186, 187, 188, 204, 234, 344
Angiosperms, 205, **210,** 222, 347
Anther cultures, 5, 9, 171, 203, **205,** 416, 423, 429
Anther response, relation to bud size, 220
Anther wall, 234
Anthers, 211, 219, 298, 353, 416, 430
Anthesis, 227
Anthoceros, 142
Anthocyanin, 147, 187, 202, 262, 294
Anthranilic acid, 267
Antiobiotics, 21, 32, 261, 295, 360, 416
Antigens, 261, 371
Antimetabolite, 203, 321, 411, 427
Antimicrobial substance, 387
Antirrhinum majus, 9, 269, 270, 394, 396, 397
Antiserum, 371, 373
α-oxoglutaric acid, 327
Aphelenchoides ritzemabosi, 418, 420
Aphids, 409
Apical meristem culture, *see* Shoot meristem culture
Aplanobacter populi, 416

Appressoria, 396
Arabidopsis thaliana, 428
Arabinose, 292
Arachis hypogea, see Peanut
Arginase, 372, 379
Arginine, 236, 238, 297, 353, 379
Arisaema triphyllum, 395, 398, 399, 402
Armoracea rusticana, 345
Ascomycetes, 394
Ascorbic oxidase, 56
Aseptic technique, 4, 11, **21**, 257, 338
Aseptic transfer rooms, **23**
Asparagine, 194, 265, 353, 362, 376, 377
Asparagus officinalis, 211, 213, 215
Aspartate transcarbamylase, 307, 310
Aspartic acid, 327, 411
Aspen, 139
Aster, 409
Aster yellows virus (AYV), 409
Astragalus, 261
Asynchronous cell cultures, 311
    study of cell cycle in, **299**
Atmosphere above cultures, 196, 198
ATP, 281, 284
Atropa belladonna, see Belladonna
Atropine, 261, 268
Aureogenus magnivena, 356, 382, 389
Aureomycin, 360
Autoclaving, **19**
Autolysis, 245
Automatic sampling valve, 73
Autonomous growth, 356, 363, 366, 375
Autoradiography, 50, 301
Autotrophic cultures, 37, 262, 427
Auxin, 6, 41, 83, 102, 162, 171, 184, 185, 187,
    195, 198, 201, 213, 218, 262, 265, 266,
    271, 344, 345, 346, 351, 354, 359, 362,
    365, 366, 367, 374, 375, 376, 377, 378,
    380, 381, 383, 386, 387, 388, 389, 390,
    399, 401, 419, 420, 424
    metabolism, of **264**
Auxin/kinin balance, 263
Auxin oxidase, *see* IAA oxidase
Auxin protectors, 376, 378
Auxin transport, 419
Auxophyton, 47, **60**
Auxotrophic mutants, 201, 203, 237
Avena coleoptile curvature test, 386, 387
Avena coleoptiles, 102
Axenic culture studies, 401, 420
Axenic isolates, 402
Azauracil, 384

B chromosomes, 163, 172, 178
β rays, 384
B vitamins, 6

Bacteria, 298, 393, **416**, 422
Bacteria-free tumour cells, **357**, **375**, 380
Bacterial DNA, **364**, 428
Bactericidal, ultraviolet tubes, 23, 25
Bacteriophages (*see also* Phages isolated
    from crown-gall bacteria), 428
Balanced growth, 309, 317, 318, 336
Banana, 122
Banding patterns, 177
Bar magnet, 69
Base analogues of nucleic acid, 346
Basidiomycetes, 394
Basidiospore, 396, 397, 402
Batch culture unit, **72**
Batch cultures, 269, 334
    growth of, 85, 272, 310
    growth patterns in, **274**
    large, 67
    limiting factors in growth of, **297**
Belladonna, 150, 158, 202, 211, 261, 266,
    268, 271, 348, 353, 429
Benazolin, 390
6-benzylaminopurine, 110, 215, 304, 307
Benzylviologen, 290
Beryllium, 38
Beta vulgaris, 366, 394, 399, 401
β-glucosidase, 291
Biochemical mutants, 236, 238, 426
Biochemical techniques, **54**
Biogels, 44, 103, 197
Biomass, concentration of, 336
Biomass production, 315, 322, 323, 325, 326,
    328
Biosynthesis, 280
    pathways of, 263, 378
Biotin, 238, 362
Bipolar embryoids, 348
Bivalents, 385
Black shank disease, 406
Blackman's formula, 259
Blow pipe, 21
β-mercaptoethanol, 56, 384
Body cell, 209
Brassica, 384, 394, 398, 399, 400, 405
Brassica oleracea, 210, 211, 213, 214, 215,
    217, 219, 221
Brassica oleracea × *B. alboglabra*, 210, 215
Brassica napus, 401
Brassicaceae, 353, 354
Breakage–fusion–bridge cycle, 179, 181
Breaking-in glassware, 13
Bremia lactucae, 395, 397
Bridge formation, 178
Broad bean, 110, 162, 163, 260, 299, 300
Bromine water, 21, 33
Bromodeoxyuridine (BUdR), **202**, 237, 238

Bromus inermis, 348
Bruising of cells, 245
Bryophyllum, 384
β-sitosterol, 262, 267, 295
Bud formation, 8, 201, 341, 348, 381, 424, 429, 430
Budding, 152, 210

C value, 169, 209
^{14}C-acetate, 287, 288
Cacodylate buffer, 95
Caffeine, 267, 268
Calendula officinalis, see Marigold
Callistephus chinensis, see Aster
Callose, 120, 206
Callus cultures, 2, **31**
  cytological characteristics of, 137
  establishment of, **240**
  friability, 8
  growth of, **257**
  metabolic patterns in, **260**
  morphology of, 258, **259**
  species from which derived, 31
  techniques for, **43**
Cambium, 31, 41, 137, 341, 366
^{14}C-amino acids, 55
Campe-sterol, 262, 267
Canavanine, 379
Capsicum annuum, 218
Carbohydrate accumulation, 286
Carbon, sources of, 37, 362, 427
Carbon dioxide
  concentration of, as a factor in conditioning, 196, 197
  determination of, 94
  fixation, 262
  release of, 243, 287, 288
Carbon dioxide electrode, 57
Carborundum disc, 30
Carborundum powder, 366, 407
Cardenolides, 295
Cardiac glycosides, 267
Carnation, 280, 413, 414, 415
  viruses infecting, 413
Carnation mottle virus, 413, 415
Carnation ring spot virus, 413
Carnoy's fixative, 165
Carotene, 122, 125
Carotenoids, 266
Carrot, 6, 8, 35, 37, 41, 49, 50, 60, 81, 104, 122, 127, 137, 140, 145, 148, 152, 162, 189, 245, 247, 252, 263, 264, 271, 273, 339, 340, 341, 342, 343, 344, 345, 346, 348, 351, 359, 362, 366, 367, 388, 416, 418, 419, 424, 429
Casein hydrolysate, 18, 43, 237, 288, 354

Castor bean, 261
Catalase, 154, 264, 270
Catechins, 333
Cathalanceine, 262
Catharanthus roseus, 194, 262, 267, 295, 360, 361, 362, 363, 375, 376, 377, 378, 416
Caulocaline, 348
Cavincine, 262
CCC, 420
Cell aggregates, 95, **143**, 152, 158, 269, 349, 425
Cell association, 426
Cell counting
  semi-automated, 52
  techniques for, **51**, **90**
Cell cultures, 3, **59**
  aggregation of, **270**, 312
  cell composition of, **278**, 311, 313, 332, 333
  cell separation in, **271**
  growth patterns in, **269**
  heterogeneity of, **270**
  metabolism of, **286**
  synchronization of, **303**
Cell cycle, 168, 169, 186, 278, 299, 336, 423
  duration of, 300, 301
  duration of phases of, 300, 301, 302
Cell density, 84, 85, 320, 335
Cell division
  changes in rate of, **276**, 280, 339, 423
  induction of in explants, **127**, 240, 243, 371
  inhibition of, 242, 411
  maintenance of, **254**
Cell expansion, 152, 423
Cell fusion, 101
Cell generations, number of, 313
Cell hybridization, 120, 171, 429
Cell lineage, 425
Cell morphology, 274, 326, 423
  change during growth cycle, 149
  uniformity of, 200
Cell number, 51, 90, 253, 275, 276, 277, 281, 282, 283, 308, 309, 311, 313, 316, 319, 322, 325, 326, 327, 328, 331
Cell plate, 133, 135, 139, 141, 153, 157, 207, 228, 234
Cell populations, 423, 427
Cell separation, 170, 189, 198, 200, 430
  changes during growth cycle, **272**
  selection pressure in favour of, **271**
Cell size, 144, 250, 252, 254, 257
Cell types, diversity of, 145
Cell volume, 318, 326, 330, 332
Cell wall degrading enzymes, 100, 198, 274
  purification of, 103, 198

Cell wall formation, 9, 135, 206, 243, 263, 425
Cell wall polysaccharides, 260, 281, 291
Cell wall regeneration, 104, 117
Cell wall thickenings, 159
Cellophane, 21
Cellular contact, 425
Cellular differentiation, 7, 137, 297, 298, 317, 334, 352, 424, 425
  onset of, **255**
Cellular physiology, 422
Cellular units, 95, 193
Cellulase, 9, 103, 104, 106, 108, 112, 115, 116, 118, 166, 198, 199
Cellulose, 206, 260
Centrifugation, 407
Centromeres, 163, 164, 175, 177, 183
  breakage at, 168
Cereal leaves, protoplasts from, 109
Cereals, 420
Cetavlon, 111
Chamaecyparis lawsoniana, 41
Chelates, 37
Chemical mutagens, **237**
Chemostat theory, 323
Chemostats, **75**, 268, 300, 318, **321**, 334
Chemotherapy, 413
Chicory, 41, 340, 341, 342, 345
Chimeras, 236, 427
(2-chloro-ethyl)trimethylammonium chloride, 374
4-chlorophenoxyisobutyric acid, 374
Chlorophyll (*see also* Chloroplasts), 202, 210, 262, 266, 405
Chlorophyllase, 266
Chloroplasts, 109, 123, 125, **139**, 159, 202, 266, 428
Cholesterol, 262, 267
Chromatid bridges, 178, 179
Chromatids, 138, 164, 168, 169, 171, 182, 183
Chromatin, 125, 131, 135, 140
Chromatophores, 122, 376
Chromic acid, 30, 51, 90
Chromic acid–sulphuric acid mixture, 13
Chromosomal instability, 189, 190
Chromosomal mutations, 161, **166**, 204, 389, 427, 428
Chromosome aberrations, **175**, 182, 184, 210, 236, 386, 389
Chromosome breakage, 163, 166, 186
Chromosome numbers, 172, 173, 174, 178, 181, 186, 201, 235, 270, 285, 302, 387, 389, 405, 427, 429, 430
Chromosomes 134, 153, 156, 161, 270, 385
  loss of, **172**, 186

Chrysanthemum frutescens, 358, 359
Chrysanthemum motifolium, 413, 415
Chrysanthemum nematode, 418
Chrysanthemum-infecting viruses, 413, 415
Chytridiales 394
Cichorium intybus, see Chicory
Cinemicrography, 131
Cinnamic acid, 289, 404
Clips for shakers, 27
Clonal maintenance, 84
Clonal variants, **199**
  selection of, **202**
Clone, 3, **191**, 205
Closed continuous cultures, **315**, 334
Clover, 382, 383, 418
Club root, 394, 399
Cobalt source, 236
Coconut milk, 8, 18, 37, 41, 122, 129, 145, 201, 213, 214, 215, 218, 244, 246, 252, 260, 271, 338, 349, 390, 399, 400, 405, 411
Colchicine, 164, 166, 203, 235
Cold shocks, 166
Cold-requiring plants, 342, 344
Collenchyma, 339
Collodion membrane, 396
Colonies
  colour of, 427
  growth of, 95, 192, 194, 198, 211, 427
  texture of, 427
Compact cultures, 260, 425
Compressed air, 12
Compressions, 425
Conditioning
  apparatus for, 88
  of host cells to crown-gall bacterium, 363
  of media, **87**, 97, 193, 194
Conidia, 395, 396, 399, 401, 403
Continuous cultures, 4, **67**, 270, **309**, 334, 422
  closed, **315**, 334
  open, **317**, 334
Convolvulus arvensis, 81, 195, 201, 271, 273, 277, 341
Conway microdiffusion technique, 55
Cormel stem-tips, 416
Corpo opaco, 125, 140
Cotton, 137
Cotton wool plugs, 20
Cotyledons, 32, 211, 346, 395
Coulter counter, 51
Coumarin, 262, 267, 295
Crassulacean acid metabolism, 262
Crataegus globosum, 402
Crepis capillaris, 162, 170, 175, 177, 189, 340, 343, 389

Criss-cross bridge-formation, 179, 180, 182
Critical dilution rate, 77, 328, 329
Critical initial density, 86, 193
Cronartium ribicola, 395, 398, 399, 402, 403
Cross-conditioning, 89, 193
Cross-walls, 108
Crown-gall disease, **356**, 416
Crown-gall tissues, 7, 8, 45, 56, 104, 147, 170, 188, 190, 194, 195, 200, 264, 342, **356**, 410, 411, 428
  nutritional requirements of, **362**
Crown-gall tumours, 428
  growth form of, 357
  induction of, **357**
Crucifers, 399
Crystal violet, 52
Crystal-containing bodies, 123, 127, 129, 130, 133
Culture filtrates, 416
Culture media
  agitated liquid, **46**
  defined, 338
  for pollen and anther culture, **213**
  preparation of, **16**, **37**, **79**, **217**
  solid, **44**
  stationary liquid, **45**
Culture systems, **60**
  closed, continuous, 74
  continuous, **67**
  open, continuous, 75
Culture vessels
  cylindrical, 68
  for platform shakers, **64**
  round-bottomed, 70
  two-tier, **65**
  V-shaped, 68
Cultured cells, general cytology of, **121**
^{14}C-uridine, 297
Cuticle, 396, 397
Cutter for preparing explants, 35
Cuvette, 77, 83, 318
Cyanide, 264
Cycads, 348
Cyclamen, 345
Cyclic AMP (c-AMP), 377
Cyclic AMP phosphodiesterases, 377
Cycloheximide, 264
Cymbidium spp., 415
Cynipids, 417
Cysteine, 43, 56
Cysteine hydrochloride, 6
Cytidylic acid, 194, 362, 376, 377
Cytochemistry, 375
Cytochrome oxidase, 270
Cytodifferentiation, *see* Cellular differentiation

Cytokinesins, 376, 377
Cytokinesis, 153, 157, 171, 184, 249, 282, 299, 306, 336
Cytokinin, 37, 41, 42, 83, 195, 201, 213, 265, 304, 305, 345, 346, 349, 359, 362, 365, 366, 367, 374, 375, 376, 377, 378, 381, 387, 388, 390, 400, 401, 411, 420, 424
Cytological stability, **186**, 202, 427
  effect of nutrient media on, 186, 204
Cytology
  influence of media on, 427
  nuclear, **161**
  of cultured cells, 121, 375, 423, 427

2,4-D, 41, 122, 129, 140, 145, 147, 160, 170, 171, 175, 201, 202, 214, 216, 241, 244, 245, 246, 251, 262, 264, 266, 271, 280, 282, 293, 294, 304, 306, 362, 387, 399, 400, 415, 420
Dahlia, 412
Damage to cells, 245, 257
Datura innoxia, 205, 214, 217, 295, 354
Datura metel, 214, 233, 234
Datura meteloides, 214
Datura muricata, 214
Datura spp., 211, 214, 217, 225, 267
Datura stramonium, 214, 362, 379
Daucus carota, *see* Carrot
Daughter nuclei, 153, 157, 171, 207
DEAE-dextran, 114
DEAE-sephadex, 371
Decodon verticillatus, 395
Dedifferentiation, 137, 240, 261, 339, 425
Deep-freezing, 189, 204
Deletion, 168, 175, 187
Density detector, 77, 80, 81, 317, 319
Deoxyisoflavones, 267
Derepressor, 428
Detergents, 13, 21, 33, 109, 110
Developmental biology, 423
Developmental phase of crown-gall tumours, **375**
Developmental programme, 354
Dextran sulphate, 109, 112
Dialysis tubing, 88, 195
Dianthus, 262
Dicentric, 168, 177
2,4-dichlorophenoxyacetic acid, *see* 2,4-D
Dicotyledons, 31, 339, 357
Dictyosomes, 123, 125, 129, 133, 135, 144, 156, 157, 208
Differentiation
  cellular, 7, 137, **255**
  in callus, 240
  vascular, 8
Digitalis purpurea, 267, 295

Dilution rate (D), 75, 320, 327, 329, 332, 333, 336
  relationship to biomass, 327
Dimethyl sulphoxide, 204
Dinitrofluorobenzine, 50
Dioon, 298
Dioscorea deltoidea, 295
Diosgenin, 267, 295
Diphenylamine reaction, 93
Diplochromosomes, 167, 184
Diploid, 148, 163, 167, 170, 184, 189, 201, 234, 238, 343, 354, 375, 429, 430
Disease resistance, study of, 292, **403**
Dispersion of suspensions, 198, 199
Disposable petri dishes, 95, 110
Distilled water, preparation of, **14**
Dithiothreitol, 56
Ditylenchus destructor, 418
Ditylenchus dipsaci, 419
Diurnal fluctuations, 262
Dividing zones in callus, 166
Division machinery, 242
Division walls, 152
DNA, 33, 92, 227, 250, 276, 279, 281, 282, 284, 285, 297, 311, 325, 364, 365, 366, 367, 369, 370, 372, 373, 390
  inhibition of replication of, 303, 304
  replication of, 166, 169, 171, 179, 184, 202, 207, 209, 238, 242, 247, 307, 336
DNA polymerase, 56, 247
DNA precursors, 301
DNAase, 56, 365
Dodder, 380
Double helix, 372
Doubling time (*td*), 278, 280, 282, 312, 324, 326, 331, 335, 400
Downy mildews, 393, 394, 395, 396, 397, 398, 399, 403
'Drain and refill' technique, 69, 271
Dry weight, 275, 279, 280, 281, 282, 286, 288, 294, 296, 313, 316, 318, 319, 321, 322, 324, 325, 326, 327, 330, 332
  determination of, 30, 50, 91
dTMP kinase, 56
Dual cultures of parasites and tissue cultures, **393**
  experimental studies with, **401**
  maintenance of, **398**
  of viruses and plant tissue cultures, **410**
Duration of culture, effect on embryogenesis, 352

EDTA, 37, 38, 43, 51, 83, 218
Effective yield, 333
Efficiency, 333
Efflux of metabolites, 193, 195

Egg cell, 348
Electricity supply, 12, 27
Electron microscopy, **48**, 94, 127, 154, 228, 349, 361, 372, 373, 374, 423, 425
Electronic control circuit, 74
Electrophoresis, 247, 248, 261
Embedding for electron microscopy, 49
Embryo cultures, 2
Embryogenesis, 4, 8, 221, **348**, 424, 425, 430
  factors controlling, **351**
  from haploid cells, **353**
Embryogenic callus, 141, 147
Embryogenic pollen grains, 226, 227
Embryogenic potential, 145, 425
Embryoid formation, maintenance of the ability for, 352
Embryoids, 5, 138, 211, 212, 220, 224, 234, 338, 425, 430
  abortive, **231**, 233
  induction of formation of, 348, 351
  origin from single cell, 349
Embryology, segmentations of, 425
Embryos, *see* Embryoids
Embryo-sac, 298
EMP pathway, 281
Endergonic, 281
Endocytosis, 101, 120
Endomitosis, 167, 169, 183, 234, 235, 344, 355
Endoplasmic reticulum, 123, 127, 133, 138, 154, 157
Endopolyploidy, 33, 204
Endoreduplication, 309
Endosperm, 32, 261, 266, 298, 314
Endo-symbiotic associations, 120
Enzymatic adaptation, 388
Enzyme activities, cytochemical methods, 50
Enzyme techniques, **56**
Enzymes, 378, 421
  changes in activity of, 242, 247, 265, 392
  secretion of, 392
Ephedra foliata, 209, 214, 217, 225
Epidermal peeling, 111, 115
Epidermal strip, 112
Epidermis, 396, 397
Epigenetic modification, 390
Equatorial plate, 135, 138
Equilibrium substrate concentration, 323, 325, 326, 328, 329, 330, 331
Erlenmeyer flasks, 27
Erwinia caratovorum, 416
Erysiphales, 393, 394, 403
Erysiphe cichoracearum, 394, 396
Escherichia coli, 370
Established (permanent) cultures, 4

Esterases, 261, 262
Ethyl ether sterilization, 19
Ethyl methane sulphonate, 237
Ethylene, 197, 287, 288
Etioplasts, 139
Eucalyptus camaldulensis, 154
Eukaryotes, 303, 423
Exine, 206, 208, 223, 224, 225, 231
Exogenous factors, 338
Explants, 2, 4, 33,
    preparation of, **35**
Exponential growth, 85, 259, 275, 278, 281, 301, 313
Exponential phase, 149, 198, 253, 297, 300, 335
Extinction values, 319
Extracellular macromolecules, 291

F_1 hybrid, 234
Face mask, 25
'Feeder' cells, 87, 202, 273
Fermenter, 335
    V-type, 69, 271, 314
Ferns, 31
Festuca arundinacea, 216
Feulgen staining, 50, 53, **93**, 165, 166, 184, 219, 220, 224, 227, 285, 302
Fibres, 339
Filaments of cells, 152, 223
Filter paper support, 45, 185, 211, 414
Filter sterilization, 19, 194, 407
Filterable forms of *Agrobacterium tumefaciens*, 373
Fixatives, **48**
'Fixed tension', 153
Flavoprotein, 264
Flax, 150, 395, 298, 401, 402
Flow rate, 75, 333, 336
Flower buds, 210, 211, 236, 342
Flower formation in tissue cultures, 340, 342, 344, 346
Flowers, 338, 339, 344, 382, 424
Fluorescent technique, 177
Fluorodeoxyuridine (FUdR), 242, 303
Fluorophenylalanine, 427
Folic acid, 362
Foreign DNA, 364, 371, 428, 430
Fraction collector, 74
Fragaria xananassa, see Strawberry
Free cells, 201, 271
Free-cell cultures, 271, 273
    by use of enzymes, **273**
Free-cell fraction, 273
Freesia, 218
Freeze-dryer, 30
Freezing preservation, 189, 204, 428, 430

French bean, 245
French press, 30
Fresh weight, 252, 254, 255, 276, 279, 289, 313, 324
    determination of, 50, 90
Friable cultures, 83, 202, 260, 270, 388, 425, 427
Fructose-1-6-diphosphatase, 291
Fruit protoplasts, 102
Fruits, 34, 409, 424
Fuchs-Rosenthal counting chamber, 96, 109, 111, 114
Fungal hyphae, 398, 399
Fungal obligate parasites grown in host tissue cultures, **394**
Fungi
    as vectors of virus, 409
    parasitic, 392, **393**
Fungistatic substance, 397, 401
Fungitoxin, 397

G1 phase, 168, 242, 249, 282, 285, 300, 301, 303, 336
G2 phase, 169, 249, 283, 285, 300, 301, 336
γ rays, 384
Galactose, 201, 292
Galactosidases, 291
Galacturonic acid, 292
Gall callus, 398, 402, 417
Gallocyanin–chrome alum, 50, 226
Gametophytes, 206, 222, 298
Gaseous exchange, 60, 246
Gelatin, 44
Gelman filters, 20
Gene mutations, 161, 204, 236, 427
Gene regulation, 388
Gene resistance, **403**, 406
Generative cell, 207, 208, 221, 223
Generative nucleus, 207, 226
Genes, 186, 205, 371, 388, 403
Genetic information, transfer of, 428
Genetic stability, **186**, 202, 355
Genetic tumour disease, 356, 367
Genetic tumours, 344, 356, 367, 379, **384**, 429
Genome, 187, 190, 205, 343, 353, 361, 370, 385, 387, 390, 427, 428, 429
Gentamycin, 110
Germ tubes, 396, 397
Giant cells, 145, 375, 419
Giant pollen grains, 231
Gibberellic acid, 102, 195, 374, 399, 420
Gibberellins, 37, 102, 201, 263, 346, 424
Ginkgo biloba, 209, 214, 217, 236, 271, 343
Ginkgo-type growth pattern, 222
Gladiolus, 416

Glass homogenizer, 59
Glass rings, 97
Globular stage of embryoids, 346, 348
Glucose, 292, 325, 328, 329, 332, 362, 376, 414
  transport of, 297
Glucose-6-phosphate dehydrogenase, 56, 247, 279
Glutamate-oxaloacetate transaminase, 327
Glutamate-pyruvate transaminase, 327
Glutamic acid, 411
Glutamine, 265, 351, 353, 362, 375, 377
Glutaraldehyde, 48, 95
Glycerol, 189, 291
Glycine, 43, 373
Glycine max, see Soybean
Glycosidases, 291
Glycosides, 355
Glyoxylate cycle, 261
Gnotobiotic techniques, 393
Golgi bodies see Dictyosomes
Gooseberry, 413
Gooseberry vein banding virus, 413
Gradients in callus, 256
Grafting, 8, 367, 369, 380, 381, 382, 383, 384, 387, 388, 389, 390
Gramineae, 353
Grana, 266
Grape, 201, 388, 417
Green light, effect on cell division, 243
Growth
  induction of, **241**
  measurement of, **50, 89**
Growth cycle, 85, 197, 273, 274, 335
  cell division during, **276**
  metabolic changes during, **286**
Growth factor synthesizing systems, 377
Growth factors, **38**, 217, 244, 362, 363, 375, 376, 377, 378, 406, 420
Growth parameters, 50, 363, 427
Growth phases in batch culture, 85
Guanidine bases of tumours, 358, 379
Guanine, 383, 345
Guanylic acid, 194, 362, 376, 377
Gymnosperms, 31, 205, **209**, 221, 222, 225, 230, 236, 357
Gymnosporangium globosum, 402
Gymnosporangium juniperi-virgininae, 395, 397, 398, 401, 402
Gymnosporangium spp., 395

Habituated tissue cultures, **388**, 411
Habituation, 188, 367, 379, 382, 384, **388**
Haemocytometer, 51, 90
Hanging drop cultures, 97, 210
Haploid mutants, **236**, 237

Haploids, 5, 9, 170, 172, 189, 190, 201, 203, 205, 234, 269, 353, 423, 427, 429, 430
Haplopappus gracilis, 103, 150, 153, 162, 163, 164, 165, 167, 170, 172, 174, 178, 182, 183, 187, 189, 202, 260, 266, 270, 278, 287, 294, 300, 301, 303, 304, 305, 312, 318, 319, 324, 343
Haustoria, 398, 399
Heart-shaped stage of embryoids, 346, 351
Heat treatment, 412, 413, 415
  of tumour tissues, **360**
Helianthus annuus, see Sunflower
Helianthus tuberosus, see Jerusalem artichoke
Herbicides, 420
Heterodicentric, 168, 178, 180, 182
Heterokaryons, 120
Heterozygous species, 222
Hexokinase, 247
Histochemical techniques, **48**, 94, 425
Histogenesis, 7
Histological techniques, **48**, 94, 423
History of plant tissue culture, **5**
Hollyhock, 360
Homoarginine, 379
Homogenizer, 30
Homokaryons, 120
Homozygous lines, production of, **235**
Homozygous mutants, 203
Hop, 394, 413
  viruses infecting, 413
Hordeum vulgare, 209, 213, 215, 217, 221, 225, 233
Hormones, 271, 338, 344, 353, 370, 374, 375, 390, 423, 427
  endogenous, 244, 264, 424
  for pollen and anther culture, **213**
  interaction with nematodes, 419
Host cultures, infection of, **396**
Host–parasite interaction, 392, **403**
Humulus lupulus, see Hop
Hybrid tumour tissue cultures, 386
Hybrid tumours, **384**
  genetic background of, **385**
  physiology and biochemistry of, **386**
Hydrogen peroxide, 32
Hydroxamic acid, 289
Hydroxyphaseollin, 405
8-hydroxyquinoline, 166
Hydroxyurea, 303, 305
Hyphal branching, 239
Hyphal-like cells, 210
Hypochaeris, 266
Hypochlorite, 21, 32, 109, 110, 111, 115, 359
Hypocotyl, 212, 269
Hypoxanthine, 238, 383

IAA, 6, 18, 41, 214, 215, 235, 237, 262, 263, 264, 346, 362, 374, 384, 387, 420
IAA oxidase, 202, 264, 376
Illumination of cultures, 26, 222
Immuno-diffusion techniques, 371, 374, 391
Immuno-electrophoresis, 261
Impatiens, 262
Inbreeding species, 222
Incubation of cultures, **25**
Incubator shakers, 26, 28
Indole, 267
Indole alkaloids, 267
Indoleglycerol phosphate, 267
Indolylacetaldehyde, 264
Indol-3yl-acetic acid, *see* IAA
Indolylpyruvic acid, 264
Induced doubling of chromosome number, 235
Induced fusion of protoplasts, 120
Induction of callus development, 240
Induction period, 363
Inductive treatments, 345, 352
Infection hyphae, 396
Infra-red carbon dioxide analyser, 94
Inhibitors, 203, 242, 364, 401, 415, 420
Inoculum, 4
  size of, 84, 276, 293, 305
Inoculum of fungus, contaminant-free, **393**
Inoculum of virus, contaminant-free, **407**
Input concentration of substrate (S_R), 323, 326, 328, 329, 330, 331
Insect vectors, 382, 407, **409**
Insects, in gall induction, **417**
Insoluble nitrogen, 254, 313
Inter-relationships between lines of experimentation, 430
Intine, 206, 207, 208, 223
Intrinsic factors, 321
Inulin, 121, 261
Inversion, 168
Invertases, 291
Ion uptake, 377, 378
Ionic status, 194, 195, 198, 378
Ionizing radiations, 384
Ipomoea, 267, 271, 286, 394
Iron, 38, 83, 194, 217, 218, 387
Irradiation, **236**
Isochromosome, 168, 175
Isodicentric, 168, 178, 180, 182
Isoenzymes, 261, 262, 264
Isogenic cells, 355
Isolated cells, special growth requirements of, **191**
Isoleucine, 267
Isotope competition, 267
Isotopic procedures, **55**, 93, 378, 401

Jerusalem artichoke, 33, 35, 36, 37, 41, 42, 48, 50, 55, 56, 123, 125, 127, 131, 134, 135, 140, 143, 162, 241, 242, 243, 245, 246, 247, 248, 249, 250, 251, 252, 253, 256, 261, 343, 344, 348, 359, 362, 379
Juglans regia, 358
Juniperus communis, 267
Juniperus spp., 395
Juniperus virginiana, 395, 398, 401

^{42}K, 378
Kalanchoë, 266, 373
Kalanchoë daigremontiana, 357, 364, 365, 366
Karyosome-like bodies, 131
Karyotype analysis, 162, 177, 202
Kinetin, 8, 37, 42, 145, 160, 162, 175, 184, 194, 195, 214, 215, 216, 294, 304, 306, 307, 345, 346, 354, 362, 377, 382, 384, 387, 389, 390, 399, 400
Knop's solution, 6, 37
Kunitz units, 112
Kurz system, 67

Labelling index, 299, 300
Laboratory layout, **12**
Laboratory organization, **11**
Lactic dehydrogenase, 327
Lactuca tatarica, 395
Lag phase, 149, 274, 281, 335
Lagging of chromatids, 172, 173
Laminair cabinets, **22**
Lanceine, 262
Lead citrate, 49, 95
Leaf hoppers, 382, 383, 409
Leaf mesophyll, 32
Leaf protoplasts, **108**
  culture of, **116**
  for virus infection studies, **111**
  from cereal leaves, **109**
  from leguminous plants, **110**
Leakage, 193
Leaves, 34, 212, 338, 339, 341, 346, 366, 381, 387, 395, 396, 398, 399, 402, 407, 408, 424
Leguminous leaves, protoplasts from, **110**
Leguminous plants, 417
Lemon, 41, 246
Leptohormone, 244
Lettuce, 397
Leucoanthocyanins, 333
Light
  chloroplast differentiation, 139, 210
  effect on stability of isolated protoplasts, 100
  green light, effect on cell division, 243

inhibition of colony formation by, 97, 203, 238
responses to, **266**, 321, 344, 348, 354, 389, 401, 415
Light sensitive resistors, 83, 318
Lignified cells, 7, 137, 147, 160, 339
Lignin, 104, 147, 160, 267, 293, 294
released into the medium, 160
Ligno-suberin, 245
Lilium, 299, 384
Linear growth, 85, 336
Linear regression, 281, 282, 309
Lineweaver/Burk plot, 329, 330
Linum usitatissimum, see Flax
Lipid bodies, 127, 229, 231
Lipids, 261, 288
Liquid emulsion methods, 50
Logarithmic growth, 108, 274, 335, 357
Lolium multiflorum, 211, 213, 214, 216, 217, 233
Lolium perenne, 162, 178
Lucerne, 418
Lunaria annua, 342, 346
Lupin, 110
Lupinus polyphyllus, see Lupin
Lycopersicon, 384
Lycopersicon esculentum, see Tomato
Lyophilized tissue, 30
Lysine, 238, 267
Lysis, 372
Lysopine, 369, 378, 379, 380
Lysosomes, 129, 143

Macerozyme, 104, 110, 112, 115, 116, 118, 198, 199
Macro-nutrients, **39**, 217, 218, 414
Macrosteles fascifrons, 409
Magnetic stirrers, 36, 46, 67
Magnetic valve, 67
Maintenance of stock cultures, **47, 84**
Maize, 179, 271, 314, 315
Malachite green, 415
Malate, 287
Malic dehydrogenase, 257, 327
Mallow, 388
Malus spp., 358
Mammalian cells, 298, 303
Manganese, 264
Mannitol, 104, 105, 109, 112, 115, 116, 118
Mannose, 292
Marigold, 144, 200, 269, 416, 418
Maximov double cover-slip, 97
Maximum specific growth rate (μ_{max}), 323, 335
Mean generation time (*g*), 75, 278, 311, 336
Medicago sativa, 343

Meicelase P, 104, 105, 109, 110
Meiosis, 205, 206, 235, 299
Melampsora lini, 395, 398, 401, 402
Melilotus spp., 201, 287, 383
Meloidogyne incognita, 418, 419
Mercuric chloride, 21, 33, 359
Meristem tip culture, 392, **412**, 321
technique of, 414
Meristematic cells, 143, 223, 241, 273, 375
Meristematic dome, 413
Meristematic nodules, 137, 256, 260
Meristems, 338, 339, 410, 425, 427
Meso-inositol, *see Myo*-inositol
Mesophyll, 108, 110, 112, 396, 397, 409
Messenger RNA, 369
Metabolic consequences of induction, **243**
Metabolism, **89**, 281, 334, 388
Metabolites
interchange of, 425
release of, **87**, 97, 193, 291, 397
Metaphase, 135, 153, 164, 165, 166, 170, 182, 184, 188
Metaphase index, 303
Metaphase labelling, 301
Methacrylate, 95
Methionine, 267
Mevalonic acid, 266
Microbodies, 123, 129, 154
Microchambers, growth of cells in, **97**, 191, 193, 199
Microdensitometry, 120, 285, 301
Micro-elements, 38, **39**
Micro-environment, 426
Microfibrils, 125, 157, 158
Micro-injection methods, 407
Micro-metering pump, 69, 75
Micro-nutrients, 38, **39**, 217, 218, 414
Micro-organisms, uptake by protoplasts, 120
Micropropagation, 424, 430
Microspectrophotometry, *see* Microdensitometry
Microspores, 206, 207, 299
Microsporogenesis, 206, 219
stages in, 219
Microtubules, 125, 130, 133, 135, 141, 157
Middle lamella, 425
Millipore filters, 19, 112, 115, 407
Mineral nutrients, **37**, 377
Mineral oil, 204, 418
Minimal medium, 195, 202, 204
Minimum effective density, 87, 193, 195, 196
Mist propagator, 212
Mitochondria, 123, 127, 129, 133, 137, **141**, 144, 145, 147, 154
Mitomycin C, 372

Mitosis, 134, 153, 167, 169, 170, 184, 207, 209, 222, 223, 236, 246, 249, 277, 298, 336, 339
  duration of, 299, 300
Mitotic cycle, *see* Cell cycle
Mitotic index, 246, 272, 277, 278, 299, 300, 303, 304
  determination of, **51**, 91
Mitotic proteins, 265
Molecular biology, 423
Molecular events before division, **246**
Monocentric, 179, 180
Monocotyledons, 31, 162, 357, 399
Monopolar organs, 341, 348
Monoxenic stocks, 418
Morphactins, 389, 390
Morphogenesis, **338**, 424, 429
Morphogenetic potential, 187, 201, 204, 210, 338, 352, 388, 424, 425
  decrease with increasing duration of culture, 342
Morphogenetic stimuli, 263
'Moruloid' masses, 152
Mosses, 31
Multicellular organisms, 423
Multi-disciplinary approach, 422
Multinucleate cells, 152, 171, 231, 234, 375
Multinucleate protoplasts, 107, 114
Multinucleation, 107
Multipolar spindle, 167, 172, 173
Multistage treatments, 67
Multivesicular bodies, 135, 158, 160
Mutagenic treatment, 202, 203, 236, 237, 403, 429
Mutants, 4, 187, 190, 202, **236**, 387, 416, 422, 426, 429, 430
  selection of, **202**, 296, 423, 426
Mutations, 161, 162, **202**, 204, 236, 355, 388, 389, 405, 423, 426
Mycelium, 396, 397, 398, 399
  peninsulas of, 403
  systemic, 397
Myelin-like bodies, 135, 140
Myo-inositol, 194, 195, 217, 265, 362, 375, 377, 414

NAA, 41, 122, 159, 160, 170, 194, 195, 201, 214, 215, 216, 217, 235, 260, 264, 266, 270, 271, 288, 294, 362, 374, 399, 400, 411, 414
NADH$_2$, 290, 379
NADPH$_2$, 281
Naked cells, 100
Naphthalene acetic acid, *see* NAA
Naphthol yellow S, 50, 227
Narcissus spp., 415

Natural gas, 12
Necrosis, 427
Negative acceleration of growth, 84
Nematode bank, 419
Nematodes
  infection of callus by, **418**
  plant parasitic, 392, 409, **417**
Neoplasms, 356
Neoplastic growth, 388
Nicotiana alata, 214, 387
Nicotiana debneyi-tabacum, 385, 386
Nicotiana debneyi-tabacum × N. longiflora, 387
Nicotiana glauca, 385, 386, 387, 415
Nicotiana glauca × N. langsdorffii, 7, 339, 356, 385, 386, 387, 429
Nicotiana glutinosa, 214
Nicotiana hybrid tumours, 384
Nicotiana langsdorffii, 385, 386, 387
Nicotiana longiflora, 385, 386
Nicotiana plumbaginifolia, 387
Nicotiana rustica, 215, 415
Nicotiana sauveolus × N. langsdorffii, 387
Nicotiana sylvaticum, 150
Nicotiana sylvestris, 215
Nicotiana tabacum, *see* Tobacco
Nicotiana tabacum × N. glutinosa, 97, 200
Nicotiana-type growth pattern, 222, 223
Nicotine, 261, 262, 267, 295
Nicotinic acid, 43, 362
Nile blue, 150
Nipple flask, 60, 61, 62
N-6-(isopentenyl)adenosine, 304
Nitrate, 290, 291, 325, 326, 329, 331, 351, 377
Nitrate reductase, **290**, 291, 326, 327
Nitrite, 290
Nitrogen/auxin ratio, 351
Nitrogen, soluble, 284
Nitrogen metabolism, abnormal, 371, 374
Nitrogen source, 38, 83, 145, 150, 195, 243, 274, 346, 351, 353
Nitrogen utilization, 286, 326, 370
Nitrogenase, 417
Nitrogen-fixing bacteria, 120, 417, 421, 428
Nitrogen-limited cultures, 327
Nitrosoguanidine, 202, 237
N-methyl-N'-nitro-nitrosoguanidine, *see* Nitrosoguanidine
N-3-nitrophenyl-N'-phenylurea, 238
Nobel agar, 44
Nomarski interference, 106, 147
Non-disjunction, 167, 172, 174
Non-embryogenic pollen grains, 226, 227, 230, 232
Non-histone protein, 375

Nonpolysomatic species, 162
Nopaline, 358, 367, 369, 372, 379, 380
Nopaline dehydrogenase, 372, 379
Nucellus, 137
Nuclear aberrations, 234, 235, 386
Nuclear cytology, **161**, 361, 424, 425, 428
  techniques in, 163
Nuclear/cytoplasmic ratio, 206
Nuclear DNA, assay of, **93**, 285, 302
Nuclear envelope, 123, 169
Nuclear extensions, 133
Nuclear fusion, 167, **171**, 234
Nuclear stability, 271, 343, 429
Nuclei, 93, 125, 131, 133, 137, 139, 150, 207,
  208, 285, 339, 375
Nucleic acid determination, 54, **92**
Nucleic acid hybridization, 370, 371, 374,
  391, 428
Nucleic acids, 226, 227, 251, 254, 265, 268,
  299, 307, 346, 364, 366, 367
Nucleoli, 123, 125, 131, 133, 137, 140, 208,
  241, 275
Nucleotides, 239, 281, 284, 297
Nurse culture, 87, 191, 210, 218, 403
Nurse technique, 9, 87, 97, 191, 192
Nutrient media, **37**
Nutrient requirements of cultured cells, **37**,
  375

Oat, 418
Obligate parasites, 393, 398
  dual cultures of, **393**
  grown in host tissue cultures, **394**, 400
Octopine, 358, 367, 369, 372, 379, 380, 381
Octopine dehydrogenase, 372, 379
Octopinic acid, 369, 379
Octoploid, 148, 167, 169
Oil droplets, 123
Olpidium brassicae, 409
Omega phage, 372
Onozuka, 104, 105, 115, 116, 118
Oomycetes, 403
Oospores, 399
Open continuous cultures, **317, 321**, 334
Operon, 428
Optical density of suspensions, 75, 81, 83,
  317, 325
Opuntia spp., 201, 379
Orbital incubators, 27
Orbital shakers, **62**
Orchids, 424
Organ cultures, 2, 5
Organ differentiation, 201, **339**, **342**,
  345
Organic acids, 262, 287, 288, 386
Organic carbon, 94

Organic growth factors, **38**, 217, 218, 377,
  378
Organic nitrogen, 346, 351
Organization, emergence of, 425
Organogenesis, 4, 213, 238, 263, 334, **339**,
  352, 387, 397, 424, 430
  factors controlling, **342**
  influence of physical factors on, 344
  theories of the regulation of, 345
Orotic acid, 346
Oryza sativa, 211, 213, 215, 219, 221, 233,
  234, 261, 267, 354
Oryza sativa japonica, 216
Osmiophilic globules, 160
Osmium tetroxide, 48, 95
Over-conditioning, 89
Oxalis, 49, 266
Oxygen availability, 243, 317
Oxygen electrode, 57, 67, 71, 94
Oxygen uptake, 247, 254, 255, 325, 328

Packed cell volume, 30, 90, 316, 321, 322,
  325, 327
Paeonia, 218, 219
p-aminobenzoic acid, 238
Pantothenic acid, 43, 362
Paper raft technique, 9, 185, 191, 192, 199
Paraffin-oil-nitrocellulose membrane, 397
Parafilm, 197
Paramural bodies, 125, 135, 160
Parenchyma cells, 339
Paris daisy, 359
Parsnip, 35
Parthenocissus tricuspidata, 45, 90, 104,
  264, 312, 376, 388, 409
'Partial synchrony', 303, 305
Partial transformation, 364
Pastinaca sativa, 366
Pathogen-free stocks, 421
Pattern of gene activation, 355
p-coumarate: CoA ligase, 287, 289
Pea, 97, 110, 162, 165, 171, 175, 185, 271,
  339, 340, 343, 345, 405
Peanut, 144, 152, 270, 315
Pectinase, 51, 108, 166
Pectinol, 109
Pectins, 260
Pelargonium hortorum, 416
Pennisetum typhoides, 394, 399, 403
Pentaploid, 171, 234
Pentose phosphate pathway, 281
Peptone, 402
Periderm, 137
Perinuclear cytoplasm, 131, 133
Periodic acid-Schiff, 129
Peripheral cells, 243, 258

Peristaltic flow inducer, 75
Peristaltic pump, 75
Periwinkle, 418
Permanganate, 49
Permeability, 194, 195, 377, 426
Peronospora farinosa, 394, 395, 399, 400, 401, 403, 405
Peronospora parasitica, 394, 395, 398, 405
Peronospora parasitica var. *nicotianae*, 406
Peronospora tabacina, 394
Peronosporaceae, 393
Peronosporales, 394, 403
Peroxidase, 262, 264, 270, 279, 291, 376, 386
Perspex boxes, 29
Petri dish plating, 95
Petunia, 110, 114, 115, 118, 211, 215, 234
pH
    electrode, 67, 71
    of media, 30, 83, 194, 195, 325, 326, 411
Phages isolated from crown-gall bacteria, 358, 364, 366, 367, 369, 372, 373, 381, 390
Phase contrast, 99, 148, 149, 150, 153
Phaseolus vulgaris, 8, 9, 144, 269, 271, 278, 312, 315, 367, 374, 416
Phases in microspore development, 206, 220
Phenols, 106, 243, 263, 264, 279, 289, 293, 404
Phenotype, 161, 213, 236
Phenylalanine ammonia-lyase (PAL), 287, 289
Phloem, 122, 123, 341, 365
Phosphate, 56, 325, 326, 329, 330, 345, 377, 378, 383, 387
Phosphoenol pyruvate carboxylase, 262
Phosphorous, *see* Phosphate
Photocell, 77, 82, 318
Photographic document paper, 96
Photoperiodism, 342, 344, 346
Photosynthesis of cultures, 262, 266
Photosynthetic lamellae, 139
Phragmoplast, 153
Phragmosome, 133, 135, 153
Phycomycetes, 394
p-hydroxybenzoic acid, 404
Phyllocaline, 348
Phylloxera gall tissues, 201, 417
Physiological isolation for embryogenesis, 349
Physiological races, 392
Phytoalexin, 405
Phytochrome, 267
Phytoferritin, 123, 129, 139, 160
Phytohormones, *see* Hormones
Phytol, 266
Phytophthora infestans, 403, 404, 405

Phytophthora megasperma var. *sojae*, 405
Phytostat, 323, 324, 335
Picea glauca, 43, 260, 269
Pigmented cells, 270
Pinocytosis, 113, 408, 409
Pinto bean, 367, 373, 374, 380
Pinus monticola, 395, 397, 398, 399, 402, 403
Pipetting unit, for suspension cultures, **86**
Pisatin, 405
Pisum sativum, *see* Pea
Pith parenchyma, 345, 346
Planes of division, 256
Plant breeding, 205, 355, 429
Plant cultures, 2
Plant parasites in tissue culture, **392**
Plantlets from cultures, 201, 211, 212, 218, 219, 220, 346, 414, 430
    abnormal, **233**
Plasmalemma, 101, 120, 130, 135, 256, 428
Plasmalemmasomes, 160
Plasmatic inheritance, 355
Plasmodesmata, 125, 135, 158, 351, 408, 412, 425
Plasmodiophora brassicae, 394, 398, 399, 400, 401
Plasmodiophorales, 394
Plasmodium, multinucleate, 399, 400
Plasmolysis, 101, 106
Plasmolyticum, 100
Plasmopara viticola, 393, 398, 401
Plasticity of cell morphology, 339
Plastids, 125, 127, 129, 133, 137, 144, 147, 208, 228, 231, 233
    greening of, 139, 147, 266
Platform shakers, 27, 29, 46, **62**
Plating efficiency (PE), 96, 194, 195, 198, 211
Plating (Bergmann) technique, 9, 95, 191, 192, 197, 199, 209, 238
Ploidy level, 165, 170, 171, 175, 177, 343
    of pollen plants **234**
Plumbago indica, 342, 346
Podocarpus macrophyllus, 209
Polarity, 44, 46, 228, 233, 339, 425
Pollen, **205**, 348, 353
    initial behaviour of, **222**
    stages in embryoid formation in, 224
    storage of, 209
    tricellular, 221
Pollen age, **219**
Pollen culture, **205**, 343
Pollen protoplasts, **118**
Pollen tetrads, 120, 206
Pollen tubes, 205, 238
    septate, 223
Pollination, 207
Polyacrylamide gel electrophoresis, 55

Poly-L-lysine, 114
Poly-L-ornithine, 112, 113, 114, 365, 409
Polylysogenic strains of *Agrobacterium tumefaciens*, 372
Polymers, non-dialysable, 291, 292
Polynomial functions, 286
Polyoma, 369
Polyphenol oxidase, 56
Polyphenols, 243, 244, 245, 293
Polyploidy, 145, 148, 162, 165, 169, 170, 173, 175, 184, 186, 190, 210, 343, 375
Polyribosomes, *see* Polysomes
Polysaccharides, 292
Polysomatic plants, 183, 186, 190, 343
Polysomes, 123, 127, 129, 133, 156, 246, 343
Polytene, 167, 182, 309, 375
Polyvinyl chloride, 13
Polyvinylpyrolidone (PVP), 56, 106
Population dependence, 191
Populus cardicus, 416
Populus nigra, 6
Populus tremuloides, *see* Aspen
Post-exponential phase, 252
Post-fixation, 49
Potato, 35, 41, 122, 144, 218, 244, 341, 394, 403, 404, 405, 412, 413, 416, 418
viruses infecting, 413
Potato virus X, 410, 413, 415
Potato virus Y, 410, 413
Potato wart disease, 394
Potting on of plantlets, 211
Powdery mildews, 393, 394, 396, 403
^{32}P-phosphate, 56, 378
Pratylenchus minyus, 417
Pratylenchus vulvus, 418
Pre-prophase band, 130
Preservation of cells, 189
Preservation of cultures, 48, 189, 204, 428, 430
Preservation of genotypes, 189
Pre-thylakoid body, 140
Primary tumours, 359, 380
Primordia, 339, 340, 341, 344, 347
Pro-embryoids, 147, 346, 348
Progesterone, 296
Progressive deceleration phase, 85, 336
Prokaryotes, 303, 423
Prolamellar body, 140
Proliferations, frothy, 256
Proline, 238
Propagules of plant parasites, 392
Prophase, 123, 131, 133, 137, 153, 207
Prophase: metaphase ratio, 306
Propionic orcein, 91, 165, 309
Proplastids, 139, 141, 156, 266

Protective clothing, 13
Protein, 156, 227, 228, 242, 245, 247, 254, 263, 265, 279, 281, 287, 288, 292, 311, 321, 322, 325, 327, 328, 332, 371, 390, 409, 411
determination of, 55, 91, 92
Protein output, 333
Protein synthesis, 93, 127, 223, 270, 314, 375, 411
Proteosomes, 127
Prothallial cells, 209
Prothallus, 201
Protoplasmic continuity, 3
Protoplasmic streaming, 153
Protoplasts, **100**, 428, 430
fusion of, 9, **100**, 114, 428
isolation of, **100**, **108**, 409, 421
isolation of, from leaves, **108**, 111
isolation of, from pollen, **118**
isolation of, from tissue cultures, **103**
pinocytosis in, 408
regeneration of cell wall by, **100**, 428
Protoxylem, 341
Protozoa, 298
Provascular tissue, 32
Pseudomonas spp., 416
Pseudoperonospora humuli, 394, 403
Pseudotsuga menziesii, 403
Pteridium, 142
Pterocarpan, 405
PTFE, 36, 46, 71
Puccinia antirrhini, 394, 396, 397
Puccinia graminis tritici, 402
Puccinia helianthi, 394, 396
Puccinia malvacearum, 402
Puccinia minutissima, 395
Puccinia recondita f.sp. *tritici*, 396
Puccinia tatarica, 395
Pulse chase technique, 249, 301
Pulse labelling, 249, 301
Purine synthesis, 265
Pycnia, 399
Pyridoxine, 43, 362
Pyrimidine synthesis, 265
Pyrus spp., 402
Pyruvate, 379
Pyruvate kinase, 279

Qo$_2$, 281, 330, 332
Quadruplochromosomes, 167, 184
Quiescent cells, structure and physiology of, **121**, 154, 241, 248

Radioactive constituents, 18
Radioactive precursors, 45, 55, 93
Radioactivity, determination of, 94

Ranunculus sceleratus, 139, 140, 141, 147, 349
Regenerated plants, 187, 188, 348, 354, **415**, 421, 424
Regressive changes in explants, 240, **247**
Regulator genes, 426
Relative growth rate, *see* Specific growth rate
Repressors, 355, 426
Resistance
 to parasitic nematodes, **419**, 421
 to pathogenic fungi, 403, 421
Respiration, 123, 129, 286, 299, 325
Respirometry, 56, 94
Restitution nucleus, 171
Reverse mutation, 402
R-genes, 403
Rheum raponticum, see Rhubarb
Rhizobium spp., 417
Rhizocaline, 346
Rhizogenesis, 162, 171, 339, 344
Rhododendron, 41
Rhubarb, 413
 viruses infecting, 413
Ribes uva-crispa, see Gooseberry
Riboflavin, 264, 362
Ribonuclease, 112
Ribosomal RNA, 247, 250, 369
Ribosomes, 123, 131, 138, 154, 229, 250, 369
Ricinus communis, 366
Rifampicin, 365
Ring dicentric, 168, 177
Rishitin, 404
RNA, 92, 156, 223, 227, 228, 229, 241, 243, 245, 247, 250, 251, 263, 265, 280, 283, 297, 307, 311, 325, 330, 332, 365, 369, 370, 382, 383, 409
RNA polymerase, 365, 370
RNAase, 227, 365
Root cultures, 2, 6, 392, 417, 418, 424, 430
Root formation, **339**, 340
Root initiation, 201, 346, 348, 424
Root nodules, 417
Roots, 138, 211, 236, 244, 245, 250, 254, 261, 263, 299, 300, 339, 341, 345, 346, 357, 399, 414, 430
 adventitious, 341
Root-tips, 165
Rose, 396, 418
 Paul's Scarlet, 105, 106, 107, 144, 270, 275, 276, 278, 279, 280, 282, 284, 287, 288, 289, 293, 318, 323, 324
Rubus fruticosus, 41, 277
Rubus occidentalis, 358
Rumex acetosa, 383
Rusts, 393, 394, 395, 396, 397, 398, 399, 401, 403

Ruta graveolens, 262, 287, 295
Rye, 109, 110

^{35}S, 93
S phase, 169, 186, 203, 247, 248, 249, 299, 300, 336
 duration of, 299, 300, 301
Saccharum, 173, 174, 175
Safety devices, 12, 27
Salicylic acid, 404
Salivary gland chromosomes, 182
Salix capraea, 6
Saprophytic growth of obligate parasites, 398, 401
Sartorius filters, 20, 109, 407
Satellite, 163, 164, 177
Satellite DNA, 367, 370, 371
Saturation constant, (K_s) 323, 329, 330
Scintillation counting, 94
Sclerospora graminicola, 394, 399, 403
Scopoletin, 262, 263, 264, 267, 268, 386
Scopolia japonica, 295
Scopolin, 262, 263, 267, 268, 386
Scorzonera spp., 41, 342, 362, 367, 370, 378, 379, 388
Secondary cell wall, 160, 243, 254
Secondary plant products, **267**, 289, **293**, 334, 355, 405, 426
Secondary single-cell clones, 200
Secondary tumours, 359, 380
Sectors, 187
Seeds, 34
Selection pressure, 427
Selenium tolerance, 261
Semi-continuous cultures, 271, **309**, 324, 334
Semi-logarithmic plot, 275, 276, 277, 279, 281, 282, 283, 308, 311
Senescent cells, 145, 147
Sequential treatments, 351, 426
Sequoia, 341
Sesquiterpenoid compounds, 404
Sets of chromosomes (*n*), 168
Settled volume, 314, 315
Shadowgraph prints, 96
Shake cultures, **62**
Shakers, reciprocal, 8, 112
Shaking speed, 63
Sheet meristems, 256
Shikimic acid, 267, 297
Shoot meristem culture, **406**, **412**, 414, 424
Shoot-root relationship, 7
Shoots, 138, 250, 263, 339, 341, 345, 347, 357, 366, 389, 406
Silicone rubber, 21, 75
Silver nitrate, 32

Single-cell clones, 3, 4, 8, 95, **191**, 269, 271, 381, 412, 423, 427, 428, 430
  derivation of, **198**
  differences between, **199**
Single-cell variants, 4
'Sink', 426
Sintered glass filters, 18, 19, 195, 407
Siphon, 74
Sister-chromatids, 169
Sodium nitrate, 101
Soft-rot bacteria, 416
Solanaceae, 353
Solanum menongema, 348
Solanum nigrum, 216
Solanum spp., 213, 234, 267, 403
Solanum tuberosum, see Potato
Solanum xanthocarpum, 295
Solenoid valves, **69**
Solid buffers, 83
Somatic endoreduplication, 167, **181**
Somatic hybridization, 101, 429, 430
Somatic mutation, 344, 389
Sophoro angustifolia, 296
Sorbitol, 109, 199, 274
Soybean, 67, 103, 117, 156, 267, 271, 287, 289, 291, 324, 405, 417
Specific growth rate (μ), 75, 279, 281, 282, 317, 320, 321, 322, 323, 330, 335, 336
'Spent' medium, 316
Sperms, 205, 207, 221
Sphaerotheca pannosa, 396
Spheroids, 139
Spheroplasts, 374
Spherosomes, 142, 147
Spindle, 134, 169, 171, 186
Spindle abnormalities, 166
Spindle inhibitor, 165, 235
Spinning cultures, 27, **65**
Spontaneous doubling of chromosome number, 235
Spontaneous fusion of protoplasts, 114, 116
Spontaneous tumours, 385
Spontaneous variation, 202, 204
Sporangia, 209, 225
Spore mother cell, 206, 299
Sporopollenin, 206
Sporulation, 401
Stability of clones, **204**
'Staling', 89
Stalk cell, 209
Stamens, 210, 212
Starch, 122, 125, 139, 156, 159, 208, 221, 226, 231, 263, 266
Stationary phase, 84, 97, 149, 159, 198, 273, 281, 304, 336

Steady-state cultures, 297, 299, 300, 318, 321, 322, 325, 326, 327, 329, 330, 325, 326, 336, 337, 423, 430
'Step-down' transition, 320
'Step-up' transition, 320, 328
Sterility, test for, 11
Sterilization, **19**
  procedures of, 34
Sterilizing agents, effectiveness of, 33
Steroids, 261, 262, 267, 296
Steward apparatus, **60**
Stigma-sterol, 262, 267, 295
Stirred cultures, **67**
Stock solutions, 18
Storage organs, 34
Storage parenchyma, 31, 35
Strain, 3
Strawberry, 413
  viruses infecting, 413
Streptanthus tortuosus, 140
Streptomycin, 360
Stripping-film technique, 302
Subculture, **47**, 84, 257
  frequent serial, **310**
  generation-interval, **312**
Sub-micron filters, 22
Sub-protoplasts, 118
Substrate limitation, 320
Sucrose, 43, 105, 110, 111, 115, 116, 118, 121, 214, 215, 216, 217, 237, 286, 362, 375, 387
Sucrose/sorbitol density gradient, 104
Sugar phosphates, 262
Sugar-beet, 400, 403, 405
Sugars, autoclaving of, 19
Sulphate incorporation, 93
Sulphydryl groups, 299
Sunflower, 142, 144, 147, 190, 264, 359, 362, 365, 366, 367, 369, 373, 375, 376, 378, 380, 388, 394, 396
Surface sterilization, 21, 32
Suspension cultures, 3, 8, 191, 430
  aggregation in, 270, 312
  cell composition of, **278**, 311, 373
  cell form and structure in, **143**
  cell separation in, **271**
  electron microscope studies of, **154**
  growth patterns in, **269**
  heterogeneity of, **270**
  metabolism of, 286
  techniques for, **59**
Suspensor cells, 182, 346
Switch mechanism in development, 239
Sycamore, 45, 65, 90, 93, 123, 129, 144, 145, 147, 149, 150, 152, 153, 154, 156, 157, 159, 160, 162, 192, 194, 195, 198, 199,

202, 270, 271, 272, 273, 274, 277, 278, 279, 280, 281, 282, 283, 284, 287, 290, 291, 293, 294, 300, 302, 303, 304, 307, 308, 309, 310, 312, 313, 314, 316, 318, 319, 320, 321, 322, 324, 325, 326, 327, 328, 329, 330, 331, 333
Symbiotic nitrogen-fixation, 417
Synchronization of cultures, 303
Synchronous cultures, **297**, 336
  transition to asynchronous, 311
Synchronous mitoses, 171, 305, 306, 307, 309
Synchrony of division, 54, 71, 117, 134, 238, 242, **246**, 261, 273, 278, **297**, 336, 423, 430
  naturally occurring, 298
Synchrony percentage, 303
Synchytrium endobioticum, 394
Syncytia, 420
Synergistic action, 344, 351
'Synthetic' conditioned medium, 97, 193

Tagetes erecta, 8
Tannins, 267
Tapetal cells, 182, 234
Taraxacum, 341
Taxus brevifolia, 209, 210, 214, 217
TdR kinase, 247, 249
Tea, 267, 268
Teflon, 69
Telial host, 397, 402
Telocentric, 168, 175
Telophase, 135, 139
Telutospore, 398
Temperature, 203, 221, 250, 252, 253, 262, 265, 344, 348, 354, 363, 370, 372, 384, 401, 411, 415
Temperature-controlled rooms, 26, 29
Templates, 370
Tensions, 425
Teratomas, 341, 381, 382, 389
Terms defined, **2**
Tetrads of spores, 206, 207, 221
Tetraploid, 148, 163, 167, 169, 170, 184, 185, 190, 201, 236, 238, 343
Thermostats, 26
Thermotherapy, *see* Heat treatment
Thiamine (vit B$_1$), 6, 43, 217, 267, 362, 383, 414
Thiouracil, 415
Thorium dioxide, 101
Threonine, 267
Threonine deaminase, 289, 291
Thylakoidal system, 139
Thylakoids, 140, 266

Thymidine, 55, 184, 185, 203, 237, 301, 303, 384
Thymidine kinase, 56, 307, 310
Thymidine triphosphate, 242
Time-lapse photography, 147
Tissue cultures, 2, **31**
  metabolic patterns in, **260**
  short-term, 344, 352
Titanium, 38
TMP kinase, 247, 249
Tobacco, 8, 9, 33, 38, 41, 43, 56, 93, 104, 108, 110, 111, 113, 114, 116, 117, 127, 131, 144, 147, 153, 162, 169, 171, 172, 173, 184, 185, 186, 187, 188, 201, 203, 206, 208, 209, 211, 215, 217, 218, 291, 220, 221, 224, 225, 226, 227, 231, 232, 233, 234, 235, 236, 237, 238, 239, 257, 258, 259, 261, 262, 263, 264, 266, 267, 268, 269, 270, 274, 275, 278, 290, 291, 292, 295, 296, 304, 306, 318, 339, 342, 343, 344, 345, 346, 353, 354, 357, 359, 362, 365, 369, 370, 371, 377, 378, 379, 381, 382, 384, 388, 389, 394, 406, 407, 408, 409, 410, 411, 412, 415, 416, 418, 420, 424, 427, 429
Tobacco mosaic virus (TMV), **111**, 407, 408, 409, 410, 411, 412, 415
Tobacco necrosis virus (TNV), 409
Todea barbara, 203
Tomato, 6, 210, 213, 214, 216, 217, 271, 359, 360, 365, 366, 405, 409, 418
Tomato juice, 41
Tonoplast, 127
Torpedo-stage of embryoids, 346, 348
Torreya nucifera, 209, 210, 214, 217
Totipotency, 47, 201, 338, 348, 424, 425
Toxins, 392, 404, 416, 421
Tracheidal cells, 7
Tracheids, 137, 143, 145, 201, 253
Tradescantia paludosa, 209, 300
Transcription, 208
Transfer RNA, 247, 369
Transfer rooms, **23**
Transfer tubes, 94
Transformation, 265, 356, **361**, 388, 428
  inception phase of, 361
Transition point, 321, 322
Translocation, 168, 175, 177
Transvascuolar strands, 123, 131, 137, 150, 154
Tricentric, 177
Trichloracetic acid, 93
Tri-iodobenzoic acid, 374, 420
Trillium erectum, 299
Triploid, 171, 172, 178, 234, 298, 385, 387

Tritiated thymidine, 55, 148, 185, 249, 299, 301, 302, 307
Tritiated RNA, 370
Triticum, see Wheat
Triticum monococcum, 178, 287, 324
Tropane alkaloids, 267, 295
Tryptophane, 264, 267, 376, 386
Tumble tube, 46, 60, 62
Tumorigenesis, 356
    reversibility of, **381**
Tumorogenic principle, *see* Tumour-inducing principle
Tumour cells, 8, 265, **356**
Tumour diseases of plants, **356**
Tumour induction, **361**
Tumour tissues, 188, 194, **265**, 269, 421
Tumour-inducing ability, assay of, 365
Tumour-inducing principle (TIP), 265, **361**, 428
    transmission of, **380**
Tumour-prone hybrids, 384, **385**
Turbidity
    proportionality to biomass, 319
    units of, 319
Turbidometric measurement, 318
Turbidostats, **75**, **317**, 322, 334
Turnip, 400
Two-tier culture vessel, 65
Tyrosine, 345

UDP-glucose, 281, 284
Ultraviolet bactericidal tubes, 23, 25
Ultraviolet irradiation, 202, 372, 374
Umbelliferae, 348
Unbalanced growth, 278, 281, 297
Uncinula necator, 394, 395
'Unisat' mutation, 389
Univalents, 385
Uracil, 383
Uranyl acetate, 49, 95
Urea, 346
Urea solution, 20
Uredia, 395
Uredinales, 393, 394, 401
Uredospore, 395, 396, 397, 402

Uridine, 384
Uromyces ari-triphylli, 395, 397, 398, 399, 402

Vacuoles, 121, 127, 137, 207, 208, 229, 231
Valve
    automatic sampling, 73
    magnetic, 67
Vanillic acid, 404
Variability of cultures, 259

Variants, **199**, 402, 427
    selection of, **202**
Variegation, 236
Vascular cambium, 121, 137, 246
Vascular elements, 348, 375, 418, 424
Vegetative cell, 207, 208, 220, 222, 223, **225**, 238
    abnormal development of, 230
    cytochemical changes in, 225
    ultrastructural features of, 228
Vegetative nucleus, 207, 224, 231
Venturia inequalis, 397
Vernalization, 342, 346
Vesicles, 229, 231, 396, 409
V-fermenter, 69, 271, 314
Vicia faba, see Broad bean
Vinca rosea, see Catharanthus roseus
Vindolin, 295
Vindolinum, 295
Vine, 393, 394, 395, 398, 401, 416
Viral genome, 384
Viral nucleic acid, 408
Virginia creeper, 360, 388
Virology, 111
Virus crystals, 408
Virus inclusions, 408, 412
Virus infection, 32, 101, 111, 364, 372
    of cell and tissue cultures, **407**
    of isolated protoplasts, 101, 111, 392
Virus multiplication, 408, 409, 411, 415
Virus titre, 407, 408, 410, 411
Viruses, 392, **406**, 421
    distribution in tissue cultures, **411**
Virus-free plants, 411, **412**
Virus-free stocks, 406, 411, **412**, 424
Virus-induced tumour disease, **382**
    tumour development in, **382**
Virus-induced tumours, 339, 342, 356, **382**
Virus-resistant plants, 412, 416
Visnagin, 294, 295, 296
Vitamin B_1, *see* Thiamine
Vitamine B_2, *see* Riboflavin
Vitamin B_5, *see* Pantothenic acid
Vitamin B_6, *see* Pyridoxine
Vitamins, 37, 43, 194, 261, 267, 271, 362, 375
Vitis vinifera, see Vine
Volatile factor, 195, 196, 197
Volatile inhibitor, 243

Warburg technique, 45, 57, 94
Washing-up facilities, **13**
'Wash-out', 321, 323, 328, 337
Water melon, 264
Waterlogging, 45
Wheat, 110, 162, 396

Wheatstone bridge circuit, 318, 319, 321
White blisters, 393, 394
White turnip, 35, 41
Whole plant physiology, 423
Wound callus, 2
Wound cambium, 137, 142, 244, 256
Wound healing cycle, 361, 363, 375
Wound hormone, 244
Wound reaction, **243**, 258, 361
Wound tumour disease, 356, 379
Wound tumour tissue cultures, **383**
Wounding and virus infection, 407, 408

Xanthine, 383

X-rays, 236, 381, 384
Xylem, 380
Xylose, 292

Yeast extract, 18, 41, 162, 175, 217, 237, 260, 271, 338, 354, 402
Yield coefficient (*Y*), 323, 330, 331
Yohimbine, 262

Zea mays, *see* Maize
Zeatin, 42
Zinnia spp., 358, 372
Zoospores, 409
Zygotic embryos, 425